The
2009-10
Arriva
Bus Handbo

CU00869070

British Bus Publishing

Body codes used in the Bus Handbook series:

Type:
A	Articulated vehicle
B	Bus, either single-deck or double-deck
BC	Interurban - high-back seated bus
C	Coach
M	Minibus with design capacity of 16 seats or less
N	Low-floor bus (*Niederflur*), either single-deck or double-deck
O	Open-top bus (CO = convertible - PO = partial open-top)

Seating capacity is then shown. For double-decks the upper deck quantity is followed by the lower deck. Please note that seating capacities shown are generally those provided by the operator. It is common practice, however, for some vehicles to operate at different capacities when on certain duties.

Door position:
C	Centre entrance/exit
D	Dual doorway.
F	Front entrance/exit
R	Rear entrance/exit (no distinction between doored and open)
T	Three or more access points

Equipment:-
T	Toilet	TV	Training vehicle.
M	Mail compartment	RV	Used as tow bus or engineers' vehicle.
L	Lift for wheelchair (post 2005 express coaches are fitted with lifts as standard)		

Allocation:
s	Ancillary vehicle
t	Training bus
u	out of service or strategic reserve; refurbishment or seasonal requirement
w	Vehicle is withdrawn and awaiting disposal.

e.g. - B32/28F is a double-deck bus with thirty-two seats upstairs, twenty-eight down and a front entrance/exit.
N43D is a low-floor bus with two or more doorways.

Re-registrations:
Where a vehicle has gained new index marks the details are listed at the end of each fleet showing the current mark, followed in sequence by those previously carried starting with the original mark.

Annual books are produced for the major groups:
The Stagecoach Bus Handbook
The First Bus Handbook
The Arriva Bus Handbook
The Go-Ahead Bus Handbook
The National Express Coach Handbook (bi-annual)
Some editions for earlier years are available. Please contact the publisher.

Regional books in the series:
The Scottish Bus Handbook
The Welsh Bus Handbook
The Ireland & Islands Bus Handbook
English Bus Handbook: Smaller Groups
English Bus Handbook: Notable Independents
English Bus Handbook: Coaches

Associated series:
The Hong Kong Bus Handbook
The Malta Bus Handbook
The Leyland Lynx Handbook
The Postbus Handbook
The Mailvan Handbook
The Toy & Model Bus Handbook - Volume 1 - Early Diecasts
The Fire Brigade Handbook (fleet list of each local authority fire brigade)
The Police Range Rover Handbook

Some earlier editions of these books are still available. Please contact the publisher on 01952 255669.

2009-10 Arriva Bus Handbook

The Arriva Bus Handbook is a Major Group edition of the *Bus Handbook* series which contains the various fleets of Arriva plc, one of the major operators across Europe. The *Bus Handbook* series is published by British Bus Publishing, an independent publisher of quality books for the industry and bus enthusiasts. Further information on these may be obtained from the address below.

Although this book has been produced with the encouragement of, and in co-operation with, Arriva plc management, it is not an official group fleet list and the vehicles included are subject to variation, particularly as the vehicle investment programme continues. Some vehicles listed are no longer in regular use on services but are retained for special purposes. Also, out of use vehicles awaiting disposal are not all listed. The services operated and the allocation of vehicles to subsidiary companies are subject to variation at any time, although accurate at the time of going to print.

To keep the fleet information up to date we recommend the Ian Allan publication *Buses*, published monthly or, for more detailed information, the PSV Circle monthly news sheets.

Edited by Bill Potter and Stuart Martin

Acknowledgments:
We are most grateful to Tom Johnson, Harry Laming, Peter Marley, Colin Martin, Kevin O'Leary, the PSV Circle and the management and officials of Arriva Group plc, and their operating companies, for their kind assistance and co-operation in the compilation of this book. The front cover picture is by Chris Clegg, the rear cover views are by Bill Potter and Harry Laming; frontispiece by Harry Laming.

ISBN 9781904875291

Published by British Bus Publishing Ltd
16 St Margaret's Drive, Telford, TF1 3PH

© British Bus Publishing Ltd, November 2009

Telephone 01952 255669 - www.britishbuspublishing.co.uk

CONTENTS

Contents

Arriva ranks among the principal operators of hybrid buses in London. Volvo's new hybrid bus is the B5L with HV1 being shown at the 2008 Coach and Bus exhibition in Birmingham. Six of this model have now been supplied to Arriva and they have been placed in service alongside five Wrightbus Gemini 2 hybrid vehicles that use VDL components. HV3, LJ09KOH, is shown. *Mark Lyons*

ARRIVA plc

The Arriva Group: A Company Profile

Arriva is one of the largest transport services organisations in Europe, employing more than 44,000 people and delivering more than one billion passenger journeys across twelve European countries every year. It operates an extensive range of services including buses, trains, commuter coaches and water buses. In the UK it is also engaged in bus and coach distribution.

The group is firmly established as one of the UK's top 250 companies quoted on the London Stock Exchange. For the year ended 31 December 2008 Arriva's group revenue was £3,042.2 million.

In the UK, Arriva is one of the largest regional bus operators, with 6,500 vehicles serving customers in the North East, North West and South East of England, the Midlands, Scotland and Wales. The core activity is the operation of urban, interurban and rural local bus services, and commuter and express coach services within England, Scotland and Wales. In the capital, the group also runs buses under contract to Transport for London and in addition operates The Original Tour sightseeing buses.

Many of the operating companies in Arriva's UK Bus division are long established but the grouping emerged out of the privatisation of public sector transport companies, following the Transport Act 1985, and the subsequent consolidation of the sector. The division is organised into autonomous operating subsidiaries in order that management teams and staff can focus upon the discrete local and regional markets that they serve.

The group makes significant capital investment in the replacement and improvement of its vehicles and facilities. Arriva's UK Bus division has made significant investments in new vehicles with more than 460 new buses in 2008 and 500 in 2009, improving the passenger experience and representing a total investment of more than £230 million.

Recent years have witnessed a renewed interest in rapid transit systems and the development of intermediate technologies. Arriva has been actively positioning itself within this broadened public transport market. Fastrack, the award-winning bus rapid transit system in North Kent in the UK is operated by Arriva on behalf of Kent County Council and Kent Thameside. It is being delivered through one of the best examples of a public/private sector partnership which has proved successful in attracting people out of their cars.

Arriva Bus and Coach is the group's bus and coach distribution business with exclusive import rights for all VDL chassis supported by a wide range of bus and coach bodywork options, from Van Hool of Belgium, Ikarus from Hungary, Temsa from Turkey and Plaxton from the UK, to meet new market demands.

Arriva's UK Trains division operates two passenger rail franchises – Arriva Trains Wales and CrossCountry. Arriva Trains Wales stretches from Manchester, through north and south Wales and across to Birmingham and Gloucester. CrossCountry is the UK's most extensive rail franchise stretching from Aberdeen to Penzance and Stansted to Cardiff.

Additionally, across Europe Arriva provides passenger transport services in the Czech Republic, Denmark, Italy, Germany, Hungary, the Netherlands, Poland, Portugal, Slovakia, Spain and Sweden.

Over the last ten years the group have developed a unique position in mainland Europe, setting themselves apart from other UK public transport sector peers. Arriva's footprint has deepened and widened, with operations in twelve countries. Management teams have been strengthened, further reinforcing the group's unrivalled knowledge and experience, and the continuous development of a strong network of relationships which has been key to Arriva's success. With a proven track record, having tripled revenue from its mainland European operations in the six years from 2002, the group was awarded The Queen's Award for Enterprise: International Trade 2009.

The Cowie name - early years.

In 1931, the Cowie family started a motorcycle repair business in Sunderland. In 1934 the first sales outlet opened on what became the location of the Group Head Offices. Business came to a halt in 1942 because of the effects of the war, but reopened in 1948 and benefited from the boom in personal mobility offered by the motorcycle. A second motorcycle shop was opened in Newcastle in 1952. Further expansion occurred in 1955 with another branch in Newcastle, and new branches in Durham and Stockton-on-Tees. A move into the Scottish market was taken with the acquisition of the J R Alexander motorcycle dealerships in Edinburgh and Glasgow in 1960. However, there were signs that the market was moving against the motorcycle, and in favour of the motor car which was becoming a more mass market item in availability and price.

In 1962, Cowie acquired its first car dealership in Sunderland, and such was the pace of change that, by 1963, motor car sales constituted 80% of revenue. On the back of this change, a public company, T Cowie plc, was formed in 1965 and, in that year, a further two car dealerships were acquired, one in Redcar and a second in Sunderland. In 1967, there was investment in new car showrooms for the Ford franchise in Sunderland; there has been a long association with Ford over the years. In 1971, this was strengthened with the acquisition of Ford Blackburn and a further Ford dealership, this time in Middlesbrough. By 1971, the group turnover reached £8m per annum.

Growing on the motor business.

The first exposure to the bus and coach industry came in 1980. In that year, Cowie took over The George Ewer Group which had various motor interests including Eastern Tractors and these assets were bought with the business. It also included the long established Grey Green coach operation and this was to lead to a sea change in the make-up of the business.

In 1984 Cowie acquired the Hanger Group and this brought Ford dealerships in Nottingham and Birmingham and, significantly, Interleasing, the contract hire business. Expansion continued in 1987 when seven main dealerships were acquired from the Heron Group. On the leasing side, Marley Leasing was acquired. Further growth in this business area came in 1991 when RoyScot Drive and Ringway Leasing were also acquired, adding to the prominence of this activity for the group.

1992 was the year that Cowie very nearly took another step into the bus and coach market, narrowly failing to purchase Henlys which, by that time, had ownership of Plaxtons along with several car franchises. Consolation came with the addition of a Ford dealership in Swindon, a Peugeot dealership in Middlesbrough, and a Toyota dealership in Wakefield. In 1993, the Keep Trust Group was acquired, the dealership network being boosted by 70%.

Another notable event in 1993 was the retirement of Sir Tom Cowie, the Chairmanship of the business being taken up by Sir James McKinnon in 1994 and continuing until he retired at the end of 1999, when Gareth Cooper took over the role.

Explosion into Buses!

The Cowie group's first involvement with the bus and coach sector came with the purchase of the Ewer Group in 1980. Its subsidiary, Grey Green, was a most distinguished name in coaching but, under the Ewer Group, had already started to operate bus services for London Transport.

Grey Green, which had operating bases in Stamford Hill and Dagenham in London, and in Ipswich in East Anglia, participated in historic operations such as East Anglian Express, the Eastlander Pool, and joint services to Scotland in association with Scottish Bus Group. Private Hire coaching also played a large part in the business.

Great opportunities were perceived in the 1980s in commuter coaching after the deregulation of coach services under the 1980 Act. Commuter coach services thrived for a while, but the involvement in British Coachways was not successful. In the mid 1980s, Grey Green's involvement in the East Anglian Express Pool was taken into the National Express operation.

Early success in London Bus tendering brought Grey Green operation into to the very heart of London, notably on route 24 which passes the Palace of Westminster. After that, Grey Green gradually ceased all coach activities and concentrated entirely on tendered bus services in London.

The privatisation of the newly created subsidiaries of London Buses offered the opportunity to build on the favourable experience of Grey Green in the London bus market. Leaside was acquired in 1994 and was renamed Cowie Leaside. Later in 1995, the South London company was purchased, becoming Cowie South London. These two acquisitions made Cowie the largest single operator in the London Buses area. As the London operations were tidied up, the former Kentish Bus/Londonlinks operations at Cambridge Heath, Battersea, Croydon (Beddington Farm) and Walworth, and the East Herts & Essex operations from Edmonton were gradually incorporated into the London operation.

In 1996, Cowie purchased County Bus from the National Express Group, adding to the group's presence in the south-east of England. Then, in August 1996, Cowie completed the acquisition of British Bus plc and, at a stroke, became the second largest bus operating group in the United Kingdom. The acquisition of British Bus plc added a whole raft of bus companies across the country and very nearly brought all the disparate elements of the former London Country company, including Green Line Travel, under one ownership. There was a Monopolies and Mergers Commission inquiry into the acquisition of British Bus, with particular focus on the situation in London and the South East, but the report did not require any divestment.

The final acquisition of 1996, also in August, brought another previously divided company back under common ownership. In 1986, United Automobile Services had been split into two separate companies for privatisation: North East Bus and Northumbria. Northumbria was sold to Proudmutual, a company which had been set up to facilitate the management buyout and was acquired by Cowie in 1994. United Auto was sold in December 1987 to Caldaire who sold it on, as North East Bus, to West Midlands Travel. West Midlands Travel itself then merged with the National Express Group which decided to concentrate on its core operations and, in 1996, sold first County Bus then North East Bus to Cowie/Arriva.

The former British Bus headquarters at Salisbury, which dated back to Drawlane days, was wound down and closed at the end of 1996, with the group administration being moved to Sunderland.

The growth of British Bus plc

With the acquisition of British Bus plc, Cowie became the second largest bus operating group in the United Kingdom. This move led to the reclassification of the enlarged group from being motor trade to transport.

Drawlane Ltd

The privatisation of the National Bus Company followed the 1985 Transport Act with the National Bus Company becoming a vendor unit selling its subsidiaries to pre-qualified parties. Endless Holdings Ltd was one of those interested parties, being a group of companies based in the cleaning and building management sector with a head office on Endless Street in Salisbury. The prime mover in Endless was Ray McEnhill. Endless set up a subsidiary called Drawlane Ltd to bid for NBC companies as they were made ready for sale.

The first company bought by Drawlane was Shamrock & Rambler in July 1987. This was the major part of the coaching activities of Hants and Dorset and was based at a modern depot in Bournemouth. The business was heavily dependent on National Express contracts, although a minibus operation was set up to compete with Yellow Buses in the Bournemouth area. Shamrock & Rambler did not survive for long: its bus operations were quickly reduced in scale, and difficulties with the National Express contracts led to notice of termination being given to Shamrock &

Rambler which sealed its fate. National Express set up a local joint venture company called Dorset Travel Services Ltd to take over the workings of Shamrock & Rambler using other vehicles and based as a tenant of Yellow Buses at Mallard Road. Yellow Buses eventually purchased Dorset Travel Services. The Shamrock & Rambler vehicles were dispersed around the then Drawlane Group fleets and Shamrock & Rambler was wound up.

Drawlane was preferred bidder for three more companies: Southern National, North Devon, and London Country (South West). Each purchaser was limited to three NBC companies in the first instance. However there was concern that Drawlane might be related to another bidder called Allied Bus which had been selected as preferred bidder for another three companies: Lincolnshire Road Car, East Midland Motor Services and Midland Red North. The concern was sufficient for the preferred bidder status to be withdrawn from both and offers were re-invited.

Drawlane was successful in acquiring Midland Red (North) in January 1988 which, at the time of purchase, had 248 vehicles and 491 employees. The following month, London Country (South West) Ltd was purchased with 415 vehicles and 1250 employees, although the garages were purchased separately by Speyhawk Properties, who then leased them to the bus company with varying securities of tenure, reflecting the premium value of property in London and the South East. Finally, in March 1988, Drawlane acquired the 'new' North Western Road Car Company Ltd based in Bootle, with 340 vehicles and 870 staff.

Drawlane had also bought East Lancashire Coachbuilders from the industrial conglomerate John Brown. East Lancs was based in Blackburn and had a strong customer base in the local authority sector.

Further expansion for Drawlane would now come from acquiring bus operations from other sources. ATL (Western) Ltd had purchased Crosville Motor Services from NBC in March 1988 and in early 1989 was ready to sell. Drawlane purchased the company adding a further 470 vehicles. A quick overview of the future of Crosville is appropriate here. In an exercise to realise value from the company, the South Cheshire operations at Crewe and Etruria were transferred to Midland Red North, the Runcorn and Warrington depots were transferred to North Western, and the Macclesfield and Congleton depots were merged into Bee Line Buzz, of which more later. The remaining operations at Rock Ferry and Chester, together with the vehicles, were sold to PMT Ltd along with the 'Crosville' trading name. The original Crosville Motor Services Company was renamed North British Bus Ltd, and, though it existed for some time thereafter, it did not trade as a bus company.

Midland Fox was bought from its management team in September 1989 along with a minority share holding in the company from Stevensons of Uttoxeter. Bee Line Buzz operations in Manchester had been started up by BET, which had sold its bus operations in 1968, venturing back into bus operation in the UK. A similar operation was started in Preston. Both were sold to Ribble's management buyout team who, in turn, sold Ribble Buses to the Stagecoach Group. As part of an exchange of assets with Stagecoach in the Manchester area, Drawlane bought Bee Line Buzz from Stagecoach along with Hulme Hall Road depot in Manchester and added into the company the former Crosville operations at Macclesfield and Congleton. Bee Line had an independent existence within the group until 1993, when its operations were merged into North Western, with Macclesfield depot going to Midland Red North.

Drawlane in Transition

In 1991, Drawlane became a partner in a consortium with several banks setting up a company called Speedtheme Ltd in a bid to buy National Express Holdings from its management team. As well as the main National Express business, National Express Holdings owned Crosville Wales Ltd and its Liverpool subsidiary Amberline Ltd, Express Travel in Perth, and Carlton PSV the Neoplan coach dealer in Rotherham. Speedtheme Ltd did not want Crosville Wales and Amberline/Express Travel, and these were immediately sold to a company, called Catchdeluxe Ltd, set up by two of the main shareholders of Drawlane, Ray McEnhill and Adam Mills. Although

The first involvement with the bus and coach sector of Arriva's predecessor, The Cowie Group, came with the purchase of the Ewer Group in 1980. Its subsidiary, Grey Green, was a most distinguished name in coaching but, under the Ewer Group, had already started to operate bus services for London Transport. The Grey-Green livery has been retained on 115, F115PHM, an early Volvo Citybus double-deck. The vehicle has now been donated by Arriva to the London's Transport Museum. *Mark Lyons*

not part of Drawlane at this time, these two companies were under common management, and it was not long before they became part of the Drawlane Group.

Ray McEnhill became the Chairman and Chief Executive of National Express Group and, as this group prepared for its floatation on the Stock Exchange, Ray McEnhill and Adam Mills severed their involvement with Drawlane. The London & Country business called Speedlink Airport Services was sold by Drawlane to National Express at this time, though Drawlane retained the Green Line Travel Company, along with the Green Line trading name.

There was, in effect, a management buyout of Drawlane in the autumn of 1992 to coincide with the successful floatation of National Express and shortly thereafter Drawlane was renamed British Bus plc. (British Bus Ltd had been a dormant subsidiary of National Express Holdings, originally set up by NBC to market the Britexpress card overseas.)

British Bus grows

Throughout this period, there were various smaller acquisitions by the group companies but these are dealt with in the short histories of these companies that follow.

In 1993, British Bus purchased Southend Transport and Colchester Transport, both former municipal operations which had been offered for sale after being weakened by competition from Badgerline subsidiaries, Thamesway in Southend and Eastern National in Colchester. After acquisition, the two companies were put under common management and the supervision of London & Country. A programme of rationalisation put both back onto a firm footing, though down-sized.

In 1993, North Western acquired Liverline of Liverpool which, by then, had grown to a fifty-one vehicle company. It was run as a separate subsidiary until 1997. North Western had by this time absorbed the bus operations of Amberline.

Also in 1993, Tellings Golden Miller was sold back to its original owners. Tellings had been taken over by Midland Fox and came into the group. At the time of its sale, it had bases in Surrey and Cardiff. In the latter location, Tellings had become the joint operator of the Trawscambria service with Crosville Cymru! That role then passed to Rhondda Bus in which British Bus had a share holding for a while.

At the end of 1993, British Bus became the preferred bidder for the purchase of GM Buses North, but the position was overturned by the vendors and new bids invited. The outcome of this exercise was a winning bid from an employee-based team which was eventually completed in the spring of 1994. They subsequently sold out to First.

Further growth was funded by expanding the capital base of the group through investment by two merchant banks who took a share of the increased equity in a new parent company British Bus Group, though operational control remained with British Bus plc.

During 1994, ownership of both East Lancashire Coachbuilders Ltd and Express Travel was transferred out of the group, though they were still associated companies. East Lancs in particular was still a preferred supplier to the group for bus bodywork.

The National Greenway programme was coming to an end at this point. This programme involved the stripping down and re-engineering of Leyland National shells with new a Gardner or Volvo engine, new gearbox and new body panels mounted on the shell framework. The stripping down and mechanical overhaul work was carried out at London and Country's Reigate garage, though later some of this work was carried out by Blackburn Transport, as this was closer to East Lancs. East Lancs then did the body work with customer options as to the front design. Notable numbers were carried out both for Group companies and other operators.

Luton, Derby, Clydeside and Stevensons

In July 1994, British Bus acquired Luton and District Transport from its employees. By this time, the business included the former London Country Bus North West and a large part of the Stevenage operations bought from Sovereign Bus. This meant that a sizeable part of the former London Country company was now back in common ownership.

Luton and District had also assisted other employee buyouts such as Derby City Transport in which it had a 25% share holding, and Clydeside 2000 plc, where there was a 19% share holding. In both companies, the shareholders voted to accept offers from British Bus for the balance of the shares and they became fully owned members of the British Bus group. A third company, Lincoln City Transport, had not met with success and the employees had already agreed its sale to Yorkshire Traction-owned Lincolnshire Road Car Company Ltd before the British Bus take-over.

At the same time, there were discussions about the acquisition of Stevensons of Uttoxeter. Stevensons had grown dramatically after deregulation and operated well away from its traditional area. A strong expansion in the West Midlands was initially successful but West Midlands Travel responded to the competition and used its Your Bus acquisition to start up operations in Burton on Trent which was by then the heartland of Stevensons. A long struggle looked in prospect and a sale to British Bus was agreed. Operations in the West Midlands were scaled down and surplus vehicles were distributed around the group. After this process was complete, the geographically separated Macclesfield depot of Midland Red North was transferred to Stevensons control in January 1995.

Proudmutual and Caldaire Holdings

In the summer of 1994, British Bus also acquired the Proudmutual group. Proudmutual had been the buyout vehicle for the management team of Northumbria Motor Services to buy their business from NBC. It had also acquired some smaller businesses in the North East, including Moor-Dale.

Proudmutual had previously purchased Kentish Bus, the former London Country South East from NBC in March 1988. They had considerable success in the London Transport tendering process and further LT work was added when the LT contracts of Boro'line were purchased in

February 1992. The Proudmutual acquisition thus gave British Bus a very strong position in LT tendering when the activities of London and Country and the LDT group were taken into account. It also brought another part of the former London and Country into common ownership.

The privatisation of the London Transport Bus companies brought no success for British Bus but, as we have seen earlier, the Cowie Group was successful in acquiring two of the subsidiaries. In March of 1995, the Caldaire group was acquired. Caldaire was the buyout vehicle with which the management of West Riding Group had purchased their business from NBC in January 1987. In the December of that year they also bought United Auto from NBC. A demerger later on saw the United business being separated off again into North East Bus. The long established independent South Yorkshire Road Transport was acquired and formed one of the trading identities of the Caldaire Group, the others being West Riding, Yorkshire Woollen, and Selby & District. After acquisition, the group of companies was renamed Yorkshire Bus Group by British Bus.

Maidstone & District

What turned out to be the last major acquisition by British Bus was made in April 1995. Maidstone and District had been one of the earliest of NBC sales in late 1986. It had purchased New Enterprise of Tonbridge in 1988, and the assets of Boro'line Maidstone in 1992. Under British Bus ownership, the company was put under common management with Kentish Bus and Londonlinks as the Invictaway Group with its head office at Maidstone.

Floatation or Trade Sale?

Throughout 1995, preparations for floatation had been underway with the appointment of advisors. However, in the summer of 1995, these plans were thrown off course by reports of alleged irregularities involving support from the Bank of Boston for British Bus at an earlier point in the group's history. The timing of these allegations led to the postponement of the floatation. The alternative route for shareholders and the investing banks to realise their investment was a trade sale of the group and discussions were held with various interested parties.

Concurrently, the group was also looking to expand its interests into other modes of transport as the opportunities in the bus industry were becoming scarce due to the growth of the major players. The subsequent sales of the two former GM Buses companies by their employee owners and that of Strathclyde Buses were opportunities for growth, but British Bus was not successful. In June 1996 the Cowie Group made an offer to acquire British Bus, (which has already been covered). There was a Monopolies and Mergers Commission investigation into the take-over in view of the concentration of operation in London and the South East in the combined business. However, the resulting report made no recommendation about disinvestment, recognising the considerable presence of the other groups in the area. The acquisition by Cowie was completed in August 1996.

Trams and trains

As part of its drive to expand its business, British Bus pursued tramway operations and had become a partner in the Eurotrans consortium which bid for both the Leeds Supertram and Manchester Metrolink concessions. British Bus was also part of a consortium bid for the Croydon Tramlink project.

Eurotrans was selected as the preferred bidder for the South Leeds Supertram Project which was being promoted by West Yorkshire PTE. This was a PFI project where bidders were fighting for the concession to design, build, operate, and maintain the tramway for a period of thirty years. The Eurotrans consortium included big construction companies such as Taylor Woodrow, Morrison Construction, and Christiani & Neilsen. The tram supplier was Vevey Technologies of Switzerland, now part of the Bombardier Group. The intention was for Arriva

Yorkshire to set up a tram operating subsidiary to operate the tramway for the concession life on behalf of Eurotrans. The project stalled waiting for UK Government funding commitment, but was revived, in expanded form, in 2001 when funding was secured.

British Bus was also active in the process of franchising of the Train Operating Companies by OPRAF although none of their bids were successful. However, the exposure to the process paid off in arranging through ticketing deals with the successful bidders later on.

Arriva

In 1997 the group started rebranding of all its business interests under the new trading identity of 'Arriva'. The bus division became a separate legal identity on 1 January 1998, becoming Arriva Passenger Services (APS). The group livery and identity started to replace the former colours and names progressively from 1999.

A link with the past was broken when the Group Headquarters relocated in the latter part of 1998 from Hylton Road in Sunderland to new purpose built premises on the Doxford International Business Park on the outskirts of Sunderland near the A19.

Arriva Passenger Services shared the head office premises of Arriva Fox, first at Millstone Lane and then, from September 1997, at Thurmaston, although several staff were quartered at offices in bus depots around the country. Then, in August 1999, APS moved to purpose-built offices in the Meridian Business Park on the other side of Leicester.

In September 1997, Arriva made its first acquisition outside Britain with the purchase of Unibus Holdings in Denmark. Arriva continues to expand within Europe, and now has operations in the Czech Republic, Denmark, Germany, Italy, the Netherlands, Portugal, Spain and Sweden. From December 2007 the group will begin operating train services in Poland.

In 2000, Arriva purchased MTL, the Liverpool-based transport group, which included the major bus operator in Merseyside plus the two rail franchises. A subsidiary operation at Heysham was soon sold on and the MTL bus operation was absorbed into Arriva North West. Gilmoss depot and its operations were sold on to meet the terms required by the Office of Fair Trading for its approval of the purchase of MTL.

Arriva in London

When the Arriva name was introduced, Grey Green was renamed Arriva London North East. Cowie Leaside became Arriva London North, and Cowie South London became Arriva London South. The three companies have moved gradually closer to functioning as one unit, and the London North East operations were absorbed into London North in 2003. The head office for both London companies is at Wood Green. The Beddington Farm depot of Londonlinks transferred from Southern Counties to London South in October 1999. The Leaside Travel coach and bus contract hire fleet retained a separate livery until operations ceased at the end of 2005. Arriva London now concentrates private hire activities on its Routemaster operations, for which a number of these iconic vehicles have been retained under the title "The Arriva Heritage Fleet".

Arriva provided the first low-floor double-deck buses in London and deliveries continue apace. The first buses were DAF with Alexander bodies. Volvo chassis with Alexander, Wright and Plaxton bodies are also in fleet service. More recent orders for Wright double-decks have been on DAF chassis, while however the latest order is for a batch of ADL Trident-chassied Enviro 400 vehicles.

In line with the Mayor's strategy for transport in London, which includes the introduction of Congestion Charging, the Arriva London fleet has seen growth of over 200 extra buses together with rapid fleet renewal. The fleet is now close to being entirely low-floor. To accommodate this growth, Stamford Hill garage has been reopened, the former tram shed at Brixton was brought back into use in February 2003, and facilities elsewhere are being expanded. In March 2007, Arriva joined forces with TfL to trial the world's first Hybrid double-decker bus. Hybrid buses, which use a combination of diesel and electric power, are central to the Mayor of London and TfL's plans for a cleaner, greener fleet, and will contribute to cutting the capital's carbon dioxide emissions, with further hybrid buses added to the Arriva London fleet in 2009.

County Bus and Coach into Arriva East Herts and Essex

County Bus & Coach came into being at the beginning of 1989 to carry on the eastern operations of the former London Country North East, which in itself was one of the four parts into which London Country was divided. Operations were based at Harlow, Hertford and Grays. Ownership had progressed through the AJS group in 1988, the South of England Travel group in 1989 and the Lynton Travel group in 1990. The company was then purchased by the West Midlands Travel holding company, becoming part of the National Express Group when WMT merged into NEG. A retraction into core business led NEG to sell County to Cowie in 1996. A significant acquisition in 1989 was the bus interests and depot of Sampsons of Hoddesdon.

A restructuring of responsibilities saw County Bus take over the controlling supervision of Southend Transport and Colchester Borough Transport from London & Country and, during 1998, this responsibility passed to Arriva The Shires with overall control of what was now Arriva East Herts and Essex. The Edmonton operation passed to the Arriva London group in a rationalisation of responsibilities within London. During 1999 the fleets of these various component operations were renumbered into a single series. In a further restructuring in 2001, responsibility for Colchester, Grays and Southend passed to Arriva Southern Counties.

The Shires becomes Arriva the Shires and Essex Ltd

In 1986, United Counties Omnibus Company Ltd was divided into three parts, the southern most of these being Luton and District Transport Ltd which took over operations in Aylesbury, Dunstable, Hitchin and Luton. The new head office of the company was in Luton.

In August 1987 Luton and District became the first employee owned bus operator in the UK when its employees bought it from NBC. In the period from January 1988 to October 1990, LDT expanded the size and the area of its operations through a number of acquisitions. The assets and business of Red Rover Omnibus Ltd, operating bus services from a depot in Aylesbury, were acquired in January 1988. In June 1988, Milton Keynes Coaches was acquired, joined in May 1990 by two thirds of the bus services operated in the Stevenage area latterly Sovereign Bus Ltd.

In October 1990, LDT acquired London Country North West Ltd. LCNW operated a vehicle fleet of a similar size to LDT from a head office and depot in Garston and other depots in Hemel Hempstead, High Wycombe, Amersham, and Slough. LDT assisted in the employee buyouts of two other companies and acquired a share holding in both, Derby City Transport in 1989, and Clydeside 2000 plc in 1991.

In July 1994 LDT became part of British Bus. In October 1994, the bus operations of Stuart Palmer Travel based in Dunstable was taken over, followed in May 1995 by Buffalo Travel of Flitwick, and Motts Travel of Aylesbury in July 1995. April 1995 saw the launch of a brand new blue and yellow company livery with local trading names replacing the previous red and cream of LDT and green and grey of LCNW. The legal name was changed to LDT Ltd in May and the corporate operating name became The Shires.

In late 1997, Lucketts Garages (Watford) Ltd was acquired. In addition to local bus services in Watford there were substantial dial-a-ride operations and a commercial workshop. Lutonian Buses was acquired in March. Since then, there has been a ruling that the business should be sold under Competition regulations. The Group's challenge was unsuccessful and Lutonian was sold during 2000. The Arriva branding saw vehicles carrying the name 'Arriva serving the Shires', except at Garston garage which carries 'Arriva serving Watford'. Management responsibility for

Arriva East Herts & Essex now falls to Arriva the Shires and Essex Ltd. Responsibility for Colchester, Grays and Southend passed to Arriva Southern Counties in 2001.

In February 2006, Arriva acquired Premier Buses Limited, the holding company of MK Metro Limited (MK Metro), of Milton Keynes for £5.6 million. MK Metro operated commercial and tendered services in Milton Keynes, Bedfordshire and Northamptonshire adding two hundred and sixty employees and one hundred and twenty vehicles to the fleet.

Kentish Bus part of Arriva Southern Counties

Kentish Bus started as London Country South East on the division of London Country Bus Services in 1986. It had its Head Office at Northfleet and in April 1987 was relaunched as Kentish Bus and Coach Ltd with a new livery of cream and maroon.

In March 1988 Kentish Bus was sold to Proudmutual on privatisation. There was considerable expansion into the LT tender market for which new buses were added, many with registration indices originating in the North East. In February 1992 there was further expansion in this area when Kentish acquired the LT tendered work of the troubled Boro'line Maidstone operation, along with some 57 vehicles.

After the acquisition of the Proudmutual group by British Bus in 1994, Kentish Bus and Londonlinks were jointly managed from Northfleet. However, on the acquisition of Maidstone and District by British Bus, the management was relocated to Armstrong Road, Maidstone under the Invictaway grouping. The balance of the operation continued to be controlled from Maidstone after the reallocation of the two London depots to South London and Leaside.

Its size was significantly reduced by the transfer of operations at Battersea to the control of South London and the operations at Cambridge Heath to Leaside.

However the local Kent Thameside network was expanded and upgraded with low floor vehicles upon the opening or the Bluewater shopping centre in 1999. This has been followed by a successful Kickstart scheme on routes 495, 498 and 499 which brought in new Dennis Dart MPDs. A major development has been the operation of the award winning Fastrack services which commenced in 2006 with a new fleet of Volvo/Wrights vehicles running on significant lengths of busway. A second route started in June 2007 which features cashless operation on the bus and a dedicated bus bridge over the M25.

The Dartford operation has seen an increase in the number of London contracts operated and this now forms the vast majority of the operation there. The Arriva identity was applied as 'Arriva serving Kent Thameside' to the operations at Dartford and Northfleet.

London & Country into Arriva Surrey and West Sussex

London & Country was the trading name of London Country Bus (South West) Ltd, which was one of the four operations that London Country Bus Services Ltd was divided into prior to the privatisation of NBC. The former head office of LCBS at Reigate became that of L&C. The company was bought by Drawlane, as outlined above, in February 1988. However, the properties were leased back having been sold separately. The company was relaunched with a new livery and trading name in April 1989.

London & Country was successful in winning LT tenders and this led to the addition of new vehicles and the high profile opening of an impressive new garage at Beddington Farm in Croydon. Responsibility for this operation passed to Arriva London in 1999.

In 1990 the Woking, Guildford, and Cranleigh operations of the former Alder Valley business were purchased, and though kept as a separate operating company - Guildford and West Surrey - they were put under the same management as L&C. Also in 1990, a separate company called Horsham Buses was established for operations in the Horsham area.

Spare capacity at the Reigate garage allowed the development of the National Greenway concept in conjunction with East Lancs and many vehicles were prepared at Reigate.

Green Line coaches provide regular services between central London and Hertfordshire, Bedfordshire and Berkshire, continuing the name which is synonymous with London Transport's coach routes. Modern coaches with disability access from Van Hool are used on the London to Luton Airport service with 4065, YJ55WSW, illustrated. *Mark Lyons*

Briefly, London & Country had two subsidiaries in Dorset: Stanbridge and Crichel and Oakfield Travel. Both of these were later sold to Damory Coaches. Another subsidiary was Linkline Coaches of Harlesden in London, which specialised in coaching and corporate work. This was later sold to its management.

1993 saw the acquisition of Southend Transport and Colchester Transport and many L&C influences followed. These companies transferred to the supervision of County Bus but returned to Southern Counties in 2001.

The Croydon based operations and other LT tender operations at Walworth were transferred into a new company called Londonlinks. In 1995 Londonlinks was put under common management with Kentish Bus and Maidstone and District as part of the Invictaway Group. Reallocation of responsibilities in the enlarged group later saw the Croydon depot of Londonlinks return to L&C control before finally passing to London South.

A consequence of the property sales was the vacation of Reigate garage and its replacement by a facility at Merstham. The Head Office functions were consolidated in 1997 when the Reigate office closed as the functions moved to other premises including Crawley garage.

The three companies were renamed as Arriva Croydon and North Surrey, Arriva West Sussex, and Arriva Guildford and West Surrey, although only the latter one is still in operation. However the trading identity used is Arriva Surrey & West Sussex. In August 1998 the Countryliner coaching operation was sold to its manager. In 2001, the Crawley operation and depot was sold to Metrobus and the Merstham depot was closed.

In Guildford, significant investment has taken place with the depot completely rebuilt in 2002 and the fleet updated. Major new contracts are operated for Surrey County Council and in partnership with the University of Surrey.

Maidstone and District becomes Arriva Kent & Sussex

This is the original Maidstone and District Motor Services Ltd which was founded in 1911. Under NBC, it shared common management with East Kent from 1972 to 1983. In 1983 the Hastings and Rye area services were hived off as Hastings and District.

Maidstone and District was one of the first NBC companies to be privatised, being bought by its management team in November 1986. In 1988 New Enterprise of Tonbridge was purchased and is still kept as a separate entity. In June 1992 the assets of Maidstone Boro'line Maidstone including the premises at Armstrong Road were purchased.

In April 1995 the company was sold to British Bus, and in November the Head Office at Chatham was closed and staff moved along with Kentish Bus head office staff from Northfleet to the former Maidstone Boro'line premises at Maidstone under the Invictaway banner.

Under British Bus Cowie control, the group acquired a number of additional operators including Mercury Passenger Services of Hoo, Wealden Beeline of Five Oak Green, and the Grey Green (Medway) bus operation. In May of 1997 the Green Line operations in Gravesend and the Medway Towns were sold to the Pullman Group (London Coaches).

Recent developments have seen the substantial "Operation Overdrive" investment at Gillingham, a Kickstart scheme in Sittingbourne and further investment under the Maidstone Quality Partnership.

The Arriva branding used three identities, 'Arriva serving the Medway Towns', 'Arriva serving Kent and Sussex', and 'Arriva serving Maidstone'.

Midland Fox turns into Arriva Fox County Ltd

Midland Red East Ltd was formed in 1981 to take over the Leicestershire operations of Midland Red. In 1984 the company name was changed to Midland Fox Ltd, and there was a major relaunch of the company with a new livery and fox logo. There was also the launch of a new minibus network in Leicester under the, now discontinued, Fox Cub brand.

In 1987 the company was bought from NBC by its management with the help of the directors of Stevensons of Uttoxeter who, separately, bought the Swadlincote depot. Several smaller operators were also taken over. These included Wreake Valley of Thurmaston, Fairtax of Melton Mowbray, Astill and Jordan of Ratby, Shelton Orsborn of Wollaston, Blands of Stamford, and Loughborough Coach and Bus.

In 1989 Midland Fox was acquired by Drawlane. The following year it acquired Tellings-Golden Miller in Byfleet and this business, in turn, acquired the Coach Travel Centre in Cardiff, amongst others. Tellings bus operations eventually became part of London and Country, while Tellings was sold back to its management in 1994 before its expansion into bus operation in Cardiff.

In 1994 Pickering Transport was purchased by British Bus. Pickering build lorry bodies at their extensive site at Thurmaston, and now offer body repair and painting services which has resulted in many group vehicles appearing there. The site also houses one of Fox's three Leicester area depots. 1996 saw a launch of high quality services under the Urban Fox brand in a striking new blue livery.

Derby City Transport Ltd was a long established municipally owned bus company. In August 1989 it was sold to its employees who were assisted by Luton and District Transport. Luton and District took a 25% share holding in the business. There was a competitive interlude in Derby where Midland Red North started operations, but this ended with Derby buying out the competition in February 1990.

In 1994, after the acquisition of Luton and District by British Bus, the shareholders in Derby decided to accept an offer from British Bus for the rest of the share capital of the company. After a period of autonomy, the business was relaunched under the City Rider brand name and a yellow, red and blue livery. In January 1996, Derby City Transport was incorporated into the Midland Fox group; full integration and renumbering of the fleet took place at the start of 2000.

In 1990, "75" Taxis started as a division of Derby City Transport, building up a fleet of London style taxis. In September 1994, Midland Fox launched a new taxi service in Leicester marketed as Fox Cabs, but this operation was sold in 2001.

In September 1996, the Head Office of Midland Fox moved to the Pickering of Thurmaston premises along with a depot facility. It was later joined by the British Bus head office, now Arriva Passenger Services Ltd, which has since moved to new premises on the other side of Leicester.

The Arriva branding had vehicles carrying the 'Arriva serving the Fox County', or 'Arriva serving Derby' identities as appropriate. The taxi business remains under its previous brand. Special liveries include four vehicles in Quick Silver Shuttle for Leicester Park and Ride, Airport Car Park Shuttle and Airport Rail Link, three vehicles in a blue livery for East Midland Airport, two buses in a Corby to Kettering Rail Link for Midland Main Line, and two vehicles in a green livery for a Marks and Spencer shuttle service.

In 2003 Arriva Fox County and Arriva Midlands North merged to become Arriva Midlands.

Midland Red North & Stevensons become Arriva Midlands North

Midland Red North Ltd was founded in 1981 when the Midland Red company was divided into four operating parts by NBC. The company traded with local network names such as Chaserider for a considerable time. These were based upon the networks generated from the Viable Network Project later carried out across NBC as Market Analysis Project (MAP). The area included the then new town of Telford where a network of new services was introduced displacing many of the long-traditional operators.

The company was sold to Drawlane in January 1988 after a false start as described earlier. In 1989, it took over the Crewe and Etruria depots of fellow Drawlane subsidiary Crosville Motor Services Ltd. In 1992, Midland Red North purchased the Oswestry and Abermule operations of Crosville Wales Ltd, then an associated company. In 1993, with the dispersion of the Bee Line Buzz Company, the Macclesfield depot of that company which had traded as C-Line was taken over, having been part of Crosville for some time.

Stevensons of Uttoxeter commenced services in that part of Staffordshire in 1926 and continued as a small but successful family owned business. During the 1980s, and particularly after deregulation, significant growth occurred. In 1985 a controlling interest was acquired in the East Staffordshire Borough Council's bus operations in Burton-on-Trent. In 1987, the Swadlincote depot of Midland Fox and the Lichfield out station were purchased from NBC.

Growth in the West Midlands and the acquisition of a number of small companies including Crystal Coaches in Burslem and Viking Tours and Travel saw the company become a major independent operator in the early 1990s.

In April 1994, however, West Midlands Travel used its Your Bus subsidiary to retaliate in the Burton area against the significant level of operation Stevensons then had in the West Midlands area. This led to the sale of the company to British Bus in August 1994 and a significant scaling down of Stevensons operations in the West Midlands.

Macclesfield depot was transferred from Midland Red North into Stevensons in January 1995. From April 1995, Midland Red North and Stevensons were jointly managed. The closure of the Stevensons Head Office at Spath with the provision of central administration services from the Cannock head office became effective in 1996. A common livery was established between the two fleets though the Stevensons fleet name was retained on vehicles allocated to former Stevensons depots. Viking coaches retained a separate livery of two shades of grey until the operation was sold. The application of Arriva livery and branding saw the trading identity 'Arriva serving the North Midlands' applied to all buses.

In May 1998, the Shifnal depot of Timeline was acquired along with nineteen vehicles. The bus operations of Matthews Handybus in the Newcastle under Lyme area were acquired in February 1998, but no buses were involved. In August of the same year, the local bus operations of Williamsons of Knockin Heath were taken over and four vehicles came with the work.

1999 saw many SLF Dennis Darts arrive and these were followed by thirty Volvo B6BLE buses fitted with Wright Crusader 2 bodywork. All private hire coaching operations along with the vehicles were disposed of during the year.

In February 2003, Macclesfield, Crewe and Winsford depots were transferred from Arriva Midlands North to Arriva North West. The remaining depots of Midlands North were combined with Arriva Fox County to become Arriva Midlands.

North Western Road Car Co Ltd becomes Arriva North West

Ribble Motor Services was another NBC company divided in preparation for privatisation. The dormant Mexborough & Swinton Traction Company was renamed as above to take over the Merseyside, West Lancashire and Wigan operations of Ribble in September 1986. The head office of the new company was sited at Hawthorne Road, the Bootle area office north of Liverpool.

The company was acquired by Drawlane in March 1988. In 1989, the Runcorn and Warrington depots of Crosville were acquired. Expansion saw North Western open a depot in Altrincham, though eventually rationalisation saw the operations assumed by the Bee Line Buzz Company during its independent existence as a Drawlane subsidiary.

In 1993 Bee Line was put under the same management as North Western and a few weeks later, Liverline of Bootle was acquired along with 51 vehicles. Both were maintained as separate identities under the same management. Also acquired in 1993 was the bus operations of Express Travel, which at the time were still branded as Amberline, though this identity was not maintained.

The head office of the company later moved to Aintree depot, though a subsequent move saw the depot sold and redeveloped leaving the head office building free standing.

1995 was a busy year with two operations in the Wigan area acquired; Little White Bus and Wigan Bus Company. Also acquired in 1995 was Arrowline Travel based in Knutsford, which traded as Star Line. This brought luxury coaches on Airport related work as well as a modern fleet of mini and midi buses. The Star Line operation was later relocated to Wythenshawe, while the coaching operation was sold to Selwyns of Runcorn.

Increase in activity in the Warrington area required a new depot to be established at Haydock. The collapse of a Cheshire operator Lofty's of Mickle Trafford saw further growth in the mid-Cheshire area and new Cheshire workings took vehicles as far south as Whitchurch.

In 1997, Arriva acquired the residue of South Lancs Transport following that operator's withdrawal from Chester, and the business was put under the supervision of North Western.

In 1998, some bus operations of Timeline in the North West were purchased along with some vehicles, though the majority went to First Group. Arriva North West manages the bus and coach facilities at the Trafford Centre on behalf of the owners of this striking shopping centre which is located four miles west of Manchester. Arriva branding initially used the identity 'Arriva serving the North West'.

In 1999, the Winsford-based Nova Scotia operation was acquired, its services being integrated into the fleet. In 2000, Arriva acquired MTL, and the bus operations were put under the control of Arriva North West, those operations being branded 'Arriva serving Merseyside' and early in 2002 the head office functions of Arriva Cymru were transferred to Aintree. Upon full integration in the later months, the company was re-titled Arriva North West and Wales.

In August 2005 the vehicles and services of Blue Bus of Bolton were acquired adding 218 employees and eighty-six vehicles to the fleet and taking Arriva into the northern part of the Greater Manchester conurbation. As a result, Arriva almost doubled the buses operating in the Manchester area that hitherto covered the city centre and areas to the south of the city into Cheshire.

Yorkshire Bus Group to Arriva Yorkshire

The West Riding Automobile Company and Yorkshire Woollen District Transport were put under common management by NBC and, when privatisation happened in January 1987, the management team bought both companies. Selby and District was a trading title turned into a separate company by the new owners' Caldaire Holding company. While Caldaire became involved in the North East, the core business in West Yorkshire changed very little but there was steady investment in fleet replacement and upgrade.

There was involvement in the splitting up of National Travel East leaving a residue of operations on National Express contracts, and also competitive operations in Sheffield that led to corresponding competition in Wakefield.

The South Yorkshire Road Transport Company of Pontefract was purchased in July 1994, and maintained a separate trading identity for a time. In March 1995, the Caldaire Group was acquired by British Bus. Jaronda Travel of Selby was acquired in August 1999.

The Arriva identities used were 'Arriva serving Yorkshire' and 'Arriva serving Selby' seeing a merging together of the West Riding and Yorkshire identities for the first time. Arriva Yorkshire partnered with First Leeds in the extension to the East Leeds Guided Bus Corridor along the A64 York Road. This initiative saw both operators together contributing nearly half the scheme cost of around £9.9m with the other half coming from a partnership of Leeds City Council and West Yorkshire PTE.

United, Tees and District, and Teesside Motor Services Ltd become Arriva North East

United Automobile Services Ltd was another NBC subsidiary divided up in preparation for privatisation. In 1986, the northern part of the operating area was hived off into a new company called Northumbria. United continued to trade south of the Tyne, with its head office in Darlington. The operations in Scarborough and Pickering were transferred to a subsidiary of East Yorkshire Motor Services.

In December 1987, United was bought from NBC by Caldaire Holdings, the management buyout vehicle of the West Riding management team. In 1989, the National Express coaching activities of United were sold off to a joint venture company Durham Travel Services, set up by two former United managers with National Express Ltd.

In 1990, United was split into two parts, the Durham and North Yorkshire section continuing to trade as United, the section in Cleveland trading as Tees and District. At this time, the associated businesses of Trimdon Motor Services and Teeside Motor Services were acquired, with the Trimdon business being absorbed into United and the Teeside business continuing.

In the summer of 1992, there was a demerger of the Caldaire Group, with the North East operations passing to the Westcourt Group, and Caldaire North East becoming North East Bus.

In 1994, a new head office and engineering works in Morton Road, Darlington allowed the vacation of the Grange Road site for redevelopment. Also in 1994, the Westcourt Group sold to West Midlands Travel in the November, and North East Bus became part of the National Express Group following the merger with that group in 1995.

Eden Bus Services of Bishop Auckland was acquired in October 1995 and was absorbed into the main operation. National Express Group sold North East Bus to the Cowie Group on the last day of July 1996 and, in October, the Ripon depot operations were sold to Harrogate and District Travel.

Northumbria Motor Services Ltd into Arriva Northumbria

In 1986, the operations of United Auto were split into two parts in preparation for privatisation. The dormant Southern National Omnibus Company Ltd was renamed Northumbria and took over operations in September 1986 with a new head office in Jesmond.

In October 1987, Northumbria was acquired from NBC by its management using Proudmutual as a holding company. Proudmutual also acquired Kentish Bus in March 1988. Other acquisitions included Moor-Dale Coaches and Hunters. In 1994 the Proudmutual group was acquired by British Bus while, at the same time, Moor-Dale Coaches was sold back to former directors.

In the Arriva era, two trading identities were used: 'Arriva serving Northumbria' and 'Arriva serving the North East'.

Crosville Cymru into Arriva Cymru

Crosville Wales Ltd was, until August 1986, the Welsh and Shropshire operations of Crosville Motor Services Ltd based in Chester. In 1986, it was resolved that the Crosville company was too large to be offered for privatisation as a whole, and the then dormant Devon General Omnibus and Touring Company Ltd was revived by NBC in order to take over the assets and business of Crosville in Wales, to be renamed Crosville Wales. The management team of Crosville Wales purchased the company from NBC in December 1987.

In January 1989, the company was bought by National Express Holdings Ltd. In July of that year, it purchased a subsidiary company called Amberline, based at Speke in Liverpool, and added a bus operation to the mainly National Express coach contracts operated.

In July 1991, the National Express group was purchased by a consortium of banks led by Drawlane, as explained earlier. Ultimately, this led to Crosville Wales becoming a full member of the Drawlane Group shortly before its transformation into British Bus plc. In January 1992, the Oswestry depot and its outstation at Abermule were sold to Midland Red North.

Crosville Wales took advantage of second hand vehicles from other group companies and other operators, building a fleet of Leyland Lynx and National 2s while concurrently buying further new Mercedes minibuses and Dennis Darts.

In 1995, some of the services, but no vehicles, of Alpine Travel were acquired, leading to an operation of certain services as Alpine Bus in a red and white livery. This was superseded by a Shoreline livery of blue, white and yellow, which was phased out and replaced by route branding.

All operations were branded 'Arriva serving Wales/gwansanaethu Cymru', including those of the two acquisitions in 1998. The first was Devaway of Bretton, Chester. This brought a mixed fleet of VRTs, Nationals, and more Lynx and a depot from which to operate Chester area services. The second acquisition was Purple Motors of Bethesda. Further low floor buses arrived during 1999 to provide the Arriva share of a Quality Partnership Corridor on Deeside jointly provided with First Crosville. At the start of 2000 a large batch of low-floor Darts joined the fleet, and these displaced the last examples of the National in the fleet.

In February 2000, low-floor vehicles were introduced on low-volume rural services as north Wales authorities chose to use their share of the expanded Rural Bus Grant in improving quality rather than widening availability. Some vehicles are equipped with a bike racks.

Clydeside Buses Ltd into Arriva Scotland West Ltd

Clydeside Buses and its predecessors have been serving its core area of Renfrewshire and Inverclyde since 1928. Prior to 1985, the operation had formed the northern section of Western Scottish, part of the Scottish Bus Group (SBG). In preparation for the deregulation of local bus services, Clydeside Scottish assumed responsibility for the Glasgow, Renfrewshire and Inverclyde operations of Western Scottish in 1985. Over the next six years, there was a complex series of reorganisations between Clydeside and Western until, in 1991, Clydeside became the last SBG subsidiary to be privatised when it was purchased by its employees with assistance from Luton and District Transport Group, emerging as Clydeside 2000 plc.

After 1986, there were numerous competitors in the core area and trading proved extremely difficult. When the LDT group sold to British Bus, an offer put to the shareholders of Clydeside was accepted and Clydeside joined British Bus. There was immediate effort to update the fleet against a background of tightening up of enforcement generally in the area. Some of the competitive battles had led to the Traffic Commissioner taking steps to control the number of departures and waiting times in certain town centres.

The development of services has seen Flagship Routes introduced to raise quality levels. Additionally, opportunity was taken to acquire various smaller operators in the area such as Ashton Coaches of Greenock, and a significant share in Dart Buses of Paisley. Operations from the Greenock base were restyled as GMS-Greenock Motor Services with a separate livery. McGills Bus Service Ltd of Barrhead was acquired by the group in 1997 and for a time was kept as a separate entity from Clydeside Buses. Clydeside also acquired Bridge Coaches of Paisley which was fully absorbed into Clydeside.

During 2001, the shareholding in Dart Buses was sold to Stagecoach Western and the dormant McGills company used as a vehicle to sell off all remaining Inverclyde operations which ceased trading at the end of June. The redundant depot at Greenock and subsequently former McGill's site at Barrhead were demolished with all operations spread between remaining sites at Inchinnan and Johnstone. In a final move to consolidate the business the remote head office site located in Renfrew was vacated and all employees subsequently relocated within a refurbished facility at Inchinnan depot. The 'Arriva serving Scotland' branding covers all operations.

Arriva acquired Heathrow-based **Tellings Golden Miller** group (TGM) in January 2008. Its group of companies, made up of wholly owned subsidiaries, operates bus and coach services across England. They include; OFJ Connections, Excel Passenger Logistics, Flight Delay Services, Classic Coaches, Burtons Coaches and Link Line Coaches all of which have kept their identity so that customers can identify with local brands. TGM operates more than 250 vehicles including; front line executive coaches, local buses, and coaches operated on behalf of National Express. TGM operates from bases in the following towns or cities; London Heathrow, Harlesden, Colchester, Cambridge and Newcastle-upon-Tyne.

Expansion into Europe

September 1997	Unibus Holdings, Denmark
January 1998	Vancom Nederland
December 1998	Veonn & Hanze, Netherlands
March 1999	Bus Danmark
July 1999	Mercancias Ideal Gallego, Spain
September 1999	Transportes Finisterre, Spain
November 2000	Ami-Transportes, Portugal
December 2000	Abilio da Costa Moreira, Portugal
April 2001	Combus, Denmark
January 2002	Autocares Mallorca
June 2002	Transportes Sul ode Tejo, Portugal
July 2002	SAB Autoservizi, Italy
April 2004	Prignitzer Eisenbahn Gruppe, Germany
October 2004	Regentalbahn, Bavaria, Germany
February 2005	Sippel, Germany
July 2005	SAVDA group, Italy
2006	Verkehrsbetriebe Bils KG, Germany
	Trancentrum Bus s.r.o, Czech Republic
2007	Esfera, Madrid
	Bosak, Czech Republic

During 2003, Arriva completed its purchase of Transportes Sul do Tejo by acquiring the Barraqueiro Group's remaining 49%. It also began to operate its Danish rail franchises in Mid and North Jutland.

In 2004, the group entered the German public transport market with the acquisition of Prignitzer Eisenbahn Gruppe and in October purchased a stake in Regentalbahn AG from the

Bavarian State which was later increased to 100 per cent. A further acquisiton in Germany was the bus businness of Sippel which operates throughout the Rhine-Main area.

Arriva also grew its position in the Italian public transport market with the acquisition of Società Autoservizi FVG SpA. In July 2005 it was announced that Arriva had also agreed to acquire 80 per cent of the operations of the SAVDA Group and it has an option to acquire the remaining 20 per cent in 2008.

These acquisitions continue the strategy of developing Arriva's mainland European transport business, which is now a major contributor to the group's results. Arriva now has significant positions in the Czech Republic, Netherlands, Denmark, Portugal and Italy, Germany, Spain and Sweden.

Scandinavia

Arriva Skandinavien provides bus and rail services in Denmark and Sweden. The first acquisition in mainland Europe was Unibus in Denmark, which was acquired in 1997. Unibus operated approximately 8% of the tendered market in Copenhagen and had operations throughout Jutland and Zealand. Founded in 1985, Unibus grew by winning tendered bus operations for the Transport Authority for the Greater Copenhagen Area - HT. Unibus was the biggest coach operator in Copenhagen until in 1995, when the coaching business was sold to Lyngby Turistfart.

This involvement within the Danish bus market led to the acquisition of Bus Danmark (now Arriva Skandinavien) in 1999 and its wholly owned subsidiary Ödakra Buss based in southern Sweden. In April 2000, Arriva Skandinavien acquired the former state owned Company, COMBUS. Part of the company was sold on to Connex (Veolia) but Arriva kept the majority of its regional and provincial bus operations throughout Denmark. In 2007 Arriva Skandinavien acquired all of Veolia's public transport activities in Denmark comprising 750 buses and 1,800 employees. Arriva Skandinavien today operates a fleet of approximately 2,000 buses, including demand responsive vehicles, and 47 train sets. During 2001 the first double-deck buses for the Danish fleet were delivered.

Arriva Danmark was also successful with its bid in respect of the first rail passenger franchises to be announced in Denmark. The package consists of two rail passenger franchises connecting Mid and West Jutland with Aarhus, the second largest town in Denmark. The franchises commenced operation in January 2003. Twenty-nine new Coradia Lint 41 trains were delivered in 2004.

In August 2007, Arriva completed the acquisition of Veolia Danmark, the second largest operator in the Danish bus market. The operations have been successfully integrated enabling Arriva to build on its position as the leading bus operator in the country.

The group was the first private company to be awarded passenger rail franchises under new legislation to privatise the Danish rail industry. In January 2003 it started operating services on eight-year concessions in Jutland. Arriva carries five per cent of all passengers in the country and covers 15 per cent of the national network. Arriva is a top performing rail operator in Denmark.

Arriva's Danish operation employs 4,900 people and operates 1,930 vehicles and 47 trains. Arriva first entered the Swedish bus market when it acquired the Danish company Bus Danmark and its wholly owned subsidiary Ödakra Buss based in southern Sweden in 1999. Arriva Sverige (Ödakra Buss) activities are focused in the Skåne Län region of southern Sweden and in and around Jönköping operating a total of 340 buses. In 2005 a new contract called for a large fleet of MAN buses to operate in the city of Helsingborg. In June 2007 the group began operating train services in Sweden, with a nine-year contract to operate the Pågatåg regional train service in the Skåne region of southern Sweden. Arriva's Swedish operation employs 1,300 people and operates more than 360 buses and 26 trains.

The Netherlands

Arriva entered the Dutch transport market in 1998 when it acquired two bus companies. It bought Vancom Nederland, the first and to date the only municipal privatisation, and purchased Veonn and Hanze under a part privatisation of the Dutch state-owned transport company Connexxion.

Arriva Nederland won two large tenders to operate bus services in central Holland in 2002, the first private operator to win a franchise from the state-owned incumbent.

In recent years the company has increased its share of the rail passenger sector, with major contract wins in the northern provinces as well as the Dordrecht area, where train services operate alongside our buses.

In December 2003 Arriva Nederland was successful in bidding for a large tender to operate public transport services in the city and province of Groningen and the province of Drenthe. In June 2005, Arriva won a 120 bus, six-year contract in Waterland, north of Amsterdam. Arriva also operate buses in central Netherlands, near to Rotterdam, in the DAV area and in Brabant.

Arriva initiated a joint venture with the state-owned rail operator NS Rail in 1999, combining their trains with Arriva's buses to provide integrated services in the north of the country. The joint venture represented the first rail privatisation in the Netherlands.

In 2005 Arriva won the 15-year rail contract and is now sole provider of regional rail services in provinces of Friesland and Groningen and provides cross-border services into Germany. Arriva has also secured a 12-year rail contract as part of the DAV concession to run trains between Dordrecht and Geldermalsen. Arriva's combined businesses in the Netherlands operate 820 vehicles, 48 trains and employ 2,070 people.

Portugal

Arriva entered the Portuguese bus market in 2000 when it acquired AMI Transportes, Joao Carlos Soares and Abilio da Costa Moreira, three family-owned companies running inter-urban services in the north west of Portugal.

In June 2002 it acquired 51 per cent of Transportes Sul do Tejo (TST) from the Barraqueiro Group and exercised its option to purchase the remaining 49 per cent in September 2003.

TST is the major private operator of scheduled bus and coach services in the growing commuter region to the south of Lisbon, an area of some 600 square miles. It also operates schools and works contracts in the region. Together with Arriva's operations in the north, Arriva Portugal – Transportes LDA, it is one of Portugal's leading bus companies. The combined business has a fleet of 850 vehicles and 1,630 employees.

In 2006, Arriva acquired 21.5 per cent of Barraqueiro SGPS SA, the leading Portuguese transport operator, with bus and rail operations in and around Lisbon.

Italy

Arriva entered the Italian market in July 2002 when it acquired SAB Autoservizi Srl, a bus group with subsidiaries operating in the Lombardy, Liguria and Friuli-Venezia Giulia regions of northern Italy.

SAB and its subsidiary companies operate principally in the Lombardy region of northern Italy, providing urban, inter-urban and airport services in the Bergamo, Brescia and Lecco areas.

The group made further acquisitions in April 2004 when it acquired 49 per cent of Società Autoservizi FVG SpA (SAF) in the Udine area of the Friuli-Venezia Giulia region, and in October 2005, when it acquired 80 per cent of the SADEM operations near Turin.

SADEM and its subsidiary SAPAV operate in the Piemonte and Valle d'Aosta regions of northern Italy to the west of Arriva's other operations.

In June 2007, Arriva entered a joint venture with Ferrovie Nord Milano Group (FNM SpA) and contracted to acquire 49 per cent of Italian bus operator SPT Linea. Arriva's combined businesses in Italy have 2,100 employees and operate 1,770 vehicles.

Germany

Arriva entered the German public transport market - the largest in Europe - in April 2004, with the acquisition of rail company Prignitzer Eisenbahn Gruppe (PEG). PEG runs services in the federal states of North Rhine-Westphalia, Brandenburg and Mecklenburg-West Pomerania. It further strengthened its position with the acquisition of Regentalbahn based in eastern Bavaria. The business provides regional passenger rail services in Bavaria, Thuringia and Saxony. It also also operates rail services into the Czech Republic in co-operation with the Czech State Railway.

The group entered the German bus market with its acquisition of Sippel in February 2005 and strengthened its position by acquiring Verkehrsbetriebe Bils and Neißeverkehr in 2006.

In January 2007 Arriva acquired BB Riesen which operates bus services in Neustrelitz in Mecklenburg Vorpommern and in February 2007 acquired 80 per cent of Neißeverkehr GmbH which provides urban and interurban bus services in the Neisze region including Spremberg, Forst and Guben. Most recently, Arriva acquired a majority stake in OHE in March 2007. OHE operates bus services in and around Hannover, Braunschwieg, Bremen and Hamburg and is a shareholder in Metronom which operate two passenger rail franchises.

Now with several well established companies Arriva Deutschland provides a reliable network of bus and rail services in Bavaria, Berlin, Brandenburg, Bremen, Hamburg, Hessen, Lower Saxony, Mecklenburg-Vorpommern, Northrhine-Westphalia, Saxony and Thüringia.

The group's German operations currently employ more than 3,100 people and operate 215 trains and some 840 vehicles.

Czech Republic

Arriva entered the Czech Republic bus market in December 2006 with the acquisition of Transcentrum Bus s.r.o, a leading company operating bus services to the north of Prague. Based in Mlada Boleslav, it operates in the Stredocesky region with additional services in the regions of Liberecky and Kralovehradecky.

In January 2007 the group acquired Bosak Bus s.r.o, which operates to the south west of Prague, strengthening its position in the Czech Republic bus market.

In December 2007 Arriva began rail operations in north west Poland as part of the PCC Arriva joint venture with the first rail contact to be awarded to a private company in the country.

In July 2008 Arriva completed the purchase of 80 per cent of Eurobus Bus Invest which operates regional and contract bus services in Slovakia and Hungary.

Arriva UK Trains

Arriva operates the Arriva Trains Wales/Trenau Arriva Cymru and CrossCountry passenger rail franchises. Arriva was awarded the Welsh rail franchise in 2003 and CrossCountry in 2007.

Arriva Trains Wales/Trenau Arriva Cymru operates interurban, commuter and rural passenger services throughout Wales and the border counties.

The CrossCountry network is the most extensive rail franchise in the UK. Stretching from Aberdeen to Penzance, and from Stansted to Cardiff, it covers around 1,500 route miles and calls at over a hundred and twenty stations.

ARRIVA SCOTLAND WEST

Arriva Scotland West Ltd, Greenock Road, Inchinnan, PA4 9PG

254	IN	M104RMS	Scania L113CRL	Alexander Strider	B51F	1995
255	IN	M106RMS	Scania L113CRL	Alexander Strider	B51F	1995
256	IN	M107RMS	Scania L113CRL	Alexander Strider	B51F	1995

257-261

Scania N113CRL East Lancs European N45F 1995

| 257 | IN | M108RMS | 259 | IN | M110RMS | 260 | IN | M112RMS | 261 | IN | M113RMS |
| 258 | IN | M109RMS | | | | | | | | | |

262-269

Scania L113CRL East Lancs European N51F 1995

| 262 | IN | M114RMS | 265 | IN | M117RMS | 267 | IN | M119RMS | 269 | JO | M121RMS |
| 263 | IN | M115RMS | 266 | IN | M118RMS | 268 | IN | M120RMS | | | |

| 270 | IN | L25LSX | Scania N113CRL | East Lancs European | N51F | 1993 | Scania demonstrator, 1995 |

1001	IN	CX55EAA	Mercedes-Benz Citaro O530	Mercedes-Benz	NC42F	2006
1002	IN	CX55EAC	Mercedes-Benz Citaro O530	Mercedes-Benz	NC42F	2006
1003	IN	CX55EAE	Mercedes-Benz Citaro O530	Mercedes-Benz	NC42F	2006

1401-1407

Dennis Dart SLF 10.2m Alexander ALX200 N36F 1997 Arriva London, 2003

| 1401 | JO | P962RUL | 1403 | JO | P964RUL | 1405 | JO | P966RUL | 1407 | JO | P968RUL |
| 1402 | JO | P963RUL | 1404 | JO | P965RUL | 1406 | JO | P967RUL | | | |

1408	JO	V311NGD	Dennis Dart SLF 10.7m	Plaxton Pointer 2	N43F	1999
1409	JO	V312NGD	Dennis Dart SLF 10.7m	Plaxton Pointer 2	N43F	1999
1410	JO	V313NGD	Dennis Dart SLF 10.7m	Plaxton Pointer 2	N43F	1999

1411-1415

Dennis Dart SLF Plaxton Pointer N35F 1996

| 1411 | JO | P801RWU | 1413 | JO | P803RWU | 1414 | JO | P804RWU | 1415 | JO | P805RWU |
| 1412 | JO | P802RWU | | | | | | | | | |

1416-1425

Dennis Dart SLF Alexander ALX200 N40F 1997

1416	JO	P806DBS	1419	JO	P809DBS	1422	JO	P812DBS	1424	JO	P814DBS
1417	JO	P807DBS	1420	JO	P810DBS	1423	JO	P813DBS	1425	JO	P815DBS
1418	JO	P808DBS	1421	JO	P811DBS						

The three Mercedes-Benz Citaro buses from Arriva North West now operate from Inchinnan. Vehicles for their former route in the north-west are now being supplied by the Tellings coaching unit. In its new location, 1000, CX55EAA heads for Erskine. *Mark Doggett*

1426-1450 — Dennis Dart SLF — Plaxton Pointer — N39F — 1997

1426	JO	P816GMS	1433	JO	P823GMS	1439	JO	P829KES	1445	JO	P835KES
1427	JO	P817GMS	1434	JO	P824GMS	1440	JO	P830KES	1446	JO	P836KES
1428	JO	P818GMS	1435	JO	P825KES	1441	JO	P831KES	1447	JO	P837KES
1429	JO	P819GMS	1436	IN	HIL2148	1442	JO	P832KES	1448	JO	P838KES
1430	JO	P820GMS	1437	JO	P827KES	1443	JO	P833KES	1449	JO	P839KES
1431	IN	P821GMS	1438	JO	P828KES	1444	JO	P834KES	1450	JO	P840KES
1432	JO	P822GMS									

1451-1455 — Dennis Dart SLF — Alexander ALX200 — N40F — 1998

1451	JO	R381JYS	1453	JO	R383JYS	1454	JO	R384JYS	1455	JO	R385JYS
1452	JO	R382JYS									

1456-1465 — Dennis Dart SLF — Alexander ALX200 — N40F — 1998-99

1456	JO	S860OGB	1459	JO	S863OGB	1462	JO	S866OGB	1464	JO	S868OGB
1457	JO	S861OGB	1460	JO	S864OGB	1463	JO	S867OGB	1465	JO	S869OGB
1458	JO	S862OGB	1461	JO	S865OGB						

1466	JO	P210LKJ	Dennis Dart SLF 10.1m	Plaxton Pointer	N40F	1997	Southern Counties, 2009
1467	JO	P211LKJ	Dennis Dart SLF 10.1m	Plaxton Pointer	N40F	1997	Southern Counties, 2009
1468	JO	P212LKJ	Dennis Dart SLF 10.1m	Plaxton Pointer	N40F	1997	Southern Counties, 2009
1601	IN	M65FDS	Dennis Dart 9.8m	Plaxton Pointer	B40F	1995	
1602	IN	M67FDS	Dennis Dart 9.8m	Plaxton Pointer	B40F	1995	

1603-1610 — Dennis Dart 9.8m — Plaxton Pointer — B40F — 1995-96 — Arriva London, 2003

1603	IN	N681GUM	1606	JO	N672GUM	1608	IN	N677GUM	1610 IN N685GUM
1605	IN	P822RWU	1607	IN	N675GUM	1609	IN	N683GUM	

1614	IN	N686GUM	Dennis Dart 9.8m	Plaxton Pointer	B40F	1996	Arriva London, 2003
1615	IN	N687GUM	Dennis Dart 9.8m	Plaxton Pointer	B40F	1996	Arriva London, 2003
1619	IN	N689GUM	Dennis Dart 9.8m	Plaxton Pointer	B40F	1996	Arriva London, 2003
1620	IN	N691GUM	Dennis Dart 9.8m	Plaxton Pointer	B40F	1996	Arriva London, 2003
1624	JO	S624KHN	Dennis Dart SLF 10.1m	Plaxton Pointer 2	N39F	1999	
1628	IN	N474MUS	Dennis Dart 9.8m	Northern Counties Paladin	B39F	1995	
1642	JO	S642KHN	Dennis Dart SLF 10.1m	Plaxton Pointer 2	N39F	1999	
1643	JO	S643KHN	Dennis Dart SLF 10.1m	Plaxton Pointer 2	N39F	1999	
1693	JO	K543ORH	Dennis Dart 9m	Plaxton Pointer	B34F	1992	Arriva Midlands, 2008
1776	JO	K538ORH	Dennis Dart 9m	Plaxton Pointer	B34F	1992	Arriva London, 2002
1777	JO	K539ORH	Dennis Dart 9m	Plaxton Pointer	B34F	1992	Arriva London, 2002
1778	JO	K540ORH	Dennis Dart 9m	Plaxton Pointer	B34F	1992	Arriva London, 2002
1779	JO	K541ORH	Dennis Dart 9m	Plaxton Pointer	B34F	1992	Arriva London, 2002
1780	JO	N710GUM	Dennis Dart 9m	Plaxton Pointer	B34F	1995	Arriva London, 2003
1781	JO	N711GUM	Dennis Dart 9m	Plaxton Pointer	B34F	1995	Arriva London, 2003
1782	JO	N712GUM	Dennis Dart 9m	Plaxton Pointer	B34F	1995	Arriva London, 2003
1783	JO	P913PWW	Dennis Dart 9m	Plaxton Pointer	B34F	1996	Arriva London, 2003
1784	JO	P914PWW	Dennis Dart 9m	Plaxton Pointer	B34F	1996	Arriva London, 2003
1785	JO	P915PWW	Dennis Dart 9m	Plaxton Pointer	B34F	1996	Arriva London, 2003
1786	IN	N708GUM	Dennis Dart 9m	Plaxton Pointer	B34F	1995	Arriva London, 2003
1787	IN	N709GUM	Dennis Dart 9m	Plaxton Pointer	B34F	1995	Arriva London, 2003
1788	JO	K946SGG	Dennis Dart 9m	Plaxton Pointer	B35F	1993	
1918	IN	W78PRG	DAF SB120	Wright Cadet	N39F	2000	Arriva NE, 2005
1919	IN	W79PRG	DAF SB120	Wright Cadet	N39F	2000	Arriva NE, 2005
1921	IN	W82PRG	DAF SB120	Wright Cadet	N39F	2000	Arriva NE, 2005
1951	IN	YJ54CKG	VDL Bus SB120	Wrightbus Cadet 2	N30F	2004	
1952	IN	YJ54CKK	VDL Bus SB120	Wrightbus Cadet 2	N30F	2004	

1955-1959 — VDL Bus SB120 — Plaxton Centro — N40F — 2007

1955	IN	YJ07JSU	1957	IN	YJ07JSX	1958	IN	YJ07JSY	1959 IN YJ07JSZ
1956	IN	YJ07JSV							

Five VDL Bus SB120 vehicles joined the Arriva Scotland fleet in 2007 with ten examples of the longer SB200 arriving in 2009, this time with Wrightbus Pulsar 2 bodywork. Seen working the Paisley town service is 1956, YJ07JSV. *Richard Godfrey*

2001-2010

		VDL Bus SB200		Wrightbus Pulsar 2		N44F	2009	
2001	IN	YJ09CVH	**2004**	IN	YJ09CUV	**2007**	IN	YJ09CVO	**2009**	IN	YJ09CUW

2001	IN	YJ09CVH	2004	IN	YJ09CUV	2007	IN	YJ09CVO	2009	IN	YJ09CUW

2001 IN YJ09CVH   **2004** IN YJ09CUV   **2007** IN YJ09CVO   **2009** IN YJ09CUW
2002 IN YJ09CVK   **2005** IN YJ09CVM   **2008** IN YJ09CVR   **2010** IN YJ09CUX
2003 IN YJ09CVL   **2006** IN YJ09CVN

2601	IN	MV02XYJ	Mercedes-Benz Vario 0814	Plaxton Beaver 2	B27F	2002	Arriva NW & Wales, 2009
2602	IN	MV02XYK	Mercedes-Benz Vario 0814	Plaxton Beaver 2	B27F	2002	Arriva NW & Wales, 2009
2603	IN	MK52XNS	Mercedes-Benz Vario 0814	Plaxton Beaver 2	B27F	2002	Arriva NW & Wales, 2009
2703	IN	R703MHN	Optare MetroRider MR17	Optare	B31F	1997	Arriva North East, 2007
2740	IN	N803BKN	Optare MetroRider MR15	Optare	B29F	1996	Arriva Southern Counties, 2004
2741	IN	N804BKN	Optare MetroRider MR15	Optare	B29F	1996	Arriva Southern Counties, 2004
2757	IN	P221SGB	Optare MetroRider MR17	Optare	B29F	1996	
2801	IN	SF09LOD	Renault Master	IndCar	N15F	2009	
2802	IN	YX09EVG	Volkswagen T5	Bluebird Tucana	N14F	2009	
2803	IN	YX09EVH	Volkswagen T5	Bluebird Tucana	N14F	2009	
3001	IN	GSU347	DAF SB3000	Van Hool Alizée HE	C53F	1994	Arriva The Shires, 2009
3002	IN	WSU476	DAF SB3000	Van Hool Alizée HE	C51F	1994	Arriva The Shires, 2009

4401-4411 — Volvo B7RLE — Wrightbus Eclipse Urban — N37F — 2007

4401	JO	SF57NPK	4404	JO	SJ57DDY	4407	JO	SJ57DDU	4410	JO	SJ57DDO
4402	JO	SF57NPL	4405	JO	SJ57DDX	4408	JO	SF57NMO	4411	JO	SJ57DDN
4403	JO	SJ57DDZ	4406	JO	SJ57DDV	4409	JO	SF57NMM			

4531	JO	L201TKA	Volvo B6 9.9M	Plaxton Pointer	B38F	1994	Arriva North West & Wales, 2006
4532	JO	L210TKA	Volvo B6 9.9M	Plaxton Pointer	B38F	1994	Arriva North West & Wales, 2006
4533	JO	L216TKA	Volvo B6 9.9M	Plaxton Pointer	B38F	1994	Arriva North West & Wales, 2006
4534	JO	L306HPP	Volvo B6 9.9M	Northern Counties Paladin	N40F	1994	Arriva Midlands, 2009
4537	JO	L235TKA	Volvo B6 9.9M	Plaxton Pointer	B38F	1994	Arriva North West & Wales, 2006
4540	JO	L231TKA	Volvo B6 9.9M	Plaxton Pointer	B38F	1994	Arriva North West & Wales, 2006

4665-4670 — Scania OmniCity K230 UB — Scania — N45F — 2008

4665	IN	YR58SUH	4667	IN	YR58SUU	4669	IN	YR58SUX	4670	IN	YR58SUY
4666	IN	YR58SUO	4668	IN	YR58SUV						

7214	u	D187FYM	Leyland Olympian ONLXB/1RH	Eastern Coach Works	B42/26D	1986	Arriva London, 2007
7264	JO	M169GRY	Scania N113DRB	East Lancs	B47/33F	1994	Arriva Midlands, 2009
7265	JO	M160GRY	Scania N113DRB	East Lancs	B47/33F	1994	Arriva Midlands, 2009
7266	JO	M165GRY	Scania N113DRB	East Lancs	B47/33F	1994	Arriva Midlands, 2009
7267	JO	M831SDA	Scania N113DRB	East Lancs	BC43/29F	1995	Arriva Midlands, 2009
7268	JO	M832SDA	Scania N113DRB	East Lancs	BC43/29F	1995	Arriva Midlands, 2009
7269	JO	M833SDA	Scania N113DRB	East Lancs	BC43/29F	1995	Arriva Midlands, 2009
7273	JO	L273FVN	Leyland Olympian ON2R50C13Z4	Alexander RH	B45/29F	1993	
7274	JO	L274FVN	Leyland Olympian ON2R50C13Z4	Alexander RH	B45/29F	1993	
7301	JO	G504SFT	Leyland Olympian ONCL10/1RZ	Northern Counties Palatine	B47/30F	1989	Arriva Midlands, 2008
7304	JO	F575SMG	Leyland Olympian ONLXB/1RZ	Alexander RL	B47/32F	1988	Arriva The Shires, 2008
7305	JO	F572SMG	Leyland Olympian ONLXB/1RZ	Alexander RL	B47/32F	1988	Arriva The Shires, 2009

Previous registrations:

GSU347	M943LYR	V312NGD	V312NGD, WSU476
HIL2148	P826KES	V313NGD	V313NGD, GSU347
S869OGB	S869OGB, HIL2148	WSU476	M947LYR
V311NGD	V311NGD, WSU475		

Allocations:

Inchinnan (Greenock Road) - IN

Mercedes-Benz	2601	2602	2603					
Renault Master	2801							
Volkswagen	2802	2803						
MetroRider	2703	2740	2741	2757				
Dart	1601	1602	1603	1605	1606	1607	1608	1609
	1610	1614	1615	1619	1620	1628	1786	1787
Dart SLF	1431	1436	1460	1628				
SB120	1918	1919	1921	1951	1952	1955	1956	1957
	1958	1959						
SB200	2001	2002	2003	2004	2005	2006	2007	2008
	2009	2010						
Scania sd	254	255	256	257	258	259	260	261
	262	263	265	266	268	270	4665	4666
	4667	4668	4669	4670				
Citaro	1001	1002	1003					
DAF SB3000 coach	3001	3002						
Scania dd	7264	7265	7266	7267	7268	7269		

Six Scania OmniCity integral buses joined the Scottish operation in 2008. Pictured in Glasgow is 4670, YR58SUY, which carries the inter-urban livery that has also been applied to recent deliveries. *Richard Godfrey*

Johnstone (Cochranemill Road) - JO

Volvo B6	4531	4532	4533	4534	4537	4540		
Dart	1693	1776	1777	1778	1779	1780	1781	1782
	1783	1784	1785	1788				
Dart SLF	1401	1402	1403	1404	1405	1406	1407	1408
	1409	1410	1411	1412	1413	1414	1415	1416
	1417	1418	1419	1420	1421	1423	1424	1425
	1426	1427	1428	1429	1430	1432	1433	1434
	1435	1437	1438	1439	1440	1441	1442	1443
	1444	1445	1446	1447	1448	1449	1450	1451
	1452	1453	1454	1455	1456	1457	1458	1459
	1461	1462	1463	1464	1465	1466	1467	1468
	1624	1643						
Scania sd	269							
Volvo B7RLE	4401	4402	4403	4404	4405	4406	4407	4408
	4409	4410	4411					
Olympian	7273	7274	7301	7304	7305			

Unallocated and stored - u

Remainder

ARRIVA NORTH EAST

Arriva North East Ltd, Arriva House, Admiral Way, Sunderland, SR3 3XP

1-4			VDL Bus SB200			Plaxton Centro		N45F	2008		
1	BH	YJ58FFA	2	BH	YJ58FFB	3	BH	YJ58FFG	4	BH	YJ58FFC

271-280			Scania L113CRL			East Lancs European		NC45F	1996		
271	SN	P271VRG	275	SN	P275VRG	277	SN	P277VRG	279	SN	P279VRG
272	SN	P272VRG	276	SN	P276VRG	278	SN	P278VRG	280	DM	P814VTY
274	AS	P274VRG									

281-290			Scania L113CRL			East Lancs European		NC45F	1995		
281	AS	N281NCN	284	DM	N284NCN	287	DM	N287NCN	289	AS	N289NCN
282	DM	N282NCN	285	DM	N285NCN	288	BA	N288NCN	290	SN	N290NCN
283	DM	N283NCN									

902-923			Optare MetroRider MR15			Optare		B31F	1997-98		
902	SN	P902DRG	910	BA	R910JNL	917	AS	R917JNL	920	BA	R920JNL
903	SN	P903DRG	911	BA	R251JNL	918	AK	R918JNL	922	RR	R922JNL
904	DL	P904DRG	914	AS	R914JNL	919	AS	R919JNL	923	SN	R923JNL
905	DL	P905JNL	916	BA	R916JNL						

1201-1205			DAF SB3000			Plaxton Prima Interurban		BC51F	1997		
1201	HX	R291KRG	1203	HX	R293KRG	1204	HX	R294KRG	1205	HX	R295KRG
1202	HX	R292KRG									

1206-1214			DAF SB3000			Plaxton Prima Interurban		BC51F	1999		
1206	RR	V206DJR	1209	RR	V209DJR	1211	AK	V211DJR	1213	AK	V213DJR
1207	RR	V207DJR	1210	AK	V210DJR	1212	AK	V212DJR	1214	HX	V214DJR
1208	RR	V208DJR									

1401	BH	NK53HHX	VDL Bus SB200			Wrightbus Commander		N44F	2003		
1402	BH	NK53HHY	VDL Bus SB200			Wrightbus Commander		N44F	2003		
1403	BH	NK53HHZ	VDL Bus SB200			Wrightbus Commander		N44F	2003		

1404-1428			VDL Bus SB200			Wrightbus Pulsar 2		N44F	2009		
1404	AS	NK09BPF	1411	BH	NK09BRF	1417	BA	NK09EJF	1423	BA	NK09EJY
1405	AS	NK09BPO	1412	BH	NK09BRV	1418	BA	NK09EJG	1424	BA	NK09EJZ
1406	AS	NK09BPU	1413	BH	NK09BRX	1419	BA	NK09EJJ	1425	BA	NK09EKA
1407	AS	NK09BPV	1414	BH	NK09BRZ	1420	BA	NK09EJL	1426	BA	NK09EKB
1408	BH	NK09BPX	1415	BA	NK09EJD	1421	BA	NK09EJV	1427	BA	NK09EKC
1409	BH	NK09BPY	1416	BA	NK09EJE	1422	BA	NK09EJX	1428	BA	NK09EKD
1410	BH	NK09BPZ									

1429-1432			VDL Bus SB200			Wrightbus Pulsar 2		N44F	On order		
1429	BA	-	1430	BA	-	1431	BA	-	1432	BA	-

1504-1517			MAN 11.190			Optare Vecta		B42F	1993		
1504	SN	K504BHN	1506	SN	K506BHN	1508	SN	K508BHN	1517	SN	K517BHN
1505	SN	K505BHN	1507	SN	K507BHN	1515	SN	K515BHN			

1519-1551			MAN 11.190			Optare Vecta		B42F	1994		
1519	SN	L519FHN	1527	SN	L527FHN	1536	RR	L536FHN	1544	SN	L544GHN
1521	SN	L521FHN	1529	SN	L529FHN	1537	SN	L537FHN	1549	SN	L549GHN
1524	RR	L524FHN	1531	SN	L531FHN	1541	RR	L541FHN	1551	SN	L551GHN
1526	SN	L526FHN	1533	SN	L533FHN	1542	SN	L542FHN			

1552	u	M501AJC	MAN 11.190			Optare Vecta		B42F	1995	Arriva Cymru, 1998
1553	RR	M502AJC	MAN 11.190			Optare Vecta		B42F	1995	Arriva Cymru, 1999
1554	u	M503AJC	MAN 11.190			Optare Vecta		B42F	1995	Arriva Cymru, 1999
1555	u	M504AJC	MAN 11.190			Optare Vecta		B42F	1995	Arriva Cymru, 1999
1556	RR	UOI772	MAN 11.190			Optare Vecta		BC42F	1993	Arriva Midlands North, 1999
1557	RR	L102MEH	MAN 11.190			Optare Vecta		B42F	1994	Arriva Midlands North, 1999
1558	SN	K140RYS	MAN 11.190			Optare Vecta		BC42F	1993	Arriva Midlands North, 1999

Currently one of the main models selected by Arriva for its British fleets is the VDL Bus SB200 with Wrightbus Pulsar 2 bodywork, the revised styling introduced by Wrightbus in 2008. Darlington is the location for this view of 1428, NK09EKD. *Richard Godfrey*

1611-1626

Dennis Dart SLF 10.1m | Plaxton Pointer 2 | N39F | 1999

1611	DN	S611KHN	**1616**	PE	S616KHN	**1618**	BA	S618KHN	**1625**	BA	S625KHN
1612	DN	S612KHN	**1617**	PE	S617KHN	**1621**	RR	S621KHN	**1626**	BA	S626KHN
1613	PE	S613KHN									

1635-1640

Dennis Dart SLF 10.1m | Plaxton Pointer 2 | N39F | 1999

1635	RR	S635KHN	**1637**	DM	S637KHN	**1639**	PE	S639KHN	**1640**	PE	S640KHN
1636	PE	S636KHN	**1638**	PE	S638KHN						

1646-1656

Dennis Dart SLF 10m | Plaxton Pointer | N34F | 1997 | Arriva London, 2007

1646	SN	R421COO	**1649**	RR	R424COO	**1652**	PE	R427COO	**1655**	PE	R430COO
1647	SN	R422COO	**1650**	RR	R425COO	**1653**	PE	R428COO	**1656**	PE	R431COO
1648	SN	R423COO	**1651**	RR	R426COO	**1654**	PE	R429COO			

1657 DN K582MGT | Dennis Dart 9m | Plaxton Pointer | B34F | 1991 | Stagecoach, 2007

1658-1669

Dennis Dart 9.8m | Plaxton Pointer | B40F | 1992-93 | Stagecoach, 2007

1658	RR	K709PCN	**1661**	DN	L735VNL	**1664**	DN	L738VNL	**1667**	DN	L741VNL
1659	DN	K717PCN	**1662**	DN	L736VNL	**1665**	DN	L739VNL	**1668**	RR	L748VNL
1660	DN	L729VNL	**1663**	RR	L737VNL	**1666**	DN	L740VNL	**1669**	DN	L759VNL

1670 DN M770DRG | Dennis Dart 9.8m | Plaxton Pointer | B40F | 1994 | Stagecoach, 2007

1671-1679

Dennis Dart 9.8m | Alexander Dash | B36F | 1995-96 | Stagecoach, 2007

1671	DN	N301AMC	**1674**	DN	P615PGP	**1676**	DN	P617PGP	**1678**	DN	P637PGP
1672	DN	N303AMC	**1675**	RR	P616PGP	**1677**	DN	P634PGP	**1679**	DN	P638PGP
1673	RR	N305AMC									

1680-1684

Dennis Dart 9.8m | Alexander Dash | B40F | 1996 | Stagecoach 2007

1680	DN	P456EEF	**1682**	DN	P459EEF	**1683**	DN	P460EEF	**1684**	DN	P461EEF
1681	DN	P458EEF									

1685	DN	L141YVK	Dennis Dart 9m	Northern Counties Paladin	B35F	1994	Arriva Yorkshire, 2006
1686	DN	J220HGY	Dennis Dart 9m	Plaxton Pointer	B35F	1995	Arriva Yorkshire, 2006
1687	DN	J221HGY	Dennis Dart 9m	Plaxton Pointer	B35F	1995	Arriva Yorkshire, 2006
1688	HX	J468OKP	Dennis Dart 9.8m	Plaxton Pointer	B40F	1992	Arriva Yorkshire, 2006

1689-1692

Dennis Dart 9m • Northern Counties Paladin • B35F • 1994 • Arriva Yorkshire, 2006

1689	IN	L114YVK	1690	IN	L129YVK	1691	IN	L149BFT	1692	IN	L152YVK

1696	RR	M186YKA	Dennis Dart 9.8m	Plaxton Pointer	B40F	1995	Arriva NW&W, 2009
1697	BA	M186YKA	Dennis Dart 9.8m	Plaxton Pointer	B40F	1995	Arriva NW&W, 2009
1698	DM	M211YKD	Dennis Dart 9.8m	Plaxton Pointer	B40F	1995	Arriva NW&W, 2009
1701	DM	T701RCN	Dennis Dart SLF 8.8m	Plaxton Pointer MPD	N29F	1999	
1702	BH	T702RCN	Dennis Dart SLF 8.8m	Plaxton Pointer MPD	N29F	1999	

1703-1723

Dennis Dart SLF 8.8m • Plaxton Pointer MPD • N29F • 1999

1703	DN	V703DNL	1709	PE	V709DNL	1714	BA	V714DNL	1719	RR	V719DNL
1704	DN	V612DNL	1710	BA	V710DNL	1715	BA	V715DNL	1720	RR	V720DNL
1705	DN	V705DNL	1711	BA	V711DNL	1716	DN	V716DNL	1721	BH	V721DNL
1706	DN	V706DNL	1712	BA	V712DNL	1717	DN	V717DNL	1722	RR	V722DNL
1707	BA	V707DNL	1713	BA	V713DNL	1718	RR	V718DNL	1723	PE	V723DNL
1708	BA	V708DNL									

1724-1742

Dennis Dart SLF 8.8m • Plaxton Pointer MPD • N29F • 1999

1724	BH	V724DNL	1729	NL	V729DNL	1734	RR	V734DNL	1739	RR	V739DNL
1725	u	V725DNL	1730	NL	V730DNL	1735	DN	V735DNL	1740	BH	V740DNL
1726	BH	V726DNL	1731	NL	V731DNL	1736	BA	V736DNL	1741	BH	V741DNL
1727	RR	V727DNL	1732	NL	V732DNL	1737	BA	V737DNL	1742	BH	V742DNL
1728	BH	V728DNL	1733	BH	V733DNL	1738	RR	V738DNL			

1743-1749

Dennis Dart SLF 8.8m • Plaxton Pointer MPD • N29F • 2000

1743	PE	V743ECU	1745	DN	V745ECU	1747	DN	V747ECU	1749	HX	V749ECU
1744	RR	V744ECU	1746	DN	V746ECU	1748	PE	V748ECU			

1750-1757

Dennis Dart SLF 8.8m • Plaxton Pointer MPD • N29F • 2000

1750	NL	W751SBR	1752	NL	W753SBR	1754	RR	W756SBR	1756	RR	W758SBR
1751	NL	W752SBR	1753	RR	W754SBR	1755	RR	W757SBR	1757	PE	W759SBR

1758	NL	NK53HJA	TransBus Dart SLF 8.8m	TransBus Mini Pointer	N29F	2003	
1759	HX	NK53VKA	TransBus Dart SLF 8.8m	TransBus Mini Pointer	N29F	2004	

1760-1775

ADL Dart 8.8m • ADL Mini Pointer • NC29F • 2005

1760	HX	NK05GVX	1764	BA	NK05GWC	1768	SN	NK05GWG	1772	BA	NK05GWO
1761	DN	NK05GVY	1765	SN	NK05GWD	1769	SN	NK05GWJ	1773	BH	NK05GWU
1762	NL	NK05GVG	1766	SN	NK05GWE	1770	SN	NK05GWM	1774	BH	NK05GWV
1763	BA	NK05GWA	1767	SN	NK05GWF	1771	SN	NK05GWN	1775	BH	NK05GWW

1790	HX	NK55MYR	ADL Dart 8.8m	ADL Mini Pointer	NC29F	2006	
1791	HX	NK55MYS	ADL Dart 8.8m	ADL Mini Pointer	NC29F	2006	
1792	NE	NK55MYT	ADL Dart 8.8m	ADL Mini Pointer	NC29F	2006	
1793	DL	W166PNT	Dennis Dart SLF 8.8m	Plaxton Pointer MPD	N29F	2000	First Stop, Renfrew, 2005
1794	DL	W631RNP	Dennis Dart SLF 8.8m	Plaxton Pointer MPD	N29F	2000	First Stop, Renfrew, 2005
1795	DL	Y258KNB	Dennis Dart SLF 8.8m	Plaxton Pointer MPD	N29F	2001	First Stop, Renfrew, 2005
1796	DL	Y259KNB	Dennis Dart SLF 8.8m	Plaxton Pointer MPD	N29F	2001	First Stop, Renfrew, 2005
1797	DL	Y215BGB	Dennis Dart SLF 8.8m	Plaxton Pointer MPD	N29F	2001	First Stop, Renfrew, 2005
1798	DL	SA52MYT	Dennis Dart SLF 8.8m	Plaxton Pointer MPD	N29F	2003	First Stop, Renfrew, 2005
1799	BL	SF04RHA	TransBus Dart SLF 8.8m	TransBus Mini Pointer	N29F	2004	First Stop, Renfrew, 2005
1800	AS	NK56HKV	ADL Dart 4	ADL Enviro 200	N29F	2006	
1801	AS	NK56HKW	ADL Dart 4	ADL Enviro 200	N29F	2006	

1901-1922

DAF SB120 • Wright Cadet • N39F • 2000

1901	SN	W301PPT	1906	SN	W308PPT	1911	SN	W314PPT	1916	SN	W72PRG
1902	SN	W302PPT	1907	SN	W309PPT	1912	SN	W315PPT	1917	SN	W76PRG
1903	SN	W303PPT	1908	SN	W311PPT	1913	SN	W317PPT	1920	SN	W81PRG
1904	SN	W304PPT	1909	SN	W312PPT	1914	SN	W319PPT	1922	SN	W83PRG
1905	SN	W307PPT	1910	SN	W313PPT	1915	SN	W69PRG			

1923	AS	YJ57BVB	VDL Bus SB120	Plaxton Centro	N40F	2007	
1924	AS	YJ57BVC	VDL Bus SB120	Plaxton Centro	N40F	2007	
2504	NL	Y294PDN	Optare Solo M920	Optare	N27F	2001	Crystals, Dartford, 2008
2505	NL	Y295PDN	Optare Solo M920	Optare	N27F	2001	Crystals, Dartford, 2008
2502	BA	YN03UXW	Optare Alero	Optare	NC18F	2003	*Operated for Durham CC*
2627	SN	P627FHN	Optare MetroRider MR35	Optare	B25F	1996	

Inherited with the Proudmutual Group, Northumbria had a large number of the MetroRider integral minibus model. Many of these have now been replaced by the Optare Solo model, with Durham's 2809, YK08ESY, being one of fifty-one joining the North East fleet in 2008. *Richard Godfrey*

| 2631 | BA | P631FHN | Optare MetroRider MR35 | Optare | B25F | 1906 | |
| 2633 | RR | P633FHN | Optare MetroRider MR35 | Optare | B25F | 1997 | |

2646-2655

Mercedes-Benz Vario 0814   Alexander ALX100   B27F   2001

2646	BH	X646WTN	2649	BH	X649WTN	2652	BH	X652WTN	2654	BH	X654WTN
2647	BH	X647WTN	2650	BH	X650WTN	2653	BH	X653WTN	2655	BH	X656WTN
2648	BH	X648WTN	2651	BH	X651WTN						

2657	BH	R130GNW	Mercedes-Benz Vario 0814	Alexander ALX100	B27F	1998	Arriva North West, 2005
2658	BH	S822MCC	Mercedes-Benz Vario 0814	Plaxton Beaver 2	B27F	1998	Arriva North West, 2005
2659	BH	R113GNW	Mercedes-Benz Vario 0814	Plaxton Beaver 2	B33F	1998	Arriva North West, 2005
2660	BH	R112GNW	Mercedes-Benz Vario 0814	Plaxton Beaver 2	B33F	1998	Arriva North West, 2005
2661	BH	R129GNW	Mercedes-Benz Vario 0814	Alexander ALX100	B27F	1998	Arriva North West, 2005
2662	BH	S350PGA	Mercedes-Benz Vario 0814	Plaxton Beaver 2	B27F	1998	Arriva North West, 2005
2663	BH	S353PGA	Mercedes-Benz Vario 0814	Plaxton Beaver 2	B27F	1998	Arriva North West, 2005
2664	BH	R798DUB	Mercedes-Benz Vario 0810	Plaxton Beaver 2	B31F	1997	Arriva North West, 2005
2666	BH	S823MCC	Mercedes-Benz Vario 0814	Plaxton Beaver 2	B27F	1998	Arriva North West, 2005
2667	BH	S351PGA	Mercedes-Benz Vario 0814	Plaxton Beaver 2	B27F	1998	Arriva North West, 2005
2668	AS	R792DUB	Mercedes-Benz Vario 0810	Plaxton Beaver 2	B31F	1997	Arriva North West, 2005
2669	AS	S352PGA	Mercedes-Benz Vario 0814	Plaxton Beaver 2	B27F	1998	Arriva North West, 2005
2670	AS	R792DUB	Mercedes-Benz Vario 0810	Plaxton Beaver 2	B31F	1997	Arriva North West, 2005
2672	AS	R796DUB	Mercedes-Benz Vario 0810	Plaxton Beaver 2	B31F	1997	Arriva North West, 2005

2701-2725

Optare MetroRider MR15   Optare   B31F   1997-98

| 2701 | BA | R701MHN | 2705 | AS | R705MHN | 2711 | AS | R711MHN | 2723 | RR | R723MHN |
| 2702 | HX | R702MHN | 2710 | AS | R710MHN | 2716 | AS | R716MHN | | | |

2801-2851

Optare Solo M950   Optare   N31F   2008

2801	PE	YK08ERO	2814	DM	YK08ETJ	2827	DN	YK08XBG	2840	DN	YK08XBW
2802	PE	YK08ERU	2815	PE	YK08ETL	2828	DN	YK08XBH	2841	DN	YK08XBY
2803	DM	YK08ERV	2816	PE	YK08ETO	2829	DN	YK08XBK	2842	DN	YJ58CAA
2804	DM	YK08ERX	2817	PE	YK08ETR	2830	DN	YK08XBL	2843	DN	YJ58CAE
2805	DM	YK08ERY	2818	PE	YK08ETT	2831	DN	YK08XBM	2844	DN	YJ58CAO
2806	DM	YK08ERZ	2819	PE	YK08ETU	2832	DN	YK08XBN	2845	DN	YJ58CAV
2807	DM	YK08ESU	2820	PE	YK08ETV	2833	DN	YK08XBO	2846	DN	YJ58CAW
2808	DM	YK08ESV	2821	PE	YK08ETX	2834	DN	YK08XBP	2847	DN	YJ58CAX
2809	DM	YK08ESY	2822	PE	YK08ETY	2835	DN	YK08XBR	2848	DN	YJ58CBF
2810	DM	YK08ETA	2823	PE	YK08ETZ	2836	DN	YK08XBS	2849	DN	YJ58CBO
2811	DM	YK08ETD	2824	DN	YK08XBD	2837	DN	YK08XBT	2850	DN	YJ58CBU
2812	DM	YK08ETE	2825	DN	YK08XBE	2838	DN	YK08XBU	2851	DN	YJ58CBV
2813	DM	YK08ETF	2826	DN	YK08XBF	2839	DN	YK08XBV			

Another model chosen by Northumbria was the Optare Delta, this being based on the DAF SB220 chassis. Many of these have now been displaced while one of the last to be built, 4098, P130RWR, and which was new to Luton Airport, entered service when Arriva purchased the Blue Bus operation based in **Bolton**. *Richard Godfrey*

2852-2861

			Optare Solo SR M950			Optare		N31F	2009	

2852	BA	YJ59GJK	2855	BA	YJ59GJX	2858	BA	YJ59GKC	2860	BA	YJ59GKE
2853	BA	YJ59GJO	2856	BA	YJ59GJY	2859	BA	YJ59GKD	2861	BA	YJ59GKF
2854	BA	YJ59GJV	2857	BA	YJ59GKA						

3002-3025

Mercedes-Benz O405 — Optare Prisma — B49F — 1995

3002	RR	M302SAJ	3008	RR	N808XHN	3014	RR	N514XVN	3020	SN	N520XVN
3003	SN	M303SAJ	3009	RR	N809XHN	3015	RR	N515XVN	3022	RR	N522XVN
3004	RR	M304SAJ	3010	RR	N810XHN	3016	RR	N516XVN	3023	RR	N523XVN
3005	RR	M305SAJ	3011	RR	N511XVN	3017	RR	N517XVN	3024	RR	N524XVN
3006	RR	N806XHN	3012	RR	N512XVN	3018	RR	N518XVN	3025	RR	N525XVN
3007	RR	N807XHN	3013	RR	N513XVN	3019	SN	N519XVN			

3026	RR	L100SBS	Mercedes-Benz O405			Wright Cityranger		B51F	1993		Arriva Midlands North, 1999
4015	AS	K415BHN	DAF SB220			Optare Delta		B49F	1993		
4022	u	K422BHN	DAF SB220			Optare Delta		B49F	1993		

4023-4058

DAF SB220 — Plaxton Prestige — N45F — 1998

4023	RR	R423RPY	4033	BA	R433RPY	4041	DN	S341KHN	4050	DN	S350KHN
4024	DN	R424RPY	4034	BA	R434RPY	4042	BA	S342KHN	4051	SN	S351KHN
4025	RR	R425RPY	4035	DM	R435RPY	4043	SN	S343KHN	4052	DN	S352KHN
4026	RR	R426RPY	4036	DM	R436RPY	4044	DN	S344KHN	4053	RR	S353KHN
4027	RR	R427RPY	4037	SN	R437RPY	4045	SN	S345KHN	4054	SN	S354KHN
4030	SN	R430RPY	4038	SN	R438RPY	4046	SN	S346KHN	4056	DN	S356KHN
4031	BA	R431RPY	4039	SN	R439RPY	4048	SN	S348KHN	4057	DN	S357KHN
4032	BA	R432RPY	4040	SN	R440RPY	4049	BA	S349KHN	4058	RR	S358KHN

4059	RR	R701KCU	DAF SB220			Northern Counties Paladin		B41F	1997		

4060-4073

DAF SB220 — Plaxton Prestige — N41F — 1998

4060	DM	S702KFT	4064	DM	S706KFT	4068	DM	S710KFT	4071	DM	S713KRG
4061	DN	S703KFT	4065	DM	S707KFT	4069	DM	S711KFT	4072	DM	S714KRG
4062	DM	S704KFT	4066	DM	S708KFT	4070	DM	S712KRG	4073	DM	S715KRG
4063	SN	S705KFT	4067	DM	S709KFT						

Fourteen Scania OmniCity buses delivered in 2006 are shared between Durham and Redcar depots. Seen in Durham is 4654, NK05GXH, pictured on the Brandon service. *Richard Godfrey*

4074-4081

			DAF SB220			Ikarus CitiBus		N43F*	1999	*4080-81 are N40F
4074	BH	T74AUA	4076	BH	T76AUA	4078	BH	T79AUA	4080	RR	T81AUA
4075	BH	T75AUA	4077	BH	T78AUA	4079	BH	T83AUA	4081	RR	T82AUA

4082	RR	V653LWT	DAF SB220	Plaxton Prestige	B40F	1999	
4083	DN	J926CYL	DAF SB220	Ikarus CitiBus	B48F	1992	Arriva London, 2000
4086	BA	J413NCP	DAF SB220	Ikarus CitiBus	B48F	1992	Arriva London, 2000
4093	u	G254SRG	DAF SB220	Optare Delta	BC48F	1989	
4096	u	G257UVK	DAF SB220	Optare Delta	BC48F	1990	
4097	RR	G258UVK	DAF SB220	Optare Delta	BC48F	1990	
4098	DM	P130RWR	DAF SB220	Optare Delta	B44D	1997	Blue Bus, Bolton, 2004
4108	u	L532EHD	DAF SB220	Ikarus CitiBus	B48F	1994	North Western, 1997

4501-4515

			Volvo B10BLE			Wright Renown		N44F	1999	
4501	NL	V501DFT	4505	NL	V505DFT	4509	NL	V509DFT	4513	NL	V513DFT
4502	AS	V502DFT	4506	NL	V506DFT	4510	NL	V510DFT	4514	NL	V514DFT
4503	AS	V503DFT	4507	NL	V507DFT	4511	NL	V511DFT	4515	NL	V515DFT
4504	NL	V504DFT	4508	NL	V508DFT	4512	NL	V512DFT			

4516-4523

			Volvo B10BLE			Alexander ALX300		N44F	2000	
4516	NL	W292PPT	4518	NL	W294PPT	4520	NL	W296PPT	4522	NL	W298PPT
4517	NL	W293PPT	4519	NL	W295PPT	4521	NL	W297PPT	4523	NL	W299PPT

4524-4530

			Volvo B10BLE			Wright Renown		N44F	1999	Arriva Scotland, 2003
4524	HX	V530GDS	4526	HX	V532GDS	4528	HX	V534GDS	4530	HX	V536GDS
4525	HX	V531GDS	4527	HX	V533GDS	4529	HX	V535GDS			

4646-4659

			Scania OmniCity CN94UB			Scania		N42F	2005	
4646	DM	NK05GXW	4650	DM	NK05GXD	4654	DM	NK05GXH	4657	DM	NK05GXM
4647	DM	NK05GXA	4651	RR	NK05GXE	4655	RR	NK05GXJ	4658	RR	NK05GXN
4648	RR	NK05GXB	4652	RR	NK05GXF	4656	DM	NK05GXL	4659	RR	NK05GXO
4649	RR	NK05GXC	4653	RR	NK05GXG						

4660-4664 — Scania OmniCity CN230UB — Scania — N41F — 2007

4660	SN	NK07FZC	**4662**	SN	NK07FZE	**4663**	DM	NK07FZF	**4664**	SN	NK07FZG
4661	SN	NK07FZD									

7250-7258 — Scania N113DRB — Northern Counties Palatine — B42/29F — 1994-95 — Arriva London, 2002

7250	BH	L159GYL	**7253**	PE	M178LYP	**7255**	PE	M180LYP	**7257**	PE	N182OYH
7251	AS	L160GYL	**7254**	BH	M179LYP	**7256**	BH	N181OYH	**7258**	NL	N183OYH
7252	NL	L161GYL									

7259-7262 — Scania N113DRB — East Lancs — B47/33F — 1995 — Arriva Midlands, 2008

7259	DM	M177GRY	**7260**	DM	M170GRY	**7261**	NL	M172GRY	**7262**	NL	M176GRY

7263-7266 — Scania N113DRB — East Lancs — B45/33F — 1995 — Arriva The Shires, 2008

7263	PE	N162VVO	**7264**	PE	N164VVO	**7265**	BH	N160VVO	**7266**	DM	N163VVO

7362-7367 — Volvo Olympian — Northern Counties Palatine — B47/29F — 1998 — Arriva Midlands, 2009

7362	AS	R636MNU	**7364**	PE	S648KJU	**7366**	DN	R639MNU	**7367**	DN	R641MNU
7363	PE	R640MNU	**7365**	DN	R625MNU						

7370-7377 — Volvo Olympian — Northern Counties Palatine II — BC43/27F — 1994

7370	AK	M370FTY	**7372**	AK	M372FTY	**7374**	AK	M374FTY	**7376**	NL	M376FTY
7371	AK	M371FTY	**7373**	AK	M373FTY	**7375**	AS	M375FTY	**7377**	NL	M377FTY

7386-7393 — Scania N113DRB — East Lancs Cityzen — BC43/31F — 1996

7386	BL	N386OTY	**7388**	BL	N388OTY	**7390**	BL	N390OTY	**7392**	BL	N392OTY
7387	BL	N387OTY	**7389**	BL	N389OTY	**7391**	BL	N391OTY	**7393**	BL	N393OTY

7410-7420 — Volvo Olympian — Northern Counties Palatine II — B43/29F — 1997

7410	AS	P410CCU	**7413**	AS	P413CCU	**7416**	NL	P416CCU	**7419**	NL	P419CCU
7411	AS	P411CCU	**7414**	NL	P414CCU	**7417**	NL	P417CCU	**7420**	NL	P420CCU
7412	AS	P412CCU	**7415**	NL	P415CCU	**7418**	NL	P418CCU			

7421-7429 — Volvo Olympian — East Lancs — B44/30F — 1994 — Arriva Southern Counties, 1998

7421	NL	M685HPF	**7424**	DM	M688HPF	**7426**	DM	M690HPF	**7428**	DM	M692HPF
7422	DM	M686HPF	**7425**	DM	M689HPF	**7427**	DM	M691HPF	**7429**	BA	M693HPF
7423	NL	M687HPF									

7432-7435 — Dennis Trident — Alexander ALX400 — N51/31F — 2000

7432	NL	W397RBB	**7433**	NL	W398RBB	**7434**	NL	W399RBB	**7435**	NL	W501RBB

7436-7444 — DAF DB250 — East Lancs Lowlander — N44/29F — 2001

7436	AS	Y686EBR	**7439**	NL	Y689EBR	**7441**	BH	Y691EBR	**7443**	DN	Y693EBR
7437	AS	Y687EBR	**7440**	NL	Y685EBR	**7442**	BH	Y692EBR	**7444**	DM	Y694EBR
7438	AS	Y688EBR									

7445	NL	NK05GWX	Volvo B7TL	ADL ALX400	N45/27F	2005
7446	NL	NK05GWY	Volvo B7TL	ADL ALX400	N45/27F	2005

7447-7452 — DAF DB250 — Northern Counties Palatine 2 B43/24D — 1998 — Arriva Southern Counties, 2006

7447	BA	R204CKO	**7449**	BA	R206CKO	**7451**	BA	R208CKO	**7452**	BA	R209CKO
7448	BA	R205CKO	**7450**	BA	R207CKO						

7453-7456 — VDL Bus DB250 — East Lancs Lowlander — N / F — 2007

7453	AK	YJ57BVD	**7454**	AK	YJ57BVE	**7455**	AK	YJ57BVF	**7456**	AK	YJ57BVG

7457	SS	L94HRF	DAF DB250	Optare Spectra	B48/29F	1993	Arriva Midlands, 2009
7458	SS	L95HRF	DAF DB250	Optare Spectra	B48/29F	1993	Arriva Midlands, 2009

7501-7513 — ADL Trident 2 — ADL Enviro 400 — NC45/33F — 2007-08

7501	BH	NK57DXX	**7505**	BH	NK57GWX	**7508**	AS	NK57GXA	**7511**	AS	NK57GXD
7502	BH	NK57DXY	**7506**	BH	NK57GWY	**7509**	AS	NK57GXB	**7512**	AS	NK57GXE
7503	BH	NK57DXZ	**7507**	AS	NK57GWZ	**7510**	AS	NK57GXC	**7513**	AS	NK57GXF
7504	BH	NK57DYA									

7514-7521 — ADL Trident 2 — ADL Enviro 400 — NC45/33F — 2009

7514	BH	NK09DFMZ	**7516**	BA	NK09FNC	**7518**	BA	NK09FNE	**7520**	BA	NK09FNG
7515	BH	NK09DFNZ	**7517**	BA	NK09FND	**7519**	BA	NK09FNF	**7521**	BA	NK09FVR

Ancillary vehicles:

1001-1005		Volvo B6 9.9M		Plaxton Pointer		TV	1994	Arriva North West & Wales, 2006
1001	SN	L219TKA	**1003**	SN	L232TKA	**1004**	AS L244TKA	**1005** AS L225TKA
1002	SN	L220TKA						

9966	SN	P612FHN	Optare MetroRider MR35	Optare	B25F	1996
9967	SN	P618FHN	Optare MetroRider MR35	Optare	B25F	1996
9968	DN	P621FHN	Optare MetroRider MR35	Optare	B25F	1996

Previous registrations:

HIL2148	S869OGB	L95HRF	L95HRF, 915DYE
J353BSH	J353BSH, VLT173	UOI772	K141RYS
J354BSH	J354BSH, VLT32,WLT554	WLT954	A954SUL
L94HRF	L94HRF, 49XRF		

Depots and allocations:

Alnwick (Lisburn Street) - AK

MetroRider	918				
SB3000	1210	1211	1212	1213	
Olympian	7370	7371	7372	7373	7374
VDL DB250	7453	7454	7455	7456	

Ashington (Lintonville Terrace) - AS

Mercedes-Benz	2668	2669	2670	2672			
MetroRider	917	919	2705	2710	2711	2716	
Volvo B6	1004	1005					
DAF/VDL SB120	1923	1924					
Dart SLF	1800	1801					
DAF/VDL SB200	1404	1405	1406	1407	4015		
Volvo B10BLE	4502	4503					
Scania sd	274	281	289				
Scania dd	7251						
Olympian	7362	7375	7410	7411	7412	7413	
DB250	7436	7437	7438				
Trident 2	7507	7508	7509	7510	7511	7512	7513

Bishop Auckland (Caroline Street) - BA

MetroRider	2631	2701						
Optare Solo	2852	2853	2854	2855	2856	2857	2858	2859
	2860	2861						
Dart	1697							
Dart SLF	1618	1625	1626	1707	1708	1710	1711	1712
	1713	1714	1715	1736	1737	1763	1864	1772
DAF/VDL SB220	1415	1416	1417	1418	1419	1420	1421	1422
	1423	1424	1425	1426	1427	1428	4031	4032
	4033	4034	4042	4049				
Olympian	7429							
DAF/VDL DB250	7447	7448	7449	7450	7451	7452	7457	7458
Trident 2	7516	7517	7518	7519	7520	7521		

Blyth (Bridge Street) - BH

Mercedes-Benz	2646	2647	2648	2649	2650	2651	2652	2653
	2654	2655	2657	2658	2659	2660	2661	2662
	2663	2664	2666	2667				
Dart SLF	1702	1721	1724	1726	1728	1733	1740	1741
	1742	1773	1774	1775	1799			
DAF/VDL SB200	1401	1402	1403	1408	1409	1410	1411	1412
	1413	1414	4074	4075	4076	4077	4078	4079
Scania dd	7250	7354	7256	7265	7386	7388	7389	7390
	7391	7392	7393					
DB250	7441	7442						
Trident 2	7501	7502	7503	7504	7505	7506	7514	7515

Darlington (Faverdale) - DN

Outstation: Barnard Castle

Optare Solo	2824	2825	2826	2827	2828	2829	2830	2831
	2832	2833	2834	2835	2836	2837	2838	2839
	2840	2841	2842	2843	2844	2845	2846	2847
	2848	2849	2850	2851				
Dart	1657	1659	1660	1661	1662	1664	1665	1666
	1667	1669	1670	1671	1672	1674	1676	1677
	1678	1679	1680	1681	1682	1683	1684	1685
	1686	1687						
Dart SLF	1611	1612	1703	1704	1705	1706	1716	1717
	1735	1745	1746	1747	1761	1793	1794	1795
	1796	1797	1798					
DAF/VDL SB220	4024	4041	4044	4050	4052	4056	4057	4061
	4083							
Olympian	7365	7366	7367					
DB250	7443							

Ancillary   9968

Durham (Waddington Street) - DU

Solo	2803	2804	2805	2806	2807	2808	2809	2810
	2811	2812	2813	2814	2815			
Dart	1698							
Dart SLF	1637	1701						
DAF/VDL SB220	4035	4036	4060	4062	4064	4065	4066	4067
	4068	4069	4070	4071	4072	4073	4098	
Scania OmniCity	4646	4647	4650	4652	4654	4657	4663	
Scania dd	7259	7260	7266					
Olympian	7422	7424	7425	7426	7427	7428		
DAF DB250	7444							

Hexham (Burn Lane) - HX

MetroRider	2702						
Dart	1688						
Dart SLF	1749	1759	1760	1790	1791		
DAF Interurban	1201	1202	1203	1204	1205	1214	
Volvo B10BLE	4524	4525	4526	4527	4528	4529	4530

Newcastle (Jesmond Road) - NL

Optare Solo	2504	2505						
Dart SLF	1729	1730	1731	1732	1750	1751	1752	1758
	1762	1792						
VDL SB200	2	3	4					
Volvo B10BLE	4501	4504	4505	4506	4507	4508	4509	4510
	4511	4512	4513	4514	4515	4516	4517	4518
	4519	4520	4521	4522	4523			
Scania dd	7252	7258	7261	7262				
Olympian	7376	7377	7414	7415	7416	7417	7418	7419
	7420	7421	7423					
Trident	7432	7433	7434	7435				
DAF DB250	7439	7440						
Volvo B7TL	7445	7446						

Peterlee (Davey Drive) - PE

Optare Solo	2801	2802	2815	2816	2817	2818	2819	2820
	2821	2822	2823					
Dart	1613	1616	1617	1636	1638	1639	1640	1652
	1653	1654	1655	1656	1709	1723	1743	1748
	1757							
Scania dd	7253	7255	7257	7263	7264			
Olympian	7363	7364						

Redcar (Ennis Road, Dormanstown) - RR

MetroRider	904	916	2633	2723				
Dart	1658	1663	1668	1673	1675	1696		
Dart SLF	1621	1635	1649	1650	1651	1718	1719	1720
	1722	1727	1734	1738	1744	1753	1754	1755
	1756							
MAN Vecta	1524	1536	1541	1553	1556	1557		
DAF Interurban	1206	1207	1208	1209				
Mercedes-Benz O405	3002	3004	3005	3006	3007	3008	3009	3010
	3011	3012	3013	3014	3015	3016	3017	3018
	3022	3023	3024	3025	3026			
DAF/VDL SB220	4023	4025	4026	4027	4053	4058	4059	4080
	4081	4082	4086	4097				
Scania OmniCity	4648	4649	4651	4652	4653	4655	4658	4659

Stockton (Boathouse Lane) - SN

MetroRider	902	923	2627					
Volvo B6	1001	1002	1003					
MAN Vecta	1504	1505	1506	1507	1508	1515	1517	1519
	1521	1526	1527	1529	1531	1533	1537	1542
	1544	1549	1551	1558				
Scania interurban	271	272	275	276	277	278	279	290
Dart SLF	1646	1647	1648	1765	1766	1767	1768	1769
	1770	1771						
DAF/VDL SB120	1901	1902	1903	1904	1905	1906	1907	1908
	1909	1910	1911	1912	1913	1914	1915	1916
	1917	1920	1922					
Mercedes-Benz O405	3003	3019	3020					
DAF/VDL SB220	4030	4037	4038	4039	4040	4043	4045	4046
	4048	4051	4054	4063				
Scania OmniCity	4660	4661	4662	4664				
Ancillary	*9966*	*9967*						

Unallocated or stored - u/w

Remainder

ARRIVA YORKSHIRE

Arriva Yorkshire Ltd; Arriva Yorkshire North Ltd,
24 Barnsley Road, Wakefield, West Yorkshire, WF1 5JX
Arriva Yorkshire West Ltd, Mill Street East, Dewsbury, West Yorkshire, WF12 9AG

11	DY	R10WAL	DAF SB220			Ikarus CitiBus		B49F	1997	K-Line Travel, 2000
13	DY	R69GNW	DAF SB220			Ikarus CitiBus		B49F	1998	K-Line Travel, 2000

22-29
			DAF SB220			Ikarus CitiBus		B49F	1994	K-Line Travel, 2000
22	u	M812RCP	24	u	M814RCP	26	u	M816RCP	28	u	M818RCP
23	u	M813RCP	25	u	M815RCP	27	u	M817RCP	29	u	M819RCP

53	DY	N51FWU	DAF SB220	Ikarus CitiBus	B49F	1995	Arriva Bus & Coach, 2004
54	DY	P202RUM	DAF SB220	Ikarus CitiBus	B49F	1996	Arriva Bus & Coach, 2004
56	DY	N52FWU	DAF SB220	Ikarus CitiBus	B49F	1995	Arriva Bus & Coach, 2004
57	DY	YD02RJO	DAF SB220	Ikarus CitiBus	N44F	2002	Arriva Bus & Coach, 2004
58	DY	YD02RJJ	DAF SB220	Ikarus CitiBus	N44F	2002	Arriva Bus & Coach, 2004

102-109
			Volvo B10BLE	Wright Renown	NC44F	2000
102	CD	W102EWU	104	CD	W104EWU	107	CD	W107EWU	109	SB	W109EWU
103	CD	W103EWU	106	CD	W106EWU	108	SB	W108EWU			

136-146
			Dennis Dart 9m	Northern Counties Paladin	B35F	1994	Arriva Southern Counties, 2001
136	u	L136YVK	138	w	L128YVK	143	w	L159BFT	146	w	L146YVK
137	u	L137YVK	140	w	L140YVK						

156	u	K320CVX	Dennis Dart 9m	Plaxton Pointer	B35F	1992	Arriva Southern Counties, 2004
157	u	J465MKL	Dennis Dart 9.8m	Plaxton Pointer	B40F	1991	Arriva Southern Counties, 2004
161	WF	P326HVX	Dennis Dart 9m	Plaxton Pointer	B34F	1996	
165	SB	W165HBT	Dennis Dart SLF	Alexander ALX200	N40F	2000	
166	SB	W166HBT	Dennis Dart SLF	Alexander ALX200	N40F	2000	
167	SB	TWY7	Dennis Dart SLF	Alexander ALX200	N40F	2000	

170-199
			Dennis Dart SLF	Alexander ALX200	N40F	1997
170	SB	P170VUA	179	SB	P179VUA	186	DY	P186VUA	193	DY	P193VUA
171	WF	P171VUA	180	SB	P180VUA	187	DY	P187VUA	194	DY	P194VUA
172	SB	P172VUA	181	SB	P181VUA	188	DY	P188VUA	195	DY	P195VUA
173	WF	P173VUA	182	CD	P182VUA	189	DY	P189VUA	196	DY	P196VUA
175	CD	P175VUA	183	CD	P183VUA	190	DY	P190VUA	197	DY	P197VUA
176	CD	P176VUA	184	CD	P184VUA	191	DY	P191VUA	198	DY	P198VUA
177	SB	P177VUA	185	CD	P185VUA	192	DY	P192VUA	199	DY	P199VUA

200	SB	R103GNW	Dennis Dart SLF	UVG Urbanstar	N40F	1998	Jaronda Travel, Cawood, 1999

201-229
			Dennis Dart SLF	Plaxton Pointer MPD	N29F	2000
201	WF	V201PCX	208	WF	V208PCX	215	WF	V215PCX	223	CD	V223PCX
202	WF	A1YBG	209	WF	V209PCX	216	WF	V216PCX	224	CD	V224PCX
203	WF	V203PCX	210	WF	V210PCX	217	WF	V217PCX	225	WF	V225PCX
204	WF	V204PCX	211	WF	V211PCX	218	WF	V218PCX	226	CD	V226PCX
205	WF	V205PCX	212	WF	V212PCX	219	WF	A4YBG	227	CD	V227PCX
206	WF	V206PCX	213	WF	V213PCX	220	WF	V220PCX	228	CD	V228PCX
207	WF	V207PCX	214	WF	A2YBG	221	WF	V221PCX	229	CD	V229XUB

230	WF	W244SNR	Dennis Dart SLF	Plaxton Pointer MPD	N29F	2000	Arriva Midlands, 2009

Carrying the Inter-urban Arriva livery is Yorkshire's 442, R442KWT, from the first batch of ALX300s to carry Alexander bodywork. The vehicle is based in Wakefield. *Andy Jarosz*

260	SB	SN55HTX	ADL Dart 9m		ADL Pointer		N34F	2006		
261	SB	SN55HTY	ADL Dart 9m		ADL Pointer		N34F	2006		
262	SB	SN55HTZ	ADL Dart 9m		ADL Pointer		N34F	2006		

411-424
Volvo B10B — Alexander Strider — B51F — 1994

| 411 | CD | M411UNW | 415 | u | M415UNW | 419 | u | M419UNW | 423 | u | M423UNW |
| 413 | u | M413UNW | 417 | u | M417UNW | 420 | | M420UNW | 424 | u | M424UNW |

435	WF	P10LPG	DAF SB220		Northern Counties Paladin		B42F	1997	Arriva Bus & Coach, 2004
436	WF	R989FNW	DAF SB220		Northern Counties Paladin		B42F	1997	Arriva Bus & Coach, 2004
437	WF	R985FNW	DAF SB220		Northern Counties Paladin		B42F	1997	Arriva Bus & Coach, 2004
438	WF	R28GNW	DAF SB220		Northern Counties Paladin		B41F	1998	Arriva Bus & Coach, 2004
439	WF	R29GNW	DAF SB220		Northern Counties Paladin		B41F	1998	Arriva Bus & Coach, 2005

440-471
DAF SB220 — Alexander ALX300 — N42F — 1998

440	WF	R440GWY	449	WF	R449KWT	457	WF	R457KWT	465	HE	S465GUB
441	WF	R441KWT	450	CD	R450KWT	458	WF	R458KWT	466	HE	S466GUB
442	WF	R442KWT	451	WF	R451KWT	459	WF	R459KWT	467	HE	S467GUB
443	WF	R443KWT	452	WF	R452KWT	460	HE	R460KWT	468	HE	S468GUB
445	WF	R445KWT	453	WF	R453KWT	461	HE	R461KWT	469	HE	S469GUB
446	WF	R446KWT	454	WF	R454KWT	462	HE	S462GUB	470	HE	S470GUB
447	WF	R447KWT	455	WF	R455KWT	463	HE	S463GUB	471	HE	S471GUB
448	WF	R448KWT	456	WF	R456KWT	464	HE	S464GUB			

472-491
DAF SB220 — Alexander ALX300 — N42F — 1998

472	WF	S472ANW	477	WF	S477ANW	482	CD	S482ANW	487	CD	S487ANW
473	WF	S473ANW	478	WF	S478ANW	483	CD	S483ANW	488	CD	S488ANW
474	WF	S474ANW	479	CD	S479ANW	484	CD	S484ANW	489	CD	S489ANW
475	WF	S475ANW	480	WF	S480ANW	485	CD	S485ANW	490	CD	S490ANW
476	WF	S476ANW	481	CD	S481ANW	486	CD	S486ANW	491	CD	S491ANW

495-499
VDL Bus SB200 — Wrightbus Commander — N44F — 2004

| 495 | WF | YJ04HJC | 497 | WF | YJ04HJE | 498 | WF | YJ04HJF | 499 | WF | YJ04HJG |
| 496 | WF | YJ04HJD | | | | | | | | | |

Castleford is the location for this view of Volvo T7L 653, W653CWX, one of the 2000 intake with Alexander ALX400 bodywork. *David Little*

511-514

		Volvo Olympian YN2RV16Z4		East Lancs			B44/30F	1994	Arriva London, 2004-05
511	SB	M694HPF	512	SB	M702HPF	513	SB	M696HPF	514	SB	M697HPF

621	SB	N621KUA	Volvo Olympian YN2RV18Z4	Northern Counties Palatine II	B43/30F	1996
622	SB	N622KUA	Volvo Olympian YN2RV18Z4	Northern Counties Palatine II	B43/30F	1996
623	SB	N623KUA	Volvo Olympian YN2RV18Z4	Northern Counties Palatine II	B43/30F	1996

624-641

		DAF DB250		Optare Spectra			B48/29F	1999	
624	WF	T624EUB	629	WF	T629EUB	634	WF	T634EUB	638	WF	T638EUB
625	WF	T625EUB	630	WF	T630EUB	635	WF	T635EUB	639	WF	T639EUB
626	WF	T626EUB	631	WF	T631EUB	636	WF	T636EUB	640	WF	V640KVH
627	WF	T627EUB	632	WF	T632EUB	637	WF	T637EUB	641	WF	V641KVH
628	WF	T628EUB	633	WF	T633EUB						

651-674

		Volvo B7L		Alexander ALX400			N47/28F	2000	
651	CD	W651CWX	657	CD	W657CWX	663	CD	W663CWX	669	CD	W669CWX
652	CD	W652CWX	658	CD	W658CWX	664	CD	W664CWX	671	CD	W671CWX
653	CD	W653CWX	659	CD	W659CWX	665	CD	W665CWX	672	CD	W672CWX
654	CD	W654CWX	661	CD	W661CWX	667	CD	W667CWX	673	CD	W673CWX
656	CD	W656CWX	662	CD	W662CWX	668	CD	W668CWX	674	CD	W674CWX

675-696

		Volvo B7L		Plaxton President			N47/28F	2001	
675	SB	X675YUG	681	SB	X681YUG	686	SB	X686YUG	692	HE	X692YUG
676	SB	X676YUG	682	SB	X682YUG	687	HE	X687YUG	693	HE	X693YUG
677	SB	X677YUG	683	SB	X683YUG	688	HE	X688YUG	694	HE	X694YUG
678	SB	X678YUG	684	SB	X684YUG	689	HE	X689YUG	695	HE	X695YUG
679	SB	X679YUG	685	SB	X685YUG	691	HE	X691YUG	696	HE	X696YUG

Optare's current full-size single-deck bus is the Tempo which uses either the Mercedes-Benz OM906LA SCR Euro 4 engine or the MAN D0836 engine for the Euro 4 EGR Solution. Delivered in 2009, 1300, YJ09EYA is the first of thirteen allocated to Castleford. *Mark Lyons*

700-723

DAF DB250 — Optare Spectra — N47/27F — 2002

700	WF	YD02PXW	706	CD	YG52CFE	712	CD	YG52CFN	718	HE	YD02PYU
701	WF	YD02PXX	707	WF	YG52CFF	713	CD	YG52CFO	719	HE	YD02PYV
702	WF	YD02PXY	708	CD	YG52CFJ	714	CD	YG52CFP	720	HE	YD02PYW
703	WF	YD02PXZ	709	CD	YG52CFK	715	CD	YG52CFU	721	HE	YD02PYX
704	WF	YG52CFA	710	CD	YG52CFL	716	HE	YG52CFV	722	HE	YD02PYY
705	WF	YG52CFD	711	CD	YG52CFM	717	HE	YG52CFX	723	HE	YD02PYZ

1050-1066

ADL Dart 4 — ADL Enviro 200 — N38F — 2009

1050	DY	YJ09CSU	1055	DY	YJ09CUA	1059	DY	YJ09CUK	1063	DY	YJ09CVB
1051	DY	YJ09CTV	1056	DY	YJ09CUC	1060	DY	YJ09CUO	1064	DY	YJ09CVC
1052	DY	YJ09CTX	1057	DY	YJ09CUG	1061	DY	YJ09CUY	1065	DY	YJ09CVD
1053	DY	YJ09CTY	1058	DY	YJ09CUH	1062	DY	YJ09CVA	1066	DY	YJ09CVE
1054	DY	YJ09CTZ								

1100-1112

Volvo B7RLE — Wrighbus Eclipse Urban — N44F — 2008

1100	HE	YJ08DVA	1104	HE	YJ08DVG	1107	HE	YJ08DVN	1110	HE	YJ08DVR
1101	HE	YJ08DVB	1105	HE	YJ08DVH	1108	HE	YJ08DVO	1111	HE	YJ08DVT
1102	HE	YJ08DVC	1106	HE	YJ08DVK	1109	HE	YJ08DVP	1112	HE	YJ08DVU
1103	HE	YJ08DVF								

1300-1312

Optare Tempo X1100 — Optare — N43F — 2009

1300	CD	YJ09EYA	1304	CD	YJ09EYG	1307	CD	YJ09EYL	1310	CD	YJ09EYP
1301	CD	YJ09EYB	1305	CD	YJ09EYH	1308	CD	YJ09EYM	1311	CD	YJ09EYR
1302	CD	YJ09EYC	1306	CD	YJ09EYK	1309	CD	YJ09EYO	1312	CD	YJ09EYS
1303	CD	YJ09ETF								

1401-1408 VDL Bus SB200 Wrightbus Commander N44F 2006

1400	WF	YJ56JYE	1403	WF	YJ56JYH	1405	WF	YJ56JYL	1407	WF	YJ56JYO
1401	WF	YJ56JYF	1404	WF	YJ56JYK	1406	WF	YJ56JYN	1408	WF	YJ56JYP
1402	WF	YJ56JYG								

1409-1415 VDL Bus SB200 Wrightbus Pulsar N44F 2007

1409	DY	YJ57BVT	1411	DY	YJ57BVV	1413	DY	YJ57BVX	1415	DY	YJ57BVZ
1410	DY	YJ57BVU	1412	DY	YJ57BVW	1414	DY	YJ57BVY			

1450-1453 VDL Bus SB200 Wrightbus Pulsar 2 N44F 2009

1450	WF	YJ59BUH	1451	WF	YJ59BRY	1452	WF	YJ59BRZ	1453	WF	YJ59BSO

1500-1507 VDL Bus DB300 Wrighbus Eclipse Gemini 2 N--/--F 2009

1500	SB	YJ59BTO	1502	SB	YJ59BTV	1504	SB	YJ59BTY	1506	SB	YJ59BUA
1501	SB	YJ59BTU	1503	SB	YJ59BTX	1505	SB	YJ59BTZ	1507	SB	YJ59BUE

1600-1613 VDL Bus DB250 East Lancs Lowlander N47/27F 2006

1600	DY	YJ06WLX	1604	DY	YJ06WMD	1608	DY	YJ06WMK	1611	DY	YJ06WWX
1601	DY	YJ06WLZ	1605	DY	YJ06WME	1609	DY	YJ06WML	1612	DY	YJ06WWY
1602	DY	YJ06WMA	1606	DY	YJ06WMF	1610	DY	YJ06WWV	1613	DY	YJ06WWZ
1603	DY	YJ06WMC	1607	DY	YJ06WMG					

1800-1815 Volvo B9TL Darwen Olympus N51/30F 2008

1800	WA	YJ57BEO	1804	WA	YJ08EEB	1808	WA	YJ08EEM	1812	WA	YJ08EES
1801	WA	YJ57BEU	1805	WA	YJ08EEF	1809	WA	YJ08EEN	1813	WA	YJ08EET
1802	WA	YJ08ECY	1806	WA	YJ08EEG	1810	WA	YJ08EEP	1814	WA	YJ08EEU
1803	WA	YJ08EEW	1807	WA	YJ08EEH	1811	WA	YJ08EER	1815	WA	YJ08EEV

1900-1913 ADL Trident 2 ADL Enviro 400 N47/33F 2008

1900	HE	YJ58FHA	1904	HE	YJ58FHE	1908	HE	YJ58FHK	1911	HE	YJ58FHN
1901	HE	YJ58FHB	1905	HE	YJ58FHF	1909	HE	YJ58FHL	1912	HE	YJ58FHO
1902	HE	YJ58FHC	1906	HE	YJ58FHG	1910	HE	YJ58FHM	1913	HE	YJ58FHP
1903	HE	YJ58FHD	1907	HE	YJ58FHH					

Ancillary vehicles:

T52	DYt	J23GCX	DAF SB220	Optare Delta	B49F	1991	K-Line Travel, 2000
T401	CDt	K401HWW	Volvo B10B	Alexander Strider	B51F	1993	
T402	HEt	K402HWW	Volvo B10B	Alexander Strider	B51F	1993	
T403	WFt	K403HWW	Volvo B10B	Alexander Strider	B51F	1993	

Previous registrations:

A1YBG	V202PCX	P10LPG	P10LPG, 99D73675
A2YBG	V214PCX	TWY7	W167HBT
A4YBG	V219PCX		

Depots and Allocations:

Castleford (Wheldon Road) - CD

Dart SLF	176	182	183	184	185	223	224	226
	227	228	229					
DAF/VDL SB200	450	479	481	482	483	484	485	486
	487	488	489	490	491			
Optare Tempo	1300	1301	1302	1303	1304	1305	1306	1307
	1308	1309	1310	1311	1312			
Volvo B7TL	651	652	653	654	656	657	658	659
	661	662	663	664	665	667	668	669
	671	672	673	674				
DAF/VDL DB250	706	708	709	710	711	712	713	714
	715							

Ancillary T401

Dewsbury (Mill Street East) - DY

Dart	186	187	189	190	191	192	193	194
	195	196	197	198	199	1050	1051	1052
	1053	1054	1055	1056	1057	1058	1059	1060
	1061	1062	1063	1064	1065	1066		
DAF/VDL SB220	11	13	29	53	54	56	57	58
	1409	1410	1411	1412	1413	1414	1415	
DB250	1600	1601	1602	1603	1604	1605	1606	1607
	1608	1609	1610	1611	1612	1613		

Ancillary   T52

Heckmondwike (Beck Lane) - HE

DAF/VDL SB220	460	461	462	465	466	467	468	469
	470	471						
Volvo B7RLE	1100	1101	1102	1103	1104	1105	1106	1107
	1108	1109	1110	1111	1112			
Olympian	520	521						
Volvo B7TL	687	688	689	691	692	693	694	695
	696							
DAF/VDL DB250	716	717	718	719	720	721	722	723
Trident 2	1900	1901	1902	1903	1904	1905	1906	1907
	1908	1909	1910	1911	1912	1913		

Ancillary   T402

Selby (Cowie Drive, Ousegate) - SB

Dart SLF	165	166	167	170	172	177	179	180
	181	200	260	261	262			
Volvo B10BLE	108	109						
Olympian	512	513	514	616	621	622	623	
Volvo B7L	675	676	677	678	679	681	682	683
	684	685	686					
VDL SB300	1500	1501	1502	1503	1504	1505	1506	1507

Volvo B9TL 1802, YJ08ECY, is pictured in Leeds Road, Lofthouse in September 2009. One of sixteen with East Lancs Olympus bodywork constructed during the short period when East Lancs was part of the Darwen Group.
Mark Lyons

Arriva colours are carried by Optare Spectra 704, YG52CFA. One of the last batches of buses from the Yorkshire builder it was one of the low-floor batch supplied in 2002. It is seen in Leeds. *Mark Lyons*

Wakefield (Belle Isle, Barnsley Road) - WF

Dart SLF	161	171	173	201	202	203	204	205
	206	207	208	209	210	211	212	213
	214	215	216	217	218	219	220	221
	225	230						
DAF/VDL SB120	1400	1401	1402	1403	1404	1405	1406	1407
	1408							
Volvo B10B	411							
DAF/VDL SB220	435	436	437	438	439	440	441	442
	443	445	446	447	448	449	451	452
	453	454	455	456	457	458	459	472
	473	474	475	476	477	478	480	495
	496	497	498	499				
VDL SB200	1400	1401	1402	1403	1404	1405	1406	1407
	1408	1409	1410	1411	1412	1413	1414	1415
	1450	1451	1452	1453				
DAF/VDL DB250	624	625	626	627	628	629	630	631
	632	633	634	635	636	637	638	639
	640	641	700	701	702	703	704	705
Volvo B9TL	1800	1801	1802	1803	1804	1805	1806	1807
	1808	1809	1810	1811	1812	1813	1814	1815

Ancillary	*T403*

Unallocated or stored - u/w

Remainder

ARRIVA NORTH WEST & WALES

Arriva North West Ltd, Arriva Merseyside Ltd,
Arriva Cymru Ltd, Arriva Manchester Ltd, Arriva Liverpool Ltd,
73 Ormskirk Road, Aintree, Liverpool, L9 5AE

391	WX	W191CDN	Mercedes-Benz Vario 0814	Alexander ALX100	B27F	2000	Arriva Bus & Coach, 2000
392	WX	W192CDN	Mercedes-Benz Vario 0814	Alexander ALX100	B27F	2000	Arriva Scotland, 2001
393	LJ	W193CDN	Mercedes-Benz Vario 0814	Alexander ALX100	B27F	2000	Arriva Scotland, 2001
394	LJ	W194CDN	Mercedes-Benz Vario 0814	Alexander ALX100	B27F	2000	Arriva Scotland, 2001
601	u	M001YBG	Neoplan N4009	Neoplan	N23F	1995	
602	u	M302YBG	Neoplan N4009	Neoplan	N23F	1995	
603	u	M303YBG	Neoplan N4009	Neoplan	N23F	1995	
619	WX	YN04XZH	Optare Alero	Optare	N12F	2004	*Operated for Flintshire CC*
621	CH	YK05CAE	Optare Solo M920	Optare	N23F	2005	*Operated for Flintshire CC*
622	CH	YK05CBX	Optare Solo M850	Optare	N23F	2005	*Operated for Flintshire CC*
623	CH	YJ06ATK	Optare Solo M850	Optare	N23F	2006	*Operated for Flintshire CC*
624	CH	YJ56ATK	Optare Solo M710 SE	Optare	N23F	2006	*Operated for Flintshire CC*

660-672

			Optare Solo M880	Optare	N28F	2007-08

660	MA	CX57CYO	663	MA	CX57CYT	667	MA	CX57CYW	670	MA	CX57CZA
661	MA	CX57CYP	664	MA	CX57CYU	668	MA	CX57CYX	671	CH	CX58ETY
662	MA	CX57CYS	665	MA	CX57CYV	669	MA	CX57CYY	672	CH	CX58ETZ

673	CH	CX58EUA	Optare Solo M950 SL	Optare	N32F	2008
674	CH	CX58EUB	Optare Solo M950 SL	Optare	N32F	2008
675	CH	CX58EUA	Optare Solo M950 SL	Optare	N32F	2008

676-680

			Optare Solo M880 SL	Optare	N28F	2008-09

676	WY	CX58FYU	678	WY	CX58FYW	679	WY	CX58FYY	680	WY	CX58FYZ
677	WY	CX58FYV									

Pwllheli is the location for this view of Optare Solo 688, CX09BFZ, one of the batch supplied to North Wales depots in 2009. As shown in the view, the slimline version of the Solo has been chosen. *Richard Hughes*

681-697   Optare Solo M950 sl   Optare   N32F   2009

681	BG	CX09BFM	686	BG	CX09BFV	690	BG	CX09BGF	694	BG	CX09BGV
682	BG	CX09BFN	687	BG	CX09BFY	691	BG	CX09BGK	695	BG	CX09BGW
683	BG	CX09BFO	688	BG	CX09BFZ	692	BG	CX09BGO	696	RH	CX09BGZ
684	BG	CX09BFP	689	BG	CX09BGE	693	BG	CX09BGU	697	RH	CX09BHA
685	BG	CX09BFU									

801	LJ	R546ABA	Dennis Dart SLF 8.8m	Plaxton Pointer MPD	N28F	1997	

802-809   Dennis Dart SLF 8.8m   Plaxton Pointer MPD   N25F   1998

802	AB	S872SNB	804	LJ	S874SNB	806	BO	S876SNB	808	RU	S878SNB
803	BO	S873SNB	805	WX	S875SNB	807	WX	S877SNB	809	RU	S879SNB

810-813   Dennis Dart SLF 8.8m   Plaxton Pointer MPD   N29F   1999   Nova Scotia, Winsford, 2000

810	BO	T62JBA	811	BO	T63JBA	812	LJ	T64JBA	813	BO	T65JBA

814-820   Dennis Dart SLF 8.8m   Plaxton Pointer MPD   N27F   1999

814	LJ	T564JJC	816	LJ	T566JJC	818	LJ	T568JJC	820	AB	T570JJC
815	LJ	T565JJC	817	LJ	T567JJC	819	LJ	T569JJC			

821-852   Dennis Dart SLF 8.8m   Plaxton Pointer MPD   N27F   2000-01

821	LJ	W269NFF	831	RH	X271RFF	841	AB	Y541UJC	847	RH	Y547UJC
822	LJ	W394OJC	832	AB	X272RFF	842	AB	Y542UJC	848	WX	Y548UJC
823	AB	V553ECC	833	BG	X273RFF	843	RH	Y543UJC	849	WX	Y549UJC
824	AB	V554ECC	834	AB	X274RFF	844	RH	Y544UJC	851	WX	Y551UJC
826	AB	V556ECC	838	WX	Y538VFF	846	RH	Y546UJC	852	WX	Y552UJC
827	BG	V557ECC	839	WX	Y539VFF						

856-859   Dennis Dart SLF 8.8m   Plaxton Pointer MPD   N29F   1999   Arriva Midlands North, 2003

856	WI	T526AOB	857	WI	T527AOB	858	WI	T528AOB	859	WI	T529AOB

860-886   Dennis Dart SLF 8.8m   Plaxton Pointer MPD   N29F   2000-01

860	WI	X209JOF	865	WI	X215JOF	869	WI	X32KON	878	WI	Y38TDA
861	WI	X211JOF	866	WI	X216JOF	872	WI	Y32TDA	879	WI	Y39TDA
862	WI	X212JOF	867	WI	X217JOF	876	WI	Y36TDA	882	WI	Y42TDA
863	WI	X213JOF	868	WI	X218JOF	877	WI	Y37TDA	886	WI	Y46TDA
864	LJ	X214JOF									

890	MA	T10BLU	Dennis Dart SLF	Plaxton Pointer MPD	N29F	2002	Blue Bus, Bolton, 2005
891	WY	T11BLU	Dennis Dart SLF	Plaxton Pointer MPD	N29F	2002	Blue Bus, Bolton, 2005
892	WY	W12LUE	Dennis Dart SLF	Plaxton Pointer MPD	N29F	2002	Blue Bus, Bolton, 2005
893	WY	X13LUE	Dennis Dart SLF	Plaxton Pointer MPD	N29F	2002	Blue Bus, Bolton, 2005
894	WY	X14LUE	Dennis Dart SLF	Plaxton Pointer MPD	N29F	2002	Blue Bus, Bolton, 2005

1035-1040   Scania L113CRL   East Lancs Flyte   B47F   1996

1035	JS	P135GND	1037	JS	P137GND	1039	JS	P139GND	1040	JS	P140GND
1036	JS	P136GND	1038	JS	P138GND						

1041-1061   Scania L113CRL   Northern Counties Paladin   B47F   1997

1041	JS	P41MVU	1047	RU	R47XVM	1052	JS	P52MVU	1057	JS	R57XVM
1042	RU	P42MVU	1048	RU	R48XVM	1053	JS	P53MVU	1058	JS	P58MVU
1043	JS	P43MVU	1049	RU	P49MVU	1054	JS	R54XVM	1059	JS	R59XVM
1044	RU	P244NBA	1050	JS	P250NBA	1055	JS	R255WRJ	1060	JS	P260NBA
1045	RU	P45MVU	1051	JS	R51XVM	1056	JS	P56MVU	1061	JS	P61MVU
1046	RU	P46MVU									

1062	JS	M102RMS	Scania L113CRL	Northern Counties Paladin	B51F	1995	Arriva Scotland West, 2002
1063	JS	M103RMS	Scania L113CRL	Northern Counties Paladin	B51F	1995	Arriva Scotland West, 2002
1065	JS	M105RMS	Scania L113CRL	Alexander Strider	B51F	1995	Arriva Scotland West, 2002
1068	JS	L588JSG	Scania L113CRL	Northern Counties Paladin	B51F	1994	Arriva Scotland West, 2002
1126	u	N676GUM	Dennis Dart 9.8m	Plaxton Pointer	B40F	1995	Arriva London, 2001
1127	WX	N671GUM	Dennis Dart 9.8m	Plaxton Pointer	B40F	1995	Arriva London, 2001
1128	u	N682GUM	Dennis Dart 9.8m	Plaxton Pointer	B40F	1995	Arriva London, 2001
1129	AB	N704GUM	Dennis Dart 9m	Plaxton Pointer	B34F	1995	Arriva London, 2001
1130	RH	N707GUM	Dennis Dart 9m	Plaxton Pointer	B34F	1995	Arriva London, 2001
1141	BD	L115YVK	Dennis Dart 9m	Northern Counties Paladin	B35F	1994	Arriva Southern Counties, 2003

The Dennis Dart has been the principal midi-size bus for Arriva North West and Wales with many body styles being supplied. Most common has been the Pointer, represented here by 891, T11BLU, its index mark a clue to its use by Blue Bus of Bolton in which town it was seen working a Blackburn Road service. *Richard Godfrey*

1157	u	M157WKA	Dennis Dart 9.8m	East Lancs	B40F	1995	
1167	u	M167WKA	Dennis Dart 9.8m	East Lancs	B40F	1995	
1172	u	M172YKA	Dennis Dart 9.8m	Plaxton Pointer	B40F	1995	

1189-1198   Dennis Dart 9.8m   Plaxton Pointer   B40F   1995

1189	w	M189YKA	1191	w	M191YKA	1195	u	M195YKA	1198	GL M198YKA
1190	w	M190YKA	1192	w	M192YKA					

1201	u	M201YKA	Dennis Lance 11m	Plaxton Verde	B49F	1995
1212	BD	M212YKD	Dennis Dart 9.8m	Plaxton Pointer	B40F	1995
1213	BD	M213YKD	Dennis Dart 9.8m	Plaxton Pointer	B40F	1995
1214	BD	M214YKD	Dennis Dart 9.8m	Plaxton Pointer	B40F	1995
1215	BD	M215YKD	Dennis Dart 9.8m	Plaxton Pointer	B40F	1995
1216	BD	M216YKD	Dennis Dart 9.8m	Plaxton Pointer	B40F	1995

1217-1264   Dennis Dart 9.8m   East Lancs   B40F   1995

1217	u	M217AKB	1230	u	M230AKB	1242	MA	N242CKA	1254	MA N254CKA
1218	u	M218AKB	1231	u	M231AKB	1243	u	N243CKA	1255	WY N255CKA
1219	w	M219AKB	1232	u	M232AKB	1244	MA	N244CKA	1256	WY N256CKA
1220	WI	M220AKB	1233	u	N233CKA	1245	WY	N245CKA	1257	WY N257CKA
1221	u	M221AKB	1234	u	N234CKA	1246	WY	N246CKA	1258	BN N258CKA
1223	u	M223AKB	1235	u	N235CKA	1247	WY	N247CKA	1259	BN N259CKA
1224	u	M224AKB	1236	u	N236CKA	1248	WY	N248CKA	1260	MA N260CKA
1225	u	M225AKB	1237	WY	N237CKA	1249	MA	N249CKA	1261	MA N261CKA
1226	u	M226AKB	1238	u	N238CKA	1250	MA	N250CKA	1262	WY N262CKA
1227	w	M227AKB	1239	u	N239CKA	1251	MA	N251CKA	1263	WY N263CKA
1228	u	M228AKB	1240	WI	N240CKA	1252	BN	N252CKA	1264	BN N264CKA
1229	u	M229AKB	1241	u	N241CKA	1253	WY	N253CKA		

1276	u	J6SLT	Dennis Dart 9.8m	Plaxton Pointer	B40F	1996	South Lancashire, 1997
1277	u	J7SLT	Dennis Dart 9.8m	Plaxton Pointer	B38F	1996	South Lancashire, 1997
1287	GL	M20GGY	Dennis Dart 9.8m	Plaxton Pointer	B40F	1994	David Ogden, Haydock, 1995
1288	u	M30GGY	Dennis Dart 9.8m	Plaxton Pointer	B40F	1994	David Ogden, Haydock, 1995
1289	LJ	N678GUM	Dennis Dart 9.8m	Plaxton Pointer	B40F	1995	Arriva London, 2003

1290-1299   Dennis Lance 11m   Plaxton Verde   B49F   1994   Clydeside, 1996

1290	u	M930EYS	1293	u	M933EYS	1296	u	M936EYS	1298	u M928EYS
1291	u	M931EYS	1294	u	M934EYS	1297	u	M927EYS	1299	u M929EYS
1292	u	M932EYS								

1300	u	P3SLT	Dennis Dart 9.8m	Plaxton Pointer	B40F	1996	South Lancashire, 1997

In the 1990s East Lancs Coachbuilders was within the same organisation as British Bus, one of Arriva's acquisitions, and thus many of the vehicles purchased during that time carry East Lancs bodywork. The body supplied on the Dart is shown by 1309, N529SPA, which was new to London & Country. Recent moves have seen the bus placed into reserve. *Richard Godfrey*

1301-1310

			Dennis Dart 9.8m		East Lancs EL2000		B34F	1995	Arriva Southern Counties, 2001
1303	u	M523MPF	1308	u	N528SPA	1309	u	N529SPA	1310	u	N530SPA
1304	u	M524MPF									

1301					East Lancs EL2000		B34F	1995	Arriva Southern Counties, 2001
1303	u	M523MPF	1308	u	N528SPA	1309	u	N529SPA	1310	u	N530SPA
1304	u	M524MPF									

1311-1318

			Dennis Dart 9m		Plaxton Pointer		B34F	1995-96	Arriva London, 2005
1311	WX	N705GUM	1313	WX	N703GUM	1316	WX	P916PWW	1318	WX	P918PWW
1312	WX	N706GUM	1314	WX	N701GUM	1317	WX	P917PWW			

1323-1338

			Dennis Dart 9.8m		Plaxton Pointer		B40F	1996	Arriva London, 2001-03
1323	AB	P823RWU	1327	WX	P827RWU	1331	WX	P831RWU	1334	u	P834RWU
1325	AB	P825RWU	1328	AB	P828RWU	1332	BD	P832RWU	1338	u	P838RWU
1326	AB	P826RWU	1330	BD	P830RWU	1333	WX	P833RWU			

1340	WX	M160SKR	Dennis Dart 9m	Plaxton Pointer	B35F	1995	Arriva Southern Counties, 1999
1341	WX	M161SKR	Dennis Dart 9m	Plaxton Pointer	B35F	1995	Arriva Southern Counties, 1999
1342	u	M162SKR	Dennis Dart 9m	Plaxton Pointer	B35F	1995	Arriva Southern Counties, 1999
1343	AB	M163SKR	Dennis Dart 9m	Plaxton Pointer	B35F	1995	Arriva Southern Counties, 1999
1344	u	K1BLU	Dennis Dart 9.8m	East Lancs EL2000	B40F	1993	Blue Bus, Bolton, 2005
1346	u	N4BLU	Dennis Dart 9.8m	Alexander Dash	B40F	1995	Blue Bus, Bolton, 2005
1772	AB	E52UNE	Leyland Tiger TRBTL11/3ARZA	Alexander N	B55F	1988	Arriva Midlands North, 2003
1776	AB	H278LEF	Leyland Tiger TRCL10/3ARZA	Alexander Q	B70F	1990	Arriva Midlands North, 2003
1777	AB	H279LEF	Leyland Tiger TRCL10/3ARZA	Alexander Q	B70F	1990	Arriva Midlands North, 2003

1778-1788

			Volvo B10M-50 Citybus		Alexander Q		B70F	1992	Timeline, Leigh, 1998
1778	AB	H78DVM	1785	AB	H85DVM	1787	AB	H87DVM	1788	AB	H588DVM
1779	AB	H79DVM	1786	AB	H86DVM						

1795	u	N25FWU	DAF SB220	Northern Counties Paladin	B49F	1995	West Coast Motors, 1996
1796	u	N24FWU	DAF SB220	Northern Counties Paladin	B49F	1995	West Coast Motors, 1996
1797	u	M847RCP	DAF SB220	Northern Counties Paladin	B49F	1995	Citybus, Southampton, 1996
1799	u	M849RCP	DAF SB220	Northern Counties Paladin	B49F	1995	Citybus, Southampton, 1996

Pictured in Aberystwyth and carrying School Bus yellow is 1787, H87DVM, a Volvo B10M that carries the Alexander Q design generally assembled at the Belfast plant. The bus was new to Shearings who moved into the service network shortly after deregulation of bus services was introduced. *Laurie Rufus*

1940-1949

		Dennis Lance 11m		East Lancs			B49F	1996	Arriva London, 2001

1940	SK	N210TPK	1943	SK	N213TPK	1946	SK	N216TPK	1948	SK	N218TPK
1941	SK	N211TPK	1944	SK	N214TPK	1947	SK	N217TPK	1949	u	N219TPK
1942	SK	N212TPK	1945	u	N215TPK						

2001-2005

		Scania L113CRL		Wright Axcess-ultralow			N42F	1996	

2001	JS	N101YVU	2003	JS	N103YVU	2004	JS	N104YVU	2005	SK	N105YVU
2002	GL	M2SLT									

2006-2034

		Scania L113CRL		Wright Axcess-ultralow			N43F	1996	

2006	GL	N106DWM	2014	JS	N114DWM	2021	JS	N121DWM	2028	GL	N128DWM
2007	JS	N107DWM	2015	JS	N115DWM	2022	JS	N122DWM	2029	JS	N129DWM
2008	GL	N108DWM	2016	SK	N116DWM	2023	SK	N123DWM	2030	GL	N130DWM
2009	GL	N109DWM	2017	SK	N117DWM	2024	GL	N124DWM	2031	GL	N131DWM
2010	GL	N110DWM	2018	JS	N118DWM	2025	GL	N125DWM	2032	GL	N132DWM
2011	GL	N211DWM	2019	JS	N119DWM	2026	GL	N126DWM	2033	GL	N133DWM
2013	JS	N113DWM	2020	JS	N120DWM	2027	JS	N127DWM	2034	GL	N134DWM

2042-2054

		Scania N113CRL		Wright Pathfinder			B39F	1994	

2042	w	RDZ1702	2049	u	RDZ1709	2051	u	RDZ1711	2053	u	RDZ1713
2045	w	RDZ1705	2050	u	RDZ1710	2052	u	RDZ1712	2054	u	RDZ1714
2048	u	RDZ1708									

2061	SP	CX05EOV	Scania OmniCity CN94UB	Scania	N34F	2005
2062	SP	CX05EOW	Scania OmniCity CN94UB	Scania	N34F	2005
2063	SP	CX05EOY	Scania OmniCity CN94UB	Scania	N34F	2005

2101-2109

		Dennis Dart SLF 10.1m		Plaxton Pointer 2			N39F	1997	Arriva North East, 2003

2101	WI	R601MHN	2104	WX	R604MHN	2106	WX	R606MHN	2108	AB	R608MHN
2102	WI	R602MHN	2105	WX	R685MHN	2107	WX	R607MHN	2109	BD	R609MHN
2103	SK	R603MHN									

2110-2134 — Dennis Dart SLF 10.1m — Plaxton Pointer 2 — N39F — 1999 — Arriva North East, 2005-06

2110	CH	S610KHN	2122	CH	S622KHN	2129	CH	S629KHN	2132	WX	S632KHN
2114	BD	S614KHN	2123	BD	S623KHN	2130	CH	S630KHN	2133	WX	S633KHN
2119	BD	S619KHN	2127	JS	S627KHN	2131	BG	S631KHN	2134	BG	S634KHN
2120	CH	S620KHN	2128	CH	S628KHN						

2141-2144 — ADL Dart 4 — ADL Enviro 200 — N38F — 2008

2141	SO	CX08DJJ	2142	SO	CX08DJK	2143	SO	CX08DJO	2144	SO	CX08DJU

2201-2262 — Dennis Dart SLF 10.1m — Plaxton Pointer 2 — N36F — 2000-01

2201	GL	X201ANC	2217	BO	X217ANC	2234	BO	X234ANC	2248	MA	X248HJA
2202	GL	X202ANC	2218	BO	X218ANC	2235	BO	X235ANC	2249	MA	X249HJA
2203	GL	X203ANC	2219	BO	X219ANC	2236	BO	X236ANC	2251	MA	X251HJA
2204	GL	X204ANC	2221	BO	X221ANC	2237	BO	X237ANC	2252	MA	X252HJA
2207	GL	X207ANC	2223	BO	X223ANC	2238	MA	X238ANC	2253	MA	X253HJA
2208	GL	X208ANC	2224	BO	X224ANC	2239	MA	X239ANC	2254	MA	X254HJA
2209	GL	X209ANC	2226	BO	X226ANC	2241	MA	X241ANC	2256	MA	X256HJA
2211	JS	X211ANC	2227	BO	X227ANC	2242	MA	X242ANC	2257	CH	X257HJA
2212	JS	X212ANC	2228	BO	X228ANC	2243	MA	X243HJA	2258	CH	X258HJA
2213	JS	X213ANC	2229	BO	X229ANC	2244	MA	X244HJA	2259	MA	X259HJA
2214	JS	X214ANC	2231	BO	X231ANC	2246	MA	X246HJA	2261	WY	X261OBN
2215	BO	X215ANC	2232	BO	X232ANC	2247	MA	X247HJA	2262	WY	X262OBN
2216	BO	X216ANC	2233	BO	X233ANC						

2263-2272 — Dennis Dart SLF 10.2m — Alexander ALX200 — N40F — 2000-01

2263	BO	X263OBN	2266	BO	X266OBN	2268	BO	X268OBN	2271	BO	X271OBN
2264	BO	X264OBN	2267	BO	X267OBN	2269	BO	X269OBN	2272	BO	X272OBN
2265	BO	X265OBN									

2273	LJ	S558MCC	Dennis Dart SLF 10.2m	Alexander ALX200	N40F	1998
2274	LJ	S559MCC	Dennis Dart SLF 10.2m	Alexander ALX200	N40F	1998

2276-2279 — Dennis Dart SLF 10.2m — Alexander ALX200 — N36F — 1997 — Arriva London, 2002-03

2276	LJ	P953RUL	2277	LJ	P959RUL	2278	LJ	P960RUL	2279	LJ	P961RUL

2280-2284 — Dennis Dart SLF 10.1m — Plaxton Pointer — N40F — 1996 — Arriva Southern Counties, 2004

2280	GL	P180LKL	2282	GL	P182LKL	2283	GL	P183LKL	2284	GL	P214LKJ

2285-2288 — Dennis Dart SLF 10.1m — Plaxton Pointer — N39F — 1996 — Arriva Southern Counties, 2004

2285	GL	P419HVX	2286	GL	P420HVX	2287	GL	P422HVX	2288	GL	P430HVX

2291-2294 — Dennis Dart SPD 11.3m — Plaxton Super Pointer — N41F — 1996

2291	SO	S248UVR	2292	SO	S249UVR	2293	SO	S250UVR	2294	SO	S251UVR

2296-2300 — Dennis Dart SLF 10.1m — Plaxton Pointer — N34F — 1997 — Arriva London, 2003

2296	RH	R416COO	2298	RH	R418COO	2299	RH	R419COO	2300	RH	R420COO
2297	RH	R417COO									

2301	WX	R301PCW	Dennis Dart SLF 10.1m	Plaxton Pointer 2	N39F	1998

2302-2313 — Dennis Dart SLF 10.2m — Alexander ALX200 — N40F — 1998

2302	WY	R302CVU	2305	WY	R305CVU	2309	WY	R309CVU	2312	WY	R312CVU
2303	WY	R303CVU	2306	WY	R606FBU	2310	WY	R310CVU	2313	WY	R313CVU
2304	WY	R304CVU	2308	WY	R308CVU	2311	WY	R311CVU			

2314-2324 — Dennis Dart SLF 10.1m — Plaxton Pointer 2 — N36F — 1999

2314	BO	T314PNB	2317	BO	T317PNB	2320	BO	T320PNB	2323	BO	T323PNB
2315	BO	T315PNB	2318	BO	T318PNB	2321	BO	T821PNB	2324	BO	T324PNB
2316	BO	T316PNB	2319	BO	T319PNB	2322	BO	T322PNB			

2325	AB	R521UCC	Dennis Dart SLF 10.1m	Plaxton Pointer	N39F	1997	
2326	AB	R522UCC	Dennis Dart SLF 10.1m	Plaxton Pointer	N39F	1997	
2328	CH	S848RJC	Dennis Dart SLF 10.1m	Plaxton Pointer 2	N39F	1998	Ieuan Williams, Deiniolen, 1999
2330	BG	T560JJC	Dennis Dart SLF 10.1m	Plaxton Pointer 2	N39F	1999	
2331	BG	T561JJC	Dennis Dart SLF 10.1m	Plaxton Pointer 2	N39F	1999	
2332	AB	T562JJC	Dennis Dart SLF 10.1m	Plaxton Pointer 2	N39F	1999	

The Plaxton Pointer was produced in several lengths. Here a 10.1 metre model, 521, R521UCC, is seen leaving Pwllheli for Porthmadog. *Richard Godfrey*

2341-2361

| | | | Dennis Dart SLF 10.1m | | | Plaxton Pointer 2 | | | N39F* | 1999-2000 | *2341/2 are N33F |

2341	CH	V571DJC	2347	RH	V577DJC	2352	RH	V582DJC	2357	CH	V587DJC
2342	CH	V572DJC	2348	LJ	V578DJC	2353	RH	V583DJC	2358	CH	V588DJC
2343	BG	V573DJC	2349	RH	V579DJC	2354	RH	V584DJC	2359	CH	V580ECC
2344	BG	V574DJC	2350	RH	V580DJC	2355	RH	V585DJC	2360	CH	V590DJC
2345	RH	V575DJC	2351	RH	V581DJC	2356	CH	V586DJC	2361	CH	V591DJC
2346	BG	V576DJC									

2391	BO	M517KPA	Dennis Lance SLF 11m	Wright Pathfinder	N40F	1995	Arriva Southern Counties, 2002
2392	BO	M518KPA	Dennis Lance SLF 11m	Wright Pathfinder	N40F	1995	Arriva Southern Counties, 2002
2393	BO	M519KPA	Dennis Lance SLF 11m	Wright Pathfinder	N40F	1995	Arriva Southern Counties, 2002
2394	BO	N527SPA	Dennis Lance SLF 11m	Wright Pathfinder	N39F	1995	Arriva Southern Counties, 2002
2395	BO	M761JPA	Dennis Lance SLF 11m	Wright Pathfinder	N39F	1995	Arriva Southern Counties, 2004
2396	BO	M762JPA	Dennis Lance SLF 11m	Wright Pathfinder	N39F	1995	Arriva Southern Counties, 2004
2397	BO	M763JPA	Dennis Lance SLF 11m	Wright Pathfinder	N39F	1995	Arriva Southern Counties, 2004
2400	MA	R91GNW	DAF SB220	Plaxton Prestige	N40F	1998	Blue Bus, Bolton, 2005
2401	SP	R151GNW	DAF SB220	Plaxton Prestige	N38F	1998	Arriva London, 1999
2402	SP	R152GNW	DAF SB220	Plaxton Prestige	N38F	1998	Arriva London, 1999
2403	SP	R153GNW	DAF SB220	Plaxton Prestige	N38F	1998	Arriva London, 1999

2404-2415

| | | | DAF SB220 | | | Alexander ALX300 | | | N42F | 2000 | |

2404	BD	V404ENC	2407	SP	V407ENC	2410	SP	V410ENC	2413	SP	V413ENC
2405	SP	V405ENC	2408	SP	V408ENC	2411	BD	V411ENC	2414	BD	V414ENC
2406	SP	V406ENC	2409	BD	V409ENC	2412	BD	V412ENC	2415	SP	V415ENC

2416-2449

| | | | DAF SB120 | | | Wright Cadet | | | N39F | 2000-01 | |

2416	SP	X416AJA	2426	SP	X426AJA	2434	SP	X434HJA	2442	BD	X442HJA
2417	SP	X417AJA	2427	BD	X427AJA	2435	BD	X435HJA	2443	BD	X443HJA
2418	SP	X418AJA	2428	BD	X428HJA	2436	BD	X436HJA	2445	BD	X445HJA
2419	SP	X419AJA	2429	BD	X429HJA	2437	BD	X437HJA	2446	BD	X446HJA
2421	SP	X421AJA	2431	SP	X431HJA	2438	BD	X438HJA	2447	BD	X447HJA
2422	SP	X422AJA	2432	SP	X432HJA	2439	BD	X439HJA	2448	BD	X448HJA
2423	SP	X423AJA	2433	SP	X433HJA	2441	BD	X441HJA	2449	BD	X449HJA
2424	SP	X424AJA									

Birkenhead is the location for this view of DAF SB220 2460, Y243KBU, which features an East Lancs Myllennium body. The batch with Arriva North West is now shared between this depot and Southport. *Alan Blagburn*

2450	BN	V33BLU	DAF SB220			East Lancs Myllennium	N42F	1999	Blue Bus, Bolton, 2005

2451-2474   DAF SB220   East Lancs Myllennium   N44F   2001

2451	BD	Y451KBU	2457	BD	Y457KBU	2463	SP	Y463KNF	2469	SP	Y469KNF
2452	BD	Y452KBU	2458	BD	Y458KBU	2464	BN	Y464KNF	2470	SP	Y733KNF
2453	BD	Y453KBU	2459	BD	Y459KBU	2465	SP	Y465KNF	2471	SP	Y471KNF
2454	BN	Y454KBU	2460	BD	Y243KBU	2466	SP	Y466KNF	2472	SP	Y472KNF
2455	BD	Y241KBU	2461	BD	Y461KNF	2467	SP	Y467KNF	2473	SP	Y473KNF
2456	BD	Y242KBU	2462	SP	Y462KNF	2468	SP	Y468KNF	2474	BN	Y744KNF

2475	SP	T917KKM	DAF SB220			Plaxton Prestige	N39F	1999	Arriva Southern Counties, 2004
2476	SP	T920KKM	DAF SB220			Plaxton Prestige	N39F	1999	Arriva Southern Counties, 2004
2477	SP	T922KKM	DAF SB220			Plaxton Prestige	N39F	1999	Arriva Southern Counties, 2004
2479	SP	X782NWX	DAF SB120			Wright Cadet	N39F	1999	Selwyns, Runcorn, 2005

2480-2485   VDL Bus SB120   Wrightbus Cadet 2   N39F   2004

2480	BG	CX04AXW	2482	BG	CX04AXZ	2484	BG	CX04AYB	2485	BG	CX04AYC
2481	BG	CX04AXY	2483	BG	CX04AYA						

2489-2503   VDL Bus SB120   Wrightbus Cadet 2   N39F   2004

2489	JS	CX54DKD	2493	JS	CX54DKK	2497	WI	CX54DKU	2501	WI	CX54DLF
2490	JS	CX54DKE	2494	JS	CX54DKL	2498	WI	CX54DKV	2502	WI	CX54DLJ
2491	JS	CX54DKF	2495	JS	CX54DKN	2499	WI	CX54DKY	2503	WI	CX54DLK
2492	JS	CX54DKJ	2496	JS	CX54DKO	2500	WI	CX54DLD			

2504-2513   VDL Bus SB200   Wrightbus Commander   NC43F   2005

2504	AB	CX54EPJ	2507	AB	CX54EPN	2510	AB	CX05AAF	2512	AB	CX05AAK
2505	BG	CX54EPK	2508	AB	CX54EPO	2511	AB	CX05AAJ	2513	WX	CX05AAN
2506	AB	CX54EPL	2509	AB	CX05AAE						

A popular model with Arriva is the VDL Bus SB120 with Wrightbus Cadet bodywork, a model Arriva Bus and Coach sales also sells to the wider market. Forty-two were added to the North West fleet in 2005 with 2520, DK55FXE, shown with local lettering for the Runcorn to Liverpool network. Following recent reallocations this bus has moved to Southport. *Richard Godfrey*

2514-2555   VDL Bus SB120   Wrightbus Cadet 2   N39F   2005

2514	RU	DK55FWY	2525	RU	DK55FXL	2536	RU	DK55FXZ	2546	RU	DK55FYL			
2515	RU	DK55FWZ	2526	RU	DK55FXM	2537	RU	DK55FYA	2547	RU	DK55FYM			
2516	RU	DK55FXA	2527	RU	DK55FXO	2538	RU	DK55FYB	2548	RU	DK55FYN			
2517	BO	DK55FXB	2528	RU	DK55FXR	2539	RU	DK55FYC	2549	SP	DK55FYO			
2518	BO	DK55FXC	2529	RU	DK55FXS	2540	RU	DK55FYD	2550	SP	DK55FYP			
2519	BO	DK55FXD	2530	RU	DK55FXT	2541	RU	DK55FYE	2551	SP	DK55FYR			
2520	SP	DK55FXE	2531	RU	DK55FXU	2542	RU	DK55FYF	2552	RU	DK55FYS			
2521	RU	DK55FXF	2532	RU	DK55FXV	2543	RU	DK55FYG	2553	RU	DK55FYT			
2522	RU	DK55FXG	2533	RU	DK55FXW	2544	SP	DK55FYH	2554	RU	DK55FYV			
2523	RU	DK55FXH	2534	RU	DK55FXX	2545	RU	DK55FYJ	2555	RU	DK55FYW			
2524	RU	DK55FXJ	2535	RU	DK55FXY									

2556	BN	Y20BLU	DAF SB120	Wrightbus Cadet	N36F	2001	Blue Bus, Bolton, 2005
2557	BN	Y21BLU	DAF SB120	Wrightbus Cadet	N36F	2001	Blue Bus, Bolton, 2005
2558	BN	MM02ZVH	DAF SB120	Wrightbus Cadet	N39F	2002	Blue Bus, Bolton, 2005
2559	BN	MM02ZVJ	DAF SB120	Wrightbus Cadet	N39F	2002	Blue Bus, Bolton, 2005
2560	BN	V34ENC	DAF SB220	Ikarus Citibus 481	N42F	1999	Blue Bus, Bolton, 2005
2561	BN	V35ENC	DAF SB220	Ikarus Citibus 481	N42F	1999	Blue Bus, Bolton, 2005
2562	BN	W174CDN	DAF SB220	Ikarus Citibus 481	N42F	2000	Blue Bus, Bolton, 2005
2563	MA	Y36KNB	DAF SB220	Ikarus Polaris	N42F	2001	Blue Bus, Bolton, 2005
2564	MA	Y37KNB	DAF SB220	Ikarus Polaris	N42F	2001	Blue Bus, Bolton, 2005
2565	BN	Y38KNB	DAF SB220	Ikarus Polaris	N42F	2001	Blue Bus, Bolton, 2005
2566	MA	MK52XNN	DAF SB220	Ikarus Polaris	N44F	2002	Blue Bus, Bolton, 2005
2567	MA	MK52XNO	DAF SB220	Ikarus Polaris	N44F	2002	Blue Bus, Bolton, 2005
2568	MA	MK52XNP	DAF SB220	Ikarus Polaris	N44F	2002	Blue Bus, Bolton, 2005
2569	MA	MK52XNR	DAF SB220	Ikarus Polaris	N44F	2002	Blue Bus, Bolton, 2005

2570-2573   VDL Bus SB120   Wrightbus Cadet 2   N39F   2006

2570	CH	CX06BGU	2571	CH	CX06BGV	2572	CH	CX06BGY	2573	CH	CX06BGZ

2574-2607 VDL Bus SB200 Wrightbus Commander 2 N44F 2006

2574	GL	CX06BHA	2583	SP	CX06BHP	2592	SP	CX06BJK	2600	SP	CX06BKE
2575	GL	CX06BHD	2584	SP	CX06BHU	2593	SP	CX06BJO	2601	SP	CX06BKF
2576	GL	CX06BHE	2585	SP	CX06BHV	2594	SP	CX06BJU	2602	SP	CX06BKG
2577	GL	CX06BHF	2586	SP	CX06BHW	2595	SP	CX06BJV	2603	SP	CX06BKJ
2578	GL	CX06BHJ	2587	SP	CX06BHY	2596	SP	CX06BJY	2604	SP	CX06BKK
2579	GL	CX06BHK	2588	SP	CX06BHZ	2597	SP	CX06BJZ	2605	SP	CX06BKL
2580	SP	CX06BHL	2589	SP	CX06BJE	2598	SP	CX06BKA	2606	SP	CX06BKN
2581	SP	CX06BHN	2590	SP	CX06BJF	2599	SP	CX06BKD	2607	SP	CX06BKO
2582	SP	CX06BHO	2591	SP	CX06BJJ						

2608-2619 VDL Bus SB120 Plaxton Centro N40F 2006

2608	WI	CX56CDY	2611	WI	CX56CEF	2614	WI	CX56CEN	2617	WI	CX56CEV
2609	WI	CX56CDZ	2612	WI	CX56CEJ	2615	WI	CX56CEO	2618	WI	CX56CEY
2610	WI	CX56CEA	2613	WI	CX56CEK	2616	WI	CX56CEU	2619	WI	CX56CFA

2620-2662 VDL Bus SB200 Wrightbus Pulsar N44F 2007

2620	LJ	CX07COJ	2631	CH	CX07CRF	2642	RH	CX07CSZ	2653	LJ	CX07CUG
2621	BN	CX07COU	2632	CH	CX07CRJ	2643	LJ	CX07CTE	2654	LJ	CX07CUH
2622	BN	CX07CPE	2633	CH	CX07CRK	2644	RH	CX07CTF	2655	LJ	CX07CUJ
2623	BN	CX07CPF	2634	CH	CX07CRU	2645	LJ	CX07CTK	2656	RH	CX07CUK
2624	SO	CX07CPK	2635	CH	CX07CRV	2646	RH	CX07CTO	2657	LJ	CX07CUU
2625	BN	CX07CPN	2636	CH	CX07CRZ	2647	LJ	CX07CTU	2658	RH	CX07CUV
2626	BN	CX07CPO	2637	CH	CX07CSF	2648	RH	CX07CTV	2659	LJ	CX07CUW
2627	BN	CX07CPU	2638	BO	CX07CSO	2649	LJ	CX07CTY	2660	RH	CX07CUY
2628	BN	CX07CPV	2639	CH	CX07CSU	2650	LJ	CX07CTZ	2661	LJ	CX07CVA
2629	CH	CX07CPY	2640	CH	CX07CSV	2651	LJ	CX07CUA	2662	RH	CX07CVB
2630	CH	CX07CPZ	2641	CH	CX07CSY	2652	RH	CX07CUC			

2663-2700 VDL Bus SB200 Wrightbus Pulsar N44F 2008

2663	GL	CX58EUD	2673	GL	CX58EUP	2683	BN	CX58EVD	2692	JS	CX58EWE
2664	GL	CX58EUE	2674	GL	CX58EUR	2684	BN	CX58EVF	2693	JS	CX58EWF
2665	GL	CX58EUF	2675	GL	CX58EUT	2685	BN	CX58EVG	2694	JS	CX58EWG
2666	GL	CX58EUH	2676	BN	CX58EUU	2686	BN	CX58EVH	2695	JS	CX58EWH
2667	GL	CX58EUJ	2677	BN	CX58EUV	2687	BN	CX58EVJ	2696	JS	CX58EWJ
2668	GL	CX58EUK	2678	BN	CX58EUW	2688	BN	CX58EVK	2697	JS	CX58EWK
2669	GL	CX58EUL	2679	BN	CX58EUY	2689	JS	CX58EWB	2698	JS	CX58EWL
2670	GL	CX58EUM	2680	BN	CX58EUZ	2690	JS	CX58EWC	2699	JS	CX58EWM
2671	GL	CX58EUN	2681	BN	CX58EVB	2691	JS	CX58EWD	2700	JS	CX58EWN
2672	GL	CX58EUO	2682	BN	CX58EVC						

2701-2730 Volvo B10BLE Wrightbus Renown N44F 2001

2701	JS	X701DBT	2709	JS	X709DBT	2717	SP	Y717KNF	2724	SO	Y724KNF
2702	JS	X702DBT	2710	JS	X956DBT	2718	SP	Y718KNF	2725	SO	Y475KNF
2703	JS	X703DBT	2711	JS	Y711KNF	2719	SP	Y719KNF	2726	SO	Y726KNF
2704	JS	X704DBT	2712	JS	Y712KNF	2720	SP	Y457KNF	2727	SO	Y727KNF
2705	JS	X705DBT	2713	JS	Y713KNF	2721	SP	Y721KNF	2728	SO	Y728KNF
2706	JS	X706DBT	2714	JS	Y714KNF	2722	SO	Y722KNF	2729	JS	Y729KNF
2707	JS	X707DBT	2715	JS	Y715KNF	2723	SO	Y723KNF	2730	JS	Y458KNF
2708	JS	X708DBT	2716	SP	Y716KNF						

2732	BN	V22BLU	Volvo B10BLE	Wright Renown	N42F	1999	Blue Bus, Bolton, 2005
2733	BN	X23BLU	Volvo B10BLE	Wright Renown	N42F	2000	Blue Bus, Bolton, 2005

2740-2749 ADL E300 ADL Enviro 300 N45F 2008

2740	MA	CX58EVL	2743	MA	CX58EVR	2746	MA	CX58EVV	2748	MA	CX58EVY
2741	MA	CX58EVN	2744	MA	CX58EVT	2747	MA	CX58EVW	2749	MA	CX58EWA
2742	MA	CX58EVP	2745	MA	CX58EVU						

2793-2799 Volvo B7RLE Wrightbus Eclipse Urban N44F 2004-07 KMP, 2008

2793	BG	CX04HRN	2795	BG	CX04HRP	2797	BG	CX55FAF	2799	BG	CX57BZO
2794	BG	CX04HRR	2796	BG	CX05JVD	2798	BG	CX55FAJ			

2800	BN	Y22CJW	Volvo B7L	Wrightbus Eclipse	N41F	2001

In 2008 ten ADL Enviro 300s were added to the Manchester operation. These feature a chassis built in Guildford with bodywork constructed in Falkirk. Pictured in Piccadilly is 2749, CX58EWA. *Chris Clegg*

2801-2822
Volvo B6BLE — Wright Crusader 2 — N39F — 2000

2801	SO	X801AJA	2806	JS	X806AJA	2812	JS	X812AJA	2817	SO	X817AJA
2802	SO	X802AJA	2807	JS	X807AJA	2813	SO	X813AJA	2818	SO	X818AJA
2803	SO	X803AJA	2808	JS	X808AJA	2814	SO	X814AJA	2819	SO	X819AJA
2804	SO	X804AJA	2809	JS	X809AJA	2815	SO	X815AJA	2821	SO	X821AJA
2805	JS	X805AJA	2811	JS	X811AJA	2816	SO	X816AJA	2822	SO	X822AJA

2825	u	T222MTB	MAN 18.220	Alexander ALX300	N42F	1999	Blue Bus, Bolton, 2005

2826-2829
MAN 14.220 — East Lancs Myllennium — N39F — 2002 — Blue Bus, Bolton, 2005

2826	BN	MF52LYY	2827	BN	MF52LYZ	2828	BN	MF52LZA	2829	BN	MF52LZB

2860-2867
Optare Tempo X1130 — Optare — N41F — 2005-06 — *operated for Ceredigion*

2860	AB	YJ55BKG	2862	AB	YJ55BKL	2864	AB	YJ55BKO	2866	AB	YJ55BKV
2861	AB	YJ55BKK	2863	AB	YJ55BKN	2865	AB	YJ55BKU	2867	AB	YJ06YRY

2868	AB	YJ06YRZ	Optare Tempo X1200	Optare	NC42F	2006	*operated for Ceredigion*
2869	AB	YJ06YRO	Optare Tempo X1200	Optare	NC42F	2006	*operated for Ceredigion*

2900-2929
VDL Bus SB200 — Wrightbus Pulsar 2 — N44F — 2008

2900	GL	CX58EWO	2908	WX	CX58EWY	2916	BO	CX58EXH	2923	BO	CX58EZE
2901	GL	CX58EWP	2909	WX	CX58EWZ	2917	BO	CX58EXJ	2924	BO	CX58EZF
2902	GL	CX58EWR	2910	WX	CX58EXA	2918	BO	CX58EXK	2925	BO	CX58EZG
2903	GL	CX58EWS	2911	WX	CX58EXB	2919	BO	CX58EXL	2926	BO	CX58EZH
2904	GL	CX58EWT	2912	WX	CX58EXC	2920	BO	CX58EZA	2927	BO	CX58EZJ
2905	GL	CX58EWU	2913	WX	CX58EXE	2921	BO	CX58EZB	2928	BO	CX58EZK
2906	WX	CX58EWV	2914	WX	CX58EXF	2922	BO	CX58EZC	2929	BO	CX58EZL
2907	WX	CX58EWW	2915	WX	CX58EXG						

Pictured in Orrell, not far from the Northern Counties works in Wigan where the Palatine II body on 3285, N285CKB, was built. The chassis is a Volvo Olympian. *Richard Godfrey*

2930-2999 VDL Bus SB200 Wrightbus Pulsar 2 N44F 2009

2930	WY	MX09EKK	2948	RU	MX09LXT	2966	JS	MX09OOU	2983	JS	MX09OPO
2931	WY	MX09EKL	2949	RU	MX09LXU	2967	JS	MX09OOV	2984	JS	MX59
2932	WY	MX09EKM	2950	RU	MX09LXV	2968	JS	MX09OOW	2985	JS	MX59
2933	WY	MX09EKN	2951	RU	MX09LXW	2969	JS	MX09OOY	2986	JS	MX59JZA
2934	WY	MX09EKO	2952	RU	MX09LXY	2970	JS	MX09OPA	2987	JS	MX59JZC
2935	WY	MX09EKP	2953	RU	MX09LXZ	2971	JS	MX09OPB	2988	JS	MX59JZD
2936	WY	MX09EKR	2954	GL	MX09JHH	2972	JS	MX09OPC	2989	JS	MX59JZE
2937	WY	MX09EKT	2955	GL	MX09JHK	2973	JS	MX09OPD	2990	JS	MX59JZF
2938	WY	MX09EKU	2956	GL	MX09JHL	2974	JS	MX09OPE	2991	JS	MX59JZC
2939	WY	MX09EKW	2957	GL	MX09JHO	2975	JS	MX09OPF	2992	JS	MX59JZH
2940	WY	MX09EKY	2958	GL	MX09JHU	2976	JS	MX09OPG	2993	JS	MX59JZJ
2941	BO	MX09LYA	2959	GL	MX09JHV	2977	JS	MX09OPH	2994	GL	MX59FHB
2942	BO	MX09LYC	2960	GL	MX09JHY	2978	JS	MX09OPJ	2995	BN	MX59FGD
2943	BO	MX09LYD	2961	GL	MX09JHZ	2979	JS	MX09OPK	2996	MA	MX59FGE
2944	BO	MX09LYF	2962	GL	MX09JJE	2980	JS	MX09OPL	2997	MA	MX59FGF
2945	BO	MX09LYG	2963	GL	MX09JJF	2981	JS	MX09OPM	2998	MA	MX59FGG
2946	BO	MX09LYH	2964	GL	MX09JTY	2982	JS	MX09OPN	2999	MA	MX59FGJ
2947	BO	MX09LYJ	2965	JS	MX09OOJ						

3000-3022 VDL Bus SB200 Wrightbus Pulsar 2 N44F 2009

3000	BD	MX59AAE	3006	BD	MX59AAU	3012	GL	MX59ABN	3018	BO	MX59FFW
3001	BD	MX59AAF	3007	BD	MX59AAV	3013	GL	MX59FFR	3019	BO	MX59FFX
3002	BD	MX59AAJ	3008	BD	MX59AAY	3014	GL	MX59FFS	3020	BO	MX59FFZ
3003	BD	MX59AAK	3009	BD	MX59AAZ	3015	BO	MX59FFT	3021	BO	MX59FGA
3004	BD	MX59AAN	3010	BD	MX59ABF	3016	BO	MX59FFU	3022	BO	MX59FGB
3005	BD	MX59AAO	3011	BD	MX59ABK	3017	BO	MX59FFV			

3023-3045 VDL Bus SB200 Wrightbus Pulsar 2 N44F 2009

3023	SK	MX59JJE	3029	SK	MX59JJV	3035	-	MX59	3041	-	MX59
3024	SK	MX59JJF	3030	GL	MX59JJY	3036	-	MX59	3042	-	MX59
3025	SK	MX59JJK	3031	GL	MX59JJZ	3037	-	MX59	3043	-	MX59
3026	SK	MX59JJL	3032	GL	MX59JKZ	3038	-	MX59	3044	-	MX59
3027	SK	MX59JJO	3033	-	MX59	3039	-	MX59	3045	-	MX59
3028	SK	MX59JJU	3034	-	MX59	3040	-	MX59			

| 3125 | u | D242FYM | Leyland Olympian ONLXB/1RH | Eastern Coach Works | B42/30F | 1986 | Arriva Midlands, 2007 |

3271-3308
Volvo Olympian YN2RV18Z4 — Northern Counties Palatine II D17/00F — 1995-96

3271	SP	N271CKB	3281	SP	N281CKB	3290	SK	N290CKB	3299	BD	N299CKB
3270	SK	N272CKB	3282	SP	N282CKB	3291	SK	N291CKB	3301	BD	N301CKB
3273	SK	N273CKB	3283	SK	N283CKB	3292	SK	N292CKB	3302	BD	N302CKB
3274	SK	N274CKB	3284	SK	N284CKB	3293	BD	N293CKB	3303	BD	N303CLV
3275	SK	N275CKB	3285	SK	N285CKB	3294	BD	N294CKB	3304	BD	N304CLV
3276	SP	N276CKB	3286	SK	N286CKB	3295	BD	N295CKB	3305	BD	N305CLV
3277	SP	N277CKB	3287	SK	N287CKB	3296	BD	N296CKB	3306	BD	N306CLV
3278	SP	N278CKB	3288	SK	N288CKB	3297	BD	N297CKB	3307	BD	N307CLV
3279	SP	N279CKB	3289	SK	N289CKB	3298	BD	N298CKB	3308	BD	N308CLV

3309-3337
Volvo Olympian YN2RV18Z4 — Northern Counties Palatine II B47/30F — 1998

3309	BO	R309WVR	3315	BQ	R315WVN	3326	SP	R326WVR	3332	SP	R332WVR
3310	BQ	R310WVN	3317	SP	R317WVR	3327	SP	R327WVR	3334	SP	R334WVR
3311	BO	R311WVR	3319	BO	R319WVR	3329	SP	R329WVR	3335	SP	R335WVR
3312	BO	R312WVR	3321	SP	R321WVR	3330	SP	R330WVR	3336	BO	R336WVR
3313	BO	R313WVR	3322	SP	R322WVR	3331	SP	R331WVR	3337	BO	R337WVR
3314	BO	R314WVR	3324	BO	R324WVR						

3338	SP	M218YKC	Volvo Olympian YN2RV18Z4	Northern Counties Palatine II	B47/29F	1995	
3339	SP	M219YKC	Volvo Olympian YN2RV18Z4	Northern Counties Palatine II	B47/29F	1995	
3340	w	M921PKN	Volvo Olympian YN2RV50C16Z4	Northern Counties Palatine	B47/30F	1994	
3341	w	L211SBG	Volvo Olympian YN2RV18Z4	Northern Counties Palatine II	B47/29F	1993	

3343-3349
Volvo Olympian — Northern Counties Palatine — B47/29F — 1998

3343	WI	R233AEY	3345	AB	R235AEY	3347	WX	R237AEY	3349	WI	R239AEY
3344	SK	R234AEY	3346	WI	R236AEY	3348	WX	R238AEY			

3350-3354
Volvo Olympian YN2RV18Z4 — Northern Counties Palatine — B47/30F — 1996 — Arriva Southern Counties, 2004

3350	WI	N705TPK	3352	WI	N707TPK	3353	WI	N708TPK	3354	WI	N709TPK
3351	WI	N706TPK								

3355-3360
Volvo Olympian — Northern Counties Palatine — B45/30F — 1997 — Arriva Southern Counties, 2004

3355	BG	P938MKL	3357	BG	P940MKL	3359	BG	P942MKL	3360	WI	P943MKL
3356	BG	P939MKL	3358	BG	P941MKL					

3361	SK	V41DJA	Volvo Olympian	East Lancs Pyoneer	B47/30F	1999	Blue Bus, Bolton, 2005
3362	SO	T42PVM	Volvo Olympian	East Lancs Pyoneer	B47/30F	1999	Blue Bus, Bolton, 2005
3363	SK	S43BLU	Volvo Olympian	East Lancs Pyoneer	B47/30F	1998	Blue Bus, Bolton, 2005
3364	SK	R44BLU	Volvo Olympian	East Lancs Pyoneer	BC45/30F	1998	Blue Bus, Bolton, 2005
3365	SO	S45BLU	Volvo Olympian	East Lancs Pyoneer	B47/30F	1998	Blue Bus, Bolton, 2005

3601-3613
DAF DB250 — Northern Counties Palatine II B47/30F — 1995 — Arriva London, 2001

3601	MA	N601DWY	3605	MA	N605DWY	3608	MA	N608DWY	3611	SP	N611DWY
3602	MA	N602DWY	3606	u	N606DWY	3609	u	N609DWY	3612	SP	N612DWY
3603	MA	N603DWY	3607	MA	N607DWY	3610	MA	N610DWY	3613	MA	N613DWY
3604	MA	N604DWY								

3614-3618
DAF DB250 — Northern Counties Palatine II B43/28F — 1998 — Arriva London/SC, 2003/04

3614	SP	R213CKO	3616	SP	R201CKO	3617	SP	R202CKO	3618	SP	R203CKO
3615	SP	V715LWT								

3837	BO	N716TPK	Dennis Dominator DDA2006	East Lancs	B45/31F	1996	
3976	RH	GYE456W	MCW Metrobus DR101/12	MCW	O43/28D	1980	Original Sightseeing Tour, 2004
3977	RH	KYV663X	MCW Metrobus DR101/14	MCW	PO43/28D	1981	Original Sightseeing Tour, 2004
3979	RH	KYV689X	MCW Metrobus DR101/14	MCW	O43/28D	1981	Original Sightseeing Tour, 2004
3980	SO	GKA449L	Leyland Atlantean AN68/1R	Alexander AL	O43/32F	1973	
3984	RH	E224WBG	Leyland Olympian ONCL10/1RZ	Alexander RL	O43/30F	1988	
3987	RH	E227WBG	Leyland Olympian ONCL10/1RZ	Alexander RL	O43/30F	1988	
3992	RH	YMB512W	Bristol VRT/SL3/6LXB	Eastern Coach Works	O43/31F	1981	Crosville, 1986
3995	LJ	G35HKY	Scania N113DRB	Northern Counties	O47/33F	1990	Arriva Fox County, 2002
3996	LJ	D170FYM	Leyland Olympian ONLXB/1RH	Eastern Coach Works	O42/30F	1986	Arriva Midlands, 2007
3997	LJ	D171FYM	Leyland Olympian ONLXB/1RH	Eastern Coach Works	O42/30F	1986	Arriva Midlands, 2007
3998	LJ	C212GTU	Leyland Olympian ONLXB/1R	Eastern Coach Works	O42/27F	1985	Crosville, 1986
4000	SP	Y46ABA	Dennis Trident	Plaxton President	N47/28F	2001	Blue Bus, Bolton, 2005
4001	SP	Y47ABA	Dennis Trident	Plaxton President	N47/28F	2001	Blue Bus, Bolton, 2005
4002	SP	Y48ABA	Dennis Trident	Plaxton President	N47/28F	2001	Blue Bus, Bolton, 2005

4025-4031 — DAF DB250 10.6m — Alexander ALX400 — N45/23F — 1999 — Arriva London, 2006

4025	SP	S265JUA	4027	SP	S267JUA	4029	SP	S269JUA	4031	SP	S271JUA
4026	SP	S266JUA	4028	SP	S268JUA	4030	SP	S270JUA			

4100-4129 — Volvo B7TL — ADL ALX400 — BC45/27F — 2006

4100	SP	CX55EAF	4108	SP	CX55EAY	4116	SP	CX06EAM	4123	SP	CX06EBD
4101	SP	CX55EAG	4109	SP	CX55EBA	4117	SP	CX06EAO	4124	SP	CX06EBF
4102	SP	CX55EAJ	4110	SP	CX55EBC	4118	SP	CX06EAP	4125	SP	CX06EBG
4103	SP	CX55EAK	4111	SP	CX55EBD	4119	SP	CX06EAW	4126	SP	CX06EBJ
4104	SP	CX55EAM	4112	SP	CX55EBF	4120	SP	CX06EAY	4127	SP	CX06EBK
4105	SP	CX55EAO	4113	SP	CX55EBG	4121	SP	CX06EBA	4128	SP	CX06EBL
4106	SP	CX55EAP	4114	SP	CX55EBJ	4122	SP	CX06EBC	4129	SP	CX06EBM
4107	SP	CX55EAW	4115	SP	CX06EAK						

4400-4434 — ADL Trident 2 — ADL Enviro 400 — BC47/33F — 2008-09

4400	BD	CX58FZM	4409	BD	CX58FZW	4418	BD	CX58GBU	4427	BO	MX09LXJ
4401	BD	CX58FZN	4410	BD	CX58FZY	4419	BD	CX58GBV	4428	BO	MX09LLXK
4402	BD	CX58FZO	4411	BD	CX58FZZ	4420	BD	CX58GBY	4429	BO	MX09LLXL
4403	BD	CX58FZP	4412	BD	CX58GAA	4421	BD	CX58GBZ	4430	BO	MX09LLXM
4404	BD	CX58FZR	4413	BD	CX58GAO	4422	BD	CX58GCF	4431	BO	MX09LLXN
4405	BD	CX58FZS	4414	BD	CX58GAU	4423	BO	MX09LXE	4432	BO	MX09LLXO
4406	BD	CX58FZT	4415	BD	CX58GBE	4424	BO	MX09LXF	4433	BO	MX09LLXR
4407	BD	CX58FZU	4416	BD	CX58GBF	4425	BO	MX09LXG	4434	BO	MX09LLXS
4408	BD	CX58FZV	4417	BD	CX58GBO	4426	BO	MX09LXH			

5301-5320 — Scania L113CRL — Wright Axcess-ultralow — N40F — 1996

5301	GL	P301HEM	5307	GL	P307HEM	5312	GL	P312HEM	5317	GL	P317HEM
5302	GL	P302HEM	5308	GL	P308HEM	5313	GL	P313HEM	5318	GL	P318HEM
5303	GL	P303HEM	5309	GL	P309HEM	5314	GL	P314HEM	5319	GL	P319HEM
5305	GL	P305HEM	5310	GL	P310HEM	5315	GL	P315HEM	5320	GL	P320HEM
5306	GL	P306HEM	5311	GL	P311HEM	5316	GL	P316HEM			

6514-6543 — Volvo B10B — Wright Endurance — BC49F — 1994

6514	u	M514WHF	6524	u	M524WHF	6531	u	M531WHF	6537	u	M537WHF
6515	u	M515WHF	6525	u	M525WHF	6532	u	M532WHF	6538	u	M538WHF
6516	u	M516WHF				6533	u	M533WHF	6540	u	M540WHF
6518	u	M518WHF	6527	SO	M527WHF	6534	u	M534WHF	6541	u	M541WHF
6520	SO	M520WHF	6528	u	M528WHF	6535	u	M535WHF	6542	u	M542WHF
6521	u	M521WHF	6529	u	M529WHF	6536	SO	M536WHF	6543	u	M543WHF

6544-6623 — Volvo B10B — Wright Endurance — BC49F — 1994-96

6544	u	M544WTJ	6565	BO	M565YEM	6584	GL	N584CKA	6605	u	N605CKA
6545	SO	M545WTJ	6566	GL	M566YEM	6585	GL	N585CKA	6606	GL	N606CKA
6546	SO	M546WTJ	6567	GL	M567YEM	6586	BD	N586CKA	6607	BO	N607CKA
6547	u	M547WTJ	6568	GL	M568YEM	6587	GL	N587CKA	6608	BO	N608CKA
6548	u	M548WTJ	6569	GL	M569YEM	6588	GL	N588CKA	6609	BO	N609CKA
6549	u	M549WTJ	6570	BD	M570YEM	6589	GL	N589CKA	6610	SO	N610CKA
6550	u	M550WTJ	6571	GL	M571YEM	6590	BD	N590CKA	6611	SO	N611CKA
6551	u	M551WTJ	6572	GL	M572YEM	6591	BD	N591CKA	6612	SO	N612CKA
6552	u	M552WTJ	6573	GL	M573YEM	6592	BD	N592CKA	6613	SO	N613CKA
6553	u	M553WTJ	6574	SO	M574YEM	6593	BO	N593CKA	6614	BN	N614CKA
6554	u	M554WTJ	6575	GL	M575YEM	6594	SK	N594CKA	6615	SO	N615CKA
6556	GL	M556WTJ	6576	GL	N576CKA	6595	BO	N595CKA	6616	SO	N616CKA
6557	u	M557WTJ	6577	GL	N577CKA	6596	SK	N596CKA	6617	SO	N617CKA
6558	BO	M558WTJ	6578	GL	N578CKA	6597	SO	N597CKA	6618	SO	N618CKA
6559	SO	M559WTJ	6579	GL	N579CKA	6598	SK	N598CKA	6619	SO	N619CKA
6561	BO	M561WTJ	6580	GL	N580CKA	6599	GL	N599CKA	6620	SO	N620CKA
6562	BO	M562WTJ	6581	GL	N581CKA	6601	GL	N601CKA	6621	SO	N621CKA
6563	u	M563WTJ	6582	GL	N582CKA	6603	GL	N603CKA	6622	SO	N622CKA
6564	u	M564YEM	6583	GL	N583CKA	6604	GL	N604CKA	6623	GL	N623CKA

7531-7545 — Dennis Dart SLF 9.8m — Plaxton Pointer — N38F — 1996-97

7531	RH	N531DWM	7535	RH	P535MBU	7539	RH	P539MBU	7543	BD	P543MBU
7532	RH	N532DWM	7536	CH	P536MBU	7540	RH	P540MBU	7544	JS	P544MBU
7533	RH	P533MBU	7537	CH	P537MBU	7541	BG	P541MBU	7545	BN	P545MBU
7534	RH	P534MBU	7538	CH	P538MBU	7542	BG	P542MBU			

Newly into service is a batch of Enviro 400 double-decks. Illustrating the batch is 4429, MX09LXL, which is seen on the trunk service from Manchester to Altrincham. *Richard Godfrey*

7547-7571

Dennis Dart SLF 9.8m — Plaxton Pointer — N38F — 1998

7547	BD	R547ABA	7553	WX	R553ABA	7560	WX	R560ABA	7566	LJ	R566ABA
7548	GL	R548ABA	7554	BD	R554ABA	7561	AB	R561ABA	7567	GL	R567ABA
7549	GL	R549ABA	7556	BD	R556ABA	7562	RH	R562ABA	7568	GL	R568ABA
7550	WX	R550ABA	7557	BD	R557ABA	7563	AB	R563ABA	7569	GL	R569ABA
7551	BD	R551ABA	7558	LJ	R558ABA	7564	RH	R564ABA	7570	JS	R570ABA
7552	BN	R552ABA	7559	WX	R559ABA	7565	AB	R565ABA	7571	JS	R571ABA

7612-7623

Dennis Dart SLF 10.5m — Marshall Capital — N38F — 1999

7612	WI	T612PNC	7615	WY	T615PNC	7618	WX	T618PNC	7621	SK	T621PNC
7613	SK	T613PNC	7616	WY	T616PNC	7619	GL	T619PNC	7622	BN	T622PNC
7614	WI	T614PNC	7617	SK	T617PNC	7620	WI	T620PNC	7623	LJ	T623PNC

7624-7676

Dennis Dart SLF 10.5m — Marshall Capital — N38F — 1999-2000

7624	u	V624DBN	7637	SK	V637DVU	7650	SK	V650DVU	7663	BO	V663DVU
7625	LJ	V625DVU	7638	SK	V638DVU	7651	SO	V651DVU	7664	GL	V664DVU
7626	BO	V626DVU	7639	SK	V639DVU	7652	SO	V652DVU	7665	GL	V665DVU
7627	WI	V627DVU	7640	SK	V640DVU	7653	SO	V653DVU	7667	BO	V667DVU
7628	WI	V628DVU	7641	SK	V641DVU	7654	SO	V654DVU	7668	GL	V668DVU
7629	LJ	V629DVU	7642	JS	V642DVU	7655	SO	V655DVU	7669	SK	V669DVU
7630	LJ	V630DVU	7643	SK	V643DVU	7656	SO	V656DVU	7670	GL	V670DVU
7631	GL	V631DVU	7644	WI	V644DVU	7657	GL	V657DVU	7671	LJ	V671DVU
7632	SK	V632DVU	7645	JS	V645DVU	7658	GL	V658DVU	7672	GL	V672DVU
7633	SK	V633DVU	7646	SK	V646DVU	7659	BO	V659DVU	7673	WI	V673DVU
7634	SK	V634DVU	7647	WI	V647DVU	7660	SK	V660DVU	7674	SK	V674DVU
7635	SK	V635DVU	7648	SK	V648DVU	7661	JS	V661DVU	7675	SK	V675DVU
7636	SK	V636DVU	7649	SK	V649DVU	7662	BO	V662DVU	7676	BN	V676DVU

Ancillary vehicles:

8175	WI	K27EWC	Leyland Lynx LX2R11C15Z4R	Leyland Lynx 2	TV	1992	Colchester, 1994
8182	SP	J251KWM	Leyland Lynx LX2R11G15Z4R	Leyland Lynx	TV	1991	
8201	SP	L301TEM	Volvo B10B	Alexander Strider	TV	1994	
8202	SP	L302TEM	Volvo B10B	Alexander Strider	TV	1994	
8203	MA	L303TEM	Volvo B10B	Alexander Strider	TV	1994	
8204	GL	M109XKC	Volvo B10B	Northern Counties Paladin	TV	1993	Liverbus, 1995
8205	WX	M110XKC	Volvo B10B	Northern Counties Paladin	TV	1993	Liverbus, 1995
8206	BD	M112XKC	Volvo B10B	Northern Counties Paladin	TV	1993	Liverbus, 1995
8207	BO	M113XKC	Volvo B10B	Northern Counties Paladin	TV	1993	Liverbus, 1995
8212	SK	L502TKA	Volvo B10B	Wright Endurance	BC49F	1994	
8215	CH	L505TKA	Volvo B10B	Wright Endurance	BC49F	1994	
8217	SO	L507TKA	Volvo B10B	Wright Endurance	BC49F	1994	
8218	JS	L508TKA	Volvo B10B	Wright Endurance	BC49F	1994	
8219	BO	M519WHF	Volvo B10B	Wright Endurance	BC49F	1994	
8223	RU	M523WHF	Volvo B10B	Wright Endurance	BC49F	1994	
8230	JS	M530WHF	Volvo B10B	Wright Endurance	BC49F	1994	
8232	BN	M532WHF	Volvo B10B	Wright Endurance	BC49F	1994	
8245	GL	N605CKA	Volvo B10B	Wright Endurance	BC49F	1996	

Previous registrations:

CX04HRN	L77KMP	L508JSG	94D28205
CX04HRP	L777KMP	M2SLT	N102YVU
CX04HRR	N777KMP	M5SLT	M20CLA
CX05JVD	M7KMP	R91GNW	R91GNW, R33GNW
CX55FAE	A7KMP	T10BLU	MF51TVV
CX55FAJ	K7KMP	T11BLU	MF51TVW
CX57BZO	N77KMP	V580ECC	V589DJC
J6SLT	N192BNB	W12LUE	MF51TVX
K250CBA	K133TCP	X13LUE	MF51TVY
L411UFY	L175THF	X14LUE	MV02XYH
L412UFY	L176THF		

Depots and Allocations:

Aberystwyth (Park Avenue) - AB

Outstations: Dolgellau, Lampeter, Machynlleth and New Quay

Dart	802	820	823	824	826	832	834	841
	842	1129	1323	1325	1326	1328	1343	2108
	2325	2326	2332	7561	7563	7565	7624	
Volvo B10M bus	1778	1779	1785	1786	1787	1788		
DAF/VDL SB200	2504	2506	2507	2508	2509	2510	2511	2512
Optare Tempo	2860	2861	2862	2863	2864	2865	2866	2867
	2868	2869						
Tiger bus	1772	1776	1777					
Olympian	3345							

Bangor (Llandegai Industrial Estate) - BG

Outstations: Amlwch, Holyhead and Pwllheli

Optare Solo	681	682	683	684	685	686	687	688
	689	690	691	692	693	694	695	
Dart	827	833	2131	2134	2330	2331	2343	2344
	2346	7541	7542					
DAF/VDL SB120	2480	2481	2482	2483	2484	2485		
VDL Bus SB200	2505							
Volvo B7RLE	2793	2794	2795	2796	2797	2798	2799	
Olympian	3355	3356	3357	3358	3359			

Birkenhead (Laird Street) - BD

Dart	1212	1213	1214	1215	1216	1330	1332	2109
	2114	2119	2123	7543	7547	7551		
	7554	7556	7557					
DAF/VDL SB120	2427	2428	2429	2435	2436	2437	2438	2439
	2441	2442	2443	2445	2446	2447	2448	2449
Volvo B10B	6570	6586	6590	6591	6592			
DAF/VDL SB220	2404	2409	2411	2412	2414	2451	2452	2453
	2455	2456	2457	2458	2459	2460	2461	
VDL Bus SB200	3000	3001	3002	3003	3004	3005	3006	3007
	3008	3009	3010	3011				
Olympian	3292	3293	3294	3295	3296	3297	3298	3299
	3301	3302	3303	3304	3305	3306	3307	3308
Trident 2	4400	4401	4402	4403	4404	4405	4406	4407
	4408	4409	4410	4411	4412	4413	4414	4415
	4416	4417	4418	4419	4420	4421	4422	
Ancillary	*8206*							

Bolton (Folds Road) - BN

Dart	890	1252	1258	1259	1264	7545	7552	7622
	7676							
DAF/VDL SB120	2556	2557	2558	2559				
DAF/VDL SB220	2450	2454	2464	2474	2560	2561	2562	2565
VDL Bus SB200	2621	2622	2623	2625	2626	2627	2628	2676
	2677	2678	2679	2680	2681	2682	2683	2684
	2685	2686	2687	2688	2994			
MAN	2826	2827	2828	2829				
Volvo B10BLE	2732	2733						
Volvo B7L	2800							
Volvo B10B	6614							
Ancillary	*8232*							

Bootle (Hawthorne Road) - BO

Dart	803	806	810	811	813	2215	2216	2217
	2218	2219	2221	2223	2224	2226	2227	2228
	2229	2231	2232	2233	2234	2235	2236	2237
	2263	2264	2265	2266	2267	2268	2269	2271
	2272	2314	2315	2316	2317	2318	2319	2320
	2321	2322	2323	2324	7626	7659	7662	7663
	7667							
DAF/VDL SB120	2517	2518	2519	2638				
Lance	2391	2392	2393	2394	2395	2396	2397	
Volvo B10B	6558	6561	6562	6565	6593	6595	6607	6608
	6609							
VDL SB200	2916	2917	2918	2919	2920	2921	2922	2923
	2924	2925	2926	2927	2928	2929	2941	2942
	2943	2944	2945	2946	2947	3015	3016	3017
	3018	3019	3020	3021	3022			
Dominator	3837							
Olympian	3272	3309	3310	3311	3312	3313	3314	3315
	3319	3324	3336	3337				
Trident 2	4423	4424	4425	4426	4427	4428	4429	4430
	4431	4432	4433	4434				
Ancillary	*8207*	*8219*						

Chester (Manor Lane, Hawarden) - CH

Solo	621	623	624	671	672	673	674	675
Dart	2110	2120	2122	2128	2129	2130	2257	2258
	2328	2341	2342	2355	2356	2357	2358	2359
	2360	2361	7536	7537	7538			
DAF/VDL SB120	2570	2571	2572	2573	2629	2630	2631	2632
	2633	2634	2635	2636	2637	2639	2640	2641
Ancillary	*8215*							

Liverpool (Green Lane) - GL

Dart	2201	2202	2203	2204	2207	2208	2209	2280
	2282	2283	2284	2285	7548	7549	7567	7568
	7569	7619	7631	7657	7658	7664	7665	7668
	7670	7671	7672					
Volvo B10B	6556	6566	6567	6568	6569	6570	6571	6572
	6573	6575	6576	6577	6578	6579	6580	6581
	6582	6583	6584	6585	6587	6588	6589	6599
	6603	6604	6606	6623				
Scania L113	2002	2006	2008	2009	2010	2011	2024	2025
	2026	2028	2030	2031	2032	2033	2034	5301
	5302	5303	5305	5306	5307	5308	5309	5310
	5311	5312	5313	5314	5315	5316	5317	5318
	5319	5320						
VDL Bus SB200	2574	2575	2576	2577	2578	2579	2663	2664
	2665	2666	2667	2668	2669	2670	2671	2672
	2673	2674	2675	2900	2901	2902	2903	2904
	2905	2954	2955	2956	2957	2958	2959	2960
	2961	2962	2963	2964	3012	3013	3014	
Ancillary	*8204*	*8245*						

Liverpool (Shaw Road, Speke) - SP

DAF/VDL SB120	2416	2417	2418	2419	2421	2422	2423	2424
	2425	2426	2431	2432	2433	2434	2549	2550
	2551							
Volvo B10B	2716	2717	2718	2719	2720	2721		
DAF/VDL SB220	2401	2402	2403	2405	2406	2407	2408	2410
	2413	2415	2462	2463	2465	2466	2467	2468
	2469	2470	2471	2472	2473	2475	2476	2477
VDL Bus SB200	2580	2581	2582	2583	2584	2585	2586	2587
	2588	2589	2590	2591	2592	2593	2594	2595
	2596	2597	2598	2599	2600	2601	2602	2603
	2604	2605	2606	2607				
OmniCity	2061	2062	2063					
DB250	3611	3612	3614	3615	3616	3617	3618	4025
	4026	4027	4028	4029	4030	4031		
Olympian	3271	3276	3277	3278	3279	3281	3282	3317
	3321	3322	3326	3327	3329	3330	3331	3332
	3334	3335	3338	3339	3341			
Trident	4000	4001	4002					
Volvo B7TL	4100	4101	4102	4103	4104	4105	4106	4107
	4108	4109	4110	4111	4112	4113	4114	4115
	4116	4117	4118	4119	4120	4121	4122	4123
	4124	4125	4126	4127	4128	4129		
Ancillary	*8182*	*8201*	*8202*					

Llandudno Junction (Glan-y-mor Road) - LJ

Mercedes-Benz	393	394						
Dart	801	804	812	814	815	816	817	818
	819	821	822	864	1289	2273	2274	2276
	2277	2278	2279	2348	7558	7623	7625	7629
	7630							
VDL Bus SB200	2620	2643	2645	2647	2649	2650	2651	2653
	2654	2655	2657	2659	2661			
Scania open-top	3995							
Olympian	3996	3997	3998					

Manchester (St Andrew's Square, Piccadilly) - MA

Optare Solo	660	661	662	663	664	665	667	668
	669	670						
Dart	1244	1249	1250	1251	1254	1260	1261	2238
	2239	2241	2242	2243	2244	2246	2247	2248
	2249	2251	2252	2253	2254	2256	2259	
DAF/VDL SB220	2400	2563	2564	2566	2567	2568	2569	
VDL SB200	2995	2996	2997	2998	2999			
DB250	3601	3602	3603	3604	3605	3607	3608	3610
	3613							
Ancillary	*8203*							

Rhyl (Ffynnongroew Road) - RH

Optare Solo	696	697						
Dart	831	843	844	846	847	1130	2296	2297
	2298	2299	2300	2341	2345	2347	2349	2350
	2351	2352	2353	2354	7531	7532	7533	7534
	7535	7539	7540	7562	7564			
VDL Bus SB200	2642	2644	2646	2648	2652	2656	2658	2660
	2662							
Bristol VR open-top	3992							
Metrobus open-top	3976	3977	3979					
Olympian open-top	3984	3987						

Runcorn (Beechwood) - RU

Dart	808	809						
DAF/VDL SB120	2514	2515	2516	2521	2522	2523	2524	2525
	2526	2527	2528	2529	2530	2531	2532	2533
	2534	2535	2536	2537	2538	2539	2540	2541
	2542	2543	2545	2546	2547	2548	2552	2553
	2554	2555						
VDL Bus SB200	2948	2949	2950	2951	2952	2953		
Scania sd	1042	1044	1045	1046	1047	1048	1049	
Ancillary	*8223*							

St Helens (Jackson Street) - JS

Dart	2127	2211	2212	2213	2214	7544	7570	7571
	7642	7645	7661					
Volvo B6	2805	2806	2807	2808	2809	2811	2812	
SB120	2489	2490	2491	2492	2493	2494	2495	2496
Volvo B10B	6542	6543	6549	6551	6552	6553	6554	6597

The North Wales countryside is popular with tourists and with that in mind several open-top buses are used in the area. Northern Counties-bodied Scania 3995, G35HKY, is seen at Penygwryd. *Richard Godfrey*

Scania sd	1035	1036	1037	1038	1039	1040	1041	1043
	1050	1051	1052	1053	1054	1055	1056	1057
	1058	1059	1060	1061	1062	1063	1068	2001
	2003	2004	2007	2013	2014	2015	2017	2018
	2019	2020	2021	2022	2027	2029		
Volvo B10BLE	2701	2702	2703	2704	2705	2706	2707	2708
	2709	2710	2711	2712	2713	2714	2715	2729
	2730							
VDL Bus SB200	2689	2690	2691	2692	2693	2694	2695	2696
	2697	2698	2699	2700	2965	2966	2967	2968
	2969	2970	2971	2972	2973	2974	2975	2976
	2977	2978	2979	2980	2981	2982	2983	
Ancillary	*8218*	*8230*						

Skelmersdale (Neverstitch Road) - SK

Dart	2103	7613	7617	7621	7632	7633	7634	7635
	7636	7637	7638	7639	7640	7641	7642	7643
	7646	7648	7649	7650	7660	7665	7669	7674
Lance	1940	1941	1942	1943	1944	1946	1947	1948
Scania sd	2005	2016	2023					
Volvo B10B	6594	6596	6598					
VDL SB200	3023	3024	3025	3026	3027			
Olympian	3272	3273	3274	3275	3283	3284	3285	3286
	3287	3288	3289	3290	3291	3343	3344	3361
	3363	3364						
Ancillary	*8212*							

Southport (Cobden Road) - SO

Dart	2286	2287	2288	2291	2292	2293	2294	7651
	7652	7653	7654	7655	7656			
Volvo B6	2801	2802	2803	2804	2813	2814	2815	2816
	2817	2818	2819	2821	2822			
Dart 4	2141	2142	2143	2144				
Volvo B10B	6520	6527	6536	6545	6546	6559	6574	6597
	6610	6611	6612	6613	6615	6616	6617	6618
	6619	6620	6621	6622	6622			
Volvo B10BLE	2722	2723	2724	2725	2726	2727	2728	
VDL Bus SB200	2624							
Open-top	3980							
Olympian	3362	3365						
Ancillary	*8217*							

Winsford (Winsford Industrial Estate) - WI

Outstation: Macclesfield

Dart	1220	1233						
Dart SLF	856	857	858	859	860	861	862	863
	865	866	867	868	869	872	876	877
	878	879	882	886	2101	2102	7612	7614
	7620	7627	7628	7644	7647	7673		
SB120	2497	2498	2499	2500	2501	2502	2503	2608
	2609	2610	2611	2612	2613	2614	2615	2616
	2617	2618	2619					
Olympian	3343	3346	3349	3350	3351	3352	3353	3354
	3360							

Wrexham (Berse Road, Caego) - WX

Alero	619							
Solo	622							
Mercedes-Benz	391	392						
Dart	1127	1311	1312	1313	1314	1316	1317	1318
	1327	1331	1333	1340	1341			
Dart SLF	805	807	838	839	848	849	851	852
	2104	2105	2106	2107	2132	2133	2301	7518
	7550	7553	7559	7560				
SB200	2513	2906	2907	2908	2909	2910	2911	2912
	2914	2915						
Olympian	3347	3348						
Ancillary	*8205*							

Wythenshawe (Greeba Road) - WY

Optare Solo	676	677	678	679	680			
Dart SLF	891	892	893	894	2261	2262	2302	
	2303	2304	2305	2306	2308	2309	2310	
	2311	2312	2313	7615	7616			
Dart	1234	1237	1246	1247	1248	1253	1255	1256
	1257	1262	1263					
VDL Bus SB200	2930	2931	2932	2933	2934	2935	2936	2937
	2938	2939	2940	2994	2995	2996	2997	2998
Ancillary	*8175*							

Unallocated/stored - u

remainder

ARRIVA MIDLANDS

Arriva Midlands North Ltd; Arriva Derby Ltd; Stevensons of Uttoxeter Ltd; Arriva Fox County Ltd, 852 Melton Road, Thurmaston, Leicester, LE4 8BT

1159	SY	R159UAL	Mercedes-Benz Vario 0814	Alexander ALX100	B27F	1998	
1357	SY	N357OBC	Mercedes-Benz 709D	Alexander Sprint	B27F	1996	Arriva Fox County, 2002
1358	u	N358OBC	Mercedes-Benz 709D	Alexander Sprint	B27F	1996	Arriva Fox County, 2002

2007-2022 Dennis Dart 9m East Lancs EL2000 B33F 1994

| 2007 | u | L507BNX | 2014 | u | L514BNX | 2016 | SY | L516BNX | 2022 | CK | L522BNX |
| 2013 | u | L513BNX | 2015 | u | L515BNX | | | | | | |

2024	TF	P824RWU	Dennis Dart 9.8m	Plaxton Pointer	B40F	1996	Arriva London, 2001
2025	CK	N680GUM	Dennis Dart 9.8m	Plaxton Pointer	B40F	1995	Arriva London, 2002
2026	u	N673GUM	Dennis Dart 9.8m	Plaxton Pointer	B40F	1995	Arriva London, 2002
2027	SD	N674GUM	Dennis Dart 9.8m	Plaxton Pointer	B40F	1995	Arriva London, 2002
2028	u	N679GUM	Dennis Dart 9.8m	Plaxton Pointer	B40F	1995	Arriva London, 2002
2029	u	L300SBS	Dennis Dart 9.8m	Plaxton Pointer	B40F	1994	
2030	u	J327VAW	Dennis Dart 9.8m	Carlyle Dartline	B40F	1991	Williamsons, Shrewsbury, 1998
2032	SD	M802MOJ	Dennis Dart 9.8m	Marshall C37	B40F	1994	
2033	SD	M803MOJ	Dennis Dart 9.8m	Marshall C37	B40F	1994	
2034	BT	M804MOJ	Dennis Dart 9.8m	Marshall (2001)	B35F	1994	
2035	BT	P835RWU	Dennis Dart 9.8m	Plaxton Pointer	B40F	1996	Arriva London, 2001
2036	CK	P836RWU	Dennis Dart 9.8m	Plaxton Pointer	B40F	1996	Arriva London, 2001
2037	SD	P837RWU	Dennis Dart 9.8m	Plaxton Pointer	B40F	1996	Arriva London, 2001
2038	BT	M100PHA	Dennis Dart 9.8m	Marshall C37	BC40F	1995	Arriva Southern Counties, 1999

2039-2055 Dennis Dart 9.8m Plaxton Pointer B40F 1996 Arriva London, 2001-02

2039	u	P839RWU	2044	BT	P844PWW	2048	BT	P848PWW	2052	SY	P852PWW
2040	SD	P840PWW	2045	BT	P845PWW	2049	BT	P849PWW	2053	u	P853PWW
2041	SD	P841PWW	2046	u	P846PWW	2050	CK	P850PWW	2054	LE	P854PWW
2042	CK	P842PWW	2047	CK	P847PWW	2051	SY	P851PWW	2055	u	P855PWW
2043	BT	P843PWW									

2060	u	M20MPS	Dennis Dart 9.8m	Marshall C37	B40F	1994	Arriva Southern Counties, 1999
2062	u	K542ORH	Dennis Dart 9m	Plaxton Pointer	B34F	1992	Arriva London, 2002
2064	SY	K544ORH	Dennis Dart 9m	Plaxton Pointer	B34F	1992	Arriva London, 2002
2065	u	K542ORH	Dennis Dart 9m	Plaxton Pointer	B34F	1992	Arriva London, 2002

2071-2074 Dennis Dart 9m Northern Counties Paladin B35F 1994 Arriva Yorkshire, 2006

| 2071 | u | L127YVK | 2072 | CK | L157YVK | 2073 | CK | L131YVK | 2074 | CK | L130YVK |

2081	u	K551ORH	Dennis Dart 9m	Plaxton Pointer	B34F	1992	Arriva London, 2001
2086	u	L142YVK	Dennis Dart 9m	Northern Counties Paladin	B35F	1994	Arriva The Shires, 2002
2087	CK	L144YVK	Dennis Dart 9m	Northern Counties Paladin	B35F	1994	Arriva The Shires, 2002
2089	u	N689GUM	Dennis Dart 9.8m	Plaxton Pointer	B40F	1995	Arriva London, 2002
2090	BT	N690GUM	Dennis Dart 9.8m	Plaxton Pointer	B40F	1995	Arriva London, 2002
2093	u	L303NFA	Dennis Dart 9.8m	Plaxton Pointer	B40F	1994	
2096	TF	N806EHA	Dennis Dart 9.8m	East Lancs	B40F	1995	
2097	TF	N807EHA	Dennis Dart 9.8m	East Lancs	B40F	1995	
2098	SY	N808EHA	Dennis Dart 9.8m	East Lancs	B40F	1995	
2099	SD	M805MOJ	Dennis Dart 9.8m	Marshall C37	B40F	1994	
2106	u	L156UKB	Dennis Dart 9m	Plaxton Pointer	B34F	1994	Arriva North West & Wales, 2005
2112	CK	K73SRG	Dennis Dart 9.8m	Plaxton Pointer	B43F	1993	Arriva North West & Wales, 2007
2115	CK	L308HPP	Volvo B6 9.9M	Northern Counties Paladin	N40F	1994	Arriva The Shires, 2007
2117	CK	L311HPP	Volvo B6 9.9M	Northern Counties Paladin	N40F	1994	Arriva The Shires, 2007
2118	u	L601EKM	Volvo B6 9.9m	Plaxton Pointer	B40F	1994	Arriva Southern Counties, 2007
2119	CK	L516CPJ	Volvo B6 9.9m	Plaxton Pointer	B41F	1994	Arriva Southern Counties, 2007
2120	u	L203YCU	Volvo B6 9.9m	Northern Counties Paladin	B39F	1994	Arriva Southern Counties, 2007
2123	u	L135YVK	Dennis Dart 9m	Northern Counties Paladin	B35F	1994	Arriva Southern Counties, 2007
2124	u	L143YVK	Dennis Dart 9m	Northern Counties Paladin	B35F	1994	Arriva Southern Counties, 2007
2125	CK	L155YVK	Dennis Dart 9m	Northern Counties Paladin	B35F	1994	Arriva Southern Counties, 2007

Initially used on Telford town services, several of the 1994 batch of East Lancs-bodied Darts have now left the fleet. One those that remain is 2016, L516BNX, currently allocated to Shrewsbury where it is seen on route 22.
Richard Godfrey

2130-2138

Dennis Dart SLF 8.8m — Plaxton Pointer MPD — N29F — 2000 — Arriva London, 2009

2130	LE	V423DGT	2133	CV	V426DGT	2135	LE	V428DGT	2137	CV	V430DGT
2131	LE	V424DGT	2134	CV	V427DGT	2136	CV	V429DGT	2138	CV	V431DGT
2132	LE	V425DGT								

2200	BT	R929RAU	Dennis Dart SLF 10.1m	Plaxton Pointer 2	B41F	1997	Trent Buses, Derby, 2007

2201-2206

Dennis Dart SLF 10.1m — Plaxton Pointer — N39F — 1997

2201	CV	P201HRY	2203	CV	P203HRY	2205	CV	P205HRY	2206	CV	P206HRY
2202	BT	P202HRY	2204	CV	P204HRY					

2207	BT	S207DTO	Dennis Dart SLF 10.1m	Plaxton Pointer 2	N39F	1998
2208	CV	S208DTO	Dennis Dart SLF 10.1m	Plaxton Pointer 2	N39F	1998

2209-2212

TransBus Dart SLF 8.8m — TransBus Mini Pointer — N29F — 2003

2209	WG	SN03LGC	2210	WG	SN03LGD	2211	WG	SN03LGE	2212	WG	SN03LGF

2214	LE	P954RUL	Dennis Dart SLF 10.2m	Alexander ALX200	N36F	1997	Arriva London, 2002
2215	BT	R45VJF	Dennis Dart SLF 10.2m	Alexander ALX200	N40F	1997
2216	BT	R46VJF	Dennis Dart SLF 10.2m	Alexander ALX200	N40F	1997

2217-2224

Dennis Dart SLF 9.8m — Plaxton Pointer 2 — N33F — 1999

2217	DE	T47WUT	2219	CK	T49JJF	2222	DE	T52JJF	2224	DE	T54JJF
2218	DE	T48WUT	2221	DE	T51JJF	2223	DE	T53JJF			

2226-2238

Dennis Dart SLF 10.2m — Alexander ALX200 — N40F — 2000

2226	DE	W226SNR	2229	DE	W229SNR	2233	DE	W233SNR	2236	DE	W236SNR
2227	DE	W227SNR	2231	DE	W231SNR	2234	DE	W234SNR	2237	DE	W237SNR
2228	DE	W228SNR	2232	DE	W232SNR	2235	DE	W235SNR	2238	DE	W238SNR

A popular model within the Dart range is the Mini Pointer, a model that competes with the Optare Solo. Oswestry, on the Welsh border, is the location for this view of 2296, **BF52NZO**. *Richard Godfrey*

2239-2251

Dennis Dart SLF 8.8m — Plaxton Pointer MPD — N29F — 2000

2239	SY	W239SNR	2243	BT	W243SNR	2247	LE	W247SNR	2249	CK	W249SNR
2241	SY	W241SNR	2246	CK	W246SNR	2248	SY	W248SNR	2251	SY	W251SNR
2242	BT	W242SNR								

2252	DE	X252HBC	Dennis Dart SLF 10.2m	Alexander ALX200	N40F	2000

2253-2267

Dennis Dart SLF 8.8m — Plaxton Pointer MPD — N29F — 2001

2253	WG	Y253YBC	2258	WG	Y258YBC	2262	LE	Y262YBC	2265	LE	Y265YBC
2254	WG	Y254YBC	2259	WG	Y259YBC	2263	LE	Y263YBC	2266	LE	Y266YBC
2256	WG	Y256YBC	2261	WG	Y261YBC	2264	LE	Y264YBC	2267	LE	Y267YBC
2257	WG	Y257YBC								

2268-2275

Dennis Dart SLF 8.8m — Plaxton Pointer MPD — N29F — 2002

2268	LE	SK52MLE	2270	LE	SK52MLJ	2272	LE	SK52MLN	2274	SY	FK52MML
2269	LE	SK52MLF	2271	LE	SK52MLL	2273	LE	SK52MLO	2275	SY	FL52MML

2276-2280

TransBus Dart SLF 8.8m — TransBus Mini Pointer — N29F — 2003

2276	CK	SN53ESG	2277	CK	SN53ESO	2279	TF	SN03LDV	2280	TF	SN03LDX

2281-2288

Dennis Dart SLF 8.8m — Plaxton Pointer MPD — N29F — 1999

2281	SD	V201KDA	2283	SD	V203KDA	2285	SD	V205KDA	2287	CK	V207KDA
2282	SD	V202KDA	2284	SD	V204KDA	2286	CK	V206KDA	2288	BT	V208KDA

2289-2297

Dennis Dart SLF 8.8m — Plaxton Pointer MPD — N29F — 2001-02

2289	CK	BU51KWJ	2292	CK	BU51KWL	2294	CK	Y184TUK	2296	OS	BF52NZO
2290	CK	BU51KWN	2293	CK	BU51KWK	2295	OS	BF52NZN	2297	OS	BF52NZP
2291	CK	BU51KWM								

2299	BT	T61JBA	Dennis Dart SLF 10.6m	Marshall Capital	N37F	1999	Arriva North West, 2000

Supplied new to Gilligan & Wilson of Winsford, Dart 2299, T61JBA, latterly operated alongside four Plaxton examples. The Marshall Capital-bodied example now carries inter-urban livery, seen here at Burton-upon-Trent in August 2009. *Richard Godfrey*

2301-2305
Dennis Dart SLF 10.6m   Plaxton Pointer   N37F   1996

2301	CK	N301ENX	**2303**	SY	N303ENX	**2304**	SY	N304ENX
2302	SY	N302ENX						

(2305 SY N305ENX)

2306-2310
Dennis Dart SLF 10.6m   Plaxton Pointer   NC37F   1996

2306	SY	P306FEA	**2308**	SY	P308FEA	**2309**	SY	P309FEA
2307	SY	P307FEA						

(2310 SY P310FEA)

2311-2315
Dennis Dart SLF 10.6m   East Lancs Spryte   N41F   1996

2311	SD	P311FEA	**2313**	SD	P313FEA	**2314**	SD	P314FEA
2312	SD	P312FEA						

(2315 SD P315FEA)

2316-2327
Dennis Dart SLF 10.6m   Plaxton Pointer   NC39F   1997

2316	BT	P316FEA	**2319**	BT	P319HOJ	**2322**	BT	P322HOJ
2317	BT	P317FEA	**2320**	BT	P320HOJ	**2323**	CK	P323HOJ
2318	CK	P318FEA	**2321**	BT	P321HOJ	**2324**	BT	P324HOJ

(2325 CV P325HOJ / 2326 BT P326HOJ / 2327 BT P327HOJ)

2329-2344
Dennis Dart SLF 10.6m   Plaxton Pointer 2   NC39F   1997-98

2329	CK	R329TJW	**2334**	SD	R334TJW	**2338**	SD	R338TJW
2330	SD	R330TJW	**2335**	SD	R335TJW	**2339**	LE	R339TJW
2331	SD	R331TJW	**2336**	SD	R336TJW	**2340**	LE	R340TJW
2332	SD	R332TJW	**2337**	SD	R337TJW	**2341**	LE	R341TJW

(2342 SY R342TJW / 2343 SY R343TJW / 2344 CK R344TJW)

2345-2353
Dennis Dart SLF 10.6m   Plaxton Pointer 2   NC44F   1999

2345	SY	S345YOG	**2348**	OS	S348YOG	**2350**	CK	S350YOG
2346	OS	S346YOG	**2349**	OS	S349YOG	**2351**	CK	S351YOG
2347	OS	S347YOG						

(2352 OS S352YOG / 2353 SY S353YOG)

2354-2358
Dennis Dart SLF 10.2m   Alexander ALX200   N36F   1997   Arriva London, 2002

2354	LE	P952RUL	**2356**	LE	P956RUL	**2357**	LE	P957RUL
2355	LE	P955RUL						

(2358 LE P958RUL)

2359-2366		Dennis Dart SLF 9.5m	East Lancs Spryte	N31F	1996	Arriva Southern Counties, 2002	

2359	SY	N238VPH	**2361**	TF	N241VPH	**2363**	CK	N243VPH	**2365**	TF N248VPH
2360	SD	N240VPH	**2362**	SY	N242VPH	**2364**	SD	N244VPH	**2366**	TF N249VPH

2367-2379		ADL Dart 10.7m	ADL Pointer	N41F*	2004-05	*2371-7 are NC41F

2367	OS	FJ54OTN	**2371**	SD	FJ55BWA	**2374**	SD	FJ55BWD	**2377**	SD FJ55BWG
2368	OS	FJ54OTP	**2372**	SD	FJ55BWB	**2375**	SD	FJ55BWE	**2378**	OS FJ55BVT
2369	OS	FJ54OTT	**2373**	SD	FJ55BWC	**2376**	SD	FJ55BWF	**2379**	OS FJ55BVU
2370	OS	FJ54OTR								

2380	CK	PSU969	Dennis Dart SLF	UVG CitiStar	N40F	1998	Chase Coaches, Burntwood, 2007
2381	CK	PSU988	Dennis Dart SLF	UVG CitiStar	N40F	1998	Chase Coaches, Burntwood, 2007
2382	CK	PSU989	Dennis Dart SLF	UVG CitiStar	N40F	1998	Chase Coaches, Burntwood, 2007
2390	BT	V338MBV	Dennis Dart SLF 10.5m	East Lancs Spryte	N33F	1999	Tellings-Golden Miller, 2009
2391	BT	V337MBV	Dennis Dart SLF 10.5m	East Lancs Spryte	N33F	1999	Tellings-Golden Miller, 2009
2392	BT	R920RAU	Dennis Dart SLF	Plaxton Pointer 2	N40F	1997	Trent Barton, 2009
2393	CK	P81MOR	Dennis Dart SLF	UVG CitiStar	N38F	1997	Ensign Bus, 2009
2394	CK	R561UOT	Dennis Dart SLF	UVG CitiStar	N43F	1997	Ensign Bus, 2009
2395	LE	P82MOR	Dennis Dart SLF	UVG CitiStar	N37F	1997	Ensign Bus, 2009
2396	LE	R503MOT	Dennis Dart SLF	UVG CitiStar	N37F	1997	Ensign Bus, 2009
2397	LE	R504MOT	Dennis Dart SLF	UVG CitiStar	N34F	1997	Ensign Bus, 2009
2398	LE	R505MOT	Dennis Dart SLF	UVG CitiStar	N37F	1997	Ensign Bus, 2009
2400	BT	P174VUA	Dennis Dart SLF	Alexander ALX200	N40F	1997	Arriva Yorkshire, 2009
2401	LE	P514CVO	Dennis Dart SLF	East Lancs Spryte	N44F	1996	City of Nottingham, 2009

2613-2642		Volvo B6BLE	Wright Crusader 2	N40F	1999-2000	

2613	TH	V213KDA	**2621**	TH	V221KDA	**2629**	TH	V229KDA	**2636**	TF V236KDA
2614	TH	V214KDA	**2622**	TH	V212KDA	**2630**	TH	V230KDA	**2637**	TF V237KDA
2615	TH	V215KDA	**2623**	TH	V223KDA	**2631**	TF	V231KDA	**2638**	TF V238KDA
2616	TH	V216KDA	**2624**	TH	V224KDA	**2632**	TF	V232KDA	**2639**	TF V239KDA
2617	TH	V217KDA	**2625**	TH	V225KDA	**2633**	TF	V233KDA	**2640**	TH V210KDA
2618	TH	V218KDA	**2626**	TH	V226KDA	**2634**	TF	V234KDA	**2641**	TH V211KDA
2619	TH	V219KDA	**2627**	TH	V227KDA	**2635**	TF	V235KDA	**2642**	TH V209KDA
2620	TH	V220KDA	**2628**	TH	V228KDA					

2701	CK	YJ54CKE	DAF SB120 9.4m	Wrightbus Cadet	N30F	2004
2702	CK	YJ54CKF	DAF SB120 9.4m	Wrightbus Cadet	N30F	2004

2703-2707		DAF SB120 10.8m	Wrightbus Cadet	N39F	2002	

2703	SY	BU02URX	**2705**	SY	BU02URZ	**2706**	SY	BU02USB	**2707**	SY BU02USC
2704	SY	BU02URY								

2708-2727		DAF SB120 10.8m	Wrightbus Cadet	N39F	2001	

2708	SY	Y348UON	**2715**	TF	Y365UON	**2720**	TF	Y347UON	**2724**	TF Y364UON
2711	TF	Y351UON	**2716**	TF	Y356UON	**2721**	SY	Y361UON	**2725**	TF Y346UON
2712	TF	Y352UON	**2717**	TF	Y357UON	**2722**	TF	Y362UON	**2726**	TF Y366UON
2713	TF	Y353UON	**2718**	TF	Y358UON	**2723**	TF	Y363UON	**2727**	TF Y367UON
2714	TF	Y354UON	**2719**	TF	Y349UON					

2728-2736		DAF SB120 10.8m	Wrightbus Cadet	N39F	2002-03	

2728	SY	BF52OAG	**2731**	SD	BU03HRD	**2733**	SD	BU03HRF	**2735**	CK BU03HRJ
2729	SY	BF52NZM	**2732**	SD	BU03HRE	**2734**	TF	BU03HRG	**2736**	CK BU03HRK
2730	SY	BU03HRC								

2737	OS	CX04EHZ	DAF SB120 10.8m	Wrightbus Cadet	N39F	2004	Arriva North West & Wales, 2004
2738	TF	X781NWX	DAF SB120 9.4m	Wrightbus Cadet	N30F	2001	Arriva North West & Wales, 2005
2739	TF	X783NWX	DAF SB120 9.4m	Wrightbus Cadet	N30F	2001	Arriva North West & Wales, 2005
2740	OS	CX04EHV	DAF SB120 10.8m	Wrightbus Cadet	N39F	2004	Arriva North West & Wales, 2006
2741	OS	CX04EHW	DAF SB120 10.8m	Wrightbus Cadet	N39F	2004	Arriva North West & Wales, 2006
2742	OS	CX04EHY	DAF SB120 10.8m	Wrightbus Cadet	N39F	2004	Arriva North West & Wales, 2006

2900	CV	YJ57EKA	Optare Solo M880	Optare	N29F	2007	
2901	CV	YJ57EKB	Optare Solo M880	Optare	N29F	2007	
2902	CV	YJ57EKC	Optare Solo M880	Optare	N29F	2007	
2903	SY	YJ57EKD	Optare Solo M920	Optare	N34F	2007	
2904	SY	YJ07VRU	Optare Solo M850	Optare	N29F	2007	Optare demonstrator, 2008

Twenty-four Optare Versa buses operate for Arriva Midlands, most based at the Shropshire depots for the long X5/892 link to Wolverhampton and rural service connecting Bridgnorth and Rodington. Oswestry's 2993, YJ09MKG, is shown, and 2996 is shown on the cover. *Chris Clegg*

2905-2922

			Optare Solo SR M890			Optare		N26F		2008-09		
2905	DE	YJ58CCA	2910	DE	YJ58CCN	2915	DE	YJ09MLO	2919	DE	YJ09MLZ
2906	DE	YJ58CCD	2911	DE	YJ58CCO	2916	DE	YJ09MLV	2920	DE	YJ09MMA
2907	DE	YJ58CCE	2912	DE	YJ58CCU	2917	DE	YJ09MLX	2921	DE	YJ09MME
2908	DE	YJ58CCF	2913	DE	YJ09MLL	2918	DE	YJ09MLY	2922	DE	YJ09MMF
2909	DE	YJ58CCK	2914	DE	YJ09MLN						

2923-2930

			Optare Solo M920			Optare		N31F		2009		
2923	LE	YJ09MJE	2925	LE	YJ09MJK	2927	LE	YJ09MJX	2929	LE	YJ09OUA
2924	LE	YJ09MJF	2926	LE	YJ09MJV	2928	LE	YJ09MJY	2930	LE	YJ09OUB

2974-2995

			Optare Versa V1100			Optare		N39F*		2009		*2985-7 are N35F
2974	TF	YJ09MKM	2980	TF	YJ09MKX	2986	BT	YJ09LBL	2991	SY	YJ09MKE
2975	TF	YJ09MKN	2981	SY	YJ09MKZ	2987	BT	YJ09LBN	2992	OS	YJ09MKF
2976	TF	YJ09MKO	2982	SY	YJ09MLE	2988	SY	YJ09OTW	2993	OS	YJ09MKG
2977	TF	YJ09MKP	2983	SY	YJ09MLF	2989	SY	YJ09MKC	2994	SY	YJ09MKK
2978	TF	YJ09MKU	2984	SY	YJ09MLK	2990	SY	YJ09MKD	2995	OS	YJ09MKL
2979	TF	YJ09MKV	2985	BT	YJ09LBK						

2996	SY	YJ58PHX	Optare Versa V1100			Optare		N38F	2008
2997	SY	YJ57EKE	Optare Versa V1100			Optare		N35F	2007
3201	DE	YJ55KZS	VDL Bus SB4000			Van Hool T9 Alizée		C49FT	2005
3202	DE	YJ54CPE	VDL Bus SB4000			Van Hool T9 Alizée		C49FT	2004
3203	WG	YJ53VFY	DAF SB4000			Van Hool T9 Alizée		C49FT	2003
3204	WG	YJ03PFX	DAF SB4000			Van Hool T9 Alizée		C49FT	2003
3206	WG	YJ04BKF	VDL Bus SB4000			Van Hool T9 Alizée		C49FT	2004
3207	DE	YJ54CPF	VDL Bus SB4000			Van Hool T9 Alizée		C49FT	2004
3208	WG	YJ05PVT	VDL Bus SB4000			Van Hool T9 Alizée		C49FT	2005
3209	WG	T209XVO	DAF SB3000			Van Hool T9 Alizée		C51FT	1999

3250-3255

			Scania K340 EB			Caetano Levante		C49FT	2006
3250	DE	FJ56PCX	3252	DE	FJ56PCZ	3254	WG	FJ56PDO	3255	WG	FJ56OBP
3251	DE	FJ56PCY	3253	WG	FJ56PDK						

3305	TF	H74DVM	Volvo Citybus B10M-50	Alexander Q	B55F	1991	Timeline, Leigh, 1998
3310	TF	H81DVM	Volvo Citybus B10M-50	Alexander Q	B55F	1991	Timeline, Leigh, 1998

3415-3429

			Scania L113CRL		Plaxton Paladin	NC45F*	1998	*3415-19 are NC47F

3415	BT	R415TJW	3419	BT	R419TJW	3423	BT	R423TJW	3427	SY	R427TJW
3416	TH	R416TJW	3420	BT	R420TJW	3424	BT	R424TJW	3428	SY	R428TJW
3417	TH	R417TJW	3421	BT	R421TJW	3425	SY	R425TJW	3429	u	R429TJW
3418	SY	R418TJW	3422	BT	R422TJW	3426	SY	R426TJW			

3466-3479

Scania L113CRL — East Lancs European — NC51F* — 1996 — *3476-9 are NC49F

3466	u	N166PUT	3471	u	N171PUT	3475	u	N175PUT	3478	u	N178PUT
3468	BT	N168PUT	3473	u	N173PUT	3477	u	N177PUT	3479	BT	N179PUT
3470	u	N170PUT	3474	u	N174PUT						

3489	TH	N429XRC	Scania L113CRL	East Lancs European	N51F	1996
3491	u	N429XRC	Scania L113CRL	East Lancs European	N51F	1996
3492	u	N429XRC	Scania L113CRL	East Lancs European	N51F	1996

3501-3504

Scania N113CRL — East Lancs European — BC42F — 1995

3501	SY	M401EFD	3502	SY	M402EFD	3503	SY	M403EFD	3504	SY	M404EFD

3550	TH	YN56NNA	Scania OmniCity CN230 UD	Scania	N36F	2007
3551	TH	YN56NNB	Scania OmniCity CN230 UD	Scania	N36F	2007

3552-3566

Scania OmniCity CN230 UB — Scania — N41F — 2008

3552	DE	YR59SRO	3556	DE	YR59SRY	3560	DE	YR59SSO	3564	DE	YR59SSZ
3553	DE	YR59SRU	3557	DE	YR59SRZ	3561	DE	YR59SSU	3565	DE	YR59STX
3554	DE	YR59SRV	3558	DE	YR59SSJ	3562	DE	YR59SSV	3566	DE	YR59STY
3555	DE	YR59SRX	3559	DE	YR59SSK	3563	DE	YR59SSX			

3567-3577

Scania OmniCity CN230 UB — Scania — N42F — 2009

3567	DE	YT09ZBL	3570	DE	YT09ZBP	3573	DE	YT09ZBV	3576	DE	YT09ZBY
3568	DE	YT09ZBN	3571	DE	YT09ZBR	3574	DE	YT09ZBW	3577	DE	YT09ZBZ
3569	DE	YT09ZBO	3572	DE	YT09ZBU	3575	DE	YT09ZBX			

3578	WG	YN58RCF	Scania OmniCity CN94 UB	Scania	N42F	2005	Rotala, 2009
3579	WG	YN05HCG	Scania OmniCity CN94 UB	Scania	N42F	2005	Rotala, 2009
3580	DE	YN04AHA	Scania OmniCity CN94 UB	Scania	N41F	2004	Rotala, 2009

3601-3612

Volvo B10BLE — Alexander ALX300 — N44F — 2000

3601	CK	V601DBC	3604	CV	V604DBC	3607	CV	V607DBC	3610	CK	V610DBC
3602	CK	V602DBC	3605	CV	V605DBC	3608	CV	V608DBC	3611	CV	V611DBC
3603	CK	V603DBC	3606	BT	V606DBC	3609	CK	V609DBC	3612	CV	V612DBC

3701-3704

DAF SB200 — Wrightbus Commander — N44F — 2003

3701	CK	FD52GGO	3702	CK	FD52GGP	3703	CK	FD52GGU	3704	CK	FD52GGV

3705-3718

DAF SB200 — Wrightbus Commander — N44F — 2002

3705	TF	BF52NZR	3709	TF	BF52NZV	3713	TF	BF52NZZ	3716	TF	BF52OAC
3706	TF	BF52NZS	3710	TF	BF52NZW	3714	TF	BF52OAA	3717	TF	BF52OAD
3707	TF	BF52NZT	3711	TF	BF52NZX	3715	TF	BF52OAB	3718	TF	BF52OAE
3708	TF	BF52NZU	3712	TF	BF52NZY						

3719-3726

VDL Bus SB200 — Wrightbus Commander — N44F — 2006

3719	WG	FJ06ZTE	3721	WG	FJ06ZTG	3723	WG	FJ06ZTK	3725	LE	FJ06ZTM
3720	WG	FJ06ZTF	3722	WG	FJ06ZTH	3724	WG	FJ06ZTL	3726	LE	FJ06ZTN

3727-3731

VDL Bus SB200 — Wrightbus Pulsar — N44F — 2007

3727	CK	YJ57BUA	3729	CK	YJ57BPZ	3730	CK	YJ57BRF	3731	CK	YJ57BRV
3728	CK	YJ57BUE									

3732-3739

VDL Bus SB200 — Plaxton Centro — N44F — 2007

3732	LE	YJ57AZD	3734	LE	YJ57AZG	3736	LE	YJ57AZN	3738	SD	YJ57AZP
3733	LE	YJ57AZF	3735	LE	YJ57AZL	3737	SD	YJ57AZO	3739	SD	YJ57AZR

3740	SD	YJ57AZT	VDL Bus SB200	Wrightbus Commander	N44F	2007
3741	SD	YJ57AZU	VDL Bus SB200	Wrightbus Commander	N44F	2007

Arriva Midlands now operates thirty-one Scania OmniCity integral buses, all to the eastern side of the operation. Seen in Swadlincote while heading back to Derby is 3575, **YT09ZBX**. *Richard Godfrey*

3742-3755

VDL Bus SB200 — Wrightbus Pulsar 2 — N44F — 2009

3742	TF	YJ59BVA	**3746**	TF	YJ59BVH	**3750**	TF	YJ59BVN	**3753** TF YJ59BUU
3743	TF	YJ59BVB	**3747**	TF	YJ59BVK	**3751**	TF	YJ59BUO	**3754** TF YJ59BUV
3744	TF	YJ59BVF	**3748**	TF	YJ59BVL	**3752**	TF	YJ59BUP	**3755** TF YJ59BUW
3745	TF	YJ59BVG	**3749**	TF	YJ59BVM				

3800-3812

Scania OmniLink K230 — Scania — NC45F — 2008

3800	TH	YN08HZK	**3804**	TH	YN08HZR	**3807**	TH	YN08HZU	**3810** WG YN08HZX
3801	TH	YN08HZL	**3805**	TH	YN08HZS	**3808**	WG	YN08HZV	**3811** WG YN08HZY
3802	TH	YN08HZM	**3806**	TH	YN08HZT	**3809**	WG	YN08HZW	**3812** WG YN08HZZ
3803	TH	YN08HZP							

3900-3913

Volvo B7RLE — Wrightbus Eclips Urban — N45F — 2008

3900	LE	FY58HYH	**3904**	LE	FY58HYN	**3908**	LE	FY58HYS	**3911** LE FY58HYV
3901	LE	FY58HYK	**3905**	LE	FY58HYO	**3909**	LE	FY58HYT	**3912** LE FY58HYW
3902	LE	FY58HYL	**3906**	LE	FY58HYP	**3910**	LE	FY58HYU	**3913** LE FY58HYX
3903	LE	FY58HYM	**3907**	LE	FY58HYR				

4001-4014

Volvo B7TL — Wrightbus Eclipse Gemini — N41/29F — 2006

4001	WG	FJ06ZPX	**4005**	WG	FJ06ZRL	**4009**	WG	FJ56OBG	**4012** WG FJ56OBL
4002	WG	FJ06ZPW	**4006**	WG	FJ56OBC	**4010**	WG	FJ56OBH	**4013** WG FJ56OBM
4003	WG	FJ06ZPV	**4007**	WG	FJ56OBE	**4011**	WG	FJ56OBK	**4014** WG FJ56OBN
4004	WG	FJ56OBD	**4008**	WG	FG56OBF				

4200-4207

Volvo B9TL — Wrightbus Eclipse Gemini — N43/29F — 2008

4200	TH	FJ08LVL	**4202**	TH	FJ08LVN	**4204**	TH	FJ08LVP	**4206** TH FJ08LVS
4201	TH	FJ08LVM	**4203**	TH	FJ08LVO	**4205**	TH	FJ08LVR	**4207** TH FJ08LVT

4208-4224

Volvo B9TL — Wrightbus Eclipse Gemini — N43/27F — 2008

4208	DE	FJ58KXF	**4213**	DE	FJ58KXM	**4217**	DE	FJ58KXR	**4221** DE FJ58KXV
4209	DE	FJ58KXG	**4214**	DE	FJ58KXN	**4218**	DE	FJ58KXS	**4222** DE FJ58KXW
4210	DE	FJ58KXH	**4215**	DE	FJ58KXO	**4219**	DE	FJ58KXT	**4223** DE FJ58KXX
4211	DE	FJ58KXK	**4216**	DE	FJ58KXP	**4220**	DE	FJ58KXU	**4224** DE FJ58KXY
4212	DE	FJ58KXL							

4607-4638 — Volvo Olympian — Northern Counties Palatine — B47/29F — 1996-98

4607	CV	P607CAY	4616	WG	R616MNU	4619	LE	R619MNU	4622	CV	R622MNU
4614	CV	R614MNU	4617	LE	R617MNU	4620	CV	R620MNU	4624	LE	R624MNU
4615	CV	R615MNU	4618	LE	R618MNU	4621	CV	R621MNU	4638	WG	R638MNU

4644-4653 — Volvo Olympian — Northern Counties Palatine — B47/29F — 1998

4644	WG	S644KJU	4647	LE	S647KJU	4650	BT	S650KJU	4652	BT	S652KJU
4645	u	S645KJU	4649	LE	S649KJU	4651	WG	S651KJU	4653	CK	S653KJU
4646	LE	S646KJU									

4701-4716 — DAF DB250 — East Lancs Lowlander — N44/29F — 2001

4701	WG	Y701XJF	4705	WG	Y705XJF	4709	WG	Y709XJF	4714	WG	FE51YWM
4702	WG	Y702XJF	4706	WG	Y706XJF	4711	WG	FE51YWJ	4715	WG	FE51WSU
4703	WG	Y703XJF	4707	WG	Y707XJF	4712	WG	FE51YWK	4716	WG	FE51WSV
4704	WG	Y704XJF	4708	WG	FE51YWH	4713	WG	FE51YWL			

4717-4733 — DAF DB250 — East Lancs Lowlander — N44/29F — 2002

4717	WG	FD02UKB	4722	LE	FN52XBG	4726	LE	PN52XBF	4730	CV	FD02UKR
4718	LE	FD02UKC	4723	LE	FD02UKJ	4727	LE	FD02UKN	4731	CV	FD02UKS
4719	WG	FD02UKE	4724	LE	FD02UKK	4728	LE	FD02UKO	4732	CV	FD02UKT
4720	WG	PN52XBH	4725	LE	FD02UKL	4729	LE	FD02UKP	4733	CV	FD02UKU
4721	WG	FD02UKG									

4734-4745 — DAF DB250 — East Lancs Lowlander — N44/29F — 2003

4734	CV	PN52XRJ	4737	CV	PN52XRM	4740	CV	PN52XRR	4743	WG	PN52XRU
4735	CV	PN52XRK	4738	CV	PN52XRO	4741	CV	PN52XRS	4744	WG	PN52XRV
4736	CV	PN52XRL	4739	WG	PN52XRP	4742	WG	PN52XRT	4745	WG	PN52XRW

4746-4777 — DAF DB250 — Wrightbus Pulsar Gemini — N44/29F — 2006

4746	WG	FJ06ZTO	4754	WG	FJ06ZSZ	4762	WG	FJ06ZSK	4770	LE	FJ06ZRP
4747	WG	FJ06ZTP	4755	WG	FJ06ZTB	4763	WG	FJ06ZSL	4771	LE	FJ56KFC
4748	WG	FJ06ZST	4756	WG	FJ06ZTC	4764	WG	FJ06ZSN	4772	LE	FJ56KFD
4749	WG	FJ06ZSU	4757	WG	FJ06ZTD	4765	LE	FJ06ZSO	4773	LE	FJ56KFE
4750	WG	FJ06ZSV	4758	WG	FJ06ZSD	4766	LE	FJ06ZSP	4774	LE	FJ56KFF
4751	WG	FJ06ZSW	4759	WG	FJ06ZSE	4767	LE	FJ56KFA	4775	LE	FJ56KFG
4752	WG	FJ06ZSX	4760	WG	FJ06ZSF	4768	LE	FJ06ZRN	4776	LE	FJ56KFK
4753	WG	FJ06ZSY	4761	WG	FJ06ZSG	4769	LE	FJ06ZRO	4777	LE	FJ56KFL

Tamworth and Derby depots have received several new double-deck buses recently, the Tamworth ones being used on the services into Birmingham. Pictured in Tamworth is 4206, FJ08LVS.
Dave Heath

6000-6006 Optare Solo M850 · Optare · N24F · 2003 · Operated for Shropshire CC

6000	SY	BU03HRL	**6002**	SY	BU03HPX	**6004**	SY	BU03HPZ	**6006**	SY	FJ04PFX
6001	SY	BU03HPV	**6003**	SY	BU03HPY	**6005**	SY	BU03HRA			

6007	TF	FN04AFJ	Optare Solo M920	Optare	N33F	2004	Operated for Shropshire CC
6008	SY	FJ54OTV	Optare Solo M850	Optare	N29F	2004	Operated for Shropshire CC
6009	SY	FJ54OTW	Optare Solo M850	Optare	N29F	2004	Operated for Shropshire CC
6010	TF	FJ54OTX	Optare Solo M920	Optare	N30F	2004	Operated for Shropshire CC

Ancillary vehicles:

9516	LE	K108OHF	Volvo B10B	Northern Counties Paladin	TV	1995	Arriva North West & Wales, 2007
9517	CK	K107OHF	Volvo B10B	Northern Counties Paladin	TV	1995	Arriva North West & Wales, 2007
9520	DE	K102OHF	Volvo D10D	Northern Counties Paladin	TV	1995	Arriva North West & Wales, 2007
9521	TF	K105OHF	Volvo B10B	Northern Counties Paladin	TV	1995	Arriva North West & Wales, 2007
9522	LE	K101OHF	Volvo B10B	Northern Counties Paladin	TV	1995	Arriva North West & Wales, 2007
9525	LE	M422UNW	Volvo B10B	Alexander Strider	TV	1994	Arriva Yorkshire, 2009
9526	CK	M430UNW	Volvo B10B	Alexander Strider	TV	1994	Arriva Yorkshire, 2009
9527	u	M429UNW	Volvo B10B	Alexander Strider	TV	1994	Arriva Yorkshire, 2009
9528	u	M431UNW	Volvo B10B	Alexander Strider	TV	1994	Arriva Yorkshire, 2009
9530	u	M421UNW	Volvo B10B	Alexander Strider	TV	1994	Arriva Yorkshire, 2009

Previous registrations:

PSU969	S407JUA	PSU988	S406JUA
PSU989	S402JUA		

Depots and Allocations:

Burton-on-Trent (Wetmore Road) - BT

Dart	2034	2035	2043	2044	2045	2040	2049	2090
Dart SLF	2200	2202	2207	2215	2216	2242	2243	2244
	2288	2299	2316	2317	2319	2320	2321	2322
	2324	2326	2327	2390	2391	2392	2400	
Optare Versa	2985	2986	2987					
Scania L113	3415	3419	3420	3421	3422	3424	3468	3479
Volvo B10BLE	3606							
Olympian	4650	4652						

Cannock (Delta Way) - CK

Dart	2022	2025	2036	2042	2047	2050	2072	2073
	2074	2087	2112	2115	2125			
Dart SLF	2219	2246	2249	2276	2277	2286	2287	2289
	2290	2291	2292	2293	2294	2301	2318	2323
	2329	2344	2350	2351	2363	2380	2381	2382
	2393	2394						
Volvo B6	2117	2119						
DAF/VDL SB120	2701	2702	2735	2736				
Volvo B10BLE	3601	3602	3603	3609	3610			
VDL Bus SB200	3701	3702	3703	3704	3727	3728	3729	3730
	3731							
Olympian	4653							
Ancillary	*9517*	*9526*						

Coalville (Ashby Road) - CV

Solo	2900	2901	2902					
Dart SLF	2133	2134	2136	3137	2138	2201	2203	2204
	2205	2206	2208	2325				
Volvo B10BLE	3604	3605	3607	3608	3611	3612		
Olympian	4607	4614	4615	4620	4621	4622		
DAF/VDL DB250	4735							

Derby (Ascot Drive) - DE

Solo	2905	2906	2907	2908	2909	2910	2911	2912
	2913	2914	2915	2916	2917	2918	2919	2920
	2921	2922						
Dart SLF	2217	2218	2221	2222	2223	2224	2226	2227
	2228	2229	2231	2232	2233	2234	2235	2236
	2237	2238	2252					
VDL Bus SB4000	3201	3202	3207					
Scania coach	3250	3251	3252					
Scania OmniCity	3552	3553	3554	3555	3556	3557	3558	3559
	3560	3561	3562	3563	3564	3565	3566	3567
	3568	3569	3570	3571	3572	3573	3574	3575
	3576	3577	3580					
Volvo B9TI	4208	4209	4210	4211	4212	4213	4214	4215
	4216	4217	4218	4219	4220	4221	4222	4223
	4224							
Ancillary	*9507*	*9520*						

Leicester (Melton Road, Thurmaston) - LE

Optare Solo	2923	2924	2925	2926	2927	2928	2929	2930
Dart	2041	2054						
Dart SLF	2130	2131	2132	2135	2214	2247	2262	2263
	2264	2265	2266	2267	2268	2269	2270	2271
	2272	2273	2340	2341	2354	2355	2356	2357
	2358	2395	2396	2397	2398	2401		
VDL Bus SB200	3725	3726						
Volvo B7RLE	3900	3901	3902	3903	3904	3905	3906	3907
	3908	3909	3910	3911	3912	3913		
Olympian	4617	4618	4619	4624	4626	4627	4629	
VDL Bus DB250	4718	4722	4723	4724	4725	4726	4727	4728
	4729	4765	4766	4767	4768	4769	4770	4771
	4772	4773	4774	4775	4776			
Ancillary	*9516*	*9522*	*9525*					

Oswestry (Salop Road) - OS

Optare Solo	2903							
Dart	2295	2296	2297	2346	2347	2348	2349	2352
	2367	2368	2369	2370	2378	2379		
Optare Versa	2992	2993	2995					
VDL Bus SB120	2737	2740	2741	2742				

Shrewsbury (Spring Gardens) - SY

Outstation: Bridgnorth

Mercedes-Benz	1159	1357						
Solo	2904	6000	6001	6002	6003	6004	6005	6006
	6008	6009						
Versa	2982	2983	2984	2988	2989	2990	2991	2994
	2996	2997						
Dart	2015	2016	2051	2052	2064	2082	2098	
Dart SLF	2241	2239	2248	2251	2274	2275	2302	2303
	2304	2305	2306	2307	2308	2309	2310	2340
	2342	2343	2344	2345	2346	2348	2353	2359
	2362							
DAF/VDL SB120	2703	2704	2705	2706	2707	2708	2721	2728
	2729	2730						
Scania L113	3418	3425	3426	3427	3428			
Scania N113	3501	3502	3503	3504				

Stafford (Dorrington Park Industrial Estate, Common Road) - SD

Dart	2027	2032	2033	2037	2040	2041	2099	
Dart SLF	2281	2282	2283	2284	2285	2311	2312	2313
	2314	2315	2330	2331	2332	2334	2335	2336
	2337	2338	2360	2364	2371	2372	2373	2374
	2375	2376	2377					
DAF/VDL SB120	2731	2732	2733					
VDL Bus SB200	3737	3738	3739	3740	3741			

Tamworth (Aldergate) - TH

Volvo B6	2613	2614	2615	2616	2617	2618	2619	2620
	2621	2622	2623	2624	2625	2626	2627	2628
	2629	2640	2641	2642				
Scania L113	3416	3417	3489					
Scania OmniCity	3550	3551						
Scania OmniLink	3800	3801	3802	3803	3804	3805	3806	3807
Volvo B9TL	4200	4201	4202	4203	4204	4205	4206	4207

Telford (Charlton Street, Wellington) - TF

Solo	6007	6010						
Dart	2024	2096	2097					
Dart SLF	2279	2280	2361	2365	2366			
Volvo B6	2630	2631	2632	2633	2634	2635	2636	2637
	2638	2639						
DAF/VDL SB120	2711	2712	2713	2714	2715	2716	2717	2718
	2719	2720	2722	2723	2724	2725	2726	2727
	2734	2738	2739					
Optare Versa	2974	2975	2976	2977	2978	2979	2980	2981
Volvo B10M	3305	3310						
DAF/VDL SB200	3705	3706	3707	3708	3709	3710	3711	3712
	3713	3714	3715	3716	3717	3718	3742	3743
	3744	3745	3746	3747	3748	3749	3750	3751
	3752	3753	3754	3755				
Ancillary	9521							

Wigston (Station Street, South Wigston) - WG

Dart	2209	2210	2211	2212	2253	2254	2256	2257
	2258	2259	2261					
DAF/VDL SB200	3719	3720	3721	3722	3723	3724		
DAF/VDL coach	3203	3204	3206	3208	3209			
Scania coach	3253	3254	3255					
Scania OmniCity	3578	3579						
Scania OmniLink	3808	3809	3810	3811	3812			
Olympian	4616	4638	4644	4651				
Volvo B7TL	4001	4002	4003	4004	4005	4006	4007	4008
	4009	4010	4011	4012	4013	4014		
DAF/VDL DB250	4701	4702	4703	4704	4705	4706	4707	4708
	4709	4711	4712	4713	4714	4715	4716	4717
	4719	4720	4721	4739	4742	4743	4744	4745
	4746	4747	4748	4749	4750	4751	4752	4753
	4754	4755	4756	4757	4758	4759	4760	4761
	4762	4763	4764	4765	4766	4767	4768	4769
	4770	4771	4772	4773	4774	4775	4776	4777

Unallocated, stored and withdrawn - u

Remainder

ARRIVA THE SHIRES & ESSEX

Arriva The Shires Ltd; Arriva East Herts & Essex Ltd
487 Dunstable Road, Luton, LU4 8DS

0300-0329 — Mercedes-Benz Sprinter 515 cdi / Optare/Ferqui Soroco — C19F — 2007-08 — Operated for easyBus

0300	ST	YX57CCO	0305	ST	YX57CAE	0312	ST	YX57AOE	0316	ST	YX57AOS
0301	ST	YX57CCU	0306	ST	YX57CAO	0313	ST	YX57AOF	0317	ST	YX57AOT
0302	ST	YX57CCV	0307	ST	YX57CAU	0314	ST	YX57AOG	0329	ST	YX08HWP
0303	ST	YX57CCY	0308	ST	YX57CAV	0315	ST	YX57AOR			

418-424 — Scania K340 EB4 / Caetano Levanté — C49FT — 2008

418	MK	FJ08DXG	420	MK	FJ08DXL	422	MK	FJ08DXO	424	MK	FJ08DXR
419	MK	FJ08DXK	421	MK	FJ08DXM	423	MK	FJ08DXP			

Fleet		Reg	Chassis	Body	Type	Year	Notes
442	AY	Y42HBT	Optare Solo M850	Optare	N23F	2001	Op'd for Buckinghamshire CC
444	HA	YS02UBY	Optare Alero	Optare	N14F	2002	Operated for Essex CC
446	AY	Y46HBT	Optare Solo M850	Optare	N23F	2001	Op'd for Buckinghamshire CC
447	AY	Y47HBT	Optare Solo M850	Optare	N23F	2001	Op'd for Buckinghamshire CC
1258	HA	KS05JJE	Mercedes-Benz Vito	Mercedes-Benz	M8	2005	Op'd for Buckinghamshire CC
2128	LU	N908ETM	Mercedes-Benz 709D	Plaxton Beaver	B27F	1995	
2132	LU	N912ETM	Mercedes-Benz 709D	Plaxton Beaver	B27F	1995	
2133	LU	N913ETM	Mercedes-Benz 709D	Plaxton Beaver	B27F	1995	

2173-2180 — Mercedes-Benz Vario 0810 / Plaxton Beaver 2 — B27F — 1997-98

2173	AY	R173VBM	2177	HA	R177VBM	2179	HH	R179VBM	2180	HH	R180VBM

Fleet		Reg	Chassis	Body	Type	Year	Notes
2196	HW	R196DNM	Mercedes-Benz Vario 0814	Plaxton Beaver 2	B31F	1998	
2248	HA	R758DUB	Mercedes-Benz Vario 0810	Plaxton Beaver 2	B27F	1997	Arriva Yorkshire, 1999-2003
2249	HH	R759DUB	Mercedes-Benz Vario 0810	Plaxton Beaver 2	B27F	1997	Arriva Yorkshire, 1999-2003
2273	HH	R943VPU	Mercedes-Benz Vario 0810	Plaxton Beaver 2	BC25F	1998	
2401	HW	YJ57EKF	Optare Versa V1110	Optare	N38F	2007	
2402	HW	YJ57EKG	Optare Versa V1110	Optare	N38F	2007	

Arriva The Shires operates services from Luton Airport for easyJet airline under the easyBus name. Showing the scheme on a Van Hool Acron T917 is 4385, YJ58FFV, pictured in London's Park Lane. *Mark Lyons*

Spring 2009 in Chelmsford and Optare Versa 2407, YJ58PFZ, is seen leaving the bus station on route 59 to Harlow. It one of five allocated to the route. *Richard Godfrey*

2403-2407

		Optare Versa V1110			Optare		N34F	2008	
2403	WA	YJ58PFU	**2405**	WA	YJ58PFX	**2406**	WA	YJ58PFY	**2407** WA YJ58PFZ
2404	WA	YJ58PFV							

2418-2423

		Optare Solo M850			Optare		N31F	1999	MK Metro, 2005
2418	MK	T405ENV	**2420**	MK	T407ENV	**2422**	MK	T409ENV	**2423** MK T410ENV
2419	MK	T406ENV	**2421**	MK	T408ENV				

2424-2430

		Optare Solo M850			Optare		N31F	1999-2001	MK Metro, 2005
2424	MK	V412UNH	**2427**	MK	W415KNH	**2429**	MK	T45KAW	**2430** MK X351AUX
2425	MK	V413UNH	**2428**	MK	W416KNH				

2431-2434

		Optare Solo M920			Optare		N35F	1999	MK Metro, 2005
2431	MK	S401ERP	**2432**	MK	S402ERP	**2433**		S403ERP	**2434** S404ERP

2435	MK	S903DUB	Optare Solo M920	Optare	N33F	1998	MK Metro, 2005
2436	MK	X417BBD	Optare Solo M920	Optare	N35F	2000	MK Metro, 2005
2437	MK	X418BBD	Optare Solo M920	Optare	N35F	2000	MK Metro, 2005
2438	MK	X419BBD	Optare Solo M920	Optare	N35F	2000	MK Metro, 2005
2439	MK	V82EVU	Optare Solo M920	Optare	N37F	1999	MK Metro, 2005
2440	MK	MK02BUS	Optare Solo M920	Optare	N33F	2002	MK Metro, 2005
2441	MK	KJ02JXT	Optare Solo M920	Optare	N33F	2002	MK Metro, 2005
2442	MK	W681DDN	Optare Solo M920	Optare	N33F	2000	MK Metro, 2005
2443	MK	YN53SVG	Optare Solo M920	Optare	N33F	2003	MK Metro, 2005
2444	MK	YN04LXM	Optare Solo M920	Optare	N33F	2004	MK Metro, 2005
2445	MK	YN03NEF	Optare Solo M920	Optare	N31F	2003	MK Metro, 2005
2446	MK	YN03NCF	Optare Solo M920	Optare	N31F	2003	MK Metro, 2005
2447	MK	YJ05JXU	Optare Solo M1020	Optare	N37F	2005	MK Metro, 2005
2448	MK	YJ05JXV	Optare Solo M1020	Optare	N37F	2005	MK Metro, 2005
2449	MK	YJ55YGV	Optare Solo M1020	Optare	N37F	2005	MK Metro, 2005
2450	MK	YJ55YGW	Optare Solo M1020	Optare	N37F	2005	MK Metro, 2005
2451	MK	YJ08XDK	Optare Solo M950	Optare	N33F	2008	
2452	MK	YJ58PKA	Optare Solo M950	Optare	N33F	2008	

2453-2456 Optare Solo M880 Optare N33F 2006

2453	HA	YJ06FXS	2454	HA	YJ06FXT	2455	HA	YJ06FXU	2456	HA	YJ06FXV

2457-2460 Optare Solo M880 Optare N29F 2004

2457	HA	KE04PZF	2458	HA	KE04PZG	2459	HA	KE04OSU	2460	HA	KE04OSV

2461-2467 Optare Solo M880 Optare N29F 2005

2461	WR	KE55FDG	2463	WR	KE55KPG	2465	HA	KE55KTJ	2467	HA	KE55KTC
2462	WR	KE55FDF	2464	WR	KE55KPJ	2466	HA	KE55KTD			

2468-2472 Optare Solo M780SL Optare N25F 2006

2468	WD	YJ06YRP	2470	WD	YJ06YRS	2471	WD	YJ06YRT	2472	WD	YJ06YRU
2469	WD	YJ06YRR									

2473	HA	YJ56ATY	Optare Solo M880	Optare	N33F	2006
2474	HA	YJ56ATZ	Optare Solo M880	Optare	N33F	2006

2475-2492 Optare Solo M950 Optare N33F 2007

2475	WD	YJ07VPW	2480	WD	YJ07VRE	2485	HA	YJ07BEU	2489	SV	YJ57EJL
2476	WD	YJ07VPX	2482	HH	YJ07VRF	2486	HA	YJ57EJF	2490	WR	YJ57EJN
2477	WD	YJ07VPY	2483	HA	YJ07BCZ	2487	HA	YJ57EJG	2491	MK	YK07BGE
2478	WD	YJ07VRC	2484	HA	YJ07BEO	2488	HA	YJ57EJK	2492	WD	YK07BGF
2479	WD	YJ07VRD									

2493	SV	YJ57EJD	Optare Solo M880	Optare	N28F	2008
2494	SV	YJ57EJE	Optare Solo M880	Optare	N28F	2008
2495	MK	YJ57XWH	Optare Solo M950	Optare	N33F	2008
2496	HH	YK57FHH	Optare Solo M950	Optare	N33F	2008
2497	HH	YK57FHJ	Optare Solo M950	Optare	N33F	2008

2498-2508 Optare Solo M950 Optare N33F 2008

2498	MK	YJ58PKC	2501	MK	YJ58PKF	2504	MK	YJ58PKO	2507	MK	YJ58PKX
2499	MK	YJ58PKD	2502	MK	YJ58PKK	2505	MK	YJ58PKU	2508	WD	YJ58VCG
2500	MK	YJ58PKE	2503	MK	YJ58PKN	2506	MK	YJ58PKV			

2509	MK	YJ09OTY	Optare Solo M880 SL	Optare	N32F	2009
2510	MK	YJ09OTZ	Optare Solo M880 SL	Optare	N32F	2009

3002	MK	V392KVY	Optare Excel L1150	Optare	N45F	1999	Claribel, Birmingham, 2006
3003	MK	V393KVY	Optare Excel L1150	Optare	N45F	1999	Claribel, Birmingham, 2006
3085	NG	KE53KBO	TransBus Dart 8.8m	TransBus Mini Pointer	N29F	2003	Sovereign, Stevenage, 2005
3086	NG	KE53KBP	TransBus Dart 8.8m	TransBus Mini Pointer	N29F	2003	Sovereign, Stevenage, 2005

3147-3163 Scania L113CRL East Lancs European N51F* 1995 *3152-63 are NC47F

3147	LU	N697EUR	3152	AY	N702EUR	3155	AY	N705EUR	3162	AY	N712EUR
3148	AY	N698EUR	3153	AY	N703EUR	3156	HW	N706EUR	3163	HW	N713EUR
3149	AY	N699EUR	3154	AY	N704EUR	3160	AY	N710EUR			

3167	LU	N28KGS	Scania L113CRL	East Lancs European	N51F	1996
3168	LU	N29KGS	Scania L113CRL	East Lancs European	N51F	1996
3170	LU	N32KGS	Scania L113CRL	East Lancs European	N51F	1996
3171	HH	P671OPP	Dennis Dart SLF	East Lancs Flyte	N41F	1996
3172	MK	P672OPP	Dennis Dart SLF	East Lancs Flyte	N41F	1996
3173	HH	P673OPP	Dennis Dart SLF	East Lancs Flyte	N41F	1996
3174	MK	P674OPP	Dennis Dart SLF	East Lancs Flyte	N41F	1996

3175-3190 Dennis Dart SLF Plaxton Pointer N39F* 1997 *3175-8 are N41F

3175	HH	P175SRO	3179	MK	P179SRO	3183	WD	P183SRO	3187	WD	P187SRO
3176	HH	P176SRO	3180	MK	P180SRO	3184	WD	P184SRO	3188	WD	P188SRO
3177	HH	P177SRO	3181	HW	P181SRO	3185	WD	P185SRO	3189	WD	P189SRO
3178	MK	P178SRO	3182	WD	P182SRO	3186	WD	P186SRO	3190	HH	P190SRO

3191-3205 Scania L113CRL Northern Counties Paladin N51F* 1997 *3196-9, 3201-5 are NC47F

3191	LU	R191RBM	3195	LU	R195RBM	3199	LU	R199RBM	3203	HW	R203RBM
3192	LU	R192RBM	3196	AY	R196RBM	3201	HW	R201RBM	3204	HW	R204RBM
3193	LU	R193RBM	3197	AY	R197RBM	3202	HW	R202RBM	3205	HW	R205RBM
3194	LU	R194RBM	3198	LU	R198RBM						

3206-3215 — Dennis Dart SLF — Plaxton Pointer — N31F — 1997-98

3206	WD	R206GMJ	3209	HH	R209GMJ	3212	MK	R212GMJ	3214 HW R214GMJ
3207	HH	R207GMJ	3210	HW	R210GMJ	3213	HW	R213GMJ	3215 HH R215GMJ
3208	HW	R208GMJ	3211	WD	R211GMJ				

3216-3229 — Dennis Dart SLF — Plaxton Pointer 2 — N39F* — 1998-98 — *seating varies

3216	HW	S216XPP	3219	NG	T219NMJ	3225	WD	V422DGT	3228 HH T828NMJ
3217	HW	S217XPP	3220	WD	S317JUA	3226	HW	V421DGT	3229 HH T829NMJ
3218	HA	S315JUA	3221	WD	S318JUA	3227	NG	T827NMJ	

3230-3239 — Dennis Dart SLF — Plaxton Pointer MPD — N29F — 1999-2000

3230	HA	V230HBH	3233	HH	V233HBH	3236	HH	V236HBH	3238 HH V238HBH
3231	HA	V231HBH	3234	HH	V234HBH	3237	HH	V237HBH	3239 HH V239HBH
3232	HH	V232HBH	3235	HH	V235HBH				

3240	HW	P601RGS	Volvo B6LE	Wright Crusader	NC38F	1997	Sovereign, Stevenage, 2005
3241	HW	R602WMJ	Volvo B6LE	Wright Crusader	NC38F	1998	Sovereign, Stevenage, 2005
3245	HW	R603WMJ	Volvo B6LE	Wright Crusader	NC38F	1998	Sovereign, Stevenage, 2005
3248	HW	R604WMJ	Volvo B6LE	Wright Crusader	NC38F	1998	Sovereign, Stevenage, 2005
3249	HW	R605WMJ	Volvo B6LE	Wright Crusader	NC38F	1998	Sovereign, Stevenage, 2005

3250-3260 — Volvo B6BLE — Wright Crusader 2 — N40F* — 1999 — *3258-60 are N33D

3250	HA	V250HBH	3253	HA	V253HBH	3256	HA	V256HBH	3259 WD V259HBH
3251	HA	V251HBH	3254	HA	V254HBH	3257	HA	V257HBH	3260 WD V260HBH
3252	HA	V252HBH	3255	HA	V255HBH	3258	WD	V258HBH	

3261-3268 — Volvo B10BLE — Wright Renown — N44F — 1999

3261	LU	V261HBH	3263	LU	V263HBH	3265	LU	V265HBH	3267 LU V267HBH
3262	LU	V262HBH	3264	LU	V264IIDII	3266	LU	V266HBH	3268 LU V268HBH

3270-3276 — DAF SB220 LPG — Plaxton Prestige — N39F — 1999

3270	HH	V270HBH	3272	HH	V272HBH	3274	WD	V274HBH	3276 HH V276HBH
3271	HH	V271HBH	3273	WD	V273HBH	3275	WD	V275HBH	

3277	HH	T491KGB	DAF SB220 LPG	Plaxton Prestige	N42F	1999	Arriva Scotland West, 2000
3278	WD	T492KGB	DAF SB220 LPG	Plaxton Prestige	N42F	1999	Arriva Scotland West, 2000
3279	WD	T495KGB	DAF SB220 LPG	Plaxton Prestige	N42F	1999	Arriva Scotland West, 2000

3280-3297 — Dennis Dart SLF — Plaxton Pointer MPD — N29F — 1999-2000

3280	LU	V280HBH	3285	LU	V285HBH	3290	LU	V290HBH	3294 MK V294HBH
3281	SV	V281HBH	3286	LU	V286HBH	3291	LU	V291HBH	3295 AY X295MBH
3282	SV	V282HBH	3287	LU	V287HBH	3292	LU	V292HBH	3296 HA X296MBH
3283	LU	V283HBH	3288	LU	V288HBH	3293	SV	V293HBH	3297 HA X297MBH
3284	LU	V284HBH	3289	LU	V289HBH				

3298	HW	R607WMJ	Volvo B6LE	Wright Crusader	NC38F	1998	Sovereign, Stevenage, 2005
3299	HW	R608WMJ	Volvo B6LE	Wright Crusader	NC38F	1998	Sovereign, Stevenage, 2005
3300	SV	R524TWR	Volvo B10BLE	Wright Renown	BC47F	1998	Sovereign, Stevenage, 2005

3301-3310 — Volvo B10BLE — Wright Renown — NC47F — 2000 — Sovereign, Stevenage, 2005

3301	SV	W128XRO	3304	SV	W132XRO	3307	SV	W136XRO	3309 SV W138XRO
3302	SV	W129XRO	3306	SV	W134XRO	3308	SV	W137XRO	3310 SV W139XRO
3303	SV	W131XRO							

3311-3314 — Volvo B10BLE — Wright Renown — NC47F — 2002 — Sovereign, Stevenage, 2005

3311	SV	PN02HVS	3312	SV	PN02HVL	3313	SV	PN02HVM	3314 SV PN02HVO

3322	SV	PN02HVR	Volvo B10BLE	Wright Renown	NC47F	2002	Sovereign, Stevenage, 2005
3323	SV	PN02HVP	Volvo B10BLE	Wright Renown	NC47F	2002	Sovereign, Stevenage, 2005
3360	HA	K410FHJ	Dennis Dart 9.8m	Plaxton Pointer	B40F	1993	
3366	HA	J64BJN	Dennis Dart 9m	Wright Handybus	BC40F	1992	West's, Woodford Green, 1997
3371	HA	K321CVX	Dennis Dart 9m	Plaxton Pointer	B35F	1992	
3372	HA	K322CVX	Dennis Dart 9m	Plaxton Pointer	B35F	1992	
3386	NG	P256FPK	Dennis Dart SLF	Plaxton Pointer	N39F	1997	

3401-3406 — VDL Bus SB200 — Plaxton Centro — N45F — 2007

3401	WD	YJ57BWA	3403	WD	YJ57BWC	3405	WD	YJ57BWE	3406 WD YJ57BWF
3402	WD	YJ57BWB	3404	WD	YJ57BWD				

The MK Metro operation based in Milton Keynes retains its orange livery as illustrated by 3625, YN06JXL, a Scania L94 with Wrightbus Solar bodywork. Since becoming part of Arriva the number of larger buses operated has increased from a once minibus fleet. *Dave Heath*

3407-3412

VDL Bus SB200 — Wrightbus Pulsar — N44F — 2009

3407	SV	KX09KDJ	3409	SV	KX09KDN	3411	SV	KX09KDU	3412	SV	KX09KDV
3408	SV	KX09KDK	3410	SV	KX09KDO						

3413	HA	P833HVX	Dennis Dart 9m	Plaxton Pointer	B34F	1996
3414	HA	P334HVX	Dennis Dart 9m	Plaxton Pointer	B34F	1996
3416	HA	R416HVX	Dennis Dart SLF	Wright Crusader	N41F	1998
3417	HA	R417HVX	Dennis Dart SLF	Wright Crusader	N41F	1998
3418	HA	R418HVX	Dennis Dart SLF	Wright Crusader	N41F	1998
3435	HA	R165GNW	Dennis Dart SLF	Wright Crusader	N36F	1997
3439	HA	R169GNW	Dennis Dart SLF	Wright Crusader	N36F	1997
3440	HA	R170GNW	Dennis Dart SLF	Wright Crusader	N36F	1997

3441-3449

DAF SB220 — Plaxton Prestige — NC37F — 1997

3441	LU	R201VPU	3445	AY	R205VPU	3447	AY	R207VPU	3449	LU	R209VPU
3444	LU	R204VPU	3446	AY	R206VPU	3448	AY	R208VPU			

3452-3459

Volvo B10BLE — Alexander ALX300 — N44F — 2000

3452	HW	W452XKX	3454	HW	W454XKX	3458	NG	W458XKX	3459	NG	W459XKX
3453	HW	W453XKX	3457	NG	W457XKX						

3482-3498

Dennis Dart SLF — Plaxton Pointer MPD — N39F — 2000

3482	HW	W482YGS	3486	AY	W486YGS	3491	NG	W491YGS	3495	MK	W495YGS
3483	WR	W483YGS	3487	AY	W487YGS	3492	AY	W492YGS	3496	MK	W496YGS
3484	AY	W484YGS	3488	LU	W488YGS	3493	NG	W493YGS	3497	AY	W497YGS
3485	AY	W485YGS	3489	LU	W489YGS	3494	NG	W494YGS	3498	NG	W498YGS

3500	HH	KE51PSZ	Dennis Dart SLF 8.8m	Alexander Pointer MPD	N28F	2001	
3501	WD	KE51PTO	Dennis Dart SLF 8.8m	Alexander Pointer MPD	N28F	2001	
3502	HH	KE51PTU	Dennis Dart SLF 8.8m	Alexander Pointer MPD	N28F	2001	
3509	AY	KE51PTX	Dennis Dart SLF 8.8m	Alexander Pointer MPD	N28F	2001	
3510	MK	V897DNB	Dennis Dart SLF 11.3m	Plaxton Pointer SPD	N41F	1999	MK Metro, 2006
3520	MK	R809WJA	Dennis Dart SLF 10.1m	UVG UrbanStar	N38F	1997	MK Metro, 2006

A recent arrival with Arriva the Shires at Stevenage is 3553, KX09GYC, an Enviro 300, seen here in Church Road, Welwyn Garden City in June 2009. These are the first of this model for the fleet. *Mark Lyons*

3521-3524

			Dennis Dart SLF 10.7m		Plaxton Pointer		N43F	1998	MK Metro, 2006

3521	MK	HDZ2611	3522	MK	HDZ2607	3523	MK	HDZ2605	3524	MK	HDZ2604

3525	MK	W986WDS	Dennis Dart SLF 10.7m	Caetano Compass	N43F	2000	MK Metro, 2006
3526	MK	HX51LSO	Dennis Dart SLF 10.7m	Caetano Compass	N45F	2001	MK Metro, 2006
3527	MK	W3CTS	Dennis Dart SLF 10.7m	Caetano Compass	N44F	2000	MK Metro, 2006

3528-3536

			Dennis Dart SLF 10.7		Caetano Compass		N42F	1999	MK Metro, 2006

3528	MK	NDZ7935	3530	MK	NDZ7919	3533	MK	NDZ7918	3535	MK	T425LGP
3529	MK	NDZ7933	3532	MK	NDZ4521	3534	MK	T408LGP	3536	MK	T424LGP

3537	MK	HDZ2606	Dennis Dart 9.8m	UVG UrbanStar	N44F	1997	MK Metro, 2006

3550-3578

		ADL E300		ADL Enviro 300		N45F	2009

3550	SV	KX09GXZ	3558	SV	KX09GYH	3565	SV	KX09GYS	3572	MK	KX09GZA
3551	SV	KX09GYA	3559	SV	KX09GYJ	3566	SV	KX09GYT	3573	MK	KX09GZB
3552	SV	KX09GYB	3560	SV	KX09GYK	3567	HH	KX09GYU	3574	MK	KX09GZC
3553	SV	KX09GYC	3561	SV	KX09GYN	3568	HH	KX09GYV	3575	MK	KX09GZD
3554	SV	KX09GYD	3562	SV	KX09GYO	3569	HH	KX09GYW	3576	MK	KX09GZE
3555	SV	KX09GYE	3563	SV	KX09GYP	3570	MK	KX09GYY	3577	MK	KX09GXW
3556	SV	KX09GYF	3564	SV	KX09GYR	3571	MK	KX09GYZ	3578	MK	KX09GXY
3557	SV	KX09GYG									

3601-3619

		Scania L94UB		Wrightbus Solar		N43F	2005

3601	LU	KE55CTV	3606	LU	KE55CTK	3611	LU	KE55GWC	3616	LU	KE55CVL
3602	LU	KE55CTU	3607	LU	KE55CTF	3612	LU	KE55CVG	3617	LU	KE55CVM
3603	LU	KE55FBY	3608	LU	KE55GVY	3613	LU	KE55CVH	3618	LU	KE55GXR
3604	LU	KE55FBX	3609	LU	KE55GVZ	3614	LU	KE55CVJ	3619	LU	KE55CVA
3605	LU	KE55CTO	3610	LU	KE55GWA	3615	LU	KE55CVK			

3621-3628

		Scania L94UB		Wrightbus Solar		N43F	2006

3621	MK	YN55PZY	3623	MK	YN06JXJ	3625	MK	YN06JXL	3627	MK	YN06JXO
3622	MK	YN55PZZ	3624	MK	YN06JXK	3626	MK	YN06JXM	3628	MK	YN06JXP

3640	LU	KX59ACJ	Scania OmniCity CN230UB	Scania	N41F	2009
3641	LU	KX59ACO	Scania OmniCity CN230UB	Scania	N41F	2009

High-back seating is fitted to a batch of Mercedes-Benz Citaro buses with the latest E4 styling. Carrying lettering for route 321 is 3912, BV58MLF, pictured here in St Albans. *Mark Lyons*

3701	HW	KE55CKU	VDL Bus SB120 9.4m			Wrightbus Cadet 2		N35F	2005	
3702	HW	KE55CKO	VDL Bus SB120 9.4m			Wrightbus Cadet 2		N35F	2005	
3703	HW	KE55CKP	VDL Bus SB120 9.4m			Wrightbus Cadet 2		N35F	2005	

3704-3710
VDL Bus SB120 10.8m — Wrightbus Cadet 2 — N28D — 2006

3704	WD	YJ06LFE	3706	WD	YJ06LFG	3708	WD	YJ06LFK	3710	WD	YJ06LDK
3705	WD	YJ06LFF	3707	WD	YJ06LFH	3709	WD	YJ06LFL			

3711-3728
VDL Bus SB120 10.8m — Wrightbus Cadet 2 — N39F — 2006

3711	WD	YE06HRA	3716	WD	YE06HRJ	3721	WD	YE06HPK	3725	WD	YE06HPP
3712	WD	YE06HRC	3717	WD	YE06HPA	3722	WD	YE06HPL	3726	WD	YE06HPU
3713	WD	YE06HRD	3718	WD	YE06HPC	3723	WD	YE06HPN	3727	WD	YE06HNT
3714	WD	YE06HRF	3719	WD	YE06HPF	3724	WD	YE06HPO	3728	WD	YE06HNU
3715	WD	YE06HRG	3720	WD	YE06HPJ						

3729	MK	KX54AVE	VDL Bus SB120 10.8m	Wrightbus Cadet 2	N39F	2004	MK Metro, 2006
3730	MK	KX54AVD	VDL Bus SB120 10.8m	Wrightbus Cadet 2	N39F	2004	MK Metro, 2006
3731	MK	YG52CMU	VDL Bus SB120 10.8m	Wrightbus Cadet 2	N39F	2002	MK Metro, 2006

3732-3738
VDL Bus SB120 10.8m — Wrightbus Cadet 2 — N39F — 2007

3732	MK	YJ07JVU	3734	MK	YJ07JVW	3736	MK	YJ07JVY	3738	MK	YJ07JVF
3733	MK	YJ07JVV	3735	MK	YJ07JVX	3737	MK	YJ07JVZ			

3804	WD	SN56AXG	ADL Dart 4	ADL Enviro 200	N28D	2007	
3805	WD	SN56AXH	ADL Dart 4	ADL Enviro 200	N28D	2007	
3806	NG	KC03PGE	TransBus Dart 10.1m	TransBus Pointer	N37F	2003	Sovereign, Stevenage, 2005
3807	NG	KC03PGF	TransBus Dart 10.1m	TransBus Pointer	N37F	2003	Sovereign, Stevenage, 2005
3808	SV	SN54GPK	ADL Dart 10.1m	ADL Pointer	N37F	2004	Sovereign, Stevenage, 2005
3809	SV	SN54GPO	ADL Dart 10.1m	ADL Pointer	N37F	2004	Sovereign, Stevenage, 2005
3810	SV	SN54GPU	ADL Dart 10.1m	ADL Pointer	N37F	2004	Sovereign, Stevenage, 2005

3821-3827
Dennis Dart SLF — Plaxton Pointer 2 — N36F — 1996 — Wycombe Bus, 2000

3821	HW	N521MJO	3823	AY	N523MJO	3825	HW	P525YJO	3827	HW	P527YJO
3822	HW	N522MJO	3824	HW	N524MJO	3826	HW	P526YJO			

3828	HA	KE03UKK	TransBus Dart 8.8m	TransBus Mini Pointer	N29F	2003	
3829	AY	KE53NFG	TransBus Dart 8.8m	TransBus Mini Pointer	N29F	2003	

Allocated to High Wycombe, 3863, KE05GOH, is one of twelve Volvo B7RLEs added to the fleet in 2005. The High Wycombe operation now takes Arriva south-west from the town to Maidenhead and Reading. *Mark Doggett*

3830	WR	KE04CZF	VDL Bus SB120	Wrightbus Cadet	N35F	2004	
3831	WR	KE04CZG	VDL Bus SB120	Wrightbus Cadet	N35F	2004	
3832	WR	KE04CZH	VDL Bus SB120	Wrightbus Cadet	N35F	2004	

3835-3839			TransBus Dart 8.8m	TransBus Mini Pointer	N29F	2003-04	
3835	HH	KE53NFD	3837	HW	KE53NEU	3838	HW	KE53NFA	3839	HW	KE53NFC
3836	HH	KE53NFF									

| *3849-3852* | | | Volvo B10B | Wright Endurance | BC49F | 1997 | Sovereign, Stevenage, 2005 |
| 3849 | HW | R369TWR | 3850 | HW | R370TWR | 3851 | HW | R371TWR | 3852 | HW | R372TWR |

3856-3867			Volvo B7RLE	Wrightbus Eclipse Urban	N43F	2005	
3856	HA	KE54LNR	3859	HA	KE54LPJ	3862	HW	KE05FMV	3865	HW	KE05FMP
3857	HA	KE54LPC	3860	HA	KE54HHF	3863	HW	KE05GOH	3866	HW	KE05FMO
3858	HA	KE54LPF	3861	HW	KE05FMX	3864	HW	KE05FMU	3867	HW	KE05FMM

3868-3874			Volvo B7RLE	Wrightbus Eclipse Urban	N45F*	2007	*3873/4 are N44F
3868	NG	KE07EVX	3870	NG	KE07EWA	3872	NG	KE07EWC	3874	HW	KE57EPC
3869	NG	KE07EVY	3871	NG	KE07EWB	3873	HW	KE57EPA			

| 3890 | AY | LF02PVA | Volvo B7L | Wrightbus Eclipse | N41F | 2002 | *On loan from Arriva Bus & Coach* |

3901-3909			Mercedes-Benz Citaro O530	Mercedes-Benz	NC40F	2006	
3901	WR	BU06HSD	3904	WR	BU06HSG	3906	WR	BU06HSK	3908	HA	BU06HSN
3902	WR	BU06HSE	3905	WR	BU06HSJ	3907	HA	BU06HSL	3909	HA	BU06HSO
3903	WR	BU06HSF									

3910-3924			Mercedes-Benz Citaro O530	Mercedes-Benz	NC42F*	2008	*3910 is NC39F
3910	WR	BV58URL	3914	WD	BV58MLK	3918	WD	BV58MKP	3922	AY	BV58URP
3911	WD	BV58MLE	3915	WD	BV58MKO	3919	AY	BV58URM	3923	AY	BV58URR
3912	WD	BV58MLF	3916	WD	BV58MLL	3920	AY	BV58URN	3924	AY	BV58URS
3913	WD	BV58MLJ	3917	WD	BV58MLN	3921	AY	BV58URO			

4047-4056 — DAF SB3000 / Plaxton Prima Interurban / C53F / 1997

4047	HH	R447SKX	4049	HH	R449SKX	4051	HH	R451SKX	4053	HH	R453SKX
4048	HH	R448SKX	4050	HH	R450SKX	4052	HH	R452SKX	4055	HH	R455SKX

4065-4069 — VDL Bus SB4000 / Van Hool T9 Alizée / C55F / 2005

4065	NG	YJ55WSW	4067	NG	YJ55WSX	4068	NG	YJ55WSY	4069	NG	YJ55WSZ
4066	NG	YJ55WSV									

4101	MK	FJ07TKC	Scania K340 EB	Caetano Levante	C49FT	2007
4102	MK	FJ07TKE	Scania K340 EB	Caetano Levante	C49FT	2007
4103	MK	FJ07TKF	Scania K340 EB	Caetano Levante	C49FT	2007

4359-4369 — DAF SB3000 / Plaxton Prima Interurban / C53F / 2000

4359	HH	W359XKX	4363	SV	W363XKX	4365	HH	W365XKX	4368	SV	W368XKX
4361	SV	W361XKX	4364	HH	W364XKX	4367	SV	W367XKX	4369	HH	W369XKX
4362	SV	W362XKX									

4373-4387 — Van Hool Acron T917 / Van Hool / C63F / 2008-09

4373	LU	YJ58FJN	4377	LU	YJ58FJX	4381	LU	YJ58FHX	4385	LU	YJ58FFV
4374	LU	YJ58FJO	4378	LU	YJ58FJY	4382	LU	YJ58FHY	4386	LU	YJ58FFW
4375	LU	YJ58FJP	4379	LU	YJ58FJZ	4383	LU	YJ58FHZ	4387	LU	YJ09CXL
4376	LU	YJ58FJU	4380	LU	YJ58FKA	4384	LU	YJ58FJA			

4426	HW	S426MCC	DAF SB220	Plaxton Prestige	N42F	1999	Arriva North West, 2003
4427	WD	S427MCC	DAF SB220	Plaxton Prestige	N42F	1999	Arriva North West, 2003
4428	WD	S428MCC	DAF SB220	Plaxton Prestige	N42F	1999	Arriva Southern Counties, 2005
4429	HW	S429MCC	DAF SB220	Plaxton Prestige	N42F	1999	Arriva North West, 2003
4490	HW	T490KGB	DAF SB220	Plaxton Prestige	N42F	1999	Arriva Scotland, 2002
4491	HW	T494KGB	DAF SB220	Plaxton Prestige	N42F	1999	Arriva Scotland, 2002

4514-4518 — DAF SB120 9.4m / Wrightbus Cadet / N35F / 2002

4514	WD	KE51PVF	4516	NG	KE51PVK	4517	NG	KL52CWJ	4518	NG	KL52CWK
4515	WD	KE51PVZ									

4519-4525 — VDL Bus SB120 9.4m / Wrightbus Cadet / N35F / 2003

4519	NG	KE03OUN	4521	WR	KE03OUS	4523	HH	KE03OUK	4525	NG	KE03OUM
4520	NG	KE03OUP	4522	WR	KE03OUU	4524	HH	KE03OUL			

5084-5094 — Leyland Olympian ONCL10/1RZ / Alexander RL / B47/32F* / 1988 / *5091 is BC47/29F

5084	LU	F634LMJ	5087	LU	F637LMJ	5093	LU	F643LMJ	5094	LU	F644LMJ
5086	LU	F636LMJ	5091	LU	F641LMJ						

5095-5107 — Leyland Olympian ON2R50C13Z4 / Alexander RL / B47/32F / 1989-90 / *seating varies

5095	LU	G645UPP	5099	AY	G649UPP	5102	AY	G652UPP	5105	LU	G655UPP
5096	LU	G646UPP	5100	AY	G650UPP	5103	MK	G653UPP	5106	LU	G656UPP
5097	AY	G647UPP	5101	LU	G651UPP	5104	AY	G654UPP	5107	LU	G657UPP
5098	LU	G648UPP									

5109	HW	G129YEV	Leyland Olympian ONCL10/2RZ	Northern Counties	B49/33F	1989	London Country NW, 1990
5110	HW	G130YEV	Leyland Olympian ONCL10/2RZ	Northern Counties	B49/33F	1989	London Country NW, 1990

5113-5124 — Leyland Olympian ONCL10/1RZ / Leyland / B47/31F / 1989-90 / London Country NW, 1990

5113	HW	G283UMJ	5117	AY	G287UMJ	5121	AY	G291UMJ	5123	AY	G293UMJ
5116	WD	G286UMJ	5120	HW	G290UMJ	5122	HW	G292UMJ	5124	AY	G294UMJ

5127	MK	H197GRO	Leyland Olympian ON2R50C13Z4	Leyland	B47/31F	1991	
5132	HW	H202GRO	Leyland Olympian ON2R50C13Z4	Leyland	B47/31F	1991	
5134	HW	G131YWC	Leyland Olympian ONCL10/2RZ	Northern Counties	B49/33F	1989	Ensign, Purfleet, 1991
5135	LU	G132YWC	Leyland Olympian ONCL10/2RZ	Northern Counties	B49/33F	1989	London Country NW, 1990

5136-5145 — Volvo Olympian YN2RV18Z4 / Northern Counties Palatine / B47/30F / 1996

5136	HH	N36JPP	5139	LU	N39JPP	5142	LU	N42JPP	5144	AY	N35JPP
5137	SV	N37JPP	5140	LU	N46JPP	5143	LU	N43JPP	5145	AY	N45JPP
5138	LU	N38JPP	5141	HW	N41JPP						

5146-5161 — Volvo Olympian / Northern Counties Palatine II / BC39/29F / 1998

5146	WD	S146KNK	5150	WD	S150KNK	5154	WD	S154KNK	5159	AY	S159KNK
5147	HW	S147KNK	5151	HW	S151KNK	5156	AY	S156KNK	5160	AY	S160KNK
5148	HW	S148KNK	5152	HH	S152KNK	5157	AY	S157KNK	5161	AY	S161KNK
5149	HH	S149KNK	5153	HW	S153KNK	5158	AY	S158KNK			

Lettered for route 280 that links Aylesbury with Oxford is Alexander Dennis Enviro 400 number 5437, SN58EOJ. It is seen at Thame while heading north. *Mark Lyons*

5421-5433
Dennis Trident   Alexander ALX400   N47/31F   2000

5421	AY	W421XKX	**5424**	WR	W424XKX	**5427**	WR	W427XKX	**5431**	AY	W431XKX
5422	AY	W422XKX	**5425**	WR	W425XKX	**5428**	WR	W428XKX	**5432**	AY	W432XKX
5423	WR	W423XKX	**5426**	WR	W426XKX	**5429**	WR	W429XKX	**5433**	WR	W433XKX

5434-5440
ADL Trident 2   ADL Enviro 400   N47/33F   2008

5434	AY	SN58EOF	**5436**	AY	SN58EOH	**5438**	AY	SN58EOK	**5440**	AY	SN58EOO
5435	AY	SN58EOG	**5437**	AY	SN58EOJ	**5439**	AY	SN58EOM			

5442-5447
Dennis Trident   Alexander ALX400   N47/31F   2000   Arriva Southern Counties, 2005

5442	WR	W442XKX	**5445**	WR	W445XKX	**5446**	WR	W446XKX	**5447**	WR	W447XKX
5443	WR	W443XKX									

5452-5458
ADL Trident 2   ADL Enviro 400   N47/33F   2008

5452	AY	SN58ENX	**5454**	AY	SN58EOA	**5456**	AY	SN58EOC	**5458**	AY	SN58EOE
5453	AY	SN58ENY	**5455**	AY	SN58EOB	**5457**	AY	SN58EOD			

5831	LU	G231VWL	Leyland Olympian ON2R50G16Z4	Alexander RH	B47/29F	1990	Wycombe Bus, 2000
5835	LU	G235VWL	Leyland Olympian ON2R50G16Z4	Alexander RH	B47/29F	1990	Wycombe Bus, 2000

6000-6024
DAF DB250   Alexander ALX400   N45/20D   2002-03

6000	WD	KL52CWN	**6007**	WD	KL52CWW	**6013**	WD	KL52CXE	**6019**	WD	KL52CXM
6001	WD	KL52CWO	**6008**	WD	KL52CWZ	**6014**	WD	KL52CXF	**6020**	WD	KL52CXN
6002	WD	KL52CWP	**6009**	WD	KL52CXA	**6015**	WD	KL52CXG	**6021**	WD	KL52CXO
6003	WD	KL52CWR	**6010**	WD	KL52CXB	**6016**	WD	KL52CXH	**6022**	WD	KL52CXP
6004	WD	KL52CWT	**6011**	WD	KL52CXC	**6017**	WD	KL52CXJ	**6023**	WD	KL52CXR
6005	WD	KL52CWU	**6012**	WD	KL52CXD	**6018**	WD	KL52CXK	**6024**	WD	KL52CXS
6006	WD	KL52CWV									

6025	WD	YJ54CFG	VDL Bus DB250	Alexander ALX400	N45/20D	2005

6026-6036
VDL Bus DB250 10.2m   Wrightbus Pulsar Gemini   N43/21D   2006

6026	WD	YJ55WPO	**6029**	WD	YJ55WOC	**6032**	WD	YJ55WOM	**6035**	WD	YJ55WOV
6027	WD	YJ55WOA	**6030**	WD	YJ55WOD	**6033**	WD	YJ55WOR	**6036**	WD	YJ55WOX
6028	WD	YJ55WOB	**6031**	WD	YJ55WOH	**6034**	WD	YJ55WOU			

6037	WD	Y521UGC	DAF DB250 10.2m	Alexander ALX400	N43/20D	2001	Arriva London, 2006
6039	WD	Y531UGC	DAF DB250 10.2m	Alexander ALX400	N43/20D	2001	Arriva London, 2008
6100	WD	KX59AEE	VDL Bus DB300 Hybrid	Wrightbus Gemini 2	N41/24D	2009	
6101	WD	KX59AEF	VDL Bus DB300 Hybrid	Wrightbus Gemini 2	N41/24D	2009	

Previous registration:

J64BJN   J9BUS

Depots and allocations:

Aylesbury (Smeaton Close, Brunel Park) - AY

Outstation - Leighton Buzzard

Mercedes-Benz	2173	2196						
Optare Solo	442	446	447					
Dart	3295	3484	3485	3486	3487	3492	3497	3509
	3829							
Scania L113	3148	3149	3152	3153	3154	3155	3162	3196
	3197							
DAF/VDL SB220	3445	3446	3447	3448				
Volvo B7L	3890							
MB Citaro	3919	3920	3921	3922	3923	3924		
Olympian	5097	5099	5100	5101	5102	5104	5144	5145
	5156	5157	5158	5159	5160	5161		
Trident	5421	5422	5431	5432	5434	5435	5436	5437
	5438	5439	5440					

Harlow (Fourth Avenue) - HA

Outstation - Langston Road, Debden

Optare Alero	444							
Mercedes-Benz Vito	1258							
Mercedes-Benz	2177	2248						
Optare Solo	2453	2454	2455	2456	2457	2458	2459	2460
	2465	2466	2467	2473	2474	2483	2484	2485
	2486	2487	2488					
Optare Versa	2406	2407						
Dart	3218	3230	3231	3296	3297	3360	3366	3371
	3372	3413	3414	3416	3417	3418	3435	3439
	3440	3483	3828					
Volvo B6	3250	3251	3253	3254	3255	3256	3257	
Volvo B7RLE	3856	3857	3858	3859	3860	3873	3874	
MB Citaro	3907	3908	3909					

Hemel Hempstead (Whiteleaf Road) - HH

Mercedes-Benz	2179	2180	2249	2373				
Optare Solo	2482							
Dart	3171	3173	3175	3176	3177	3190	3207	3209
	3215	3228	3229	3232	3233	3234	3235	3236
	3237	3238	3239	3500	3501	3835	3836	
DAF/VDL SB120	4523	4524						
DAF/VDL SB220	3270	3271	3272	3276	3277			
Volvo B10B	3846	3847	3848					
Enviro 300	3567	3568	3569					
SB3000	4047	4048	4049	4050	4051	4052	4053	4055
	4359	4364	4365	4369				
Olympian	5136	5148	5152					
Trident	5452	5453	5454	5455	5456	5457	5458	

High Wycombe (Lincoln Road) - HW

Outstation: Old Amersham

Optare Solo	2496	2497					
Optare Versa	2401	2402					
Volvo B6	3240	3241	3245	3248	3249	3298	3299
Dart	3181	3208	3210	3213	3214	3216	3217	3226
	3482	3821	3822	3824	3825	3826	3827	3837
	3838	3839					
Scania sd	3156	3163	3201	3202	3203	3204	3205
DAF/VDL SB120	3701	3702	3703				
Volvo B10B	3849	3850	3851	3852			
Volvo B7RLE	3861	3862	3863	3864	3865	3866	3867
DAF/VDL SB220	4426	4429	4490	4491			
Olympian	5109	5110	5121	5132	5134	5141	5147	5148
	5150	5151	5153				

Luton (Dunstable Road) - LU

Mercedes-Benz	300	301	302	303	305	306	307	308
	312	313	314	315	316	317	329	2128
	2132	2133					
Dart	3280	3283	3284	3285	3286	3287	3288	3289
	3290	3291	3292	3488	3489		
Scania L113	3147	3167	3168	3170	3191	3192	3193	3194
	3195	3196	3199				
Scania L94	3601	3602	3603	3604	3605	3606	3607	3608
	3609	3610	3611	3612	3613	3614	3615	3616
	3617	3618	3619				
Scania Solar	3640	3641					
DAF/VDL SB220	3441	3444	3449				
Volvo B10BLE	3261	3262	3263	3264	3265	3266	3268
Van Hool T917	4372	4373	4374	4375	4376	4377	4378	4379
	4380	4381	4382	4383	4384	4385	4386	4387
Olympian	5084	5086	5087	5091	5093	5095	5096	5098
	5105	5106	5107	5135	5137	5138	5139	5140
	5142	5143	5831	5835			

Stevenage (Babbage Road) - SV

Optare Solo	2489	2493	2494				
Dart	3281	3282	3293	3808	3809	3810	
Enviro 300	3550	3551	3552	3553	3554	3555	3556	3557
	3558	3559	3560	3561	3562	3563	3564	3565
	3566						
VDL Bus SB200	3407	3408	3409	3410	3411	3412	
DAF SB3000	4361	4362	4363	4367	4368		

Stevenage (Norton Green Road) - NG

Dart	3085	3086	3219	3227	3386	3491	3493	3494
	3498	3806	3807				
DAF/VDL SB120	4516	4517	4518	4519	4520	4525	
Volvo B10BLE	3300	3301	3302	3303	3304	3306	3307	3308
	3309	3310	3311	3312	3313	3314	3322	3323
	3457	3458	3459				
Volvo B7RLE	3868	3869	3870	3871	3872		
DAF/VDL SB4000	4065	4066	4067	4068	4069		

Ware (Marsh Lane) - WR

Outstation - Pindar Road, Hoddesdon

Optare Solo	2461	2462	2463	2464	2490			
Optare Versa	2403	2404	2405					
DAF/VDL SB120	3830	3831	3832	4521	4522			
MB Citaro	3901	3902	3903	3904	3905	3906	3910	
Trident	5423	5424	5425	5426	5427	5428	5429	5433
	5442	5443	5445	5446	5447			

Watford (St Albans Road, Garston) - WD - *includes private hire fleet and Dial a Ride*

Optare Solo	2468	2469	2470	2471	2472	2475	2476	2477
	2478	2479	2480	2492	2508			
Dart	3182	3183	3184	3185	3186	3187	3188	3189
	3206	3211	3220	3221	3225	3502	3804	3805
Volvo B6	3258	3259	3260					
DAF/VDL SB120	3704	3705	3706	3707	3708	3709	3710	3711
	3712	3713	3714	3715	3716	3717	3718	3719
	3720	3721	3722	3723	3724	3725	3726	3727
	3728	4514	4515					
DAF/VDL SB220	3273	3274	3275	3278	3279	4427	4428	
VDL Bus SB200	3401	3402	3403	3404	3405	3406	4427	4428
Olympian	5146	5154						
Trident	5448							
DB250	6000	6001	6002	6003	6004	6005	6006	6007
	6008	6009	6010	6011	6012	6013	6014	6015
	6016	6017	6018	6019	6020	6021	6022	6023
	6024	6025	6026	6027	6028	6029	6030	6031
	6032	6033	6034	6035	6036	6037	6039	

Wolverton (Arden Park, Old Wolverton Road, Wolverton, Milton Keynes) - MK

Optare Solo	2418	2419	2420	2421	2422	2423	2424	2425
	2427	2428	2429	2430	2431	2432	2434	2435
	2436	2437	2438	2439	2440	2441	2442	2443
	2444	2445	2446	2447	2448	2449	2450	2451
	2452	2491	2495	2498	2499	2500	2501	2502
	2503	2504	2505	2506	2507	2508	2509	2510
Excel	3002	3003						
Dart	3172	3174	3178	3179	3180	3212	3294	3496
	3496	3510	3520	3521	3522	3523	3524	3525
	3526	3527	3528	3529	3530	3532	3533	3534
	3535	3536	3537					
DAF/VDL SB120	3729	3730	3731	3732	3733	3734	3735	3736
	3737	3738						
Scania L113	3160							
Scania L94	3621	3622	3623	3624	3625	3626	3627	3628
Enviro 300	3570	3571	3572	3573	3574	3575	3576	3577
	3578							
Scania coach	418	419	420	421	422	423	424	4101
	4102	4103						
Olympian	5103	5116	5117	5127				

ARRIVA LONDON

Arriva London North Ltd, 16 Watsons Road, Wood Green, London, N22 7TZ
Arriva London South Ltd, Croydon Bus Garage, Brighton Road, South Croydon, CR2 6EL

| ADL1 | EC | V701LWT | Dennis Dart SLF 10.2m | | | Alexander ALX200 | | N27D | 1999 | |

ADL2-8
Dennis Dart SLF 10.2m   Alexander ALX200   N27D   2000

2	EC	W602VGJ	4	EC	W604VGJ	6	EC	W606VGJ	8	EC	W608VGJ
3	EC	W603VGJ	5	EC	W605VGJ	7	EC	W607VGJ			

ADL9-23
Dennis Dart SLF 10.8m   Alexander ALX200   N33D*   1999   *13 is N30D

9	TC	V609LGC	13	TC	V613LGC	17	TC	V617LGC	21	TC	V621LGC
10	TC	V610LGC	14	TC	V614LGC	18	TC	V618LGC	22	TC	V622LGC
11	TC	V611LGC	15	TC	V615LGC	19	TC	V619LGC	23	TC	V623LGC
12	TC	V612LGC	16	TC	V616LGC	20	TC	V620LGC			

ADL61-81
Dennis Dart SLF 9.4m   Alexander ALX200   N23D   2000   Arriva The Shires, 2005

61	CN	W461XKX	66	EC	W466XKX	72	EC	W472XKX	77	EC	W477XKX
62	CN	W462XKX	67	EC	W467XKX	73	EC	W473XKX	78	EC	W478XKX
63	CN	W463XKX	68	EC	W468XKX	74	EC	W474XKX	79	EC	W479XKX
64	CN	W464XKX	69	EC	W469XKX	75	EC	W475XKX	81	EC	W481XKX
65	EC	W465XKX	71	EC	W471XKX	76	EC	W476XKX			

ADL969-983
Dennis Dart SLF 10.2m   Alexander ALX200   N27D   1998

969	DX	S169JUA	973	DX	S173JUA	977	DX	S177JUA	981	DX	S181JUA
970	DX	S170JUA	974	DX	S174JUA	978	DX	S178JUA	982	DX	S182JUA
971	DX	S171JUA	975	DX	S175JUA	979	DX	S179JUA	983	DX	S183JUA
972	DX	S172JUA	976	DX	S176JUA	980	DX	S180JUA			

Alexander bodywork is fitted to several batches of Dennis Darts for Arriva London including ADL8, W608VGJ, seen operating route 444. *Alan Blagburn*

DLA1-64 — DAF DB250 10.6m — Alexander ALX400 — N45/19D* — 1998-99 — *seating varies

No		Reg	No		Reg	No		Reg	No		Reg
1	Et	R101GNW	17	WN	S217JUA	33	Et	S233JUA	49	TH	S249JUA
2	TH	S202JUA	18	WN	S218JUA	34	TC	S234JUA	50	TH	S250JUA
3	TH	S203JUA	19	WN	S219JUA	35	Et	S235JUA	51	TH	S251JUA
4	TH	S204JUA	20	WN	S220JUA	36	CNt	S236JUA	52	TH	S252JUA
5	TH	S205JUA	21	Et	S221JUA	37	Et	S237JUA	53	TH	S253JUA
6	TH	S206JUA	22	TH	S322JUA	38	TH	S238JUA	54	TH	S254JUA
7	TH	S207JUA	23	Et	S223JUA	39	TH	S239JUA	55	TH	S255JUA
8	TH	S208JUA	24	Et	S224JUA	40	TH	S240JUA	56	CNt	S256JUA
9	TH	S209JUA	25	Et	S225JUA	41	TH	S241JUA	57	TH	S257JUA
10	TH	S210JUA	26	Et	S226JUA	42	TH	S242JUA	58	TH	S258JUA
11	WN	S211JUA	27	Et	S227JUA	43	TH	S243JUA	59	TH	S259JUA
12	WN	S212JUA	28	Et	S228JUA	44	TH	S244JUA	60	TH	S260JUA
13	WN	S213JUA	29	Et	S229JUA	45	CNt	S245JUA	61	TH	S261JUA
14	WN	S214JUA	30	Et	S230JUA	46	TH	S246JUA	62	CNt	S262JUA
15	WN	S215JUA	31	CNt	S231JUA	47	WN	S247JUA	63	TH	S263JUA
16	WN	S216JUA	32	Et	S232JUA	48	TH	S248JUA	64	TH	S264JUA

DLA72-92 — DAF DB250 10.6m — Alexander ALX400 — N45/19D — 1999

No		Reg	No		Reg	No		Reg	No		Reg
72	E	S272JUA	78	E	S278JUA	83	E	S283JUA	88	DX	S288JUA
73	E	S273JUA	79	E	S279JUA	84	E	S284JUA	89	DX	S289JUA
74	E	S274JUA	80	E	S280JUA	85	E	S285JUA	90	DX	S290JUA
75	E	S275JUA	81	E	S281JUA	86	DX	S286JUA	91	E	S291JUA
76	E	S276JUA	82	E	S282JUA	87	DX	S287JUA	92	E	S292JUA
77	E	S277JUA									

DLA93-125 — DAF DB250 10.6m — Alexander ALX400 — N45/17D* — 1999 — *seating varies

No		Reg	No		Reg	No		Reg	No		Reg
93	E	T293FGN	102	SF	T302FGN	110	SF	T310FGN	118	E	T318FGN
94	SF	T294FGN	103	SF	T303FGN	111	SF	T311FGN	119	E	T319FGN
95	E	T295FGN	104	SF	T304FGN	112	SF	T312FGN	120	E	T320FGN
96	SF	T296FGN	105	SF	T305FGN	113	SF	T313FGN	121	E	T421GGO
97	SF	T297FGN	106	SF	T306FGN	114	SF	T314FGN	122	E	T322FGN
98	SF	T298FGN	107	SF	T307FGN	115	SF	T315FGN	123	E	T323FGN
99	SF	T299FGN	108	SF	T308FGN	116	E	T316FGN	124	E	T324FGN
100	SF	T110GGO	109	SF	T309FGN	117	E	T317FGN	125	E	T325FGN
101	SF	T301FGN									

DLA126-189 — DAF DB250 10.2m — Alexander ALX400 — N43/18D* — 1999-2000 — *seating varies

No		Reg	No		Reg	No		Reg	No		Reg
126	TH	V326DGT	142	WN	V342DGT	158	N	V358DGT	174	N	W374VGJ
127	TH	V327DGT	143	TC	V343DGT	159	BN	V359DGT	175	TC	W432WGJ
128	TH	V628LGC	144	N	V344DGT	160	BN	V660LGC	176	TC	W376VGJ
129	TH	V329DGT	145	N	V345DGT	161	BN	V361DGT	177	TC	W377VGJ
130	TH	V330DGT	146	BN	V346DGT	162	BN	V362DGT	178	TC	W378VGJ
131	TH	V331DGT	147	BN	V347DGT	163	BN	V363DGT	179	TC	W379VGJ
132	TH	V332DGT	148	BN	V348DGT	164	BN	V364DGT	180	TC	W433WGJ
133	TC	V633LGC	149	N	V349DGT	165	BN	V365DGT	181	TC	W381VGJ
134	TC	V334DGT	150	BN	V650LGC	166	N	V366VGJ	182	TC	W382VGJ
135	TC	V335DGT	151	BN	V351DGT	167	N	W367VGJ	183	TC	W383VGJ
136	TC	V336DGT	152	BN	V352DGT	168	N	W368VGJ	184	TC	W384VGJ
137	N	V337DGT	153	BN	V353DGT	169	N	W369VGJ	185	TC	W385VGJ
138	N	V338DGT	154	N	V354DGT	170	N	W431WGJ	186	TC	W386VGJ
139	TC	V339DGT	155	N	V355DGT	171	N	W371VGJ	187	TC	W387VGJ
140	TC	V640LGC	156	TC	V356DGT	172	N	W372VGJ	188	TC	W388VGJ
141	TC	V341DGT	157	TC	V357DGT	173	N	W373VGJ	189	TC	W389VGJ

DLA190-223 — DAF DB250 10.2m — Alexander ALX400 — N43/18D* — 2000 — *seating varies

No		Reg	No		Reg	No		Reg	No		Reg
190	E	W434WGJ	199	E	W399VGJ	208	BN	W408VGJ	216	TC	X416FGP
191	E	W391VGJ	200	E	W435WGJ	209	BN	W409VGJ	217	TC	X417FGP
192	E	W392VGJ	201	E	W401VGJ	210	E	W438WGJ	218	TC	X418FGP
193	E	W393VGJ	202	E	W402VGJ	211	E	W411VGJ	219	TC	X419FGP
194	E	W394VGJ	203	E	W403VGJ	212	WN	W412VGJ	220	TC	X501GGO
195	E	W395VGJ	204	AR	W404VGJ	213	BN	W413VGJ	221	TC	X421FGP
196	E	W396VGJ	205	BN	W436WGJ	214	WN	W414VGJ	222	TC	X422FGP
197	E	W397VGJ	206	BN	W437WGJ	215	TC	X415FGP	223	TC	X423FGP
198	E	W398VGJ	207	BN	W407VGJ						

Almost four hundred DAF DB250 buses with Alexander bodywork are used by Arriva London on their services. Seen heading for Streatham Hill is DLA312, Y512UGC, showing the London version of the Arriva scheme. Recent repaints no longer incorporate the while relief. *Alan Blagburn*

DLA224-256

DAF DB250 10.2m Alexander ALX400 N43/19D* 2000-01 *seating varies

224	TC	X424FGP	233	SF	X433FGP	241	AR	X441FGP	249	E	X449FGP
225	E	X425FGP	234	AR	X434FGP	242	AR	X442FGP	250	TC	X506GGO
226	E	X426FGP	235	AR	X435FGP	243	AR	X443FGP	251	TC	X451FGP
227	E	X427FGP	236	BN	X436FGP	244	AR	X504GGO	252	TC	X452FGP
228	DA	X428FGP	237	E	X437FGP	245	AR	X445FGP	253	TC	X453FGP
229	N	X429FGP	238	E	X438FGP	246	AR	X446FGP	254	TC	X454FGP
230	AR	X502GGO	239	AR	X439FGP	247	AR	X447FGP	255	TC	X507GGO
231	AR	X431FGP	240	AR	X503GGO	248	E	X448FGP	256	TC	X508GGO
232	AR	X432FGP									

DLA270-319

DAF DB250 10.2m Alexander ALX400 N43/19D* 2000-01 *seating varies

270	BN	Y452UGC	283	AR	Y483UGC	296	SF	Y496UGC	308	EC	Y508UGC
271	BN	Y471UGC	284	AR	Y484UGC	297	SF	Y497UGC	309	EC	Y509UGC
272	BN	Y472UGC	285	AR	Y485UGC	298	SF	Y498UGC	310	EC	Y527UGC
273	AR	Y473UGC	286	AR	Y486UGC	299	FC	Y499UGC	311	BN	Y511UGC
274	AR	Y474UGC	287	AR	Y487UGC	300	EC	Y524UGC	312	BN	Y512UGC
275	AR	Y475UGC	288	AR	Y488UGC	301	EC	Y501UGC	313	BN	Y513UGC
276	BN	Y476UGC	289	AR	Y489UGC	302	EC	Y502UGC	314	BN	Y514UGC
277	AR	Y477UGC	290	BN	Y523UGC	303	EC	Y503UGC	315	BN	Y529UGC
278	AR	Y478UGC	291	SF	Y491UGC	304	EC	Y504UGC	316	BN	Y516UGC
279	AR	Y479UGC	292	SF	Y492UGC	305	EC	Y526UGC	317	BN	Y517UGC
280	AR	Y522UGC	293	SF	Y493UGC	306	EC	Y506UGC	318	BN	Y518UGC
281	AR	Y481UGC	294	SF	Y494UGC	307	EC	Y507UGC	319	BN	Y519UGC
282	AR	Y482UGC	295	SF	Y495UGC						

DLA322-336

DAF DB250 10.2m TransBus ALX400 N45/20D 2003

322	TH	LG52DAO	326	TH	LG52DBY	330	TH	LG52DCF	334	TH	LG52DCX
323	TH	LG52DAU	327	TH	LG52DBY	331	TH	LG52DCO	335	TH	LG52DCY
324	TH	LG52DBO	328	TH	LG52DBZ	332	TH	LG52DCU	336	TH	LG52DCZ
325	TH	LG52DBU	329	TH	LG52DCE	333	TH	LG52DCV			

DLA337-389　　DAF DB250 10.2m　　TransBus ALX400　　N45/20D*　2003　*338 is N45/19D

337	TH	LJ03MFX	351	EC	LJ03MKZ	364	EC	LJ03MKL	377	SF	LJ03MTK
338	TH	LJ03MFY	352	EC	LJ03MLE	365	EC	LJ03MWE	378	SF	LJ03MTU
339	TH	LJ03MFZ	353	EC	LJ03MLF	366	EC	LJ03MWF	379	SF	LJ03MTV
340	TH	LJ03MGE	354	EC	LJ03MLK	367	EC	LJ03MWG	380	SF	LJ03MTY
341	TH	LJ03MGU	355	EC	LJ03MJX	368	EC	LJ03MWK	381	SF	LJ03MTZ
342	TH	LJ03MGV	356	EC	LJ03MJY	369	EC	LJ03MWL	382	SF	LJ03MUA
343	TH	LJ03MDV	357	EC	LJ03MKA	370	SF	LJ03MUY	383	SF	LJ03MUB
344	TH	LJ03MDX	358	EC	LJ03MKC	371	SF	LJ03MVC	384	SF	LJ03MYU
345	TH	LJ03MDY	359	EC	LJ03MKD	372	SF	LJ03MVD	385	SF	LJ03MYV
346	SF	LJ03MDZ	360	EC	LJ03MKE	373	SF	LJ03MVE	386	SF	LJ03MYX
347	TH	LJ03MEU	361	EC	LJ03MKF	374	SF	LJ03MSY	387	SF	LJ03MYY
348	EC	LJ03MKU	362	EC	LJ03MKG	375	SF	LJ03MTE	388	SF	LJ03MYZ
349	EC	LJ03MKV	363	EC	LJ03MKK	376	SF	LJ03MTF	389	TH	LJ03MZD
350	EC	LJ03MKX									

DLP15-20　　DAF DB250 10.6m　　Plaxton President　　N45/19D　1999

15	E	T215XBV	17	E	T217XBV	19	u	T219XBV	20	E	T220XBV
16	E	T216XBV	18	E	T218XBV						

DLP40-75　　DAF DB250 10.6m　　Plaxton President　　N45/21D　2001

40	AD	Y532UGC	49	AD	Y549UGC	58	AD	LJ51DKF	67	WN	LJ51DLD
41	AD	Y541UGC	50	AD	LJ51DJU	59	AD	LJ51DKK	68	WN	LJ51DLF
42	AD	Y542UGC	51	AD	LJ51DJV	60	AD	LJ51DKL	69	WN	LJ51DLK
43	AD	Y543UGC	52	AD	LJ51DJX	61	AD	LJ51DKN	70	WN	LJ51DLN
44	AD	Y544UGC	53	AD	LJ51DJY	62	AD	LJ51DKO	71	AD	LJ51DLU
45	AD	Y533UGC	54	AD	LJ51DJZ	63	AD	LJ51DKU	72	WN	LJ51DLV
46	AD	Y546UGC	55	AD	LJ51DKA	64	WN	LJ51DKV	73	AD	LJ51DLX
47	AD	Y547UGC	56	AD	LJ51DKD	65	WN	LJ51DKX	74	AD	LJ51DLY
48	AD	Y548UGC	57	AD	LJ51DKE	66	WN	LJ51DKY	75	AD	LJ51DLZ

DLP76-90　　DAF DB250 10.2m　　TransBus President　　N43/19D*　2002　*seating varies

76	E	LJ51OSX	80	E	LJ51ORC	84	E	LJ51ORK	88	E	LF02PKD
77	E	LJ51OSY	81	E	LJ51ORF	85	E	LJ51ORL	89	E	LF02PKE
78	E	LJ51OSZ	82	E	LJ51ORG	86	E	LF02PKA	90	E	LF02PKJ
79	E	LJ51ORA	83	E	LJ51ORH	87	E	LF02PKC			

DLP91-110　　DAF DB250 10.6m　　TransBus President　　N45/19D*　2002　*seating varies

91	E	LF52URS	96	E	LF52URX	101	E	LF52URG	106	E	LF52URM
92	E	LF52URT	97	E	LF52URB	102	E	LF52URH	107	E	LF52UPP
93	E	LF52URU	98	E	LF52URC	103	E	LF52URJ	108	E	LF52UPR
94	E	LF52URV	99	E	LF52URD	104	E	LF52URK	109	E	LF52UPS
95	E	LF52URW	100	E	LF52URE	105	E	LF52URL	110	E	LF52UPT

DW1-50　　DAF DB250 10.3m　　Wrightbus Pulsar Gemini　　N43/21D　2003

1	TC	801DYE	14	TC	LJ03MWC	27	TC	LJ53BGK	39	CN	LJ53NHF
2	TC	LJ03MWN	15	TC	LJ03MWD	28	TC	LJ53BGO	40	CN	LJ53NHG
3	TC	LJ03MWP	16	TC	LJ03MVF	29	TC	LJ53BGU	41	CN	LJ53NHH
4	TC	LJ03MWU	17	TC	LJ03MVG	30	TC	LJ53NHV	42	CN	LJ53NHK
5	TC	LJ03MWV	18	TC	LJ53NHT	31	TC	LJ53NHX	43	CN	LJ53NHL
6	TC	LJ03MVT	19	TC	WLT719	32	TC	LJ53NHY	44	CN	VLT244
7	TC	WLT807	20	TC	LJ53BFP	33	TC	LJ53NHZ	45	CN	LJ53NHN
8	TC	LJ03MVV	21	TC	LJ53BFU	34	TC	734DYE	46	CN	LJ53NHO
9	TC	LJ03MVW	22	TC	822DYE	35	TC	LJ53NJF	47	CN	LJ53NHP
10	TC	LJ03MVX	23	TC	LJ53BFX	36	TC	LJ53NJK	48	CN	WLT348
11	TC	LJ03MVY	24	TC	LJ53BFY	37	CN	LJ53NJN	49	CN	LJ53NGU
12	TC	LJ03MVZ	25	TC	725DYE	38	CN	LJ53NHE	50	CN	LJ53NGV
13	TC	LJ03MWA	26	TC	LJ53BGF						

The Wrightbus Pulsar Gemini body was principally built for Arriva to use on the VDL DB250 chassis and, until 2009, the body was available only in highbridge form. Peckham is the location for this view of DW12, LJ03MVZ, which is based at Croydon depot. The dept code [TC] is displayed to the left of the driver's window. *Dave Heath*

DW51-93

VDL Bus DB250 10.3m   Wrightbus Pulsar Gemini   N43/22D   2004

51	CN	LJ04LDX	62	BN	LJ04LDC	73	BN	LJ04LGK	84	BN	LJ04LFX
52	CN	LJ04LDY	63	BN	LJ04LDD	74	BN	LJ04LGL	85	BN	WLT385
53	CN	LJ04LDZ	64	BN	WLT664	75	BN	LJ04LGN	86	BN	LJ04LFZ
54	CN	LJ04LEF	65	BN	LJ04LDF	76	BN	WLT676	87	BN	LJ04LGA
55	BN	LJ04LEU	66	BN	LJ04LDK	77	BN	LJ04LGV	88	BN	LJ04LGC
56	BN	656DYE	67	BN	LJ04LDL	78	BN	LJ04LGW	89	BN	LJ04LGD
57	BN	LJ04LFB	68	BN	LJ04LDN	79	BN	LJ04LGX	90	BN	LJ04LGE
58	BN	LJ04LFD	69	BN	LJ04LDU	80	BN	LJ04LGY	91	BN	LJ04LFG
59	BN	LJ04LFE	70	BN	WLT970	81	BN	LJ04LFU	92	BN	LJ04LFH
60	BN	LJ04LFF	71	BN	LJ04LGF	82	BN	LJ04LFV	93	BN	LJ04LFK
61	BN	LJ04LDA	72	BN	LJ04LGG	83	BN	LJ04LFW			

DW94-102

VDL Bus DB250 10.3m   Wrightbus Pulsar Gemini   N43/22D   2004

94	CN	LJ54BFP	97	CN	WLT997	99	CN	LJ54BFZ	101	CN	LJ54BGF
95	CN	VLT295	98	CN	LJ54BFY	100	CN	LJ54BGE	102	CN	LJ54BGK
96	CN	LJ54BFV									

DW103-134

VDL Bus DB250 10.3m   Wrightbus Pulsar Gemini   N43/22D   2005

103	BA	LJ05BJV	111	BA	LJ05BHP	119	BA	LJ05BMY	127	BN	LJ05BNL
104	BA	LJ05BJX	112	BA	LJ05BHU	120	BA	LJ05BMZ	128	BN	LJ05GKX
105	BA	LJ05BJY	113	BA	LJ05BHV	121	BA	LJ05BNA	129	BN	LJ05GKY
106	BA	LJ05BJZ	114	BA	LJ05BHW	122	BA	LJ05BNB	130	BN	LJ05GKZ
107	BA	LJ05BKA	115	BA	LJ05BHX	123	BA	LJ05BND	131	BN	LJ05GLF
108	BA	LJ05BHL	116	BA	LJ05BHY	124	BA	LJ05BNE	132	BN	LJ05GLK
109	BA	LJ05BHN	117	BA	LJ05BHZ	125	BA	LJ05BNF	133	BN	LJ05GLV
110	BA	LJ05BHO	118	BA	LJ05BMV	126	BA	LJ05BNK	134	BN	LJ05GLY

DAF single-decks with Arriva feature the Wrigthbus Cadet body, typified by DWL50, LF52UOB, seen on TfL route 84. The Cadet has been supplied to several Arriva fleets in the UK, these based in London have central exit doors. *Mark Doggett*

DW201-262

VDL Bus DB300 10.4m   Wrightbus Pulsar Gemini 2   N41/24D   2009

201	WN	LJ09KRO	217	-	LJ09STX	233	-	LJ59AEC	248	-	LJ59AAO
202	-	LJ09SUO	218	-	LJ09STZ	234	-	LJ59AED	249	-	LJ59AAU
203	-	LJ09SUU	219	-	LJ09SUA	235	-	LJ59AEE	250	-	LJ59AAV
204	-	LJ09SUV	220	-	LJ09SUF	236	-	LJ59AEF	251	-	LJ59AAX
205	-	LJ09SUX	221	-	LJ09SUH	237	-	LJ59AEG	252	-	LJ59AAY
206	-	LJ09SUY	222	-		238	-	LJ59AEK	253	-	LJ59AAZ
207	-	LJ09SVA	223	-		239	-	LJ59AEL	254	-	LJ59GVC
208	-	LJ09SVC	224	-	LJ59AET	240	-	LJ59AEM	255	-	LJ59GVE
209	-	LJ09SVD	225	-	LJ59AEU	241	-	LJ59AEN	256	-	LJ59GVF
210	-	LJ09SVE	226	-	LJ59AEV	242	-	LJ59ACU	257	-	LJ59GVG
211	-	LJ09SVF	227	-	LJ59AEW	243	-	LJ59ACV	258	-	LJ59GVK
212	-	LJ09SSO	228	-	LJ59AEX	244	-	LJ59ACX	259	-	LJ59GTF
213	-	LJ09SSU	229	-	LJ59AEY	245	-	LJ59AAF	260	-	LJ59GTU
214	-	LJ09SSV	230	-	LJ59AEZ	246	-	LJ59AAK	261	-	LJ59GTZ
215	-	LJ09SSX	231	-	LJ59AEA	247	-	LJ59AAN	262	-	LJ59GUA
216	-	LJ09SSZ	232	-	LJ59AEB						

DWL1-22

DAF SB120 10.2mAEC   Wrightbus Cadet   N27D*   2001   *seating varies

1	TH	Y801DGT	7	TH	LJ51DDK	13	E	LJ51DDX	18	WN	LJ51DFC
2	TH	Y802DGT	8	TH	LJ51DDL	14	E	LJ51DDY	19	WN	LJ51DFD
3	TH	Y803DGT	9	TH	LJ51DDN	15	WN	LJ51DDZ	20	WN	LJ51DFE
4	TH	Y804DGT	10	EC	LJ51DDO	16	WN	LJ51DEU	21	WN	LJ51DFF
5	TH	Y805DGT	11	WN	LJ51DDU	17	WN	LJ51DFA	22	WN	LJ51DFG
6	TH	Y806DGT	12	WN	LJ51DDV						

DWL23-29

DAF SB120 10.8m   Wrightbus Cadet   N30D   2002

23	E	LF02PLU	25	E	LF02PLX	27	E	LF02PMO	29	E	LF02PMV
24	E	LF02PLV	26	E	LF02PLZ						

The latest midi model from Alexander Dennis is the Enviro 200, normally matched to the Dart 4 chassis which was launched with the Euro 4 engine and now continues with that name for the Euro 5-engined version. The bodywork for the Enviro 200 is built at Scarborough while the wider Enviro 300 is built at Falkirk. *Mark Lyons*

DWL30-55   DAF SB120 10.2m   Wrightbus Cadet   N26D*   2002   *seating varies

30	CNt	LF02PMX	37	CNt	LF02PNO	44	EC	LF52UTB	50	EC	LF52UOB
31	CNt	LF02PMY	38	Et	LF02PNU	45	WN	LF52UNW	51	EC	LF52UOC
32	CNt	LF02PNE	39	Et	LF02PNV	46	EC	LF52UNX	52	WN	LF52UOD
33	CNt	LF02PNJ	40	Et	LF02PNX	47	WN	LF52UNY	53	WN	LF52UOE
34	CNt	LF02PNK	41	Et	LF02PNY	48	WN	LF52UNZ	54	WN	LF52USZ
35	Et	LF02PNL	42	EC	LF02POA	49	WN	LF52UOA	55	WN	LF52UTA
36	Et	LF02PNN	43	WN	LF02POH						

DWL56-67   DAF SB120 10.2m   Wrightbus Cadet   N26D*   2003   *seating varies

56	WN	LJ03MUW	59	CN	LJ03MZG	62	CN	LJ03MYH	65	CN	LJ03MYM
57	WN	LJ03MZE	60	CN	LJ03MZL	63	CN	LJ03MYK	66	CN	LJ53NGX
58	CN	LJ03MZF	61	CN	LJ03MYG	64	CN	LJ03MYL	67	CN	LJ53NGY

DWS1-18   DAF SB120 9.4m   Wrightbus Cadet2   N26D   2003

1	CN	LJ53NGZ	6	CN	LJ53NFT	11	CN	LJ53NFZ	15	CN	LJ53NGN
2	CN	LJ53NHA	7	CN	LJ53NFU	12	CN	LJ53NGE	16	CN	LJ53NFE
3	CN	LJ53NHB	8	CN	LJ53NFV	13	CN	LJ53NGF	17	CN	LJ53NFF
4	CN	LJ53NHC	9	CN	LJ53NFX	14	CN	LJ53NGG	18	CN	LJ53NFG
5	CN	LJ53NHD	10	CN	LJ53NFY						

EN1-13   ADL Dart 4 8.9m   ADL Enviro200   N26F   2008

1	LV	LJ57USS	5	LV	LJ57USW	8	LV	LJ57USZ	11	LV	LJ57UTC
2	LV	LJ57UST	6	LV	LJ57USX	9	LV	LJ57UTA	12	LV	LJ57UTE
3	LV	LJ57USU	7	LV	LJ57USY	10	LV	LJ57UTB	13	LV	LJ57UTF
4	LV	LJ57USV									

ENL1-9   ADL Dart 4 10.2m   ADL Enviro200   N29D   2007

1	TC	LJ07ECW	4	TC	LJ07ECZ	6	TC	LJ07EDF	8	TC	LJ07EBP
2	TC	LJ07ECX	5	TC	LJ07EDC	7	TC	LJ07EBO	9	TC	LJ07EBU
3	TC	LJ07ECY									

Currently undergoing development for London are Hybrid buses, identified by the addition of green leaves incorporated into their livery. Arriva have Volvo B5Ls alongside a Wrightbus integral model. This latter is built by Wrightbus using VDL components and has been registered as a Wrightbus DB250. Pictured on route 141 is HW3, LJ58AVK. *Mark Lyons*

ENL10-48

ADL Dart 4 10.2m   ADL Enviro200   N29D   2008-09

10	E	LJ58AVT	20	E	LJ58AVE	30	WN	LJ09KPR	40	WN	LJ09KOX
11	E	LJ58AVU	21	TC	LJ58AUV	31	WN	LJ09KPT	41	WN	LJ09KPA
12	E	LJ58AVV	22	TC	LJ58AUW	32	WN	LJ09KPU	42	WN	LJ09KPE
13	E	LJ58AVX	23	TC	LJ58AUX	33	WN	LJ09KPV	43	WN	LJ09KPF
14	E	LJ58AVY	24	TC	LJ58AUY	34	WN	LJ09KPX	44	WN	LJ09KPG
15	E	LJ58AVZ	25	TC	LJ58AVB	35	WN	LJ09KPY	45	WN	LJ09KPK
16	E	LJ58AWA	26	TC	LJ58AVC	36	WN	LJ09KPZ	46	WN	LJ09KPL
17	E	LJ58AWC	27	TC	LJ58AVD	37	WN	LJ09KRD	47	WN	LJ09KPN
18	E	LJ58AWF	28	TC	LJ58AUC	38	WN	LJ09KRE	48	WN	LJ09KPO
19	E	LJ58AWG	29	TC	LJ58AUE	39	WN	LJ09KRF			

ENS1-14

ADL Dart 4 9.3m   ADL Enviro200   N24D   2007

1	AE	LJ07EDK	4	AE	LJ07EDR	9	AE	LJ07EEA	12	AE	LJ07ECN
2	AE	LJ07EDL	5	AE	LJ07EDU	10	AE	LJ07EEB	13	AE	LJ07ECT
3	AE	LJ07EDO	6	AE	LJ07EDV	11	AE	LJ07ECF	14	AE	LJ07ECU
4	AE	LJ07EDP	7	AE	LJ07EDX						

HV1-6

Volvo B5L Hybrid 10.4m   Wrightbus Gemini 2   N39/21D   2009

1	WN	LJ09KRU	3	WN	LJ09KOH	5	WN	LJ09KOV	6	WN	LJ09KOW
2	WN	LJ09KOE	4	WN	LJ09KOU						

HW1-5

Wrightbus Gemini DB250   Wrightbus   N41/24D   2009

1	WN	LJ09KRG	3	WN	LJ58AVK	4	WN	LJ09KRK	5	WN	LJ09KRN
2	WN	LJ58AVG									

MA1-76   Mercedes-Benz Citaro O530G   AB49T   2004

1	LV	BX04MWW	20	LV	BX04MXU	39	LV	BX04NEJ	58	LV	BX04MZL
2	LV	BX04MWY	21	LV	BX04MXV	40	LV	BX04MYG	59	LV	BX04MZN
3	LV	BX04MWZ	22	LV	BX04MXW	41	LV	BX04MYH	60	LV	BX04NBK
4	LV	BX04MXA	23	LV	BX04MXY	42	LV	BX04MYJ	61	LV	361CLT
5	LV	BX04MXB	24	LV	BX04MXZ	43	LV	BX04MYK	62	LV	BX04NCF
6	LV	BX04MXC	25	LV	BX04MYA	44	LV	BX04MYL	63	LV	BX04NCJ
7	LV	BX04MXD	26	LV	BX04MYB	45	LV	BX04MYM	64	LV	BX04NCN
8	LV	BX04MXE	27	LV	BX04MYC	46	LV	BX04MYN	65	LV	BX04NCU
9	LV	BX04MXG	28	LV	BX04MYD	47	LV	BX04MYP	66	LV	BX04NCV
10	LV	BX04MXH	29	LV	BX04MYF	48	LV	BX04MYS	67	LV	BX04NCY
11	LV	BX04MXJ	30	LV	BX04MYY	49	LV	BX04MYT	68	LV	BX04NCZ
12	LV	BX04MXK	31	LV	BX04MYZ	50	LV	BX04MYU	69	LV	DX04NDC
13	LV	BX04MXL	32	LV	BX04NDD	51	LV	BX04MYV	70	LV	70CLT
14	LV	BX04MXM	33	LV	BX04NDG	52	LV	BX04MYW	71	LV	BX04NDF
15	LV	BX04MXN	34	LV	BX04NDU	53	LV	BX04MYZ	72	LV	BX04NDJ
16	LV	BX04MXP	35	LV	BX04NDV	54	LV	BX04MZD	73	EC	BX04NDK
17	LV	BX04MXR	36	LV	BX04NDY	55	LV	BX04MZE	74	LV	BX04NDL
18	LV	BX04MXS	37	LV	BX04NDZ	56	LV	BX04MZG	75	LV	BX04NDN
19	LV	BX07MXT	38	LV	BX04NEF	57	LV	BX04MZJ	76	EC	BX04NEN

MA77-157   Mercedes-Benz Citaro O530G   AB49T   2005

77	AE	BX05UWV	98	AE	398CLT	118	AE	BX55FVU	138	EC	BX55FWZ
78	AE	BX05UWW	99	AE	BX55FUW	119	AE	319CLT	139	EC	BX55FXB
79	AE	BX05UWY	100	AE	BX55FUY	120	AE	BX55FVW	140	EC	BX55FXC
80	AE	BX05UWZ	101	AE	BX55FVA	121	AE	BX55FVY	141	EC	BX55FXE
81	AE	BU05VFE	102	AE	BX55FVB	122	AE	BX55FVZ	142	EC	BX55FXF
82	AE	BX05VFF	103	AE	BX55FVC	123	AE	BX55FWG	143	EC	BX55FXG
83	AE	BX05UXC	104	AE	BX55FVD	124	AE	BX55FWH	144	EC	BX55FXH
84	AE	BU05VFG	105	AE	BX55FVF	125	AE	BX55FWJ	145	EC	BX55FXJ
85	AE	185CLT	106	AE	BX55FVG	126	AE	BX55FWK	146	EC	BX55FXK
86	AE	BU05VFH	107	AE	BX55FVH	127	EC	BX55FWL	147	EC	BX55FXL
87	AE	BU05VFJ	108	AE	BX55FVJ	128	EC	BX55FWM	148	EC	BX55FXM
88	AE	BX05UXD	109	AE	BX55FVK	129	EC	BX55FWN	149	EC	BX55FXO
89	AE	BX55FWA	110	AE	BX55FVL	130	EC	BX55FWP	150	EC	BX55FXP
90	AE	BX55FWB	111	AE	BX55FVM	131	EC	BX55FWR	151	EC	BX55FXR
91	AE	BX55FUH	112	AE	BX55FVN	132	EC	BX55FWS	152	EC	BX55FXS
92	AE	BX55FUJ	113	AE	BX55FVQ	133	EC	BX55FWT	153	EC	BX55FXYT
93	AE	593CLT	114	AE	BX55FVP	134	EC	BX55FWU	154	EC	BX55FXU
94	AE	BX55FUO	115	AE	BX55FVR	135	EC	BX55FWV	155	EC	BX55FXV
95	AE	BX55FUP	116	AE	BX55FVS	136	EC	BX55FWW	156	EC	BX55FXW
96	AE	BX55FUT	117	AE	BX55FVT	137	EC	BX55FWY	157	EC	BX55FXY
97	AE	BX55FUU									

MA162-166   Mercedes-Benz Citaro O530G   AB49T   2003   East London Group, 2008

162	LV	LX03HCE	163	LV	LX03HCG	164	LV	LX03HCL	166	LV	LX03HDE

Political decisions that highlight the intransigent British attitude have caused the number of articulated buses to be replaced in London, with several currently for sale. MA42, BX04MYJ, is seen in Park Lane. *Mark Lyons*

The PDL class are Darts with Pointer bodywork. From 2002 the model was built at Falkirk, the home of Alexanders coachbuilders. Mini Pointer PDL15, V435DGT, is seen on route W4. *Mark Doggett*

PDL12-18

Dennis Dart SLF 8.8m   Plaxton Pointer MPD   N26F   2000

12	AR	V432DGT	14	AR	V434DGT	16	AR	W136VGJ	18	AR	W138VGJ
13	AR	V433DGT	15	AR	V435DGT	17	AR	W137VGJ			

PDL19-38

Dennis Dart SLF 10.7m   Plaxton Pointer 2   N31D   2000

19	TH	X519GGO	24	DX	X524GGO	29	DX	X529GGO	34	DX	X534GGO
20	TH	X471GGO	25	DX	X475GGO	30	DX	X481GGO	35	DX	X485GGO
21	TH	X521GGO	26	DX	X526GGO	31	DX	X531GGO	36	DX	X536GGO
22	TH	X522GGO	27	DX	X527GGO	32	DX	X532GGO	37	DX	X537GGO
23	DX	X523GGO	28	DX	X478GGO	33	DX	X533GGO	38	DX	X538GGO

PDL39-49

Dennis Dart SLF 8.8m   Plaxton Pointer MPD   N23F*   2001   *seating varies

39	AR	X239PGT	42	AR	X242PGT	45	AR	X546GGO	48	AR	X248PGT
40	AR	X541GGO	43	EC	X243PGT	46	AR	X246PGT	49	AR	X249PGT
41	AR	X241PGT	44	AR	X244PGT	47	AR	X247PGT			

PDL50-69

Dennis Dart SLF 8.8m   Plaxton Pointer MPD   N23F*   2001   *seating varies

50	AR	LJ51DAA	55	DX	LJ51DBV	60	DX	LJ51DCF	65	E	LJ51DCY
51	AR	LJ51DAO	56	DX	LJ51DBX	61	DX	LJ51DCO	66	E	LJ51DCZ
52	DX	LJ51DAU	57	DX	LJ51DBY	62	E	LJ51DCU	67	E	LJ51DDA
53	DX	LJ51DBO	58	DX	LJ51DBZ	63	AR	LJ51DCV	68	E	LJ51DDE
54	DX	LJ51DBU	59	DX	LJ51DCE	64	E	LJ51DCX	69	EC	LJ51DDF

PDL70-94

TransBus Dart 8.8m   TransBus Mini Pointer   N29F*   2002   *seating varies

70	EC	LF02PTZ	77	EC	LF52UON	83	EC	LF52URZ	89	EC	LF52USJ
71	EC	LF52UOG	78	EC	LF52UOO	84	EC	LF52USB	90	EC	LF52USL
72	EC	LF52UOH	79	EC	LF52UOP	85	EC	LF52USC	91	EC	LF52URN
73	EC	LF52UOJ	80	EC	LF52UOR	86	EC	LF52USD	92	EC	LF52URO
74	EC	LF52UOK	81	EC	LF52UNV	87	EC	LF52USG	93	EC	LF52URP
75	EC	LF52UOL	82	EC	LF52URY	88	EC	LF52USH	94	EC	LF52URR
76	EC	LF52UOM									

Alexander Dennis' double deck is the Enviro 400, currently available on Volvo B9TL and Scania N230 UD as well as its own Trident 2 chassis which is built in Guildford. Arriva London received a batch based on the Trident 2 in 2008 with T38, LJ08CUH, shown at the Brent Cross shopping complex. *Dave Heath*

PDL95-116 ADL Dart 9.3m ADL Pointer N27D 2005

95	EC	LJ54BCX	101	EC	LJ54BBF	107	EC	LJ54LHG	112	EC	LJ54LHN
96	EC	LJ54BAA	102	EC	LJ54BBK	108	EC	LJ54LHH	113	EC	LJ54LHO
97	EC	LJ54BAO	103	EC	LJ54BBN	109	EC	LJ54LHK	114	EC	LJ54LHP
98	EC	LJ54BAU	104	EC	LJ54BBO	110	EC	LJ54LHL	115	EC	LJ54LHR
99	EC	LJ54BAV	105	EC	LJ54BBU	111	EC	LJ54LHM	116	EC	LJ54LGV
100	EC	LJ54BBE	106	EC	LJ54LHF						

PDL117-123 ADL Dart 10.1m ADL Pointer N29D 2005

117	TC	LJ05GOP	119	TC	LJ05GOX	121	TC	LJ05GPK	123	TC	LJ05GPU
118	TC	LJ05GOU	120	TC	LJ05GPF	122	TC	LJ05GPO			

PDL124-136 ADL Dart 9.3m ADL Pointer N24D 2006

124	CN	LJ56APZ	128	CN	LJ56ARX	131	CN	LJ56ASU	134	CN	LJ56AOW
125	CN	LJ56ARF	129	CN	LJ56ARZ	132	CN	LJ56ASV	135	CN	LJ56AOX
126	CN	LJ56ARO	130	CN	LJ56ASO	133	CN	LJ56ASX	136	CN	LJ56AOY
127	CN	LJ56ARU									

T1-65 ADL Trident 2 10.1m ADL Enviro 400 N41/26D 2008

1	AD	LJ08CVS	18	DX	LJ08CVO	34	AD	LJ08CTZ	50	TC	LJ08CTO
2	AD	LJ08CVT	19	DX	519CLT	35	AD	LJ08CUA	51	TC	LJ08CYC
3	AD	LJ08CVU	20	DX	LJ08CVR	36	AD	LJ08CUE	52	TC	LJ08CYE
4	AD	LJ08CVV	21	DX	LJ08CUU	37	AD	LJ08CUG	53	TC	LJ08CYF
5	AD	205CLT	22	DX	LJ08CUV	38	AD	LJ08CUH	54	TC	LJ08CYG
6	AD	LJ08CVX	23	DX	LJ08CUW	39	AD	LJ08CUK	55	TC	LJ08CYH
7	AD	7CLT	24	DX	324CLT	40	AD	LJ08CUO	56	TC	LJ08CYK
8	AD	LJ08CVZ	25	DX	LJ08CUY	41	AD	LJ08CSO	57	TC	LJ08CYL
9	AD	LJ08CWA	26	DX	LJ08CVA	42	AD	LJ08CSU	58	TC	LJ08CYO
10	AD	LJ08CWC	27	AD	LJ08CVB	43	AD	LJ08CSV	59	TC	LJ08CYP
11	AD	LJ08CVF	28	AD	LJ08CVC	44	AD	LJ08CSX	60	TC	LJ08CYS
12	DX	LJ08CVG	29	AD	LJ08CVD	45	AD	LJ08CSY	61	TC	LJ08CXR
13	DX	LJ08CVH	30	AD	330CLT	46	AD	LJ08CSZ	62	TC	LJ08CXS
14	DX	LJ08CVK	31	AD	LJ08CTV	47	AD	LJ08CTE	63	TC	LJ08CXT
15	DX	LJ08CVL	32	AD	LJ08CTX	48	TC	LJ08CTF	64	TC	LJ08CXU
16	DX	LJ08CVM	33	AD	LJ08CTY	49	TC	LJ08CTK	65	TC	LJ08CXV
17	DX	217CLT									

The Volvo-Alexander combination is seen on VLA24, LJ53BFN, which was pictured in the latest livery style while operating along Orchard Street on route 2 in June 2009. *Richard Godfrey*

T66-83

ADL Trident 2 10.1m   ADL Enviro 400   N41/26D   2009

66	-	LJ59ACY	71	-	LJ59ADZ	76	-	LJ59ABO	80	-	LJ59ABZ
67	-	LJ59ACZ	72	-	LJ59AEA	77	-	LJ59ABU	81	-	LJ59ACF
68	-	LJ59ADO	73	-	LJ59ABF	78	-	LJ59ABV	82	-	LJ59ACO
69	-	LJ59ADV	74	-	LJ59ABK	79	-	LJ59ABX	83	-	LJ59AAE
70	-	LJ59ADX	75	-	LJ59ABN					

VLA1-55

Volvo B7TL 10.6m   TransBus ALX400 4.4m   N49/22D   2003

1	N	LJ03MYP	15	N	LJ03MXH	29	N	LJ53BDO	43	N	LJ53BCV
2	N	LJ03MYR	16	N	LJ03MXK	30	N	LJ53BDU	44	N	LJ53BCX
3	N	LJ03MYS	17	N	LJ03MXL	31	N	LJ53BDV	45	N	LJ53BCY
4	N	LJ03MYT	18	N	LJ03MXM	32	N	LJ53BDX	46	N	LJ53BAA
5	N	LJ03MXV	19	N	LJ03MXN	33	N	LJ53BDY	47	N	LJ53BAO
6	N	LJ03MXW	20	N	LJ03MXP	34	N	LJ53BDZ	48	N	LJ53BAU
7	N	LJ03MXX	21	N	LJ53BFK	35	N	LJ53BEO	49	N	LJ53BAV
8	N	LJ03MXY	22	N	LJ53BFL	36	N	LJ53BBV	50	N	LJ53BBE
9	N	LJ03MXZ	23	N	LJ53BFM	37	N	LJ53BBX	51	N	LJ53BBF
10	N	LJ03MYA	24	N	LJ53BFN	38	N	LJ53BBZ	52	N	LJ53BBK
11	N	LJ03MYB	25	N	LJ53BFO	39	N	LJ53BCF	53	N	LJ53BBN
12	N	LJ03MYC	26	N	LJ53BCZ	40	N	LJ53BCK	54	N	LJ53BBO
13	N	LJ03MYD	27	N	LJ53BDE	41	N	LJ53BCO	55	N	LJ53BBU
14	N	LJ03MYF	28	N	LJ53BDF	42	N	LJ53BCU			

VLA56-69

Volvo B7TL 10.6m   TransBus ALX400 4.4m   N49/22D   2004   *operated by The Original Tour*

56	WD	LJ04LFL	60	WD	LJ04LFR	64	WD	LJ04YWT	67	WD	LJ04YWW
57	WD	LJ04LFM	61	WD	LJ04LFS	65	WD	LJ04YWU	68	WD	LJ04YWX
58	WD	LJ04LFN	62	WD	LJ04LFT	66	WD	LJ04YWV	69	WD	LJ04YWY
59	WD	LJ04LFP	63	WD	LJ04YWS					

VLA70-73

Volvo B7TL 10.6m   TransBus ALX400 4.4m   N49/22D   2004

70	N	LJ04YWZ	71	N	LJ04YXA	72	N	LJ04YXB	73	N	LJ04YWE

Arriva London also operates the Wrightbus double-deck on Volvo chassis with almost two hundred now in service, all being the B7TL model. VLW14, LJ51DFY, from Wood Green depot is shown heading for London Bridge rail station. *Mark Lyons*

VLA74-128 — Volvo B7TL 10.1m — ADL ALX400 — N45/19D — 2004-05

74	AR	LJ54BGO	88	AR	LJ54BDF	102	AR	LJ54BCU	116	N	LJ54BKG
75	AR	LJ54BEO	89	AR	LJ54BDO	103	AR	LJ54BCV	117	N	LJ54BKK
76	AR	LJ54BEU	90	AR	LJ54BDU	104	N	LJ05BKY	118	N	LJ54BKL
77	AR	LJ54BFA	91	AR	LJ54BDV	105	N	LJ05BKZ	119	N	LJ54BKN
78	AR	LJ54BFE	92	AR	LJ54BDX	106	N	LJ05BLF	120	N	LJ54BKO
79	AR	LJ54BFF	93	AR	LJ54BDY	107	N	LJ05BLK	121	N	LJ54BKU
80	AR	LJ54BFK	94	AR	LJ54BDZ	108	N	LJ05BLN	122	N	LJ54BKV
81	AR	LJ54BFL	95	AR	LJ54BBV	109	N	LJ05BLV	123	N	LJ54BKX
82	AR	LJ54BFM	96	AR	LJ54BBX	110	N	LJ05BLX	124	N	LJ54BJE
83	AR	LJ54BFN	97	AR	LJ54BBZ	111	N	LJ05BLY	125	N	LJ54BJF
84	AR	LJ54BFO	98	AR	LJ54BCE	112	N	LJ05BMO	126	N	LJ54BJK
85	AR	LJ54BCY	99	AR	LJ54BCF	113	N	LJ05BMU	127	N	LJ54BJO
86	AR	LJ54BCZ	100	AR	LJ54BCK	114	N	LJ05BKD	128	N	LJ54BJU
87	AR	LJ54BDE	101	AR	LJ54BCO	115	N	LJ05BKF			

VLA129-143 — Volvo B7TL 10.1m — ADL ALX400 — N45/19D — 2005

129	DX	LJ05GLZ	133	DX	LJ05GPY	137	DX	LJ05GRU	141	DX	LJ05GSU
130	DX	LJ05GME	134	DX	LJ05GPZ	138	DX	LJ05GRX	142	DX	LJ55BTE
131	DX	LJ05GMF	135	DX	LJ05GRF	139	DX	LJ05GRZ	143	DX	LJ55BTF
132	DX	LJ05GPX	136	DX	LJ05GRK	140	DX	LJ05GSO			

VLA144-179 — Volvo B7TL 10.1m — ADL ALX400 — N45/19D — 2005

144	BN	LJ55BTO	153	BN	LJ55BRV	162	BN	LJ55BUP	171	BN	LJ55BUZ
145	BN	LJ55BTU	154	BN	LJ55BRX	163	BN	LJ55BUR	172	BN	LJ55BVD
146	BN	LJ55BTV	155	BN	LJ55BRZ	164	BN	LJ55BUS	173	BN	LJ55BVE
147	BN	LJ55BTX	156	BN	LJ55BSO	165	BN	LJ55BUT	174	BN	LJ55BVF
148	BN	LJ55BTY	157	BN	LJ55BSU	166	BN	LJ55BUU	175	BN	LJ55BVG
149	BN	LJ55BTZ	158	BN	LJ55BSV	167	BN	LJ55BUV	176	BN	LJ55BVH
150	BN	LJ55BUA	159	BN	LJ55BSX	168	BN	LJ55BUW	177	BN	LJ55BVK
151	BN	LJ55BUE	160	BN	LJ55BSY	169	BN	LJ55BUX	178	BN	LJ55BVL
152	BN	LJ55BPZ	161	BN	LJ55BSZ	170	BN	LJ55BUY	179	BN	LJ55BVM

VLW1-41   Volvo B7TL   10.1m   Wrightbus Eclipse Gemini   N41/21D*   2001-02   *seating varies

No		Reg	No		Reg	No		Reg	No		Reg
1	WN	Y581UGC	12	WN	VLT12	22	WN	LJ51DGY	32	WN	VLT32
2	WN	Y102TGH	13	WN	LJ51DFX	23	WN	LJ51DGZ	33	WN	LJ51DHO
3	WN	LJ51DJF	14	WN	LJ51DFY	24	WN	LJ51DHA	34	WN	LJ51DHP
4	WN	LJ51DJK	15	WN	LJ51DFZ	25	WN	LJ51DHC	35	WN	LJ51DHV
5	WN	LJ51DJO	16	WN	LJ51DGE	26	WN	LJ51DHD	36	WN	LJ51DHX
6	WN	LJ51DFK	17	WN	LJ51DGF	27	WN	VLT27	37	WN	LJ51DHY
7	WN	LJ51DFL	18	WN	LJ51DGO	28	WN	LJ51DHF	38	WN	LJ51DHZ
8	WN	LJ51DFN	19	WN	LJ51DGU	29	WN	LJ51DHG	39	WN	LJ51DJD
9	WN	LJ51DFO	20	WN	LJ51DGV	30	WN	LJ51DHK	40	WN	LJ51DJE
10	WN	LJ51DFP	21	WN	LJ51DGX	31	WN	LJ51DHL	41	WN	LJ51OSK
11	WN	LJ51DFU									

VLW42-104   Volvo B7TL   10.1m   Wrightbus Eclipse Gemini   N41/21D*   2002-03   *seating varies

No		Reg	No		Reg	No		Reg	No		Reg
42	WN	LF02PKO	58	WN	LF02PTU	74	WN	LF52UTM	90	SF	LF52URA
43	WN	LF02PKU	59	WN	LF02PTX	75	WN	LF52USM	91	SF	LF52UPD
44	WN	LF02PKV	60	WN	LF02PTY	76	WN	LF52USN	92	SF	WLT892
45	WN	LF02PKX	61	WN	LF02PVE	77	WN	LF52USO	93	SF	LF52UPG
46	WN	LF02PKY	62	WN	LF02PVJ	78	WN	LF52USS	94	SF	LF52UPH
47	WN	VLT47	63	WN	LF02PVK	79	WN	LF52UST	95	SF	WLT895
48	WN	LF02PLJ	64	WN	LF02PVL	80	WN	LF52USU	96	SF	LF52UPK
49	WN	LF02PLN	65	WN	LF02PVN	81	WN	LF52USV	97	AR	WLT897
50	WN	LF02PLO	66	WN	LF02PVO	82	WN	LF52USW	98	AR	LF52UPM
51	WN	WLT751	67	WN	LF52UTC	83	WN	LF52USX	99	AR	LG52DDA
52	WN	LF02PSO	68	WN	LF52UTE	84	WN	LF52USY	100	AR	LG52DDE
53	WN	LF02PSU	69	WN	LF52USE	85	WN	LF52UPV	101	AR	LG52DDF
54	WN	WLT554	70	WN	LF52UTG	86	SF	LF52UPW	102	AR	LG52DDJ
55	WN	LF02PSY	71	WN	LF52UTH	87	SF	LF52UPX	103	AR	LG52DDK
56	WN	LF02PSZ	72	WN	WLT372	88	SF	WLT888	104	AR	LG52DDL
57	WN	LF02PTO	73	WN	LF52UTL	89	SF	LF52UPZ			

VLW105-179   Volvo B7TL   10.1m   Wrightbus Eclipse Gemini   N41/21D*   2002-03   *seating varies

No		Reg	No		Reg	No		Reg	No		Reg
105	DX	LJ03MHU	124	AR	LF52UOX	143	SF	LG03MFA	162	SF	LG03MRX
106	DX	LJ03MHV	125	AR	LF52UOY	144	SF	LG03MFE	163	SF	LG03MRY
107	DX	LJ03MHX	126	AR	LF52UPA	145	SF	LG03MFF	164	SF	LG03MSU
108	DX	LJ03MHY	127	AR	LF52UPB	146	SF	LG03MFK	165	SF	LG03MSV
109	DX	LJ03MHZ	128	AR	LF52UPC	147	SF	LG03MBF	166	SF	LG03MSX
110	DX	LJ03MJE	129	AR	LG52DAA	148	SF	LG03MBU	167	SF	LG03MMU
111	DX	LJ03MJF	130	AR	LJ03MGZ	149	SF	LG03MBV	168	SF	LG03MMV
112	DX	LJ03MJK	131	AR	LJ03MHA	150	SF	LG03MBX	169	AR	LG03MMX
113	DX	LJ03MJU	132	AR	LJ03MHE	151	SF	LG03MBY	170	AR	LG03MOA
114	DX	LJ03MJV	133	AR	LJ03MHF	152	SF	LG03MDE	171	AR	LG03MOF
115	DX	LJ03MGX	134	AR	LJ03MHK	153	SF	LG03MDF	172	AR	LG03MOV
116	DX	LJ03MGY	135	AR	LJ03MHL	154	SF	LG03MDK	173	AR	VLT173
117	AR	LF52UPN	136	AR	LJ03MHM	155	SF	LG03MDN	174	AR	LG03MPF
118	AR	LF52UPO	137	AR	LJ03MHN	156	SF	LG03MDU	175	AR	LG03MPU
119	AR	LF52UOS	138	AR	LJ03MFN	157	SF	LG03MPX	176	AR	LG03MPV
120	AR	LF52UOT	139	AR	LJ03MFP	158	SF	LG03MPY	177	AR	LG03MLL
121	AR	LF52UOU	140	SF	LJ03MFU	159	SF	LG03MPZ	178	AR	LG03MLN
122	AR	LF52UOV	141	SF	LJ03MFV	160	SF	LG03MRU	179	AR	LG03MLV
123	AR	LF52UOW	142	SF	LG03MEV	161	SF	LG03MRV			

VLW180-199   Volvo B7TL   10.6m   Wrightbus Eclipse Gemini   N45/24D   2003

No		Reg	No		Reg	No		Reg	No		Reg
180	AR	LJ03MLX	185	AR	LJ03MMF	190	AR	LJ03MXR	195	AR	LJ53BEU
181	AR	LJ03MLY	186	AR	LJ03MMK	191	AR	LJ03MXS	196	AR	LJ53BEY
182	AR	LJ03MLZ	187	AR	LJ03MKM	192	AR	LJ03MXT	197	AR	LJ53BFA
183	AR	LJ03MMA	188	AR	LJ03MKN	193	AR	LJ03MXU	198	AR	LJ53BFE
184	AR	LJ03MME	189	AR	LJ03MYN	194	AR	LJ03MWX	199	AR	LJ53BFF

Special event vehicles:

No		Reg	Chassis	Body	Layout	Year
RM5	N	VLT5	AEC Routemaster R2RH	Park Royal	B36/28R	1959
RM6	N	VLT6	AEC Routemaster R2RH	Park Royal	B36/28R	1959
RML901	EC	WLT901	AEC Routemaster R2RH/1	Park Royal	B40/32R	1963
RM1124	N	VYJ806	AEC Routemaster R2RH	Park Royal	B36/28R	1965
RMC1453	EC	453CLT	AEC Routemaster R2RH	Park Royal	B36/28R	1962
RMC1464	N	464CLT	AEC Routemaster R2RH	Park Royal	O36/28R	1962
RM2217	N	CUV217C	AEC Routemaster R2RH	Park Royal	B36/28R	1965
RML2355	N	CUV335C	AEC Routemaster R2RH/1	Park Royal	B40/32R	1965
RML2403	w	JJD403D	AEC Routemaster R2RH/1	Park Royal	B40/32R	1966

Ancillary vehicles:

DDL1-16   Dennis Dart SLF 10.1m   Plaxton Pointer 2   N26D   1998

1	u	S301JUA	5	u	S305JUA	9	u	S309JUA	13	u	S313JUA
2	u	S302JUA	6	u	S306JUA	10	u	S310JUA	14	u	S314JUA
3	u	S303JUA	7	u	S307JUA	11	u	S311JUA	16	u	S316JUA
4	u	S304JUA	8	u	S308JUA	12	u	S312JUA			

Previous registrations:

70CLT	BX04NDE	T324FGN	T324FGN, 99D53451
185CLT	BU05VFD	T325FGN	T325FGN, 99D53440
205CLT	LJ08CVW	V423DGT	V435DGT
217CLT	LJ08CVN	V435DGT	V423DGT
319CLT	BX55FVV	VLT12	LJ51DFV
324CLT	LJ08CUX	VLT27	LJ51DHE
330CLT	LJ08CVE	VLT32	LJ51DHN
361CLT	BX04NBL	VLT47	LF02PKZ
398CLT	BX55FUV	VLT173	LJ03MPE
480CLT	BX05UWZ	VLT244	LJ53NHM
519CLT	LJ08CVP	VLT295	LJ54BFU
593CLT	BX55FUM	VYJ806	124CLT
656DYE	LJ04LFA	WLT348	LJ53NGO
725DYE	LJ53BGE	WLT372	LF52UTJ
734DYE	LJ53NJE	WLT385	LJ04LFY
801DYE	LJ03MWM	WLT531	--
822DYE	LJ53BFV	WLT554	LF02PSX
BX04MXB	BX04MXB, 205CLT	WLT664	LJ04LDE
BX04MXR	BX04MXR, 217CLT	WLT676	from new
BX04MXT	BX04MXT, 519CLT	WLT719	LJ53NHU
BX04MXZ	BX04MXZ, 324CLT	WLT751	LF02PRZ
BX04MYY	BX04MYY, 330CLT	WLT807	LJ03MVU
BX05UWZ	BX05UWZ, 430CLT	WLT888	LF52VPY
BX55FWH	BX55FWH, 124CLT	WLT892	LF52UPE
LF52USE	LF52USE, VLT25	WLT895	LF52UPJ
LJ03MMX	LJ03MMX, VLT25	WLT897	LF52UPL
LJ53MHX	LJ53MHX, WLT531	WLT970	LJ04LDV
		WLT997	LJ54BFX

Depots and allocations:

Barking (Ripple Road) - DX

Dart	PDL23	PDL24	PDL25	PDL26	PDL27	PDL28	PDL29	PDL30
	PDL31	PDL32	PDL33	PDL34	PDL35	PDL36	PDL37	PDL38
	PDL52	PDL53	PDL54	PDL55	PDL56	PDL57	PDL58	PDL59
	PDL60	PDL61	ADL969	ADL970	ADL971	ADL972	ADL973	ADL974
	ADL975	ADL976	ADL977	ADL978	ADL979	ADL980	ADL981	ADL982
	ADL983							
Volvo B7TL	VLA129	VLA130	VLA131	VLA132	VLA133	VLA134	VLA135	VLA136
	VLA137	VLA138	VLA139	VLA140	VLA141	VLA142	VLA143	VLW105
	VLW106	VLW107	VLW108	VLW109	VLW110	VLW111	VLW112	VLW113
	VLW114	VLW115	VLW116					
DB250	DLA86	DLA87	DLA88	DLA89	DLA90			
Trident 2/Enviro 400	T12	T13	T14	T15	T16	T17	T18	T19
	T20	T21	T22	T23	T24	T25	T26	

Brixton (Streatham Hill) - BN

Outstation: Battersea - BA

DB250	DLA146	DLA147	DLA148	DLA150	DLA151	DLA152	DLA153	DLA159
	DLA160	DLA161	DLA162	DLA163	DLA164	DLA165	DLA205	DLA206
	DLA207	DLA208	DLA209	DLA213	DLA236	DAL270	DLA271	DLA272
	DLA276	DLA290	DLA311	DLA312	DLA313	DLA314	DLA315	DLA316
	DLA317	DLA318	DLA319	DW55	DW56	DW57	DW58	DW59
	DW60	DW61	DW62	DW63	DW64	DW65	DW66	DW67
	DW68	DW69	DW70	DW71	DW72	DW73	DW74	DW75
	DW76	DW77	DW78	DW79	DW80	DW81	DW82	DW83
	DW84	DW85	DW86	DW87	DW88	DW89	DW90	DW91
	DW92	DW93	DW103	DW104	DW105	DW106	DW107	DW108
	DW109	DW110	DW111	DW112	DW113	DW114	DW115	DW116
	DW117	DW118	DW119	DW120	DW121	DW122	DW123	DW124
	DW125	DW126	DW127	DW128	DW129	DW130	DW131	DW132
	DW133	DW134						

Cambridge Heath (Ash Grove) - AE

Enviro200	ENS1	ENS2	ENS3	ENS4	ENS5	ENS6	ENS7	ENS8
	ENS9	ENS10	ENS11	ENS12	ENS13	ENS14		
Citaro G	MA77	MA78	MA79	MA80	MA81	MA82	MA83	MA84
	MA85	MA86	MA87	MA88	MA89	MA90	MA91	MA92
	MA93	MA94	MA95	MA96	MA97	MA98	MA99	MA100
	MA101	MA102	MA103	MA104	MA105	MA106	MA107	MA108
	MA109	MA110	MA111	MA112	MA113	MA114	MA115	MA116
	MA117	MA118	MA119	MA120	MA121	MA122	MA123	MA124
	MA125	MA126						

Croydon (Beddington Farm Road) - CN

Dart	ADL61	ADL62	ADL63	ADL64	PDL124	PDL125	PDL126	PDL127
	PDL128	PDL129	PDL130	PDL131	PDL132	PDL133	PDL134	PDL135
	PDL136							
SB120	DWL58	DWL59	DWL60	DWL61	DWL62	DWL63	DWL64	DWL65
	DWL66	DWL67	DWS1	DWS2	DWS3	DWS4	DWS5	DWS6
	DWS7	DWS8	DWS9	DWS10	DWS11	DWS12	DWS13	DWS14
	DWS15	DWS16	DWS17	DWS18				
DB250	DW37	DW38	DW39	DW40	DW41	DW42	DW43	DW44
	DW45	DW46	DW47	DW48	DW49	DW50	DW51	DW52
	DW53	DW54	DW94	DW95	DW96	DW97	DW98	DW99
	DW100	DW101	DW102					
Ancillary	DLA31	DLA36	DLA45	DLA56	DLA62	DWL30	DWL31	DWL32
	DWL33	DWL34	DWL37					

Croydon (Brighton Road, South Croydon) - TC

Dart	ADL9	ADL10	ADL11	ADL12	ADL13	ADL14	ADL15	ADL16
	ADL17	ADL18	ADL19	ADL20	ADL21	ADL22	ADL23	PDL117
	PDL118	PDL119	PDL120	PDL121	PDL122	PDL123		
Dart 4 / Enviro200	ENL1	ENL2	ENL3	ENL4	ENL5	ENL6	ENL7	ENL8
	ENL9							
DB250	DLA133	DLA134	DLA135	DLA136	DLA139	DLA140	DLA14	DLA143
	DLA156	DLA157	DLA175	DLA176	DLA177	DLA178	DLA179	DLA180
	DLA181	DLA182	DLA183	DLA184	DLA185	DLA186	DLA187	DLA188
	DLA189	DLA215	DLA216	DLA217	DLA218	DLA219	DLA220	DLA221
	DLA222	DLA223	DLA224	DLA250	DLA251	DLA252	DLA253	DLA254
	DLA255	DLA256	DW1	DW2	DW3	DW4	DW5	DW6
	DW7	DW8	DW9	DW10	DW11	DW12	DW13	DW14
	DW15	DW16	DW17	DW18	DW19	DW20	DW21	DW22
	DW23	DW24	DW25	DW26	DW27	DW28	DW29	DW30
	DW31	DW32	DW33	DW34	DW35	DW36		

Trident 2 / Enviro 400	T48	T49	T50	T51	T52	153	T54	T55
	T56	T57	T58	T59	T60	T61	T62	T63
	T64	T65						

Edmonton (Towpath Road, Stonehill Business Park) - EC

Dart	ADL1	ADL2	ADL3	ADL4	ADL5	ADL6	ADL7	ADL8
	ADL65	ADL66	ADL67	ADL68	ADL69	ADL71	ADL72	ADL73
	ADL74	ADL75	ADL76	ADL77	ADL78	ADL79	ADL81	PDL69
	PDL70	PDL71	PDL72	PDL73	PDL74	PDL75	PDL76	PDL77
	PDL78	PDL79	PDL80	PDL81	PDL82	PDL83	PDL84	PDL85
	PDL86	PDL87	PDL88	PDL89	PDL90	PDL91	PDL92	PDL93
	PDL94	PDL95	PDL96	PDL97	PDL98	PDL99	PDL100	PDL101
	PDL102	PDL103	PDL104	PDL105	PDL106	PDL107	PDL108	PDL109
	PDL110	PDL111	PDL112	PDL113	PDL114	PDL115	PDL116	
SB120	DWL10	DWL42	DWL44	DWL46	DWL50	DWL51		
Citaro G	MA73	MA76	MA127	MA128	MA129	MA130	MA131	MA132
	MA133	MA134	MA135	MA136	MA137	MA138	MA139	MA140
	MA141	MA142	MA143	MA144	MA145	MA146	MA147	MA148
	MA149	MA150	MA151	MA152	MA153	MA154	MA155	MA156
	MA157							
Routemaster	RML901	RM1453						
DB250	DLA299	DLA300	DLA301	DLA302	DLA303	DLA304	DLA305	DLA306
	DLA307	DLA308	DLA309	DLA310	DLA348	DLA349	DLA350	DLA351
	DLA352	DLA353	DLA354	DLA355	DLA356	DLA357	DLA358	DLA359
	DLA360	DLA361	DLA362	DLA363	DLA364	DLA365	DLA366	DLA367
	DLA368	DLA369						

Enfield (Southbury Road, Ponders End) - E

Dart	PDL62	PDL64	PDL65	PDL66	PDL67	PDL68		
SB120	DWL13	DWL14	DWL23	DWL24	DWL25	DWL26	DWL27	DWL29
Dart 4 / Enviro 200	ENL10	ENL11	ENL12	ENL13	ENL14	ENL15	ENL16	ENL17
	ENL18	ENL19	ENL20					
DB250	DLA72	DLA73	DLA74	DLA75	DLA76	DLA77	DLA78	DLA79
	DLA80	DLA81	DLA82	DLA83	DLA84	DLA85	DLA91	DLA92
	DLA93	DLA95	DLA116	DLA117	DLA118	DLA119	DLA120	DLA121
	DLA122	DLA123	DLA124	DLA125	DLA190	DLA191	DLA192	DLA193
	DLA194	DLA195	DLA196	DLA197	DLA198	DLA199	DLA200	DLA201
	DLA202	DLA203	DLA210	DLA211	DLA225	DLA226	DLA227	DLA237
	DLA238	DLA248	DLA249	DLP15	DLP16	DLP17	DLP18	DLP20
	DLP76	DLP77	DLP78	DLP79	DLP80	DLP81	DLP82	DLP83
	DLP84	DLP85	DLP86	DLP87	DLP88	DLP89	DLP90	DLP91
	DLP92	DLP93	DLP94	DLP95	DLP96	DLP97	DLP98	DLP99
	DLP100	DLP101	DLP102	DLP103	DLP104	DLP105	DLP106	DLP107
	DLP108	DLP109	DLP110					
Ancillary vehicles	DLA1	DLA21	DLA24	DLA25	DLA26	DLA27	DLA28	DLA29
	DLA30	DLA32	DLA33	DLA35	DLA37	DWL35	DWL36	DWL38
	DWL39	DWL40	DWL41	DWL56				

Lea Valley (Leeside Road, Edmonton) - LV

Dart 4/Enviro 200	EN1	EN2	EN3	EN4	EN5	EN6	EN7	EN8
	EN9	EN10	EN11	EN12	EN13			
Citaro G	MA1	MA2	MA3	MA4	MA5	MA6	MA7	MA8
	MA9	MA10	MA11	MA12	MA13	MA14	MA15	MA16
	MA17	MA18	MA19	MA20	MA21	MA22	MA23	MA24
	MA25	MA26	MA27	MA28	MA29	MA30	MA31	MA32
	MA33	MA34	MA35	MA36	MA37	MA38	MA39	MA40
	MA41	MA42	MA43	MA44	MA45	MA46	MA47	MA48
	MA49	MA50	MA51	MA52	MA53	MA54	MA55	MA56
	MA57	MA58	MA59	MA60	MA61	MA62	MA63	MA64
	MA65	MA66	MA67	MA68	MA69	MA70	MA71	MA72
	MA74	MA75	MA162	MA163	MA164	MA165		

Norwood (Ernest Avenue, West Norwood) - N

Routemaster	RM5	RM6	RM1124	RMC1464	RML2217	RML2355		
DB250	DLA137	DLA138	DLA144	DLA145	DLA149	DLA154	DLA155	DAL158
	DLA166	DLA167	DLA168	DLA169	DLA170	DLA171	DLA172	DLA173
	DLA174							
Volvo B7TL	VLA1	VLA2	VLA3	VLA4	VLA5	VLA6	VLA7	VLA8
	VLA9	VLA10	VLA11	VLA12	VLA13	VLA14	VLA15	VLA16
	VLA17	VLA18	VLA19	VLA20	VLA21	VLA22	VLA23	VLA24
	VLA25	VLA26	VLA27	VLA28	VLA29	VLA30	VLA31	VLA32
	VLA33	VLA34	VLA35	VLA36	VLA37	VLA38	VLA39	VLA40
	VLA41	VLA42	VLA43	VLA44	VLA45	VLA46	VLA47	VLA48
	VLA49	VLA50	VLA51	VLA52	VLA53	VLA54	VLA55	VLA70
	VLA71	VLA72	VLA73	VLA104	VLA105	VLA106	VLA107	VLA108
	VLA109	VLA110	VLA111	VLA112	VLA113	VLA114	VLA115	VLA116
	VLA117	VLA118	VLA119	VLA120	VLA121	VLA122	VLA123	VLA124
	VLA125	VLA126	VLA127	VLA128				

Palmers Green (Regents Avenue) - AD

DB250	DLP40	DLP41	DLP42	DLP43	DLP44	DLP45	DLP46	DLP47
	DLP48	DLP49	DLP50	DLP51	DLP52	DLP53	DLP54	DLP55
	DLP56	DLP57	DLP58	DLP59	DLP60	DLP61	DLP62	DLP63
	DLP71	DLP73	DLP74	DLP75				
Trident 2 / Enviro 400	T1	T2	T3	T4	T5	T6	T7	T8
	T9	T10	T11	T27	T28	T29	T30	T31
	T32	T33	T34	T35	T36	T37	T38	T39
	T40	T41	T42	T43	T44	T45	T46	T47

Stamford Hill (Rookwood Road) - SF

DB250	DLA94	DLA96	DLA97	DLA98	DLA99	DLA100	DLA101	DLA102
	DLA103	DLA104	DLA105	DLA106	DLA107	DLA108	DLA109	DLA110
	DLA111	DLA112	DLA113	DLA114	DLA115	DLA233	DLA291	DLA292
	DLA293	DLA294	DLA295	DLA296	DLA297	DLA298	DLA346	DLA370
	DLA371	DLA372	DLA373	DLA374	DLA375	DLA376	DLA377	DLA378
	DLA379	DLA380	DLA381	DLA382	DLA383	DLA384	DLA385	DLA386
	DLA387	DLA388						
Volvo B7TL	VLW86	VLW87	VLW88	VLW89	VLW90	VLW91	VLW92	VLW93
	VLW94	VLW95	VLW96	VLW140	VLW141	VLW142	VLW143	VLW144
	VLW145	VLW146	VLW147	VLW148	VLW149	VLW150	VLW151	VLW152
	VLW153	VLW154	VLW155	VLW156	VLW157	VLW158	VLW159	VLW160
	VLW161	VLW162	VLW163	VLW164	VLW165	VLW166	VLW167	VLW168

Thornton Heath (London Road) - TH

Dart	PDL19	PDL20	PDL21	PDL22				
SB120	DWL1	DWL2	DWL3	DWL4	DWL5	DWL6	DWL7	DWL8
	DWL9							
DB250	DLA2	DLA3	DLA4	DLA5	DLA6	DLA7	DLA8	DLA9
	DLA10	DLA38	DLA39	DLA40	DLA41	DLA42	DLA43	DLA44
	DLA48	DLA49	DLA50	DLA51	DLA52	DLA53	DLA54	DLA55
	DLA57	DLA58	DLA59	DLA60	DLA61	DLA63	DLA64	DLA126
	DLA127	DLA128	DLA129	DLA130	DLA131	DLA132	DLA322	DLA323
	DLA324	DLA325	DLA326	DLA327	DLA328	DLA329	DLA330	DLA331
	DLA332	DLA333	DLA334	DLA335	DLA336	DLA337	DLA338	DLA339
	DLA340	DLA341	DLA342	DLA343	DLA344	DLA345	DLA347	DLA389

Tottenham (Philip Lane) - AB

Dart	PDL12	PDL13	PDL14	PDL15	PDL16	PDL17	PDL18	PDL39
	PDL40	PDL41	PDL42	PDL44	PDL45	PDL46	PDL47	PDL48
	PDL49	PDL50	PDL51	PLD63				
DB250	DLA204	DLA230	DLA231	DLA232	DLA234	DLA235	DLA239	DLA240
	DLA241	DLA242	DLA243	DLA244	DLA245	DLA246	DLA247	DLA273
	DLA274	DLA275	DLA277	DLA278	DLA279	DLA280	DLA281	DLA282
	DLA283	DLA284	DLA285	DLA286	DLA287	DLA288	DLA289	
Volvo B7TL	VLA74	VLA75	VLA76	VLA77	VLA78	VLA79	VLA80	VLA81
	VLA82	VLA83	VLA84	VLA85	VLA86	VLA87	VLA88	VLA89
	VLA90	VLA91	VLA92	VLA93	VLA94	VLA95	VLA96	VLA97
	VLA98	VLA99	VLA100	VLA101	VLA102	VLA103	VLW97	VLW98
	VLW99	VLW100	VLW101	VLW102	VLW103	VLW104	VLW117	VLW118
	VLW119	VLW120	VLW121	VLW122	VLW123	VLW124	VLW125	VLW126
	VLW127	VLW128	VLW129	VLW130	VLW131	VLW132	VLW133	VLW134
	VLW135	VLW136	VLW137	VLW138	VLW139	VLW169	VLW170	VLW171
	VLW172	VLW173	VLW174	VLW175	VLW176	VLW177	VLW178	VLW179
	VLW180	VLW181	VLW182	VLW183	VLW184	VLW185	VLW186	VLW187
	VLW188	VLW189	VLW190	VLW191	VLW192	VLW193	VLW194	VLW195
	VLW196	VLW197	VLW198	VLW199				

Wood Green (High Road) - WN

SB120	DWL11	DWL12	DWL15	DWL16	DWL17	DWL18	DWL19	DWL20
	DWL21	DWL22	DWL43	DWL45	DWL47	DWL48	DWL49	DWL52
	DWL53	DWL54	DWL55	DWL56	DWL57			
Dart 4 / Enviro 200	ENL30	ENL31	ENL32	ENL33	ENL34	ENL35	ENL36	ENL37
	ENL38	ENL39	ENL40	ENL41	ENL42	ENL43	ENL44	ENL45
	ENL46	ENL47	ENL48					
DB250	DLA11	DLA12	DLA13	DLA14	DLA15	DLA16	DLA17	DLA18
	DLA19	DLA20	DLA47	DLA212	DLA214	DLP64	DLP65	DLP66
	DLP67	DLP68	DLP69	DLP70	DLP72	DLP73	DLP74	DLP75
Volvo B7TL	VLW1	VLW2	VLW3	VLW4	VLW5	VLW6	VLW7	VLW8
	VLW9	VLW10	VLW11	VLW12	VLW13	VLW14	VLW15	VLW16
	VLW17	VLW18	VLW19	VLW20	VLW21	VLW22	VLW23	VLW24
	VLW25	VLW26	VLW27	VLW28	VLW29	VLW30	VLW31	VLW32
	VLW33	VLW34	VLW35	VLW36	VLW37	VLW38	VLW39	VLW40
	VLW41	VLW42	VLW43	VLW44	VLW45	VLW46	VLW47	VLW48
	VLW49	VLW50	VLW51	VLW52	VLW53	VLW54	VLW55	VLW56
	VLW57	VLW58	VLW59	VLW60	VLW61	VLW62	VLW63	VLW64
	VLW65	VLW66	VLW67	VLW68	VLW70	VLW71	VLW72	VLW73
	VLW74	VLW75	VLW76	VLW77	VLW78	VLW79	VLW80	VLW81
	VLW82	VLW83	VLW84	VLW85				
Volvo B5L Hybrid	HV1	HV2	HV3	HV4	HV5			
Wrightbus Hybrid	HW1	HW2	HW3	HW4	HW5	DW201	DW202	DW203
	DW204	DW205						

Unallocated - u/w

remainder

*DAL228 is on loan to Arriva Southern Counties at Dartford.

ARRIVA - THE ORIGINAL TOUR

The Original Tour Ltd, Jews Road, Wandsworth, SW18 1TB

EMB763	D553YNO	MCW Metrobus DR115/4	MCW	PO65/31D	1987	New World FirstBus, 2001
EMB764	E964JAR	MCW Metrobus DR115/4	MCW	PO65/31D	1987	New World FirstBus, 2001
EMB765	E965JAR	MCW Metrobus DR115/4	MCW	PO65/31D	1987	New World FirstBus, 2001
EMB766	E766JAR	MCW Metrobus DR115/4	MCW	PO65/31D	1987	New World FirstBus, 2001
EMB767	E767JAR	MCW Metrobus DR115/4	MCW	PO65/31D	1987	New World FirstBus, 2001
EMB768	E768JAR	MCW Metrobus DR115/4	MCW	PO65/31D	1987	New World FirstBus, 2001
EMB769	E769JAR	MCW Metrobus DR115/4	MCW	PO65/31D	1987	New World FirstBus, 2001
EMB770	E770JAR	MCW Metrobus DR115/4	MCW	PO65/31D	1987	New World FirstBus, 2001
EMB771	E771JAR	MCW Metrobus DR115/4	MCW	PO65/31D	1987	New World FirstBus, 2001
EMB772	E772JAR	MCW Metrobus DR115/4	MCW	PO65/31D	1987	New World FirstBus, 2001
EMB773	E773JAR	MCW Metrobus DR115/4	MCW	PO65/31D	1987	New World FirstBus, 2001
EMB774	E774JAR	MCW Metrobus DR115/4	MCW	PO65/31D	1987	New World FirstBus, 2001
EMB775	D675YNO	MCW Metrobus DR115/4	MCW	PO65/31D	1987	New World FirstBus, 2001
EMB776	UAR776Y	MCW Metrobus DR115/3	MCW	059/33D	1984	City Sightseeing, Aus., 04
EMB777	A735WEV	MCW Metrobus DR115/3	MCW	059/33D	1984	City Sightseeing, Aus., 04
EMB778	A737WEV	MCW Metrobus DR115/3	MCW	059/33D	1984	City Sightseeing, Aus., 04
EMB779	MXT179	MCW Metrobus DR115/3	MCW	059/33D	1984	City Sightseeing, Aus., 04
EMB780	A755WEV	MCW Metrobus DR115/3	MCW	059/33D	1983	New World FirstBus, 2004
EMB781	A750WEV	MCW Metrobus DR115/3	MCW	059/33D	1983	New World FirstBus, 2004
EMB782	A749WEV	MCW Metrobus DR115/3	MCW	059/33D	1983	New World FirstBus, 2004
EMB783	UAR247Y	MCW Metrobus DR115/3	MCW	059/33D	1987	New World FirstBus, 2004
EMB784	UAR250Y	MCW Metrobus DR115/3	MCW	059/33D	1987	New World FirstBus, 2004
EMB785	NKJ785	MCW Metrobus DR115/3	MCW	059/33D	1987	New World FirstBus, 2004

DLP201-214

		DAF DB250 10.6m		Plaxton President	045/21F	1999	Arriva London, 2006

201	WD	201KYD	205	WD	T205XBV	209	WD	T209XBV	212	WD	T212XBV
202	WD	T202XBV	206	WD	T206XBV	210	WD	T210XBV	213	WD	T213XBV
203	WD	T203XBV	207	WD	T207XBV	211	WD	T211XBV	214	WD	T214XBV
204	WD	T204XBV	208	WD	T208XBV						

In 2007, ten East Lancs Visionaire-bodied Volvo B9TLs were added to Arriva's Original Tour fleet. With the Palace of Westminster in the background VLE617, LJ07XEU, illustrates the model from East Lancs which is now supplied by Optare following recent business deals. *Mark Lyons*

Still a feature of the open-top tour buses are re-imported Metrobuses and Olympian Tri-axles from Hong Kong. These longer buses have higher seating capacities on the upper decks. EMB770, E770JAR, is seen passing St Paul's Cathedral. *Richard Godfrey*

OA315-352

	Leyland Olympian			Alexander RH			PO43/25D* 1992		Arriva London, 2003-05	
							321-32/8-40 are CO43/25D; 350-2 are B43/25D			

315	WD	J315BSH	325	WD	J325BSH	335	WD	J335BSH	344	WD	J344BSH
316	WD	J316BSH	326	WD	J326BSH	336	WD	J336BSH	345	WD	J345BSH
317	WD	J317BSH	327	WD	J327BSH	337	WD	J337BSH	346	WD	J346BSH
318	WD	J318BSH	328	WD	J328BSH	338	WD	J338BSH	347	WD	J347BSH
319	WD	J319BSH	329	WD	J329BSH	339	WD	J339BSH	348	WD	J348BSH
320	WD	J320BSH	330	WD	J330BSH	340	WD	J340BSH	349	WD	J349BSH
321	WD	J321BSH	331	WD	J331BSH	341	WD	J341BSH	350	WD	J350BSH
322	WD	J322BSH	332	WD	J332BSH	342	WD	J342BSH	351	WD	J351BSH
323	WD	J323BSH	333	WD	J433BSH	343	WD	J343BSH	352	WD	J352BSH
324	WD	J324BSH	334	WD	J334BSH						

VLY601-610

	Volvo B7L 10.6m			Ayats Bravo City			O51/24F	2005		

601	WD	LX05GDV	604	WD	LX05GEJ	607	WD	LX05KNZ	609	WD	EU05DVW
602	WD	LX05GDY	605	WD	LX05HRO	608	WD	LX05KOA	610	WD	EU05DVX
603	WD	LX05GDZ	606	WD	LX05HSC						

VLE611-620

	Volvo B9TL 10.9m			East Lancs Visionaire			PO49/31F	2007	

611	WD	LJ07XEN	614	WD	LJ07XER	617	WD	LJ07XEU	619	WD	LJ07XEW
612	WD	LJ07XEO	615	WD	LJ07XES	618	WD	LJ07XEV	620	WD	LJ07UDD
613	WD	LJ07XEP	616	WD	LJ07XET						

Ancillary vehicle:

MB1152 B152WUL	MCW Metrobus DR101/17		MCW	TV	1983	Arriva London, 1999

Previous registrations:

201KYD	V601LGC	E770JAR	DU8506 (HK)
A735WEV	CZ9920(HK)	E771JAR	DT9187 (HK)
A737WEV	DA2952(HK)	E772JAR	DV2896 (HK)
A749WEV	-	E773JAR	DU3481 (HK)
A750WEV	-	E964JAR	DT4549 (HK)
A755WEV	-	E965JAR	DV4883 (HK)
D553YNO	DV471 (HK)	MXT179	- (HK), UAR773Y
D675YNO	DV3433(HK)	NKJ785	- (HK), UAR772Y
E767JAR	DU3460 (HK)	UAR247Y	CZ2554(HK)
E768JAR	DU8346 (HK)	UAR250Y	CZ664(HK)
E769JAR	DT7256 (HK)	UAR776Y	

Depot: Jews Road, Wandsworth (WD)

ARRIVA SOUTHERN COUNTIES

Arriva Southern Counties Ltd, Arriva West Sussex Ltd,
Arriva Kent Thameside Ltd; Arriva Kent & Sussex Ltd; New Enterprise Ltd
Arriva Medway Towns Ltd, Arriva Guildford & West Surrey Ltd;
Arriva Southend Ltd
Invicta House, Armstrong Road, Maidstone, Kent, ME15 6TX

0304-0328
Mercedes-Benz Sprinter 515 cdi | Optare/Ferqui Soroco | C19F | 2007-08 | *Operated for Easybus*

0304	GW	YX57CAA	0318	GW	YX08HWC	0322	GW	YX08HWG	0326	GW	YX08HWM
0309	GW	YX57AOB	0319	GW	YX08HWD	0323	GW	YX08HWH	0327	GW	YX08HWO
0310	GW	YX57AOC	0320	GW	YX08HWE	0324	GW	YX08HWK	0328	GW	YX08HWN
0311	GW	YX57AOD	0321	GW	YX08HWF	0325	GW	YX08HWL			

1118-1122
Mercedes-Benz Vario 0810 | Plaxton Beaver 2 | B27F | 1998

1118	GI	R118TKO	1120	SR	R120TKO	1121	SR	R121TKO	1122	NF	R122TKO
1119	GI	R119TKO									

1172	SR	R942VPU	Mercedes-Benz Vario 0810	Plaxton Beaver 2	B27F	1998
1190	TO	P481DPE	Mercedes-Benz 711D	Plaxton Beaver 2	B27F	1997

1501-1506
Optare Solo SR M960* | Optare | N32F | 2008-09 | *1504-6 are SR M890s

1501	TW	YJ58CDZ	1503	TW	YJ58CEF	1505	NF	YJ09MMO	1506	NF	YJ09MMU
1502	TW	YJ58CEA	1504	NF	YJ09MMK						

1601-1605
Dennis Dart SLF | Plaxton Pointer MPD | N29F | 2000

1601	TW	W601YKN	1603	TW	W603YKN	1604	TW	W604YKN	1605	TW	W605YKN
1602	TW	W602YKN									

1606-1617
TransBus Dart 8.8m | TransBus Mini Pointer | N29F | 2004

1606	NF	GN04UCW	1609	SE	GN04UCZ	1612	GI	GN04UDE	1615	GI	GN04UDJ
1607	NF	GN04UCX	1610	GI	GN04UDB	1613	GI	GN04UDG	1616	GI	GN04UDK
1608	SE	GN04UCY	1611	GI	GN04UDD	1614	GI	GN04UDH	1617	GI	GN04UDL

1618-1623
ADL Dart 8.8m | ADL Mini Pointer | N29F | 2005

1618	NF	GN05ANU	1620	NF	GN05ANX	1622	NF	GN05AOB	1623	NF	GN05AOC
1619	NF	GN05ANV	1621	NF	GN05AOA						

1624-1636
ADL Dart 8.8m | ADL Mini Pointer | N23F | 2006-07

1624	DA	SN06BPE	1628	DA	SN06BPV	1631	DA	SN06BPZ	1634	SR	GN06EBF
1625	DA	SN06BPF	1629	DA	SN06BPX	1632	DA	SN06BRF	1635	SR	GN06EBG
1626	DA	SN06BPK	1630	DA	SN06BPY	1633	SR	GN06EBB	1636	SR	GN06EBH
1627	DA	SN06BPU									

1637-1649
ADL Dart 4 | ADL Enviro 200 | N26F | 2007

1637	TW	GN57BNX	1641	TW	GN57BOH	1644	DA	GN57BPK	1647	DA	GN57BPV
1638	TW	GN57BNY	1642	TW	GN57BOJ	1645	DA	GN57BPO	1648	DA	GN57BPX
1639	TW	GN57BNZ	1643	DA	GN57BPF	1646	DA	GN57BPU	1649	DA	GN57BPY
1640	TW	GN57BOF									

2000	TO	GB03TGM	Setra S415 HD	Setra	C49FT	2003	Arriva TGM, 2008
2852	TO	NM02DYA	Volvo B10M-62	Caetano Enigma	C49FT	2002	Arriva TGM, 2008
2853	TO	YN54WDE	Volvo B7R	Plaxton Profile	C53F	2004	OFJ Connections, 2008
2854	TO	YN54OCY	Volvo B7R	Plaxton Profile	C53F	2004	OFJ Connections, 2008
2855	TO	YN53VBX	TransBus Javelin	Plaxton Profile	C53F	2004	OFJ Connections, 2008
2889	TO	R456SKX	DAF SB3000	Plaxton Prima Interurban	C53F	1997	Arriva The Shires, 2009
2890	TO	M945LYR	DAF SB3000	Van Hool Alizée	C49FT	1995	London North East, 1998
2894	TO	W183CDN	DAF SB3000	Van Hool Alizée	C52F	2000	
2895	TO	SCZ9651	DAF SB3000	Van Hool T9 Alizée	C49F	1999	Eirebus, Dublin, 2003
2896	TO	SCZ9652	DAF SB3000	Van Hool T9 Alizée	C49F	1999	Eirebus, Dublin, 2003
2898	TO	W198CDN	DAF SB3000	Ikarus Blue Danube 396	C53F	2000	

2008 saw the introduction of the SR variant of the Optare Solo. One of a batch for Arriva service in Dartford is 1506, YJ09MMU. *Dave Heath*

2903-2908

DAF SB3000 — Plaxton Première 320 — C53F — 1998

2903	TO	R903BKO	2905	TO	R905BKO	2907	TO	R907BKO	2908	TO	R908BKO
2904	TO	R904BKO	2906	TO	R906BKO						

3008	NF	L510CPJ	Dennis Dart 9.8m	East Lancs EL2000	B40F	1993
3019	ME	N539TPF	Dennis Dart 9.8m	East Lancs EL2000	B40F	1995

3020-3024

Dennis Lance 11m — East Lancs — N49F — 1996

3020	TW	N220TPK	3022	TW	N322TPK	3023	GU	N223TPK	3024	GU	N224TPK
3021	TW	N221TPK									

3025-3036

Dennis Dart SLF — Plaxton Pointer 2 — N35F — 1996 — North Western (Beeline), 1998

3025	GU	N225TPK	3028	GU	N228TPK	3031	GU	N231TPK	3035	GU	N235TPK
3026	GU	N226TPK	3029	SR	N229TPK	3033	GU	N233TPK	3036	GU	N236TPK
3027	SR	N227TPK	3030	GU	N230TPK						

3037	SR	N237VPH	Dennis Dart SLF	East Lancs Spryte	N31F	1996
3038	GI	N245VPH	Dennis Dart SLF	East Lancs Spryte	N31F	1996
3039	TO	N239VPH	Dennis Dart SLF	East Lancs Spryte	N31F	1996
3046	TW	N246VPH	Dennis Dart SLF	East Lancs Spryte	N31F	1996
3047	TW	N247VPH	Dennis Dart SLF	East Lancs Spryte	N31F	1996
3050	GI	P250APM	Dennis Dart SLF	East Lancs Spryte	N31F	1996
3051	GI	P251APM	Dennis Dart SLF	East Lancs Spryte	N31F	1996
3053	GI	P253APM	Dennis Dart SLF	East Lancs Spryte	N31F	1997
3054	GI	P254APM	Dennis Dart SLF	East Lancs Spryte	N31F	1997
3055	TO	P255APM	Dennis Dart SLF	East Lancs Spryte	N31F	1997

3070-3096 — Dennis Dart SLF — Plaxton Pointer — N39F — 1997

3070	GI	P270FPK	3076	GI	P276FPK	3089	TW	P289FPK	3093	GU	P293FPK
3071	GI	P271FPK	3077	GI	P277FPK	3090	GU	P290FPK	3094	GU	P294FPK
3072	GI	P272FPK	3084	ME	P284FPK	3091	GU	P291FPK	3095	GU	P295FPK
3073	GI	P273FPK	3086	SE	P286FPK	3092	GU	P292FPK	3096	GU	P296FPK
3075	GI	P275FPK	3088	GU	P288FPK						

3097-3102 — Dennis Dart SLF — Plaxton Pointer 2 — N39F — 1997

3097	GU	R297CMV	3099	GU	R299CMV	3101	GU	R301CMV	3102	GU	R302CMV
3098	GU	R298CMV	3100	GU	R310CMV						

3109	GU	T109LKK	Dennis Dart SLF 9m	Plaxton Pointer 2	N39F	1999
3110	GU	T110LKK	Dennis Dart SLF 9m	Plaxton Pointer 2	N39F	1999
3111	NF	N523MJO	Dennis Dart SLF 9m	Plaxton Pointer 2	N36F	1996	Arriva The Shires, 2009
3112	GU	NDZ7926	Dennis Dart SLF 9m	Plaxton Pointer 2	N36F	1996	Arriva The Shires, 2009
3113	GU	NDZ7918	Dennis Dart SLF 9m	Plaxton Pointer 2	N36F	1996	Arriva The Shires, 2009
3123	SE	N543TPK	Dennis Dart 9.8m	East Lancs EL2000	B40F	1996	
3124	NF	N544TPK	Dennis Dart 9.8m	East Lancs EL2000	B40F	1996	
3172	SE	N234TPK	Dennis Dart SLF 9m	Plaxton Pointer	N35F	1996	
3175	ME	M200CBB	Dennis Dart 9.8m	Plaxton Pointer	B40F	1995	Cardiff Bluebird, 1996

3176-3191 — Dennis Dart SLF 10.1m — Plaxton Pointer — N40F — 1996

3176	ME	P176LKL	3181	SE	P181LKL	3186	NF	P186LKJ	3189	GI	P189LKJ
3177	ME	P177LKL	3184	TW	P184LKL	3187	GU	P187LKJ	3190	GI	P190LKJ
3178	GI	P178LKL	3185	TW	P185LKL	3188	NF	P188LKJ	3191	GI	P191LKJ

3192-3247 — Dennis Dart SLF 10.1m — Plaxton Pointer — N40F — 1997

3192	TW	P192LKJ	3206	TW	P206LKJ	3224	GI	P224MKL	3236	GI	P236MKN
3193	TW	P193LKJ	3207	ME	P207LKJ	3225	GI	P225MKL	3237	GI	P237MKN
3194	ME	P194LKJ	3208	ME	P208LKJ	3226	GI	P226MKL	3238	GI	P238MKN
3195	ME	P195LKJ	3209	ME	P209LKJ	3227	GI	P227MKL	3239	GI	P239MKN
3196	TW	P196LKJ	3213	ME	P213LKJ	3228	GI	P228MKL	3240	GI	P240MKN
3197	NF	P197LKJ	3215	ME	P215LKJ	3229	GI	P229MKL	3241	GI	P241MKN
3198	GI	P198LKJ	3216	SE	P216LKJ	3230	GI	P230MKL	3242	GI	P242MKN
3199	SR	P199LKJ	3218	TW	P218MKL	3231	GI	P231MKL	3243	GI	P243MKN
3201	SR	P201LKJ	3219	GI	P219MKL	3232	GI	P232MKL	3244	GI	P244MKN
3202	ME	P202LKJ	3220	GI	P220MKL	3233	GI	P233MKN	3245	GI	P245MKN
3203	ME	P203LKJ	3221	GI	P221MKL	3234	GI	P234MKN	3246	GI	P246MKN
3204	SE	P204LKJ	3223	GI	P223MKL	3235	GI	P235MKN	3247	SR	P247MKN
3205	ME	P205LKJ									

3249	ME	P279FPK	Dennis Dart SLF 10.1m	Plaxton Pointer 2	N40F	1997

3250-3259 — Scania L113CRL — Wright Axcess-ultralow — N43F — 1995

3250	SE	N250BKK	3253	SE	N253BKK	3256	SE	N256BKK	3258	SE	N258BKK
3251	SE	N251BKK	3254	SE	N254BKK	3257	SE	N257BKK	3259	SE	N259BKK
3252	SE	N252BKK	3255	SE	N255BKK						

3261-3272 — Dennis Dart SLF 10.1m — Plaxton Pointer 2 — N39F — 1998

3261	NF	R261EKO	3264	TW	R264EKO	3267	NF	R267EKO	3270	TW	R270EKO
3262	TW	R262EKO	3265	TW	R265EKO	3268	NF	R268EKO	3271	TW	R271EKO
3263	NF	R263EKO	3266	TW	R266EKO	3269	TW	R269EKO	3272	NF	R272EKO

3273-3289 — Dennis Dart SLF 10.1m — Plaxton Pointer 2 — N39F* — 1999 — *3276-81 are N37F

3273	NF	T273JKM	3278	SE	T278JKM	3282	TW	T282JKM	3286	NF	T286JKM
3274	NF	T274JKM	3279	SE	T279JKM	3283	NF	T283JKM	3287	TW	T287JKM
3275	NF	T275JKM	3280	SE	T280JKM	3284	NF	T284JKM	3288	NF	T288JKM
3276	SE	T276JKM	3281	SE	T281JKM	3285	NF	T285JKM	3289	SE	T289JKM
3277	SE	T277JKM									

3291-3303 — Dennis Dart SLF 10.1m — Plaxton Pointer 2 — N34D — 2001

3291	DA	Y291TKJ	3294	DA	Y294TKJ	3297	DA	Y297TKJ	3301	DA	Y301TKJ
3292	DA	Y292TKJ	3295	DA	Y295TKJ	3298	DA	Y298TKJ	3302	DA	Y302TKJ
3293	DA	Y293TKJ	3296	DA	Y296TKJ	3299	DA	Y299TKJ	3303	DA	Y303TKJ

Representing the large number of Darts in service with Arriva Southern Counties, 3261, R261EKO, is from the 1998 intake with Plaxton Pointer 2 bodywork. It is seen at Swanley while heading for the Bluewater Shopping Centre. *Richard Godfrey*

3304-3308

| | | | | Dennis Dart SLF 10.1m | | | Plaxton Pointer 2 | | N39F | 1997 | |

3304	GU	R304CMV	**3306**	GU	R296CMV	**3307**	SE	R307CMV	**3308**	SE	R308CMV
3305	GU	R305CMV									

3318-3326

| | | | | Dennis Dart SLF | | | Plaxton Pointer 2 | | N33F* | 1998 | * 3224-6 are N39F |

3318	SE	T218NMJ	**3321**	SE	T821NMJ	**3323**	SE	T823NMJ	**3325**	SE	T825NMJ
3320	SE	T820NMJ	**3322**	SE	T822NMJ	**3324**	SE	T824NMJ	**3326**	SE	T826NMJ

| **3384** | TW | M764JPA | Dennis Lance SLF 11m | | | Wright Pathfinder | | | N39F | 1995 | |

3387-3397

| | | | | Dennis Dart SLF | | | Plaxton Pointer | | N39F | 1997 | |

3387	SE	P257FPK	**3390**	SE	P259FPK	**3393**	SE	P263FPK	**3396**	SE	P266FPK
3388	SE	P258FPK	**3391**	SE	P261FPK	**3394**	SE	P264FPK	**3397**	SE	P267FPK
3389	SE	P259FPK	**3392**	SE	P262FPK	**3395**	SE	P265FPK			

3400	SE	R310NGM	Dennis Dart SLF	Plaxton Pointer 2	N33F	1997	Town & Country, Corringham, '00
3401	SE	R311NGM	Dennis Dart SLF	Plaxton Pointer 2	N33F	1997	Town & Country, Corringham, '00
3402	SE	R312NGM	Dennis Dart SLF	Plaxton Pointer 2	N33F	1997	Town & Country, Corringham, '00
3403	SE	R313NGM	Dennis Dart SLF	Plaxton Pointer 2	N33F	1997	Town & Country, Corringham, '00

3404-3412

| | | | | Dennis Dart | | | Plaxton Pointer | | B34F | 1996 | |

3404	NF	P324HVX	**3408**	ME	P328HVX	**3410**	ME	P330HVX	**3412**	TW	P332HVX
3407	NF	P327HVX	**3409**	ME	P329HVX	**3411**	TW	P331HVX			

3421-3431

| | | | | Dennis Dart SLF | | | Plaxton Pointer | | N39F | 1996 | |

3421	SE	P421HVX	**3425**	SE	P425HVX	**3427**	SE	P427HVX	**3429**	ME	P429HVX
3423	SE	P423HVX	**3426**	SE	P426HVX	**3428**	SE	P428HVX	**3431**	SE	P431HVX

Allocated to Northfleet is a batch of VDL Bus SB200s with Wrightbus Pulsar bodywork. The single-deck Pulsar displaced the Cadet and Commander models in 2008. Pictured at the Bluewater Shopping Centre, 3755, YJ08DZE, illustrates the Arriva interurban scheme, its high-back seating being visible. *Dave Heath*

3500-3510

DAF SB120 10.2m — Wrightbus Cadet — N31D — 2002

3500	GY	KE51PTY	3503	GY	KE51PUF	3506	GY	KE51PUK	3509	GY	KE51PUV
3501	GY	KE51PTZ	3504	GY	KE51PUH	3507	GY	KE51PUO	3510	GY	KC51NFO
3502	GY	KE51PUA	3505	GY	KE51PUJ	3508	GY	KE51PUU			

3511	GY	KE51PUY	DAF SB120 9.4m	Wrightbus Cadet	N27F	2002	
3512	GY	KC51PUX	DAF SB120 9.4m	Wrightbus Cadet	N27F	2002	
3513	TW	KE51PVD	DAF SB120 9.4m	Wrightbus Cadet	N27F	2002	
3591	GU	T591CGT	Dennis Dart SLF 10.1m	Plaxton Pointer 2	N39F	1999	Operated for Surrey CC
3592	GU	T592CGT	Dennis Dart SLF 10.1m	Plaxton Pointer 2	N39F	1999	Operated for Surrey CC
3593	TW	GJ52HDZ	TransBus Dart SLF 10.7m	TransBus Pointer	N39F	2003	Operated for Kent CC
3617	TO	M617PKP	Volvo B6-9.9M	Plaxton Pointer	B40F	1995	
3619	TO	M619PKP	Volvo B6-9.9M	Plaxton Pointer	B40F	1995	

3701-3706

Dennis Dart SLF 11.3m — Plaxton Pointer SPD — N44F — 1998

3701	ME	S701VKM	3703	ME	S703VKM	3705	ME	S705VKM	3706	ME	S706VKM
3702	ME	S702VKM	3704	ME	S704VKM						

3731-3735

Volvo B7RLE — Wrightbus Eclipse Urban — N44F — 2006

3731	GU	GN54MYO	3733	GU	GN54MYR	3734	GU	GN54MYT	3735	GU	GN54MYU
3732	GU	GN54MYP									

3751-3762

VDL Bus SB200 — Wrightbus Pulsar — NC44F — 2008

3751	NF	YJ08DZA	3754	NF	YJ08DZD	3757	NF	YJ08DZG	3760	NF	YJ08DZL
3752	NF	YJ08DZB	3755	NF	YJ08DZE	3758	NF	YJ08DZH	3761	NF	YJ08DZM
3753	NF	YJ08DZC	3756	NF	YJ08DZF	3759	NF	YJ08DZK	3762	NF	YJ08DZN

Similar in style to the Pulsar, the Wrightbus Eclipse Urban is the body fitted to Volvo B7REL chassis. Fourteen were supplied to Arriva Southern Counties in 2006 for the *Fastrack* services based in the Dartford and Gravesend area. Number 3804, GN06EWE, is shown. *Dave Heath*

3801-3814

			Volvo B7RLE			Wrightbus Eclipse Urban			NC43F	2006	
3801	NF	GN06EVG	3805	NF	GN06EVH	3809	NF	GN06EVL	3812	NF	GN06EVT
3802	NF	GN06EWC	3806	NF	GN06EVJ	3810	NF	GN06EVP	3813	NF	GN06EVU
3803	NF	GN06EWD	3807	NF	GN06EVK	3811	NF	GN06EVR	3814	NF	GN06EUU
3804	NF	GN06EWE	3808	NF	GN06EVM						

3815-3826

			Volvo B7RLE			Wrightbus Eclipse Urban			N43F	2007	
3815	NF	GN07AVB	3818	NF	GN07AVE	3821	NF	GN07AVJ	3824	NF	GN07AVN
3816	NF	GN07AVC	3819	NF	GN07AVF	3822	NF	GN07AVL	3825	NF	GN07AVO
3817	NF	GN07AVD	3820	NF	GN07AVG	3823	NF	GN07AVM	3826	NF	GN07AVP

3901	GU	BX56VTU	Mercedes-Benz Citaro O530	Mercedes-Benz	N42F	2006
3902	GU	BX56VTV	Mercedes-Benz Citaro O530	Mercedes-Benz	N42F	2006
3903	GU	BX56VTW	Mercedes-Benz Citaro O530	Mercedes-Benz	N42F	2006

3911-3921

			DAF SB220			Plaxton Prestige			N39F	1999	
3911	SE	T911KKM	3914	TW	T914KKM	3916	SE	T916KKM	3919	SE	T919KKM
3912	SE	T912KKM	3915	TW	T915KKM	3918	SE	T918KKM	3921	TW	T921KKM
3913	TW	T913KKM									

3923-3932

			DAF SB120			Wrightbus Cadet			N39F	2002	
3923	GU	GK51SYY	3926	GU	GK51SZD	3929	GU	GK51SZG	3931	GU	GK51SZL
3924	GU	GK51SYZ	3927	GU	GK51SZE	3930	GU	GK51SZJ	3932	GU	GK51SZN
3925	GU	GK51SZC	3928	GU	GK51SZF						

3933-3944

			DAF SB120			Wrightbus Cadet			N39F	2002	
3933	GU	GK52YUW	3937	GU	GK52YVB	3940	GU	GK52YVE	3943	GU	GK52YVJ
3934	GU	GK52YUX	3938	GU	GK52YVC	3941	GU	GK52YVF	3944	GU	GK52YVL
3935	GU	GK52YUY	3939	GU	GK52YVD	3942	GU	GK52YVG			

3945-3960

			VDL Bus SB120			Wrightbus Cadet 2			N39F	2004	
3945	DA	GK53AOH	3949	DA	GK53AOO	3953	DA	GK53AOU	3957	DA	GK53AOY
3946	DA	GK53AOJ	3950	DA	GK53AOP	3954	DA	GK53AOV	3958	ME	GK53AOZ
3947	DA	GK53AOL	3951	DA	GK53AOR	3955	DA	GK53AOW	3959	ME	GN04UFW
3948	DA	GK53AON	3952	DA	GK53AOT	3956	DA	GK53AOX	3960	ME	GN04UFX

The latest vehicles for Guildford depot are six Alexander Dennis Enviro 200s. Representing the batch is 4017, GN58BUA, which was pictured in Godsworth Park while heading away from Woking for the local Sainsbury's store. *Richard Godfrey*

3961-3969

VDL Bus SB200 — Wrightbus Commander 2 — N44F — 2004

3961	ME	GN04UFY	3964	ME	GN04UGB	3966	ME	GN04UGD	3968	ME	GN04UGF
3962	ME	GN04UFZ	3965	ME	GN04UGC	3967	ME	GN04UGE	3969	ME	GN04UGG
3963	ME	GN04UGA									

3970	GU	YJ06LFZ	DAF SB120 10.8m	Wrightbus Cadet 2	N39F	2006
3971	GY	YE06HPX	DAF SB120 10.8m	Wrightbus Cadet 2	N28D	2006
3972	GY	YE06HPY	DAF SB120 10.8m	Wrightbus Cadet 2	N28D	2006
3973	GY	YE06HPZ	DAF SB120 10.8m	Wrightbus Cadet 2	N28D	2006

3982-3999

ADL Dart 4 — ADL Enviro 200 — N29D — 2007

3982	DA	GN07DLE	3987	DA	GN07DLU	3992	DA	GN07DME	3996	DA	GN07DMV
3983	DA	GN07DLF	3988	DA	GN07DLV	3993	DA	GN07DMF	3997	GY	GN57BOU
3984	DA	GN07DLJ	3989	DA	GN07DLX	3994	DA	GN07DMO	3998	GY	GN57BOV
3985	DA	GN07DLK	3990	DA	GN07DLY	3995	DA	GN07DMU	3999	GY	GN57BPE
3986	DA	GN07DLO	3991	DA	GN07DLZ						

4000-4010

ADL Dart 4 — ADL Enviro 200 — N29D — 2008

4000	GY	GN08CGO	4003	GY	GN08CGX	4006	GY	GO58CHC	4009	GY	GO58CHG
4001	GY	GN08CGU	4004	GY	GN08CGY	4007	GY	GO58CHD	4010	GY	GO58CHH
4002	GY	GN08CGV	4005	GY	GN08CGZ	4008	GY	GO58CHF			

4011-4022

ADL Dart 4 — ADL Enviro 200 — N38F — 2008

4011	ME	GN58BTO	4014	ME	GN58BTX	4017	GU	GN58BUA	4020	GU	GN58BUH
4012	ME	GN58BTU	4015	ME	GN58BTY	4018	GU	GN58BUE	4021	GU	GN58BUJ
4013	ME	GN58BTV	4016	ME	GN58BTZ	4019	GU	GN58BUF	4022	GU	GN58BUO

4023-4035

ADL Dart 4 — ADL Enviro 200 — N32D* — 2009 — *seating varies

4023	DA	GN58BUP	4027	DA	GN58LVB	4030	DA	GN09AVX	4033	DA	GN09AWA
4024	DA	GN58BUU	4028	DA	GN09AVV	4031	DA	GN09AVY	4034	DA	GN09AWB
4025	DA	GN58BUV	4029	DA	GN09AVW	4032	DA	GN09AVZ	4035	DA	GN09AWC
4026	DA	GN58LVA									

Several of the Enviro 200 vehicles carry London livery for work on Transport for London contracts. Passing through Crayford is 4032, GN09AVZ. *Dave Heath*

4036-4059

			ADL Dart 4			ADL Enviro 200		N38F	2009		
4036	NF	GN09AWG	4042	NF	GN09AWR	4048	GI	GN09AWZ	4054	GI	GN09AXG
4037	NF	GN09AWH	4043	NF	GN09AWU	4049	GI	GN09AXA	4055	GI	GN09AXH
4038	NF	GN09AWJ	4044	GI	GN09AWV	4050	GI	GN09AXB	4056	GI	GN09AXJ
4039	NF	GN09AWM	4045	GI	GN09AWW	4051	GI	GN09AXC	4057	GI	GN09AXK
4040	NF	GN09AWO	4046	GI	GN09AWX	4052	GI	GN09AXD	4058	GI	GN09AXM
4041	NF	GN09AWP	4047	GI	GN09AWY	4053	GI	GN09AXF	4059	GI	GN09AXO

4060-4067

			ADL Dart 4			ADL Enviro 200		N38F	2009		
4060	ME	GN59FVB	4062	ME	GN59FVD	4064	ME	GN59FVF	4066	ME	GN59FVH
4061	ME	GN59FVC	4063	ME	GN59FVE	4065	ME	GN59FVG	4067	ME	GN59FVJ

5213	ME	N713TPK	Dennis Dominator DDA2006	East Lancs	B45/31F	1996	
5214	ME	N714TPK	Dennis Dominator DDA2006	East Lancs	B45/31F	1996	
5215	ME	N715TPK	Dennis Dominator DDA2006	East Lancs	B45/31F	1996	
5402	SE	H262GEV	Leyland Olympian ON2R50G13Z4	Leyland	B47/31F	1990	
5403	SE	H263GEV	Leyland Olympian ON2R50G13Z4	Leyland	B47/31F	1990	
5404	TO	H264GEV	Leyland Olympian ON2R50G13Z4	Leyland	BC43/29F	1990	
5405	TO	H265GEV	Leyland Olympian ON2R50G13Z4	Leyland	BC43/29F	1990	

5434-5441

			Dennis Trident			Alexander ALX400		N47/31F	2000		
5434	ME	W434XKX	5436	ME	W436XKX	5438	ME	W438XKX	5441	ME	W441XKX
5435	ME	W435XKX	5437	ME	W437XKX	5439	ME	W439XKX			

5768	SR	H768EKJ	Leyland Olympian ON2R50C13Z4	Northern Counties	B47/30F	1991	Boro'line, Maidstone, 1992
5769	SR	H769EKJ	Leyland Olympian ON2R50C13Z4	Northern Counties	B47/30F	1991	Boro'line, Maidstone, 1992

5906-5910

			Leyland Olympian ON2R50C13Z4 Northern Counties Palatine					B45/30F	1993		
5906	TO	K906SKR	5908	TW	K908SKR	5909	TW	K909SKR	5910	ME	K910SKR
5907	SR	K907SKR									

5911-5925 — Volvo Olympian — Northern Counties Palatine — B47/30F — 1994-95 — 5913 re-bodied 1995

5911	TW	M911MKM	5915	ME	M915MKM	5918	TW	M918MKM	5922	TW	M922PKN
5913	TW	M913MKM	5916	ME	M916MKM	5919	ME	M919MKM	5923	TW	M923PKN
5914	ME	M914MKM	5917	TW	M917MKM	5920	ME	M920MKM	5925	GU	M925PKN

5926-5937 — Volvo Olympian — Northern Counties Palatine — B47/30F — 1997

5926	TW	P926MKL	5929	TW	P929MKL	5932	TW	P932MKL	5935	SE	P935MKL
5927	TW	P927MKL	5930	ME	P930MKL	5933	ME	P933MKL	5936	SE	P936MKL
5928	TW	P928MKL	5931	ME	P931MKL	5934	SE	P934MKL	5937	SE	P937MKL

5938	GI	P609CAY	Volvo Olympian	Northern Counties Palatine	B47/29F	1996	Arriva Midlands, 2009
5939	NF	P610CAY	Volvo Olympian	Northern Counties Palatine	B47/29F	1996	Arriva Midlands, 2009
5940	NF	P613CAY	Volvo Olympian	Northern Counties Palatine	B47/29F	1996	Arriva Midlands, 2009
5941	TW	R637MNU	Volvo Olympian	Northern Counties Palatine	B47/29F	1998	Arriva Midlands, 2009
5942	TW	R639MNU	Volvo Olympian	Northern Counties Palatine	B47/29F	1998	Arriva Midlands, 2009
6210	TW	R210CKO	DAF DB250	Northern Counties Palatine 2	B43/24D	1998	
6211	TW	R211CKO	DAF DB250	Northern Counties Palatine 2	B43/24D	1998	
6212	TW	R212CKO	DAF DB250	Northern Counties Palatine 2	B43/24D	1998	

6213-6219 — DAF DB250 — Wrightbus Pulsar Gemini — N43/21D — 2004

6213	DA	GK53AOA	6215	DA	GK53AOC	6217	DA	GK53AOE	6219	DA	GK53AOG
6214	DA	GK53AOB	6216	DA	GK53AOD	6218	DA	GK53AOF			

6224-6236 — DAF DB250 10.2m — Alexander ALX400 — N43/20D — 2000-01 — Arriva London, 2006

6224	DA	X457FGP	6228	DA	Y461UGC	6231	DA	Y464UGC	6234	DA	Y467UGC
6225	DA	X458FGP	6229	DA	Y462UGC	6232	DA	Y465UGC	6235	DA	Y468UGC
6226	DA	X459FGP	6230	DA	Y463UGC	6233	DA	Y466UGC	6236	DA	Y469UGC
6227	DA	Y451UGC									

6237-6240 — DAF DB250 — East Lancs Olympus — N51/28F — 2008

6237	ME	YJ57BKD	6238	ME	YN57BKE	6239	ME	YN57BKF	6240	ME	YN57BKG

6401-6449 — Volvo B7TL — TransBus ALX400 — N45/27F — 2004

6401	GI	GN04UDM	6413	SR	GN04UEC	6426	GI	GN04UET	6438	TW	GN04UFG
6402	GU	GN04UDP	6414	SR	GN04UED	6427	GI	GN04UEU	6439	TW	GN04UFH
6403	GU	GN04UDS	6415	GI	GN04UEE	6428	GI	GN04UEV	6440	GI	GN04UFJ
6404	ME	GN04UDT	6417	GI	GN04UEG	6429	GI	GN04UEW	6441	GI	GN04UFK
6405	ME	GN04UDU	6418	GI	GN04UEH	6430	GI	GN04UEX	6442	GI	GN04UFL
6406	ME	GN04UDV	6419	GI	GN04UEJ	6431	GI	GN04UEY	6443	GI	GN04UFM
6407	ME	GN04UDW	6420	GI	GN04UEK	6432	GI	GN04UEZ	6444	TW	GN04UFP
6408	ME	GN04UDX	6421	GI	GN04UEL	6433	GI	GN04UFA	6445	TW	GN04UFR
6409	ME	GN04UDY	6422	GI	GN04UEM	6434	GI	GN04UFB	6446	ME	GN04UFS
6410	SR	GN04UDZ	6423	GI	GN04UEP	6435	GI	GN04UFC	6447	ME	GN04UFT
6411	SR	GN04UEA	6424	GI	GN04UER	6436	GI	GN04UFD	6448	ME	GN04UFU
6412	SR	GN04UEB	6425	GI	GN04UES	6437	GI	GN04UFE	6449	ME	GN04UFV

6450-6457 — ADL Trident 2 — ADL Enviro 400 — N47/33F — 2008

6450	GI	GN58BSO	6452	GI	GN58BSV	6454	GI	GN58BSY	6456	GI	GN58BTE
6451	GI	GN58BSU	6453	GI	GN58BSX	6455	GI	GN58BSZ	6457	GI	GN58BTF

7624-7643 — Volvo Citybus B10M-50 — Northern Counties — B45/31F — 1989 — Londonlinks, 1997

7624	GI	G624BPH	7630	GI	G630BPH	7633	GI	G633BPH	7635	GI	G635BPH
7629	GI	G629BPH	7631	GI	G631BPH	7634	GI	G634BPH	7636	GI	G636BPH

Ancillary vehicles:

T4	MEt	L506CPJ	Volvo B6-9.9M	Plaxton Pointer	TV	1994	
T5	MEt	L507CPJ	Volvo B6-9.9M	Plaxton Pointer	TV	1994	
T7	MEt	L509CPJ	Volvo B6-9.9M	Plaxton Pointer	TV	1994	
T14	MEt	M525MPM	Dennis Dart 9.8m	East Lancs EL2000	B40F	1995	

T205-212 — Volvo B6-9.9M — Northern Counties Paladin — TV — 1994 — Londonlinks, 1997

201	GI	L201YCU	208	GI	L208YCU	211	DA	L211YCU	212	GI	L212YCU
207	ME	L207YCU	210	GI	L210YCU						

T605	MEt	L605EKM	Volvo B6-9.9M	Plaxton Pointer	TV	1994	

Previous registrations:

SCZ9651	99D81499, T178AUA	SCZ9652	99D81498, T179AUA

The arrival in 2008 at Gillingham of eight Alexander Dennis Trident 2s with Enviro 400 bodywork allowed a similar number of the 2004 Volvo buses to be reallocated. The new intake included 6450, GN58BSO, seen here lettered for route 101. *Mark Lyons*

Depots and allocations:

Dartford (Central Road) - DA

Dart SLF	1624	1625	1626	1627	1628	1629	1630	1631
	1632	3291	3292	3293	3294	3295	3296	3297
	3298	3299	3301	3302	3303			
Dart 4	1643	1644	1645	1646	1647	1648	1649	3982
	3983	3984	3985	3986	3987	3988	3989	3990
	3991	3992	3993	3994	3995	3996	4023	4024
	4025	4026	4027	4028	4029	4030	4031	4032
	4033	4034	4035					
DAF/VDL SB120	3945	3946	3947	3948	3949	3950	3951	3952
	3953	3954	3955	3956	3957			
DAF/VDL DB250	6213	6214	6215	6216	6217	6218	6219	6224
	6225	6226	6227	6228	6229	6230	6231	6232
	6233	6234	6235	6236	DLA228			
Ancillary	*T211*							

Gatwick (Central Road) - GW - Easybus colours

Mercedes-Benz	304	309	310	311	312	313	314	321
	322	323	324	325	326	327	328	

Gillingham (Nelson Road) - GI

Mercedes-Benz	1118	1119						
Dart SLF	1610	1611	1612	1613	1614	1615	1616	1617
	3038	3050	3051	3053	3054	3070	3071	3072
	3073	3075	3076	3077	3178	3188	3189	3190
	3191	3198	3219	3220	3221	3223	3224	3225
	3226	3227	3228	3229	3230	3231	3232	3233
	3234	3235	3236	3237	3238	3239	3240	3241
	3242	3243	3244	3245	3246			
Dart 4	4044	4045	4046	4047	4048	4049	4050	4051
	4052	4053	4054	4055	4056	4057	4058	4059
Volvo Olympian	5938							
Volvo Citybus	7624	7629	7630	7631	7633	7634	7635	7636
Volvo B7TL	6401	6415	6417	6418	6419	6420	6421	6422
	6423	6424	6425	6426	6427	6428	6429	6430
	6431	6432	6433	6434	6435	6436	6437	6440
	6441	6442	6443					
Trident 2	6450	6451	6452	6453	6454	6455	6456	6457
Ancillary	*T207*	*T208*	*T210*	*T212*				

Grays (Europa Park, London Road) - GY

SB120 Cadet	3500	3501	3502	3503	3504	3505	3506	3507
	3508	3509	3510	3511	3512	3513		
Dart 4	3971	3972	3973	3997	3999	4000	4001	4002
	4003	4004	4005	4006	4007	4008	4009	4010

Guildford (Leas Road) - GU

Dart SLF	3023	3024	3025	3026	3028	3030	3031	3033
	3035	3036	3088	3090	3091	3092	3093	3094
	3095	3096	3097	3098	3099	3100	3101	3102
	3109	3110	3112	3113	3187	3304	3305	3306
	3591	3592						
DAF/VDL SB120	3923	3924	3925	3926	3927	3928	3929	3930
	3931	3932	3933	3934	3935	3937	3938	3939
	3940	3941	3942	3943	3944	3970		
Dart 4	4017	4018	4019	4020	4021	4022		
Mercedes-Benz O530	3901	3902	3903					
Volvo B7RLE	3731	3732	3733	3734	3735			
Volvo Olympian	5925							
Volvo B7TL	6402	6403						
Ancillary	*T605*							

Maidstone (Armstrong Road) - ME

Dart	3175	3408	3409	3410			
Dart SLF	3084	3176	3177	3194	3195	3202	3203	3205
	3207	3208	3209	3213	3215	3249	3429	3701
	3702	3703	3704	3705	3706			
Dart 4	4011	4012	4013	4014	4015	4016	4060	4061
	4062	4063	4064	4065	4066	4067		
Volvo B6	3610	3612						
DAF/VDL SB120	3958	3959	3960					
DAF/VDL SB200	3961	3962	3963	3964	3965	3966	3967	3968
	3969							
Dominator	5213	5214	5215					
OLympian	5910	5914	5915	5916	5919	5920	5930	5931
	5933							
Trident	5434	5435	5436	5437	5438	5439	5441	
DAF/VDL SB250	6237	6238	6239	6240				
Volvo B7TL	6404	6405	6406	6407	6408	6409	6446	6447
	6448	6449						

Northfleet (London Road) - NF

Mercedes-Benz	1122							
Optare Solo	1504	1505	1506					
Dart	3008	3404	3407					
Dart SLF	1606	1607	1618	1619	1620	1621	1622	1623
	3111	3124	3197	3198	3261	3263	3267	3268
	3272	3273	3274	3275	3283	3284	3285	3286
	3288							
Dart 4	4036	4037	4038	4039	4040	4041	4042	4043
VDL Bus SB200	3751	3752	3753	3754	3755	3756	3757	3758
	3759	3760	3761	3762				
Volvo B7RLE	3801	3802	3803	3804	3805	3806	3807	3808
	3809	3810	3811	3812	3813	3814	3815	3816
	3817	3818	3819	3820	3821	3822	3823	3824
	3825	3826						
Olympian	5939	5940						

Sheerness (Bridge Road) - SR

Mercedes-Benz	1120	1121	1172					
Dart SLF	1633	1634	1635	1636	3027	3029	3037	3199
	3201	3216	3247					
Olympian	5768	5769	5907					
Volvo B7TL	6410	6411	6412	6413	6414			

Southend (Short Street) - SE

Dart SLF	1608	1609	3086	3123	3172	3181	3186	3204
	3276	3277	3278	3279	3280	3281	3289	3307
	3308	3318	3320	3321	3322	3323	3324	3325
	3326	3387	3388	3389	3390	3391	3392	3393
	3394	3395	3396	3397	3400	3401	3402	3403
	3421	3423	3425	3426	3427	3428	3431	
Scania L113	3250	3251	3252	3253	3254	3255	3256	3257
	3258	3259						
DAF/VDL SB220	3911	3912	3916	3918	3919			
Olympian	5402	5403	5934	5935	5936	5937		

Ancillary   T603

Tonbridge (Cannon Lane) - New Enterprise - TO

Easybus sprinters are housed at Gatwick

Merced-Benz 709	1190							
Setra coach	2000							
DAF/VDL SB3000	2889	2890	2894	2895	2896	2898	2903	2904
	2905	2906	2907	2908				
Dart	3039	3055						
Volvo B6	3617	3619						
Volvo B7 coach	2853	2854						
Javelin coach	2855							
Volvo B10M coach	2852							
Olympian	5404	5405	5906					

Tunbridge Wells (St John's Road) - TW

Optare Solo	1501	1502	1503					
Dart	3411	3412						
Dart SLF	1601	1602	1603	1604	1605	3046	3047	3089
	3184	3185	3192	3193	3196	3206	3218	3262
	3264	3265	3266	3269	3270	3271	3282	3287
	3411	3412	3595					
Dart 4	1637	1638	1639	1640	1641	1642		
Lance	3020	3021	3022	3384				
DAF/VDL SB220	3913	3914	3915	3921				
Olympian	5908	5909	5911	5913	5918	5922	5923	5926
	5927	5928	5929	5932	5941	5942		
DAF/VDL DB250	6210	6211	6212					
Volvo B7TL	6438	6439	6444	6445				

Unallocated and stored - u/w

Remainder

TELLINGS - GOLDEN MILLER

Tellings Golden Miller Group plc; Tellings-Golden Miller Coaches Ltd,
Ensign Close, Exeter Way, London Heathrow Airport, TW6 2PQ
Burton's Coaches, Dudery Hill, Haverhill, CB9 8DR
Classic Coaches, Classic House, Morrison Road, Stanley, DA9 7RX
Excel Passenger Logistics, Monometer House, Rectory Road, Leigh-on-Sea, SS2 9HN
Flight Delay Services, Commonwealth House, Chicago Avenue, Wythenshaw, M90 3FL
Link Coaches, 1 Wrottesley Road, Harlesden, NW10 5XA
OFJ Connections, 16300 Electra Avenue, London Heathrow Airport, Hounslow, TW6 2DN
Network Colchester, 5 Whitehall Industrial Estate, Grange Way, Colchester, CO2 8GU

Tellings-Golden Miller

Reg	Chassis	Body	Seating	Year	Notes
P10TGM	Volvo B10M-62	Van Hool Alizée HE	C32FT	1997	
R957RCH	Volvo B9M	Plaxton Première 320	C43F	1997	Airlinks, Feltham, 2005
R958RCH	Volvo B9M	Plaxton Première 320	C43F	1997	Airlinks, Feltham, 2005
R959RCH	Volvo B9M	Plaxton Première 320	C43F	1997	Airlinks, Feltham, 2005
R177TKU	Volvo B10M-62	Plaxton Première 350	C49FT	1997	Link Line, Harlesden, 2003
R50TGM	Volvo B10M-62	Plaxton Première 320	C53F	1998	
W393OUF	Toyota Coaster BB50R	Caetano Optimo IV	C22F	2000	Airlinks, Feltham, 2005
W40TGM	Volvo B10M-62	Plaxton Paragon	C53F	2000	Burton's, Haverhill, 2003
W50TGM	Volvo B10M-62	Plaxton Paragon	C53F	2000	
W60TGM	Volvo B10M-62	Plaxton Paragon	C53F	2000	Burton's, Haverhill, 2006
W80TGM	Volvo B10M-62	Plaxton Première 320	C57F	2000	
X152ENJ	Mercedes-Benz Vario 0814	Plaxton Cheetah	C24F	2001	Airlinks, Feltham, 2005
X153ENJ	Mercedes-Benz Vario 0814	Plaxton Cheetah	C24F	2001	Airlinks, Feltham, 2005
X154ENJ	Mercedes-Benz Vario 0814	Plaxton Cheetah	C24F	2001	Airlinks, Feltham, 2005
X157ENJ	Mercedes-Benz Vario 0814	Plaxton Cheetah	C24F	2001	Airlinks, Feltham, 2005
X158ENJ	Mercedes-Benz Vario 0814	Plaxton Cheetah	C24F	2001	Airlinks, Feltham, 2005
X159ENJ	Mercedes-Benz Vario 0814	Plaxton Cheetah	C24F	2001	Airlinks, Feltham, 2005
X216HCD	Mercedes-Benz Vario 0814	Plaxton Cheetah	C24F	2001	Airlinks, Feltham, 2005
X217HCD	Mercedes-Benz Vario 0814	Plaxton Cheetah	C24F	2001	Airlinks, Feltham, 2005
X218HCD	Mercedes-Benz Vario 0814	Plaxton Cheetah	C24F	2001	Airlinks, Feltham, 2005
X219HCD	Mercedes-Benz Vario 0814	Plaxton Cheetah	C24F	2001	Airlinks, Feltham, 2005
X221HCD	Mercedes-Benz Vario 0814	Plaxton Cheetah	C24F	2001	Airlinks, Feltham, 2005
X223HCD	Mercedes-Benz Vario 0814	Plaxton Cheetah	C24F	2001	Airlinks, Feltham, 2005
Y501TGJ	Volvo B10M-62	Plaxton Paragon	C53F	2001	
Y504TGJ	Volvo B10M-62	Plaxton Paragon	C53F	2001	
Y506TGJ	Volvo B10M-62	Plaxton Paragon	C53F	2001	
Y507TGJ	Volvo B10M-62	Plaxton Paragon	C53F	2001	
Y10TGM	Iveco EuroRider 391E.12.35	Beulas Stergo ε	C49FT	2001	
Y15TGM	Iveco EuroRider 391E.12.35	Beulas Stergo ε	C49FT	2001	
Y20TGM	Volvo B10M-62	Plaxton Panther	C53F	2001	
Y30TGM	Volvo B10M-62	Plaxton Panther	C53F	2001	
YN51WGY	Volvo B10M-62	Plaxton Panther	C53F	2001	
YN51WGZ	Volvo B10M-62	Plaxton Panther	C53F	2001	
KP51UEV	Volvo B10M-62	Plaxton Panther	C49FT	2001	
KP51UEW	Volvo B10M-62	Plaxton Panther	C49FT	2001	
KP51UEX	Volvo B10M-62	Plaxton Panther	C49FT	2001	
KP51UEY	Volvo B10M-62	Plaxton Panther	C49FT	2001	
KP51UEZ	Volvo B10M-62	Plaxton Panther	C49FT	2001	
KP51SYF	Volvo B10M-62	Plaxton Première 350	C51FT	2002	
5141MW	Setra S315 GT-HD	Setra	C49FT	2002	
6764MW	Setra S315 GT-HD	Setra	C49FT	2002	
KU02YUF	Volvo B10M-62	Plaxton Paragon	C49FT	2002	
KU02YUG	Volvo B10M-62	Plaxton Paragon	C49FT	2002	
3262MW	Volvo B7R	Plaxton Profile	C53F	2003	
6963MW	Volvo B7R	Plaxton Profile	C53F	2003	
LB52UYK	Volvo B12M	Plaxton Panther	C49FT	2003	
GB03TGM	Setra S415 HD	Setra	C49FT	2003	
UK03TGM	Volvo B12B	Caetano Enigma	C48FT	2004	
MM03TGM	Volvo B12B	Caetano Enigma	C48FT	2004	
GB04TGM	Volvo B12M	Caetano Enigma	C48FT	2004	
UK04TGM	Volvo B12B	Caetano Enigma	C53F	2004	

Temsa was established in 1968 and builds vehicles in Turkey and has been supplying the British market with touring coaches in recent years. Tellings operates the 8.4 metre Opalin midi-coach which features an MAN engine. Pictured at the British Coach Rally is YJ08EFT. *Andrew Jarosz*

KX04HSJ	Volvo B12B	TransBus Paragon	C49FT	2004
YN54WWF	Volvo B12B	TransBus Panther	C49FT	2004
YN54WWG	Volvo B12B	TransBus Panther	C49FT	2004
GS05TGM	Volvo B12B	Van Hool T9 Alicron	C32FT	2005
YN55WSU	Volvo B12B	Plaxton Panther	C49FT	2005
YN55WSV	Volvo B12B	Plaxton Panther	C49FT	2005
YN55WSW	Volvo B12B	Plaxton Panther	C49FT	2005
YN06CJV	Scania K340 EB4	VDL Berkhof Axial 50	C55FT	2006
YN06CJX	Scania K340 EB4	VDL Berkhof Axial 50	C55FT	2006
YN06CJY	Scania K114 EB4	Irizar Century Style	C53F	2006
YN06CJZ	Scania K114 EB4	Irizar Century Style	C53F	2006
YN06PFD	Volvo B12M	Plaxton Panther	C50FT	2006
YN06PFE	Volvo B12M	Plaxton Panther	C50FT	2006
YN06PFF	Volvo B12M	Plaxton Panther	C50FT	2006
YN06PFG	Volvo B12M	Plaxton Panther	C50FT	2006
KC06EVN	Mercedes-Benz Vito 111cdi	Mercedes-Benz	M7	2006
KC06EVP	Mercedes-Benz Vito 111cdi	Mercedes-Benz	M7	2006

Web: www.tellingsgoldenmiller.co.uk

Network Colchester

101	CO	AY55DKA	Scania N94 UD	East Lancs OmniDekka	N47/33F	2005	
102	CO	YN06TFZ	Scania N94 UD	East Lancs OmniDekka	N47/33F	2006	
103	CO	YN08OBY	Scania N230 UD	East Lancs Olympus	N47/33F	2008	
104	CO	YN08OBY	Scania N230 UD	East Lancs Olympus	N47/33F	2008	
131	CO	T131AUA	DAF DB250	Plaxton President	B45/22F	1999	Wiltax, 2008
133	CO	T133AUA	DAF DB250	Plaxton President	B45/22F	1999	Wiltax, 2008
155	CO	T405SMV	Dennis Trident	East Lancs Lolyne	B45/31F	1999	Metrobus, Crawley, 2009
156	CO	T406SMV	Dennis Trident	East Lancs Lolyne	B45/31F	1999	Metrobus, Crawley, 2009
195	CO	W395RBB	Dennis Trident	Alexander ALX400	B51/31F	2000	Arriva North East, 2009
196	CO	W396RBB	Dennis Trident	Alexander ALX400	B51/31F	2000	Arriva North East, 2009
204	CO	H804RWJ	Scania N113 DRB	Northern Counties Palatine	B47/33F	1991	Excel,
213	CO	G613BPH	Volvo Citybus B10M-50	East Lancs	B49/39F	1989	Arriva Southern Counties, 2004

The Network Colchester operation falls with the Tellings group. Scania OmniDekka 102, YN06TFZ, illustrates the livery used. The OmniDekka was a joint venture between Scania and East Lancs prior to Scania suppling its own integral double-deck OmniCity. *Dave Heath*

214	CO	G614BPH	Volvo Citybus B10M-50	East Lancs	B49/39F	1989	Arriva Southern Counties, 2004	
215	CO	G615BPH	Volvo Citybus B10M-50	East Lancs	B49/39F	1989	Arriva Southern Counties, 2004	
217	CO	G617BPH	Volvo Citybus B10M-50	East Lancs	B49/39F	1989	Arriva Southern Counties, 2004	
237	CO	G37HKY	Scania N113 DRB	Northern Counties Palatine	B47/33F	1991	Excel,	
243	CO	G643BPH	Volvo B10M-50 Citybus	Northern Counties Palatine	B45/35F	1989	Arriva Southern Counties, 2008	
300	CL	W427CWX	Optare Solo M850	Optare	N27F	2000	Metrobus, Crawley, 2004	
302	CO	Y291PDN	Optare Solo M850	Optare	N27F	2001	Metrobus, Crawley, 2004	
305	CL	Y295PDN	Optare Solo M850	Optare	N27F	2001	Metrobus, Crawley, 2004	
307	CL	YJ51YWW	Optare Solo M850	Optare	N27F	2001		
429	CO	P429AHR	Optare Excel L1070	Optare	N36F	1997	Andybus, Dauntsey, 2005	
512	CO	HX51LRK	Dennis Dart SLF	Caetano Compass	N40F	2001		
513	CO	HX51LRL	Dennis Dart SLF	Caetano Compass	N40F	2001		
514	CO	HX51LRN	Dennis Dart SLF	Caetano Compass	N40F	2001		
515	CO	HX51LRO	Dennis Dart SLF	Caetano Compass	N40F	2001		

520-527			ADL Dart 10.7m	ADL Pointer	N37F	2004		

520	CO	SN54HWY	**522**	CO	SN54HXA	**524**	CO	SN54HXC	**526**	CO	SN54HXE
521	CO	SN54HWZ	**523**	CO	SN54HXB	**525**	CO	SN54HXD	**527**	CO	SN54HXF

528	CO	AY54FPZ	ADL Dart 10.7m	ADL Pointer	N37F	2004	
529	CO	AY54FRC	ADL Dart 10.7m	ADL Pointer	N37F	2004	

901-907			Volvo B10BLE	Alexander ALX300	N44F	2000	Tellings-Golden Miller, 2005

901	CO	W901UJM	**903**	CO	W903UJM	**905**	CO	W905UJM	**907**	CO	W907UJM
902	CO	W902UJM	**904**	CO	W904UJM	**906**	CO	W906UJM			

Previous registrations:

HX51LRK	HX51LRK, R60BCL	HX51LRN	HX51LRN, R70BCL
HX51LRL	HX51LRL, R80BCL	P446SWX	P446SWX
HX51LRO	HX51LRO, R90BCL		

127	BU	YK04KWF	Optare Solo M930	Optare	N33F	2004	*Operated for Addenbrookes Hospital*
128	BU	YK04KWG	Optare Solo M930	Optare	N33F	2004	*Operated for Addenbrookes Hospital*
136	BU	YM55RRX	Mercedes-Benz Vario O814	Plaxton Beaver 2	B29FL	2006	
137	BU	YM55RRY	Mercedes-Benz Vario O814	Plaxton Beaver 2	B29FL	2006	
218	BU	H683GPF	Volvo Citybus B10M-50	East Lancs	B45/35F	1991	Green Triangle, Liverpool, 2001
219	BU	G2190TV	Volvo Citybus B10M-50	Alexander RV	B47/32F	1990	Trent Barton, 1999
236	BU	F892BKK	Leyland Olympian ONCL10/1RZ	Leyland	B43/29F	1988	Arriva The Shires, 2009
23	BU	G371YUR	Leyland Olympian ONCL10/1RZ	Leyland	BC43/26F	1990	Armchair, Brentford.
239	BU	F639LMJ	Leyland Olympian ONCL10/1RZ	Leyland	B43/29F	1988	Arriva The Shires, 2009
240	BU	F640LMJ	Leyland Olympian ONCL10/1RZ	Leyland	B43/29F	1988	Arriva The Shires, 2009
246	BU	F246MTW	Leyland Olympian ONCL10/1RZ	Leyland	BC43/29F	1988	Arriva Southern Counties, 2004
255	BU	H155PVW	Leyland Olympian ON2R50C13Z4	Alexander	H47/33D	1991	Dublin Bus, 2008
307	BU	J807KHD	DAF SB220	Ikarus CitiBus	BC42F	1992	Matthews, Soytre, 2007
800	BU	N540TPF	Dennis Dart 9.8m	East Lancs EL2000	B40F	1996	Arriva Southern Counties, 2004
801	BU	N541TPF	Dennis Dart 9.8m	East Lancs EL2000	B40F	1996	Arriva Southern Counties, 2004
	BU	CCE993	MAN 24-350	Jonckheere Modulo	C53/15FT	1999	Stagecoach, 2007
	BU	X731DAU	Mercedes-Benz Vito 108	Mercedes-Benz	M8	2000	Classic, Stanley, 2007
	BU	W40BCL	Mercedes-Benz Vito 110 cdi	Mercedes-Benz	M8	2001	
	BU	Y188TDP	Citroen Relay	Citroen	M8	2001	
	BU	NL52XZV	DAF SB4000XF	Van Hool T9 Alizée	C49FT	2002	Arriva North East, 2006
	BU	NL52XZW	DAF SB4000XF	Van Hool T9 Alizée	C49FT	2002	Arriva North East, 2006
	BU	VX04JHY	LDV 400	LDV	M16	2004	
	BU	GB04BCL	Mercedes-Benz Vito 110 cdi	Mercedes-Benz	M8	2004	
	BU	LL04BCL	Bova Futura FHD14-430	Bova	C59FT	2004	
	BU	XL04BCL	Bova Futura FHD14-430	Bova	C59FT	2004	
	BU	YU04XFB	Volvo B12B	TransBus Panther Expressliner	C49FT	2004	
	BU	YU04XFC	Volvo B12B	TransBus Panther Expressliner	C49FT	2004	
	BU	YU04XFD	Volvo B12B	TransBus Panther Expressliner	C49FT	2004	
	BU	GO54BCL	Volvo B12B	Caetano Enigma	C53F	2004	
	BU	YN54DDJ	Volvo B12B	TransBus Panther	C51FT	2004	
	BU	YN54DDK	Volvo B12B	TransBus Panther	C51FT	2004	
	BU	YN54DDL	Volvo B12B	TransBus Panther	C51FT	2004	
	BU	YN54DDO	Volvo B12B	TransBus Panther	C49FT	2005	
	BU	YN54ZHK	Volvo B12B	TransBus Panther	C49FT	2005	
	BU	YN54ZHM	Volvo B12B	TransBus Panther	C49FT	2005	
	BU	YN05VRT	Volvo B12B	TransBus Panther	C49FT	2005	
	BU	YN05VRU	Volvo B12B	TransBus Panther	C49FT	2005	
	BU	GB05BCL	Volvo B12B	TransBus Panther	C49FT	2005	
	BU	EU05BCL	Volvo B12B	TransBus Panther	C49FT	2005	
	BU	UK05BCL	Volvo B12B	TransBus Panther	C49FT	2005	
	BU	LG05BCL	Bova Futura FHD14-430	Bova	C59FT	2005	
	BU	YN07LHE	Scania K124 TRB	Berkhof Axial 70	C59FT	2007	Classic, Stanley, 2009
	BU	DD08BCL	Volvo B12B	Van Hool Astrobel	C61/18CT	2008	
	BU	EU08BCL	Scania K124	Irizar Century	C49FT	2008	
	BU	GB08BCL	Scania K124	Irizar Century	C49FT	2008	
	BU	GO08BCL	Scania K124	Irizar Century	C49FT	2008	
	BU	UK08BCL	Scania K124	Irizar Century	C49FT	2008	
	BU	YN08BCL	Temsa Opalin	Temsa	C33F	2008	
	BU	YJ08EFT	Temsa Opalin	Temsa	C35FT	2008	
	BU	YJ09CYH	Van Hool T917 Astron	Van Hool	C--F	2009	

Previous registrations:

GB04BCL	AF04XGY	H155PVW	91D1094
CCE993	T57BBW	UK05BCL	WA05DFE
DD08BCL	3232MW	W40BCL	Y193HJN

Classic Coaches are based in north east England. Illustrating the Tellings-style livery applied in red is 8224, YN54AMO, a Scania K114 with Berkhof Axial 50 bodywork. *Mark Doggett*

Classic

Classic - Primrose - Hylton Castle - Moor-Dale

Classic Coaches (Continental) Ltd; Classic Buses (Stanley) Ltd; Moor-Dale Coaches Ltd; Hylton Castle Motors Ltd, Classic House, Morrison Rd, Stanley, DH9 7RX

4959	YD04MFJ	Mercedes-Benz Vito 109cdi	Traveliner	M	2004	
4960	YD04MFK	Mercedes-Benz Vito 109cdi	Traveliner	M	2004	
4962	YD04MFM	Mercedes-Benz Vito 109cdi	Traveliner	M	2004	
6110	BX54EBU	Mercedes-Benz Sprinter 416cdi	Koch	N20F	2004	
6111	BX54EBV	Mercedes-Benz Sprinter 416cdi	Koch	N20F	2004	
6133	NK04VMD	Mercedes-Benz Sprinter 411cdi	Koch	N20F	2004	
6213	T213BBR	Renault Master	Oughtred & Harrison	M10	1999	Durham Travel Services, 2002
6214	BX02CLO	Mercedes-Benz Sprinter 411cdi	Koch	N20F	2002	
6216	T216BBR	Renault Master	Oughtred & Harrison	M10	1999	Durham Travel Services, 2002
6217	T217BBR	Renault Master	Oughtred & Harrison	M10	1999	Durham Travel Services, 2002
6218	T218BBR	Renault Master	Oughtred & Harrison	M10	1999	Durham Travel Services, 2002
6219	T219BBR	Renault Master	Oughtred & Harrison	M10	1999	Durham Travel Services, 2002
6224	T224BBR	Renault Master	Oughtred & Harrison	M10	1999	Durham Travel Services, 2002
6225	T225BBR	Renault Master	Oughtred & Harrison	M10	1999	Durham Travel Services, 2002
6500	YJ02FKY	Optare Solo M920	Optare	N29F	2000	
6731	Y291PDN	Optare Solo M850	Optare	N27F	2001	Burton's, Haverhill, 2006
6733	Y293PDN	Optare Solo M850	Optare	N27F	2001	Burton's, Haverhill, 2006
6734	Y296PDN	Optare Solo M920	Optare	N27F	2001	Metrobus, Crawley, 2001
6735	Y294PDN	Optare Solo M920	Optare	N27F	2001	Metrobus, Crawley, 2001
6741	W441CWX	Optare Solo M920	Optare	N29F	2000	Burton's, Haverhill, 2006
6852	PN02LZM	LDV Convoy	Jaycas	M16	2002	
6854	PN02LZO	LDV Convoy	Jaycas	M16	2002	
6857	PN02LZR	LDV Convoy	Jaycas	M16	2002	
6865	PN02LZP	LDV Convoy	Jaycas	M16	2002	

6866	F166XCS	Mercedes-Benz 609D	Scott	C24F	1989	Hylton Castle, East Boldon, 1997
6902	J201JRP	Mercedes-Benz 811D	Plaxton Beaver	B27F	1991	MK Metro, 2000
6930	H430XGK	Mercedes-Benz 811D	Alexander Sprint	B28F	1991	MK Metro, 2003
6947	K947OEM	Mercedes-Benz 811D	Marshall C16	B27F	1993	MTL (North), 1998
6948	K948OEM	Mercedes-Benz 811D	Marshall C16	B27F	1993	MTL (North), 1998
7801	W901UJM	Volvo B10BLE	Alexander ALX300	N44F	2000	Burton's, Haverhill, 2006
7802	W902UJM	Volvo B10BLE	Alexander ALX300	N44F	2000	Burton's, Haverhill, 2006
7804	N204LCK	Optare Excel L1070	Optare	N36F	1996	Ludlows, Halesowen, 2005
7834	G234BRT	Leyland Tiger TRCTL11/3RZM	Plaxton Derwent 2	S67F	1989	Burton's, Haverhill, 2005
7835	G235BRT	Leyland Tiger TRCTL11/3RZM	Plaxton Derwent 2	S67F	1989	Burton's, Haverhill, 2005
7901	S10BCL	Volvo Olympian	East Lancs	B47/29F	1998	Burton's, Haverhill, 2006
7902	S20BCL	Volvo Olympian	East Lancs	B47/29F	1998	Burton's, Haverhill, 2006
7937	B737GCN	Leyland Olympian ONCL10/1RV	Eastern Coach Works	B46/29F	1985	Go North East, 2005
7948	GSU348	MCW Metrobus DR102/53	MCW	B42/29F	1986	Burton's, Haverhill, 2001
8003	W3CLA	Volvo B7	Plaxton Prima	C57F	2000	
8004	AA03CLA	Neoplan Skyliner N122/3	Neoplan	C57/20CT	2003	
8005	S5CLA	Volvo B10M-62	Plaxton Première 350	C49FT	1999	
8006	S6CLA	Volvo B10M-62	Plaxton Première 350	C49FT	1999	
8008	593CCE	Volvo B10M-62	Plaxton Paragon	C49FT	2000	Burton's, Haverhill, 2006
8009	W900BCL	Volvo B10M-62	Plaxton Paragon	C51FT	2000	Burton's, Haverhill, 2006
8012	Y12CLA	Volvo B7	Plaxton Prima	C57F	2000	
8013	Y13CLA	Volvo B7	Plaxton Prima	C57F	2000	
8020	G2PGL	Scania L94IB4	Irizar Century 12.35	C49FT	1999	Bus Eireann, 2003
8023	WA56ENN	Volvo B12B	Van Hool Astrobel	C61/16FT	2006	
8030	FJ06ZKK	Volvo B12B	Berkhof Axial 50	C55FT	2006	
8031	FJ06ZKL	Volvo B12B	Berkhof Axial 50	C55FT	2006	
8033	YN55KZZ	Scania K114EB6	Irizar PB	C34FT	2005	
8041	CC03HOL	Volvo B10M-62	Caetano Enigma	C49FT	2003	
8042	CC53HOL	Scania K114IB4	Irizar Century 12.35	C49FT	2004	
8054	NK51ZSR	Volvo B7	Caetano Enigma	C49FT	2001	
8056	NK51ZST	Volvo B7	Caetano Enigma	C49FT	2001	
8057	NK51ZSU	Volvo B7	Caetano Enigma	C49FT	2001	
8098	GO02STS	Volvo B10M-62	Caetano Enigma	C49FT	2002	
8110	GO02CLA	Scania K124EB4	Irizar Century 12.35	C49FT	2002	
8111	GO53CLA	Scania K114IB4	Irizar Century 12.35	C49FT	2003	
8112	FJ51JYN	Volvo B10M-62	Caetano Enigma	C49FT	2001	
8120	GO03CLA	Scania K114IB4	Irizar Century 12.35	C49FT	2003	
8185	685XHY	DAF SB3000	Van Hool Alizée H	C55F	1993	Go-Ahead, 1998
8201	CC04MAL	Scania K114IB4	Irizar InterCentury 12.32	C49FT	2004	
8204	YN54AKO	Scania K114IB4	Berkhof Axial 50	C55F	2005	
8224	YN54AMO	Scania K114IB4	Berkhof Axial 50	C55F	2005	
8230	YN54ANU	Scania K114IB4	Berkhof Axial 50	C55F	2005	
8261	YN54AGV	Scania K114IB4	Berkhof Axial 50	C55F	2005	
8290	YN06CJU	Scania K114IB4	Berkhof Axial 50	C55F	2006	
8294	YN54AJO	Scania K114IB4	Berkhof Axial 50	C55F	2005	
8433	TJI1683	Volvo B10M-61	Plaxton Supreme VI	C53F	1982	Wickson, Walsall Wood, 1997
8442	YN53OYW	Scania K114IB4	Irizar Century 12.35	C49F	2004	
8443	YN53OYX	Scania K114IB4	Irizar Century 12.35	C49F	2004	
8445	YN53OYZ	Scania K114IB4	Irizar Century 12.35	C49F	2004	
8453	YN53OZH	Scania K114IB4	Irizar Century 12.35	C49F	2004	
8455	YN53OZP	Scania K114IB4	Irizar Century 12.35	C49F	2004	
8457	YN53OZR	Scania K114IB4	Irizar Century 12.35	C49F	2004	
8701	WSV571	Volvo B10M-62	Berkhof Axial 50	C40FT	1999	
8711	WSV572	Volvo B10M-62	Berkhof Axial 50	C40FT	1997	
8887	656CCE	Volvo B10M-62	Plaxton Excalibur	C49FT	1998	
8889	R9CLA	Volvo B10M-62	Plaxton Première 350	C49FT	1998	
8942	V142EJR	DAF SB3000	Van Hool T9 Alizée	C44FT	1999	Arriva North East, 2006
8943	X143WNL	DAF SB3000	Van Hool T9 Alizée	C49FT	2000	Arriva North East, 2006
8944	X144WNL	DAF SB3000	Van Hool T9 Alizée	C49FT	2000	Arriva North East, 2006
8947	NL52XZX	DAF SB4000XF	Van Hool T9 Alizée	C49FT	2002	Arriva North East, 2006
8948	NL52XZY	DAF SB4000XF	Van Hool T9 Alizée	C49FT	2002	Arriva North East, 2006

Several of the Burton's fleet operate on National Express contracts. YU04XFB is one of three Volvo B12Bs added to the fleet on 2004. Its Panther Expressliner body was built during the time when Plaxton was part of TransBus. Carrying the latest livery it is generally found on the Cambridge service. *Mark Doggett*

Previous registrations:

593CCE	W800BCL	S10BCL	S849DGX
656CCE	R7CLA	S20BCL	S852DGX
G2PGL	99D9036, S354SET	TJI1683	UCX429X
G234BRT	03KJ43	W3CLA	W3CLA, V111ACH
G235BRT	03KJ44	WSV571	V1NFC
GSU348	C955DWJ	WSV572	V1SFC
P429AHR	P446SWX, TIL6877, N30ARJ		

Web: www.classic-coaches.co.uk

Link Line

Link Line Coaches Ltd, 1 Wrottesley Road, Harlesden, NW10 5XA

R174VBM	Mercedes-Benz Vario O810	Plaxton Beaver 2	B27F	1997	Arriva Southern Counties, 2004
T576FFC	Mercedes-Benz 308D	Mercedes-Benz	M12	1999	Mudi-Bond, Cassington, 2002
RL51ZKR	Dennis Dart SLF 9m	Caetano Nimbus	N29F	2002	
RL51ZKS	Dennis Dart SLF 9m	Caetano Nimbus	N29F	2002	
CE52UWW	Optare Solo M850	Optare	N29F	2002	Bebb, Llantwit Fardre, 2006
RN52EYH	Dennis Dart SLF 9m	Caetano Nimbus	N29F	2003	
RN52EYJ	Dennis Dart SLF 9m	Caetano Nimbus	N29F	2003	
GB03LLC	Volvo B12B	Caetano Levante	C48FT	2003	
UK03LLC	Volvo B12B	Caetano Levante	C48FT	2003	
GB04LLC	Volvo 7R	Plaxton Prima	C53F	2004	
HX04HUH	Dennis Dart SLF 9m	Caetano Nimbus	N29F	2004	
HX04HUK	Dennis Dart SLF 9m	Caetano Nimbus	N29F	2004	
MX54ZVA	Mercedes-Benz 313cdi	Mercedes-Benz	M16	2004	

ARRIVA SKANDINAVIEN

Arriva Danmark A/S; Arriva Scandinavia A/S
Herstedvang 7C, DK-2650 Albertslund, Danmark

1001-1013   Volvo B10LA   Säffle   AN44D   1998

1001	u	PC90745	1004	u	PC95836	1007	u	PC95849	1010	HD	PC95863
1002	u	PC95805	1005	KK	PC95842	1008	HD	PC95862	1011	HD	PC95873
1003	u	PC95830	1006	HD	PC95843	1009	HD	PC95872	1013	u	PC95892

1014-1027   Volvo B10BLE   Åbenrå   N43D   1998

1014	EJ	PJ88.334	1018	EJ	PJ88.338	1022	RK	PJ88.342	1025	RK	PJ88.345
1015	EJ	PJ88.335	1019	EJ	PJ88.339	1023	RK	PJ88.343	1026	RK	PJ88.346
1016	EJ	PJ88.336	1020	EJ	PJ88.340	1024	RK	PJ88.344	1027	RK	PJ88.347
1017	EJ	PJ88.337									

1028	RK	PJ97769	Volvo B10BLE	Säffle	N43D	1998
1029	RK	PJ97770	Volvo B10BLE	Säffle	N43D	1998

1084-1147   DAB Citibus S15 (LPG)   DAB   N43D   1998

1084	RG	PE97340	1096	RG	PE97352	1109	u	PJ96905	1137	u	PM95455
1086	RG	PE97342	1097	RG	PE97359	1110	HO	PJ96906	1138	u	PM95490
1087	RG	PE97343	1098	KK	PE97424	1127	RG	PL96059	1139	RG	PM95491
1088	RG	PE97344	1099	HO	PE97444	1128	RG	PL96060	1140	RG	PM95507
1089	RG	PE97345	1100	RG	PE97445	1129	RG	PL96061	1141	RG	PM95508
1090	RG	PE97346	1102	HO	PJ96813	1130	RG	PM95387	1142	RG	PM95509
1091	RG	PE97347	1103	u	PJ96814	1131	RG	PM95388	1143	RG	PP94068
1092	RG	PE97348	1104	RG	PJ96815	1132	RG	PM95408	1144	RG	PP94069
1093	ΠG	PE97349	1105	KK	PJ96860	1133	RG	PM95409	1145	RG	PP94070
1094	RG	PE97350	1106	KK	PJ96861	1134	RG	PM95410	1146	RG	PP94071
1095	RG	VZ89061	1108	KK	PJ96904	1135	u	PM95453	1147	RG	PP94072

1157-1171   Scania L113CLL   Berkhof   N43D   1998-99

1157	KK	PP94306	1160	RI	PP94326	1164	RI	PR93210	1170	GX	PR93374
1159	KK	PP94325	1162	ES	PP94356	1167	GX	PR93314	1171	RI	PR93328

1172-1181   DAB Citibus S15 (LPG)   DAB   N43D   1999

1172	KK	PP94305	1174	u	PZ90196	1177	u	PZ89838	1181	RG	PZ89894
1173	KK	PZ90195	1176	RG	PZ89723	1180	RG	PZ89893			

1187-1219   DAB Citybus S15   Silkeborg   N31D   1997-99

1187	HO	OY91905	1195	RG	OY91994	1203	RG	PE88119	1212	RG	PE88128
1188	RG	OY91906	1196	HO	OY91995	1204	RG	PE88120	1213	RG	PE97294
1189	RG	OY91770	1197	GX	OY91996	1205	RG	PE88121	1214	RG	PL96188
1190	RG	OY91908	1198	GX	OY91997	1206	RG	PE88122	1215	GX	PL96230
1191	RG	OY91910	1199	VA	OZ91742	1207	RG	PE88123	1216	GX	PM95289
1192	RG	OY91912	1200	VA	OZ91743	1208	RG	PE88124	1217	RG	PP94098
1193	RG	OY91917	1201	VA	OZ91744	1209	RG	PE88125	1218	RG	PZ89777
1194	RG	OY91918	1202	RG	PC97856	1211	RG	PE88127	1219	RG	PZ89909

1220	GX	RH90875	VDL Bus 2000LF gas	Berkhof	N33F	2000

1221-1235   VDL Bus SB4000   VDL Jonckheere   N33D   2005

1221	GX	TT94215	1225	GX	TT94338	1229	GX	TT94342	1233	GX	TT94346
1222	GX	TT94216	1226	GX	TT94339	1230	GX	TT94343	1234	GX	TT94347
1223	GX	TT94214	1227	GX	TT94340	1231	GX	TT94344	1235	GX	TT94348
1224	GX	TT94337	1228	GX	TT94341	1232	GX	TT94345			

1236	KO	UB95799	Volvo B12BLE	Volvo 8500	N33D	2006
1237	GX	UK96354	VDL Bus SB4000	VDL Jonckheere	N33D	2006

1238-1249   Volvo B7BLE   Volvo 8500   N33D   2006

1238	RG	UR90587	1241	RG	UR90657	1244	RG	UR90690	1247	RG	UR90693
1239	RG	UR90588	1242	RG	UR90669	1245	RG	UR90691	1248	RG	UR90713
1240	RG	UR90656	1243	RG	UR90670	1246	RG	UR90692	1249	RG	UR90714

Arriva Skandinavien fleet now exceeds 2430 buses since gaining contracts in Stockholm and the Storstrøm region of Denmark. The former HT branding has now been removed from 1106, PJ96861, pictured here on Copenhagen city service 10. *Bill Potter*

1250-1271 — Volvo B12BLE — Volvo — N33D — 2004

1250	RK	TD91367	1256	RK	TJ97103	1262	RK	TJ97844	1267	RK	TJ97896
1251	RK	TD91368	1257	RK	TJ97105	1263	RK	TJ97843	1268	RK	TL89534
1252	RK	TD91375	1258	RK	TJ97811	1264	RK	TJ97870	1269	RK	TL89537
1253	RK	TJ97101	1259	RK	TJ97812	1265	RK	TJ97869	1270	RK	TL89536
1254	RK	TJ97102	1260	RK	TJ97813	1266	RK	TJ97889	1271	RK	TL89635
1255	RK	TJ97104	1261	RK	TJ97842						

1272-1275 — Volvo B7BLE — Volvo 8500 — N33D — 2006

1272	RG	UR90715	1273	RG	UR90716	1274	RG	UR90710	1275	RG	UR90711

1276	RK	TJ92702	Volvo B10BLE	Säffle	N43D	1997
1277	RK	TJ92703	Volvo B10BLE	Säffle	N43D	1997
1278	AM	TX89413	VDL Bus SB4000	VDL Jonckheere	N33D	2006

1300-1313 — Volvo B10BLE — Säffle — N43D — 1997

1300	RK	TD89898	1304	RK	TD89753	1308	RK	TD89748	1311	RK	TD89708
1301	RK	TD89899	1305	RK	TD89751	1309	RK	TD89749	1312	RK	TD89752
1302	RK	TD89710	1306	RK	TD89711	1310	RK	TD89707	1313	RK	TJ92701
1303	RK	TJ92700	1307	RK	TD89750						

1314-1339 — Volvo B7RLE — Volvo — N43D — 2007

1314	RG	VN96086	1321	RG	VN92352	1328	RG	VN92397	1334	RG	VN92403
1315	RG	VN96087	1322	RG	VN92353	1329	RG	VN92398	1335	RG	VN92416
1316	RG	VN92347	1323	RG	VN92355	1330	RG	VN92399	1336	RG	VN92423
1317	RG	VN92348	1324	RG	VN92354	1331	RG	VN92400	1337	RG	VN92431
1318	RG	VN92349	1325	RG	VN92370	1332	RG	VN92401	1338	RG	VN92443
1319	RG	VN92350	1326	RG	VN92371	1333	RG	VN92402	1339	RG	VN92444
1320	RG	VN92351	1327	RG	VN92396						

1340-1381   Volvo B10BLE 12m   Åbenrå   N36D   2000

1340	GX	RM90995	1355	GX	RM91010	1361	RK	RM91016	1376	HI	RN90322		
1341	GX	RM90996	1356	GX	RM91011	1362	RK	RM91017	1377	HI	RN90323		
1342	GX	RM90997	1357	GX	RM91012	1363	RK	RM91018	1378	HI	RN90379		
1343	GX	RM90998	1358	GX	RM91013	1364	RK	RM91019	1379	HI	RN90380		
1344	GX	RM90999	1359	GX	RM91014	1365	RK	RM91020	1380	HI	RN90381		
1345	RG	RM91000	1360	RK	RM91015	1367	RK	RM91022	1381	HI	RN90382		

1385-1403   Volvo B10BLE 13.7m   Åbenrå   NC41D   2000

1385	u	SN89817	1392	RG	RN95511	1396	RG	RN95515	1400	HI	RP88736
1386	KK	RN90444	1393	RG	RN95512	1397	HI	RP88732	1401	HI	RP88737
1387	u	RN95486	1394	RG	RN95513	1398	HI	RP88733	1402	HI	RP88738
1388	MA	RN95487	1395	RG	RN95514	1399	HI	RP88735	1403	HI	RP88734
1391	RG	RN95490									

1406-1435   Volvo B10BLE 13.7m   Åbenrå   NC41D   2001

1406	HO	RX93856	1414	MA	RV92762	1425	MA	RV92764	1430	RG	RV96461
1411	FN	RV92729	1415	GX	RX93812	1426	u	RV92765	1431	MA	RV96462
1412	RG	RX93811	1418	KK	RX93859	1428	MA	RX93860	1434	RG	RV96464
1413	HO	RX93857	1419	RG	RV92763	1429	HO	RV96460	1435	RG	RX96754

1456-1473   Volvo B10BLE 13.7m   Åbenrå   N41D   2002

1456	MA	SC90611	1459	u	SC90677	1464	u	SD88212	1471	MA	SD88253
1457	MA	SC90651	1461	u	SD88164	1466	AA	SB93616	1472	u	SD88254
1458	MA	SC90652	1462	MA	SD88165	1467	MA	SB93617	1473	u	SD88255

1474-1499   DAB Citybus S15   Silkeborg   N31D   1997-99

1474	HO	OY91990	1481	GX	PL96097	1488	HO	PP94100	1494	HO	PZ89715
1475	HO	OY91991	1482	GX	PL96253	1489	w	PP94132	1495	GX	PZ89724
1476	w	OY91911	1483	HO	PL96254	1490	RG	PP94133	1496	GX	PZ89735
1477	GX	PC90596	1484	GX	PM95290	1491	RG	PZ89701	1497	GX	PZ89736
1478	GX	PE88130	1486	GX	PM95345	1492	RG	PZ89714	1498	GX	PZ89756
1479	GX	PE88132	1487	VA	PP94099	1493	RG	PZ89076	1499	w	PX89757
1480	GX	PE88139									

1500	KG	RH90371	VDL Bus 2000LF gas	Berkhof	N33F	1999
1501	w	UT97040	VDL Bus 2000LF gas	Berkhof	N33F	1999
1502	w	RE88967	VDL Bus 2000LF gas	Berkhof	N33F	1999
1503	w	RH90134	VDL Bus 2000LF gas	Berkhof	N33F	1999

1518-1580   Scania OmniCity CL94UB   Scania   A43D   2005

1518	AA	TX92484	1534	EJ	TX92615	1550	GX	TZ89326	1566	GX	UB94158
1519	EJ	TX92485	1535	EJ	TX92616	1551	GX	TZ89327	1567	GX	UB94159
1520	RG	TX92486	1536	EJ	TX92619	1552	GX	TZ89328	1568	GX	UB94223
1521	EJ	TX92487	1537	EJ	TX92649	1553	GX	TZ89399	1569	GX	UB94224
1522	EJ	TX92488	1538	EJ	TZ89150	1554	GX	TZ89423	1570	GX	UB94225
1523	EJ	TX92489	1539	EJ	TZ89151	1555	GX	TZ89448	1571	GX	UB94243
1524	EJ	TX92523	1540	EJ	TZ89152	1556	GX	TZ89463	1572	GX	UB94244
1525	EJ	TX92524	1541	EJ	TZ89203	1557	GX	UB94098	1573	GX	UB94245
1526	EJ	TX92525	1542	EJ	TZ89204	1558	GX	UB94099	1574	GX	UB94246
1527	EJ	TX92526	1543	EJ	TZ89205	1559	GX	UB94100	1575	GX	UB94284
1528	EJ	TX92527	1544	EJ	TZ89206	1560	GX	UB94117	1576	GX	UB94285
1529	EJ	TX92528	1545	EJ	TZ89232	1561	GX	UB94118	1577	GX	UB94286
1530	EJ	TX92569	1546	EJ	TZ89233	1562	GX	UB94283	1578	GX	UM94114
1531	EJ	TX92570	1547	EJ	TZ89234	1563	GX	UB94155	1579	GX	UM94113
1532	EJ	TX92571	1548	EJ	TZ89235	1564	GX	UB94156	1580	GX	UM94112
1533	EJ	TX92572	1549	EJ	TZ89236	1565	GX	UB94157			

1583-1591   DAB Citybus S15 Mk3 LPG   Silkeborg   N31D   1997

1583	GX	OZ91959	1586	HO	OZ91681	1588	GX	OZ91685	1591	GX	OZ91741
1584	GX	OY91998	1587	GX	OZ91682	1590	GX	OZ91686			

1592-1611   Volvo B10BLE 13.7m   Åbenrå   N41D   1998

1592	GX	PJ88362	1597	GX	PJ88368	1602	GX	PJ88373	1607	GX	PJ88379
1593	GX	PJ88365	1598	HO	PC92482	1603	GX	PJ88374	1608	GX	PJ88360
1594	GX	PJ88364	1599	GX	PJ88369	1604	HO	PJ88375	1609	HO	PJ88361
1595	GX	PC92476	1600	GX	PJ88370	1605	GX	TT94683	1610	GX	PJ97773
1596	GX	PC92477	1601	GX	PJ88371	1606	GX	PJ88378	1611	EJ	PJ97775

With the Roskilde cathedral in the background, 1751, RU97095, is one of three DAB Citybus S11 buses now operating from Maribo depot, one of the replacements for the Ballerup site. The trio usually works alongside the Optare Solo buses in the city. *Bill Potter*

1612-1641   DAB Citybus S15 Mk3 LPG   Silkeborg   N31D   1998-99

1612	GX	PE88117	**1620**	GX	PM95243	**1627**	GX	PZ89705	**1634**	GX	PZ89712
1613	GX	PC90580	**1621**	GX	PM95246	**1628**	GX	PZ89706	**1635**	HO	PZ89713
1614	GX	PE88129	**1622**	GX	PP94101	**1629**	GX	PZ89707	**1636**	GX	PZ89677
1615	GX	PE97295	**1623**	GX	PZ94130	**1630**	GX	PZ89708	**1637**	GX	PZ89796
1616	GX	PL96096	**1624**	GX	PZ94131	**1631**	GX	PZ89709	**1638**	GX	PZ89797
1618	HO	PL96186	**1625**	GX	PZ89703	**1632**	GX	PZ89710	**1640**	GX	RC88817
1619	GX	PL96187	**1626**	GX	PZ89704	**1633**	GX	PZ89711	**1641**	GX	RC88888

1642-1673   Volvo B12BLE   Volvo/Säffle 8500   N33D   2003

1642	RG	SR93011	**1650**	RG	ST97814	**1658**	RG	ST97831	**1666**	RG	ST97844
1643	RG	SR93012	**1651**	RG	SR93093	**1659**	RG	SR93081	**1667**	RG	ST97847
1644	RG	ST97813	**1652**	RG	SR93053	**1660**	RG	ST97837	**1668**	RG	ST97848
1645	RG	SR93034	**1653**	RG	SR93065	**1661**	RG	ST97841	**1669**	RG	ST97854
1646	RG	SR93092	**1654**	RG	SR93083	**1662**	u	ST97846	**1670**	RG	ST97855
1647	RG	SR93080	**1655**	RG	SR93094	**1663**	RG	ST97832	**1671**	RG	ST97856
1648	RG	SR93064	**1656**	RG	ST97815	**1664**	RG	ST97842	**1672**	RG	ST97869
1649	RG	SR93082	**1657**	RG	ST97830	**1665**	RG	ST97843	**1673**	RG	ST97876

1674-1682   VDL Bus SB4000   VDL Jonckheere Citybus   N33D   2005

1674	GX	TU92380	**1677**	GX	TU92377	**1679**	GX	TU92375	**1681**	GX	TU92373
1675	GX	TU92379	**1678**	GX	TU92376	**1600**	GX	TU92374	**1682**	GX	TU92372
1676	GX	TU92378									

1683-1686   Volvo B12BLE   Volvo/Säffle 8500   N33D   2005

1683	RG	TS94892	**1684**	RG	TS94893	**1685**	RG	TS94894	**1686**	RG	TS94895

1687	GX	TV88714	VDL Bus SB4000	VDL Jonckheere Citybus	N33D	2005

1700-1733   Volvo B10BLE 12m   Åbenrå   AN38D   1999

1700	HI	PZ95450	**1707**	RK	PZ95457	**1714**	RK	PZ95496	**1723**	EJ	PZ95505
1701	HI	PZ95451	**1708**	HI	PZ95458	**1715**	HI	PZ95497	**1725**	RG	PZ95546
1702	RK	PZ95452	**1709**	GX	PZ95459	**1716**	HI	PZ95498	**1728**	RG	PZ95549
1703	RK	PZ95453	**1710**	GX	PZ95492	**1717**	HI	PZ95499	**1730**	HI	PZ95551
1704	RK	PZ95454	**1711**	GX	PZ95493	**1718**	HI	PZ95500	**1731**	RG	PZ95552
1705	RK	PZ95455	**1712**	HI	PZ95494	**1719**	HI	PZ95501	**1732**	RG	PZ95553
1706	RK	PZ95456	**1713**	HI	PZ95495	**1720**	RK	PZ95502	**1733**	RG	PZ95554

1734	MA	VH91186	DAB Citybus S11		DAB		N15D	1995		
1736	MA	UY91941	DAB Citybus S11		DAB		N15D	1995		
1740	GX	PZ95592	Volvo B10BLE		Åbenrå		N68D	1999		
1741	GX	PZ95593	Volvo B10BLE		Åbenrå		N68D	1999		
1742	EJ	PZ95594	Volvo B10BLE		Åbenrå		N68D	1999		
1750	MA	RU97094	DAB Citybus S11		DAB		N15D	2001		
1751	MA	RU97095	DAB Citybus S11		DAB		N15D	2002		
1753	MA	PP94330	DAB Citybus S11		DAB		N15D	1998		

1754-1760   Optare Solo M920L   Optare   N23D   2005

1754	RK	TP97969	1756	EJ	TP97968	1758	RK	TR88011	1760	RK	TR88009
1755	RK	TP97967	1757	RK	TP97966	1759	RK	TR88010		

1761	KG	XE97902	VDL Bus ALE120		VDL Ambassadør		N27D	2009		
1762	KG	XE97901	VDL Bus ALE120		VDL Ambassadør		N27D	2009		
1763	HO	PP94331	DAB Citybus S11		DAB		N15D	1998		
1764	MA	PP94332	DAB Citybus S11		DAB		N15D	1998		
1768	HD	PJ89236	Neoplan		Neoplan		N26D	1998		
1769	HD	RV89.288	Mercedes-Benz Cito S		Mercedes-Benz		N29F	2002		
1773	EJ	PP89440	Volvo B10BLE		Åbenrå		N32D	1997		
1781	EJ	RC89417	Volvo B10BLE		Åbenrå		N32D	1998		

1783-1794 Scania OmniLink CL94UB 12m   Scania   N42D   2003

1783	EJ	SU96129	1786	EJ	SU96126	1789	EJ	SU96123	1792	EJ	SU96135
1784	EJ	SU96128	1787	EJ	SU96125	1790	EJ	SU96122	1793	EJ	SU96148
1785	EJ	SU96127	1788	EJ	SU96124	1791	EJ	SU96136	1794	EJ	SU96147

1795-1813   Scania OmniLink CL94UB 13.7m   Scania   NC43D   2003

1795	RG	SV97011	1800	RG	SX90752	1805	RG	SY89362	1810	RG	SY89417
1796	RG	SV97012	1801	RG	SX90819	1806	RG	SY89361	1811	RG	SY89416
1797	RG	SV97013	1802	RG	SX90854	1807	RG	SY89360	1812	RG	SY89479
1798	RG	SV97044	1803	RG	SX90839	1808	RG	SY89393	1813	RG	SY89496
1799	RG	SX90788	1804	RG	SX90875	1809	RG	SY89392		

1827-1858   Scania OmniLink CL94UB 13.7m   Scania   NC43D   2003-04

1827	RG	SY93825	1835	RG	SY93783	1843	RG	TB88449	1851	RG	TB88489
1828	RG	SY93784	1836	RG	SZ91745	1844	RG	TB88417	1852	RG	TB88477
1829	RG	SY93824	1837	RG	SZ91744	1845	RG	TB88418	1853	RG	TB88490
1830	RG	SY93774	1838	RG	SZ91749	1846	RG	TB88450	1854	RG	TB88491
1831	RG	SY93775	1839	RG	SZ91755	1847	RG	TB88444	1855	RG	TB88493
1832	RG	SY93776	1840	RG	SZ91756	1848	RG	TB88445	1856	RG	TB88492
1833	RG	SY93777	1841	RG	SZ91765	1849	RG	TB88483	1857	RG	TB88469
1834	RG	SY93778	1842	RG	TB88448	1850	RG	TB88484	1858	RG	TB88470

1862-1885   Volvo B10BLE 12m   Åbenrå   AN38D   1997-98

1862	KG	TX94432	1868	KG	PJ88291	1874	KG	PJ88366	1880	KG	PC92479
1863	KG	OY97165	1869	KG	PJ88300	1875	KG	TP90383	1881	KG	PC92480
1864	KG	OY97164	1870	KG	PJ88306	1876	KG	TP90384	1882	KG	PC92481
1865	KG	OY97163	1871	KG	PJ88307	1877	KG	PC92474	1883	KG	PJ88372
1866	KG	OY97251	1872	KG	UV90371	1878	KG	PC92475	1884	HO	PJ88376
1867	GX	OY97267	1873	KG	PJ88363	1879	KG	PC92478	1885	KG	PX96095

1886-1900   VDL Bus SB220 gas   Berkhof 2000LF   N31F   1999-2000

1886	KG	RE88852	1890	MA	RE88954	1894	VA	RH90133	1898	MA	TC94631
1887	MA	RE88813	1891	KG	RE88888	1895	MA	RH90763	1899	KG	UT97175
1888	KG	RE89023	1892	MA	RE88966	1896	MA	RH90847	1900	w	US96803
1889	w	RE88955	1893	KG	RE89033	1897	w	RH90848		

1901-1915   Volvo B7BLE   Volvo/Säffle 8500   N31D   2006

1901	KG	UU96912	1905	KG	UU96933	1909	KG	UU96986	1913	KG	UU97006
1902	KG	UU96913	1906	KG	UU96952	1910	KG	UU96987	1914	KG	UU97019
1903	KG	UU96914	1907	KG	UU96953	1911	KG	UU96997	1915	KG	UU97020
1904	KG	UU96932	1908	KG	UU96985	1912	KG	UU97005		

1916	RG	UJ96430	Volvo B12BLE		Volvo/Säffle 8500		N34D	2003		

Odense in the principal town on Fyn, the Island between Sjælland, where Copenhagen is situated, and Jylland. Arriva operates the FynBus network which sports a cream livery with green lettering as illustrated by 2595, PP89414, a Volvo B10M. *Bill Potter*

1917-1945   Scania OmniLink CK230 UB4   Scania   N33D   2008

1917	RG	XD92766	1925	AM	XD92776	1932	AM	XD92783	1939	AM	XD92789
1918	RG	XD92765	1926	AM	XD92777	1933	AM	XD92784	1940	AM	XJ90253
1919	RG	XD92770	1927	AM	XD92778	1934	AM	XD92785	1941	AM	XD88321
1920	RG	XD92771	1928	AM	XD92779	1935	AM	XD92786	1942	AM	XJ90254
1921	RG	XD92772	1929	AM	XD92780	1936	AM	XD92787	1943	AM	XJ90255
1922	RG	XD92773	1930	AM	XD92781	1937	AM	XJ90252	1944	AM	XJ90256
1923	RG	XD92774	1931	AM	XD92782	1938	AM	XD92788	1945	AM	XD88322
1924	RG	XD92775									

1993-1998   Volvo B10BLE   Åbenrå   N34D   1997

| 1993 | AM | OX94339 | 1995 | KK | OX94347 | 1997 | KK | OX94366 | 1998 | KK | OX94372 |
| 1994 | AM | OX94346 | 1996 | KK | OX94365 | | | | | | |

2162	OV	LS97414	Volvo B10M	Åbenrå	B--D	1989
2168	OV	LV91527	Volvo B10M	Åbenrå	B--D	1989
2169	OV	LV91557	Volvo B10M	Åbenrå	B--D	1989
2173	SN	LX94925	Volvo B10M	Åbenrå	B--D	1989
2195	OV	LY93265	Volvo B10BLE	Silkeborg	B--D	1989
2249	OV	MJ95476	Volvo B10M	Åbenrå	B38D	1991
2268	OV	MR93469	Volvo B10M	Silkeborg	B38D	1991
2287	OV	MS91113	Volvo B10M	Silkeborg	B38D	1991
2291	SN	MS97752	Volvo B10M	Åbenrå	B38D	1991

2314-2351   Volvo B10M   Silkeborg   B38D   1992

| 2314 | HV | NB89310 | 2325 | OV | NB97476 | 2344 | HG | NB97681 | 2351 | AA | NJ93727 |
| 2315 | OV | NB89311 | | | | | | | | | |

| 2358 | AA | NK96706 | Volvo B10M | Åbenrå | BC35D | 1993 |

2365-2406   Volvo B10M   Silkeborg   B38D   1993-94

| 2365 | OV | NN89572 | 2367 | OV | NN89573 | 2369 | AA | NN89575 | 2405 | U | NX93001 |
| 2366 | OV | NN89542 | 2368 | OV | NN89574 | 2379 | HG | NK96625 | 2406 | AE | NX93002 |

2462	w	OD92659	Volvo B10M			Åbenrå			B38D	1995		
2481	FN	OJ92349	Volvo B10M			DAB/Silkeborg			B38D	1996		
2482	FN	OJ92350	Volvo B10M			DAB/Silkeborg			B38D	1996		
2490	FN	OL96.926	Volvo B10M			Åbenrå			B38D	1996		
2509	FS	OM97854	Volvo B10M			Åbenrå			B38D	1996		
2521	FS	OX91625	Volvo B10M			DAB/Silkeborg			B47D	1997		
2526	FS	OX91679	Volvo B10M			DAB/Silkeborg			B47D	1997		
2534	ND	OX94494	Volvo B10M			Åbenrå			B38D	1997		

2536-2543

Scania N112CL — Scania — B47D 1997

2536	u	PB91473	2538	u	PB91475	2541	AE	PB91480	2543	u	PB91482
2537	u	PB91474	2539	u	PB91476	2542	SG	PB91481			

2546	u	PB91477	Scania N112CL	Scania	B47D	1997	
2547	u	PB91478	Scania N112CL	Scania	B47D	1997	
2552	AA	PB89095	Volvo B10M	Åbenrå	B44D	1997	
2553	AA	PB89103	Volvo B10M	Åbenrå	B44D	1997	
2554	AA	PB89102	Volvo B10M	Åbenrå	B44D	1997	
2566	AA	PC89064	Volvo B10M	Åbenrå	B44D	1997	
2568	AA	PC89080	Volvo B10M	Åbenrå	B44D	1997	
2574	u	PC97949	Scania N112CL	Scania	B47D	1998	
2575	u	PE92039	Volvo B10M	Vest	B47D	1998	
2576	u	PE92042	Volvo B10M	Vest	B47D	1998	
2578	HV	PP89431	Volvo B10M	Åbenrå	B47D	1998	

2581-2589

Volvo B10M — Vest — B47D 1998

2581	SV	PE92002	2585	OV	PP93852	2587	OV	PP93854	2589	FN	PR94089
2582	u	PE92001	2586	OV	PP93855	2588	OV	PP93852			

2595-2634

Volvo B10M — Åbenrå — B47D 1998-99

2595	FN	PP89414	2619	FS	PT95340	2633	FN	PT95382	2634	AA	PT95381
2596	FS	PP89417									

2647	HV	PX96158	Volvo B10M	Åbenrå	B47D	1999
2648	HV	PX96158	Volvo B10M	Åbenrå	B47D	1999

2658-2663

Volvo B10M — Åbenrå — B47D 1999

2658	FS	PX96253	2660	FS	PX96265	2662	FS	RC89372	2663	FS	RC89383
2659	FS	PX96266	2661	FS	PX96271						

2667	FN	RC89395	Volvo B10M	Åbenrå	B47D	1999
2668	FS	RC89396	Volvo B10M	Åbenrå	B47D	1999
2669	FN	RC89397	Volvo B10M	Åbenrå	B47D	1999

2672-2676

Volvo B10M — Åbenrå — B47D 1999

2672	ND	RD91994	2674	FN	PX96241	2675	ND	RE94201	2676	FN	RD91993
2673	ND	RD91999									

2682	SG	RH97538	Volvo B10M	Åbenrå	B47D	2000
2683	FN	RK95314	Scania L113 CLL	DAB/Silkeborg	N47D	2000

2707-2718

Scania L113 CLL — DAB/Silkeborg — N47D 2000

2707	ND	RM96994	2712	FN	RN96024	2715	FN	RN96188	2717	FN	RN95954
2710	ND	RM96996	2713	FS	RN96048	2716	FN	RN96189	2718	FN	RN95953

2745	FS	RN96076	Scania	DAB/Silkeborg	N47D	2000
2749	FN	RP91084	Volvo B10M	Åbenrå	B47D	2000

2777-2819

Scania OmniLine CL94 UB 12m — Scania — N47D 2001

2777	FN	RV95328	2790	FN	RV95341	2800	FN	RV95385	2810	FS	RV95395
2780	FS	RV95331	2791	FN	RV95342	2801	FN	RV95386	2811	FS	RV95396
2781	FS	RV95332	2792	FS	RV95343	2802	FN	RV95387	2812	FS	RV95397
2782	FS	RV95333	2793	FN	RV95344	2803	FS	RV95388	2813	FN	RV95398
2783	FS	RV95334	2794	FN	RV95345	2804	FS	RV95389	2814	FN	RV95399
2784	FS	RV95335	2795	FN	RV95382	2805	FS	RV95390	2815	FN	RV95400
2785	FS	RV95336	2796	FN	RV95383	2806	FS	RV95391	2816	FN	RV95401
2786	FS	RV95337	2797	FN	RV95346	2807	FS	RV95392	2817	FS	RV95402
2787	FN	RV95338	2798	FN	RV95384	2808	FS	RV95393	2818	FS	RV95403
2788	FN	RV95339	2799	FN	RV95347	2809	FS	RV95394	2819	FN	RV95404
2789	FN	RV95340									

In addition to the bus fleet, Arriva operates several trains in Denmark, notably around Århus, and these carry Arriva colours. Taking a break at Viborg station is AR26. *Bill Potter*

2820-2825   Scania 13.6m   Lahti Flyer   N55D   2001-02

| 2820 | FN | RV95405 | 2822 | FN | RV95407 | 2824 | FS | SJ94192 | 2825 | FS | RX97975 |
| 2821 | FN | RV95406 | 2823 | FS | RV95408 | | | | | | |

2826	HV	RX96850	Volvo B10M 13.7m	Åbenrå	NC46D	2002
2827	SV	SM97944	Scania 13.6m	Lahti Flyer	N55D	2002
2831	FN	RY88058	Scania 13.6m	Lahti Flyer	N55D	2002
2837	FN	SH93791	Volvo B10M 13.7m	Åbenrå	NC46D	2002

2847-2850   Scania OmniLine CL94UB 12m   Scania   N47D   2002

| 2847 | FS | SL95830 | 2848 | FS | SL95855 | 2849 | FS | SL95856 | 2850 | FN | SL95857 |

2853-2865   Scania OmniLink CL94UB 12m*   Scania   N47D   2003   *2864/5 are 13.5m

2853	HG	SX97361	2857	HG	SX97364	2860	HG	SX97367	2863	HG	SX97369
2854	HG	SX97362	2858	HG	SX97365	2861	HG	SX97340	2864	HG	SX97303
2855	HG	SX97363	2859	HG	SX97366	2862	HG	SX97368	2865	HG	SX97338
2856	HG	SX97339									

2866-2871   MAN 13.310   Jonckheere Modulo   C--D   2004

| 2866 | FS | SY89049 | 2868 | FS | SY89051 | 2870 | FS | SY89.081 | 2871 | FS | SY89.082 |
| 2867 | FS | SY89050 | 2869 | FS | SY89052 | | | | | | |

2883-2886   Scania L94UB 13.6m   Lahti   N--D   2004

| 2883 | FS | TK94765 | 2884 | FS | TK94766 | 2885 | FS | TK94767 | 2886 | FN | TM88197 |

| 2888 | FS | TM88244 | Scania OmniLine CL94UB 12m | Scania | NC--D | 2004 |
| 2889 | ND | TZ97307 | Scania OmniLine IL94UB 12m | Scania | C--D | 2005 |

2892-2899   Volvo B12M   Carrus   N47D   2005

| 2892 | HV | TT90158 | 2894 | HV | TT90160 | 2896 | HV | TT90162 | 2898 | HV | TT90164 |
| 2893 | HV | TT90159 | 2895 | HV | TT90161 | 2897 | HV | TT90163 | 2899 | HV | TT90165 |

| 2901 | FN | UJ92826 | Scania OmniCity L94UB 12m | Scania | NC49D | 2006 |

A few commercial routes in Denmark are operate by vehicles in Arriva colours, whereas most services are operated for the various transport authorities in their own schemes. Scania OmniLine 2848, SL95855, is shown heading for x-Ringkobing where a new base has been established. *Bill Potter*

2902-2925 — Irisbus Arway 12.8m — Irisbus — N47D — 2006

2902	FN	UR94484	2908	FN	UM93572	2914	FS	UM93583	2920	FN	UM93581
2903	FN	UR94481	2909	FN	UM93591	2915	FN	UM93584	2921	FN	UM93598
2904	FS	UR94482	2910	FN	UM93570	2916	FN	UM93599	2922	FN	UM93585
2905	FS	UR94483	2911	FS	UM93568	2917	FN	UM93586	2923	FN	UM93582
2906	FN	UM93571	2912	FS	UM93590	2918	FN	UM93588	2924	FN	US91250
2907	FN	UM93569	2913	FS	UM93580	2919	FN	UR94633	2925	FN	UM93587

2926-2931 — MAN Lion 13.7m — MAN — N53D — 2007

2926	FS	VH89850	2928	FS	VH89854	2930	FS	VH89853	2931	FS	VH89851
2927	FS	VH89855	2929	FS	VH89852						

2938-2941 — Scania OmniLine CL94UB 12m — Scania — N47D — 2007

2938	DJ	VJ93921	2939	DJ	VJ93922	2940	DJ	VJ93950	2941	DJ	VJ93951

2942-2944 — Scania OmniLine CL94UB 13.7 — Lahti Flyer — N55D — 2007

2942	DJ	VN93038	2943	DJ	VN93039	2944	DJ	VN93040

2945	AA	VY89078	Volvo B12M	-	N46D	2008
2946	AA	VY89079	Volvo B12M	-	N46D	2008
2947	AA	VY95469	Scania K230 UB	Lahti Scala	N49D	2008
2948	AA	VY95594	Scania OmniLink K230 UB	Scania	N45D	2008
2949	AA	VZ90459	Scania K230 UB	Lahti Scala	N49D	2008
2950	AA	XD90098	Scania K230 UB	Lahti Scala	N49D	2008
2951	AA	XE89989	Scania K230 UB	Lahti Scala	N49D	2008
2952	AA	XJ96062	Scania K230 UB	Lahti Scala	N49D	2008
2953	AA	XE96555	Scania OmniLink K230 UB	Scania	N45D	2008
2954	DJ	XE90177	Scania Omniline	Scania	N47D	2008
2955	DJ	XE96364	Scania Omniline	Scania	N47D	2008

2956-2965 — Volvo B12MA — Volvo/Säffle 8500 — AN64D — 2008

2956	OD	XJ89610	2959	EB	XJ89561	2962	SB	XJ94885	2964	OD	XJ94825
2957	OD	XJ89559	2960	HT	XJ89562	2963	OD	XJ89611	2965	OD	XJ94823
2958	EB	XJ89560	2961	HT	XJ89612						

Ikke i rute or 'not in service' is proclaimed by Volvo TJ97101 which has been renumbered 1253 since the last edition of the Arriva book. Copenhagen regional buses from the 3xxx series have all now been treated. *Bill Potter*

2966-2973 — Scania K230 UB — Lahti Flyer — N55D — 2008

2966	AR	XE89965	2968	AR	XE89990	2970	GN	XE90026	2972	AR	XE90193
2967	AR	XE89964	2969	EB	XE89991	2971	GN	XE90027	2973	AR	XE90199

2974-2991 — Scania OmniLine CL230 — Scania — N47D — 2008-09

2974	BO	XE96365	2979	RG	XE96428	2984	RG	XE96496	2988	UD	XE96586
2975	BO	XE96387	2980	RG	XE96494	2985	BO	XE96508	2989	OD	XE96587
2976	RG	XE96414	2981	RG	XE96495	2986	VV	XE96566	2990	OD	XJ96060
2977	RG	XE96415	2982	RG	XE96509	2987	SK	XE96567	2991	OD	XJ96061
2978	RG	XE96427	2983	SK	XE96544						

2992-2995 — Volvo B7RLE — - — N40D — 2007-08

2992	u	XD90746	2993	u	XD90747	2994	u	XD90748	2995	u	XD90749

2996	RY	XJ89717	Volvo B12MA	Volvo/Säffle 8500	AN64D	2008
2997	OD	XJ89716	Volvo B12MA	Volvo/Säffle 8500	AN64D	2008

2998-3009 — Scania K230 UB — Lahti Flyer — N55D — 2009

2998	RA	XJ96146	3001	RA	XJ96161	3004	SK	XJ96198	3007	SK	XJ96224
2999	RA	XJ96160	3002	RA	XJ96173	3005	SK	XJ96199	3008	SK	XJ96243
3000	EB	XJ96162	3003	SK	XJ96172	3006	SK	XJ96218	3009	SK	XJ96244

3106	FY	KV94917	Leyland/DAB	DAB	B--D	1986
3115	FY	LV92590	DAB	DAB	B--D	1989
3116	VI	NS88229	DAF/DAB	DAB	B--D	1989
3118	DJ	LT91490	Volvo B10M	Åbenrå	B--D	1989
3125	FY	NX88081	Leyland/DAB	DAB	B--D	1989
3135	w	ME94496	Volvo B10M	Åbenrå	B--D	1990
3146	ND	ML88147	Volvo B10M	Åbenrå	B--D	1991
3147	FY	MN92231	Volvo B10M	Åbenrå	B--D	1991
3149	DJ	MS91014	Volvo B10M	Åbenrå	B--D	1991
3152	VI	MY91487	DAF/DAB	DAB	B--D	1992
3153	DJ	OJ94187	DAF/DAB	DAB	B--D	1992

3162	ND	NJ89096	DAF/DAB			DAB		B--D	1993		
3163	ND	NJ89125	MAN/DAB			DAB Citybus S15 Mk2		B--D	1993		
3166	FY	NV91273	Volvo B10M			Åbenrå		B--D	1993		
3167	VI	OS94520	DAF/DAB			DAB		B--D	1993		
3168	HN	SJ92752	DAF/DAB			DAB Citybus S15 Mk1		B--D	1993		
3172	ND	NZ93969	DAF/DAB			DAB		B--D	1995		
3174	w	MB92888	Volvo B10M			Åbenrå		B--D	1995		
3176	ND	OB93469	MAN/DAB			DAB Citybus S15 Mk2		B--D	1995		
3184	ND	NJ89095	DAF/DAB			DAB		B--D	1996		
3185	ND	RL91182	Volvo B10M			Åbenrå		B--D	1997		
3187	ND	PE94555	Volvo B10M			Åbenrå		B--D	1998		
3188	FY	OZ92476	Volvo B10M			Vest		B--D	1998		
3189	ND	PL97691	Volvo B10M			Åbenrå		B--D	1998		
3190	ND	PE94573	Volvo B10M			Åbenrå		B--D	1998		
3191	ND	PE94583	Volvo B10M			Åbenrå		B--D	1998		
3192	ND	PE94587	Volvo B10M			Åbenrå		B--D	1998		
3193	FY	OZ92475	Volvo B10M			Vest		B--D	1998		
3194	FY	OZ92511	Volvo B10M			Vest		B--D	1998		
3195	FY	OZ92510	Volvo B10M			Vest		B--D	1998		
3196	ND	PL97595	Volvo B10BLE			Åbenrå		B--D	1998		
3197	FY	PL97652	Volvo B10M			Vest		B--D	1998		
3198	ND	PP89464	Volvo B10M			Vest		B--D	1998		
3199	ND	PX96127	Volvo B10BLE			Åbenrå		B--D	1999		
3200	ND	PX96128	Volvo B10BLE			Åbenrå		N39D	1999		
3202	DJ	PZ89729	Scania/DAB			DAB		B--D	1999		
3203	DJ	PZ89726	Scania/DAB			DAB		B--D	1999		
3204	DJ	PZ89727	Scania/DAB			DAB		B--D	1999		
3205	FY	RC89419	Volvo B10M			Vest		B--D	1999		
3206	FY	RC89418	Volvo B10M			Vest		B--D	1999		
3207	ND	RX96461	Scania/DAB			DAB		B--D	1999		

3213-3218 — Volvo B10M — Åbenrå — B36D — 2000

3213	FY	RK97522	3215	ND	RL91165	3217	ND	RL91164	3218	ND	RL91162
3214	FY	RK97521	3216	ND	RL91181						

3219-3225 — Volvo B10M — Åbenrå — B36D — 2000

3219	ND	RL91231	3221	ND	RL91230	3223	ND	RL91240	3225	ND	RL91242
3220	ND	RL91232	3222	ND	RL91197	3224	ND	RL91244			

3228-3232 — Volvo B10M — Åbenrå — B36D — 2000

3228	ND	RL91163	3230	ND	RL91229	3231	ND	RL91241	3232	ND	RL91161
3229	ND	RL91245									

3265-3284 — Volvo B12BLE — Åbenrå — N39D — 2002

3265	HN	SJ89558	3270	VJ	SJ89639	3275	VJ	SJ89651	3280	HN	SJ89664
3266	VJ	SJ89569	3271	VJ	SJ89640	3276	VJ	SJ89650	3281	HN	SJ89675
3267	VJ	SJ89581	3272	VJ	SJ89643	3277	VJ	SJ89649	3282	VJ	SJ89676
3268	VJ	SJ89633	3273	VJ	SJ89644	3278	VJ	SJ89663	3283	VJ	SJ89661
3269	VJ	SJ89634	3274	VJ	SJ89645	3279	HN	SJ89662	3284	VJ	SJ89668

3288-3307 — Volvo B12BLE — Åbenrå — N39D — 2002

3288	VJ	SJ89718	3293	HN	SL91813	3298	HN	SL91788	3303	HN	SL91807
3289	VJ	SJ89724	3294	HN	SL91821	3299	HN	SL91791	3304	HN	SL91811
3290	VJ	SJ89725	3295	HN	SL91780	3300	HN	SL91794	3305	HN	SL91881
3291	VJ	SJ89732	3296	HN	SL91786	3301	HN	SL91798	3306	HN	SL91883
3292	VJ	SJ89734	3297	HN	SL91787	3302	HN	SL91804	3307	HN	SL91993

3308-3311 — MAN — MAN — N38D — 2002

3308	VJ	SM90243	3309	VJ	SM90244	3310	VJ	SM90244	3311	VJ	SM90245

3316	ND	SX93476	Volvo B12B	Åbenrå	N47D	2004	
3317	DJ	TM88210	Scania OmniLine CL94UB 12m	Scania	N42D	2004	
3318	DJ	TM88211	Scania OmniLine CL94UB 12m	Scania	N42D	2004	
3319	DJ	TM88212	Scania OmniLine CL94UB 12m	Scania	N42D	2004	
3322	VI	SD92675	Volvo B12M	Åbenrå	NC39D	2002	
3323	ND	SU94086	Scania OmniLine CL94UB 12m	Scania	N42D	2003	
3802	FY	NE89854	Renault Durisotti	Renault	M8	1993	Handicap service
3810	DJ	OB59961	Mercedes-Benz Sprinter 312	Mercedes-Benz	M8	1995	Handicap service
3813	FY	TK94178	Renault Durisotti	Renault	M8	1996	Handicap service
3814	AH	OC93981	Iveco Ducato	Iveco	M8	1996	Handicap service
3816	OV	PZ93467	Iveco Ducato	Iveco	M8	1996	Handicap service

Aalborg is the principal town of Nordjylland where Arriva operates several of the town services alongside Connex. Volvo 4394, SJ89557, is a tri-axle B10BLE with Åbenrå bodywork. Volvo AB who owned the Åbenrå Aabenraa Karosseri plant closed the bus-making factory in 2003, when Scandinavian body production was concentrated in Sweden under the Volvo brand. *Bill Potter*

3817	FY	OS89885	Iveco Ducato	Iveco	M8	1996	Handicap service
3821	AH	OS94705	Volkswagen Kutsenits	Volkswagen	M8	1997	Handicap service
3831	AH	PZ94542	Mercedes-Benz Sprinter 416 cdi	Mercedes-Benz	M8	1999	Handicap service
3832	OV	RL96681	Iveco Ducato	Iveco	M8	2000	Handicap service
3833	FN	RP89432	Iveco Ducato	Iveco	M8	2000	Handicap service
3834	AH	RP89433	Iveco Ducato	Iveco	M8	2000	Handicap service
3837	OV	RP89436	Iveco Ducato	Iveco	M8	2000	Handicap service
3838	AH	RP89437	Iveco Ducato	Iveco	M8	2000	Handicap service
3839	AH	RP89438	Iveco Ducato	Iveco	M8	2000	Handicap service
3841	AH	RP89440	Iveco Ducato	Iveco	M8	2000	Handicap service
3842	AH	RP89441	Iveco Ducato	Iveco	M8	2000	Handicap service
3843	AH	RV91296	Mercedes-Benz Sprinter 413	Mercedes-Benz	M8	2001	Handicap service
3847	ND	SZ93361	Mercedes-Benz Vario	Mercedes-Benz	M18	2001	Handicap service
3849	AH	RX95266	Iveco Ducato	Iveco	M8	2001	Handicap service
3850	FY	SC93714	Volkswagen Kutsenits	Volkswagen	M8	2002	Handicap service

3852-3871

			Mercedes-Benz Sprinter 313			Mercedes-Benz		M8		2003	Handicap service

3852	OV	SP92838	3857	VI	SP92845	3863	VI	SP92842	3868	AH	SP93828
3853	VI	SP92848	3858	VI	SP92844	3865	VI	SP93819	3869	AH	SP93831
3854	FY	SP92846	3860	OV	SP92843	3866	VI	SP93818	3870	AH	SP93829
3855	VI	SP92839	3861	VI	SP92847	3867	AH	SP93816	3871	AH	SP93920
3856	VI	SP92840	3862	VI	SP92849						

3872	AH	SP93879	Mercedes-Benz Vario	Mercedes-Benz	M8	2003	Handicap service
3873	AA	SR92735	Citroen Jumiper	Citroen	M8	2003	Handicap service
3874	AA	SR92736	Citroen Jumiper	Citroen	M8	2003	Handicap service
3875	SK	SY92902	Ford Transit	Ford	M8	2004	Handicap service
3876	FY	SZ92876	Mercedes-Benz Vario	Mercedes-Benz	M8	2004	Handicap service
3878	HA	UK89858	Citroen Jumiper	Citroen	M8	2004	Handicap service
3879	HA	SX90495	Citroen Jumiper	Citroen	M8	2004	Handicap service

3880-3909   Mercedes-Benz Sprinter 315   Mercedes-Benz   M8   2007-09

3880	AH	UZ90025	3888	SO	VU96612	3896	SO	VP88846	3903	SO	VU96926
3881	AH	UZ90026	3889	SO	VU96608	3897	SO	VS95356	3904	SO	VU96925
3882	OV	VB96655	3890	SO	VU96641	3898	SO	VS95357	3905	SO	VU96886
3883	OV	VB96656	3891	SO	VP88723	3899	SO	VS95358	3906	SO	VU96887
3884	OV	VH94941	3892	SO	VP88726	3900	SO	VS95359	3907	SO	VU96924
3885	SO	VU96609	3893	SO	VP88722	3901	SO	VS95360	3908	SO	VU96927
3886	SO	VU96610	3894	SO	VP88725	3902	SO	VS95361	3909	SO	-
3887	SO	VU96611	3895	SO	VP88724						

3911	ST	XJ88661	Volkswagen Passat	Volkswagen	4-seat	2009	Handicap service
3912	ST	-	Volkswagen Passat	Volkswagen	4-seat	2009	Handicap service
4001	HG	OV92022	Volvo B10L	Carrus	N40D	1997	

4021-4030   Volvo B10BLE   Vest   N36D   2002

4021	FS	SN90043	4024	FS	SN90046	4027	FS	SM97699	4029	FS	SM97737
4022	FS	SN90044	4025	FS	SN90047	4028	FS	SM97714	4030	FS	SM97752
4023	FS	SN90045	4026	FS	SM97688						

4319	AA	OE92917	MAN/DAB	Silkeborg	N35D	1995
4320	AA	OE92917	MAN/DAB	Silkeborg	N35D	1995
4356	AA	OS89287	MAN/DAB	Silkeborg	N35D	1996

4361-4367   Volvo B10L   Carrus   N40D   1997

4361	AA	PB89204	4363	AA	PB89218	4365	AA	PB89226	4367	AA	PB89507
4362	AA	PB89215	4364	AA	PB89223						

4372	AA	PX96281	Volvo B7L	Åbenrå	N39D	1999

4375-4388   Volvo B10BLE   Åbenrå   N32D   2000

4375	AA	RE94369	4377	AA	RE94371	4386	AA	RN88149	4388	AA	RN88151
4376	AA	RE94370	4378	AA	RE94372	4387	AA	RN88150			

4389-4393   Scania OmniLink 13.7m   Scania   N42D   2002

4389	AA	RU97030	4391	AA	RU97078	4392	AA	RU97079	4393	AA	RU97080
4390	AA	RU97031									

4394	AA	SJ89557	Volvo B10BLE 13.7m	Åbenrå	NC42D	2002

4395-4409   Volvo B10BLE   Vest   N32D   2002

4395	AA	SJ88780	4399	AA	SJ88784	4403	AA	SJ88788	4407	AA	SJ88792
4396	AA	SJ88781	4400	AA	SJ88785	4404	AA	SJ88789	4408	AA	SJ88793
4397	AA	SJ88782	4401	AA	SJ88786	4405	AA	SJ88790	4409	AA	SJ88794
4398	AA	SJ88783	4402	AA	SJ88787	4406	AA	SJ88791			

4410-4419   Mercedes-Benz Cito   Mercedes-Benz   N17F   2002

4410	AA	SD96.842	4413	AA	SD96.845	4416	AA	SD96.848	4418	AA	SD96.850
4411	AA	SD96.843	4414	AA	SD96.846	4417	AA	SD96.849	4419	AA	SD96.851
4412	AA	SD96.844	4415	AA	SD96.847						

4420-4430   Scania OmniLink CL94UB 12m   Scania   N42D   2004

4420	AA	TB90824	4423	AA	TB90827	4426	AA	TB90830	4429	AA	TB90833
4421	AA	TB90825	4424	AA	TB90828	4427	AA	TB90831	4430	AA	TB90834
4422	AA	TB90826	4425	AA	TB90829	4428	AA	TB90832			

5047	u	KV94783	Volvo B10M	Åbenrå	B34D	1986
5150	u	LS89499	Volvo B10M	Åbenrå	B--D	1988

5401-5417   Mercedes-Benz Sprinter 312   Mercedes-Benz   M8   1997

5401	OV	OX90567	5404	OV	OX91313	5410	OV	OU96648	5416	OV	SM90263
5402	OV	OZ97632	5408	OV	OX90560	5411	OV	OU96505	5417	OV	SM90290
5403	OV	OY95589	5409	OV	OX90635	5414	OV	SM90147			

5420	OV	PE94546	Mercedes-Benz Sprinter 412	Mercedes-Benz	M8	1998
5422	OV	PE94547	Mercedes-Benz Sprinter 412	Mercedes-Benz	M8	1998
5424	OV	PE94548	Mercedes-Benz Sprinter 412	Mercedes-Benz	M8	1998
5436	OV	OU96508	Mercedes-Benz Sprinter 312	Mercedes-Benz	M8	1997
5437	OV	PE94549	Mercedes-Benz Sprinter 312	Mercedes-Benz	M8	1997
5442	OV	PE94550	Mercedes-Benz Sprinter 312	Mercedes-Benz	M8	1997

5451-5463 — Mercedes-Benz Sprinter 412 — Mercedes-Benz — M8 — 1998-99

5451	UV	SB92849	**5455**	OV	PJ96270	**5458**	OV	TB94358	**5461**	OV	TB94448
5452	OV	SM90148	**5456**	OV	SM90201	**5459**	OV	TB94359	**5462**	OV	TC93364
5453	OV	SB92850	**5457**	OV	TB94338	**5460**	OV	TB94449	**5463**	OV	TC93363
5454	OV	SB92851								

5464	OV	LS97271	Volvo B10M	Åbenrå	B38D	1989
5465	OV	SV88830	Mercedes-Benz Sprinter 412	Mercedes-Benz	M8	1997
5466	OV	TC93409	Mercedes-Benz Sprinter 412	Mercedes-Benz	M8	1997
5468	VI	TC93647	Mercedes-Benz Vario 0814	Mercedes-Benz	M8	1998
5479	OV	OJ90646	Mercedes-Benz Sprinter 400	Mercedes-Benz	M8	1996

5480-5483 — Hyundai Trajet 2.0 — Hyundai — M8 — 2006

5480	OV	YJ21272	**5481**	OV	YJ21353	**5482**	OV	YJ21354	**5483**	OV	YJ21352

5484-5493 — Mercedes-Benz Sprinter 315 — Mercedes-Benz — M8 — 2007

5484	OV	VN94176	**5487**	OV	VN94179	**5490**	OV	VN94182	**5492**	OV	VN94184
5485	OV	VN94177	**5488**	OV	VN94180	**5491**	OV	VN94183	**5493**	OV	VN94185
5486	OV	VN94178	**5489**	OV	VN94181					

5501	HA	PE94575	Volvo B10M	Åbenrå	B44D	1998
5502	HA	PJ97593	Volvo B10M	Åbenrå	B44D	1998
5503	HA	PJ97694	Volvo B10M	Åbenrå	B44D	1998
5504	HA	PB89259	Volvo B10M	Åbenrå	B44D	1998
5505	-	PB89283	Volvo B10M	Åbenrå	B44D	1998
5506	-	PE94568	Volvo B10M	Åbenrå	B44D	1998
5507	-	PE94544	Volvo B10M	Åbenrå	B44D	1998
5509	-	SN97730	Volvo B12M		B44D	2003
5510	-	SN97741	Volvo B12M		B44D	2003
5511	-	SN97734	Volvo B12M		B44D	2003
5512	NY	PE94532	Volvo B10M	Åbenrå	B44D	1998
5513	NY	VH94957	Volvo B12M		B44D	2007
5514	NY	PJ97612	Volvo B10M	Åbenrå	B44D	1998
5515	NY	PL67653	Volvo B10M 10.2m	Åbenrå	B35D	1998
5516	NY	PL97644	Volvo B10M 10.2m	Åbenrå	B35D	1998
5517	MA	OZ92514	Volvo B6BLE	-	N29D	1998
5518	NY	SZ89491	MAN 14.220	-	NC27D	2004
5519	NY	SZ89493	MAN 14.220	-	NC27D	2004
5520	NY	OU89850	Volvo B6	Åbenrå	B29D	1996
5521	NY	SZ89492	MAN 14.220	-	NC27D	2004
5522	NY	TS97877	Volvo B6	-	B29D	1996
5523	NY	UL95101	Volvo B12M	-	B47D	2006
5524	NY	UR91648	Volvo B12M	-	B47D	2006
5525	NY	UR91649	Volvo B12M	-	B47D	2006
5526	NY	PT95352	Volvo B10M	Åbenrå	BC43D	1999
5527	NY	PX96180	Volvo B10M	Åbenrå	BC43D	1999
5528	NY	UY93624	Volvo B12M	-	B47D	2006

5529-5536 — Volvo B10M — Åbenrå — BC43D — 1999

5529	NY	PX96190	**5531**	NY	PX96172	**5534**	NY	PX96176	**5536**	NY	PX9178
5530	NY	PX96192	**5532**	NY	PX96174	**5535**	NY	PT95334			

5537	NY	PL91046	Volvo B6	-	B29D	1999
5538	NY	TS94839	Volvo B12M 13.7m	-	BC47D	2005
5539	NY	TS94840	Volvo B12M 13.7m	-	BC47D	2005
5540	NY	OS89924	Volvo B10M	-	B44D	1996
5541	SK	OS89936	Volvo B10M	-	B44D	1996
5542	SK	OS89900	Volvo B10M	-	B44D	1996
5543	NY	PJ97619	Volvo B10M	-	B44D	1998
5544	SK	OS89908	Volvo B10M	-	B44D	1996
5545	PR	NM90222	DAB	-	B44D	1993
5546	PR	TD96637	Mercedes-Benz Sprinter 315	Mercedes-Benz	M8	2004
5547	PR	TD96639	Mercedes-Benz Sprinter 315	Mercedes-Benz	M8	2004
5554	SK	PB89168	Volvo B10M	-	B44D	1997
5557	EJ	PL97643	Volvo B10M 10.2m	-	B39D	1998
5559	KO	UV90046	Volvo B12M 13.7m	-	B47D	2002
5560	KO	UV90047	Volvo B12M 13.7m	-	B47D	2002

Aalborg buses use two liveries one of which is seen in this view of Scania OmniLink 4427, TB90831 while it waits time on route 2 at the rail station. Flags fly to commemorate a royal family birthday. *Bill Potter*

5562-5575

| | | Volvo B10M | | | Åbenrå | | | B44D | | 1998 |

5562	SL	RZ92525	**5566**	SK	PE94567	**5570**	SK	PJ97629	**5573**	SK	PJ97692
5563	SK	PE94533	**5567**	SK	PJ97620	**5571**	SK	PJ97636	**5574**	SK	PJ97709
5564	SK	PB89258	**5568**	SK	PJ97625	**5572**	SK	PJ97637	**5575**	SK	PE94507
5565	SK	PE94565	**5569**	SK	PJ97624						

| **5576** | UD | XE92609 | Volvo B10BLE | | - | | | N47D | | 1999 |

5577-5588

| | | Volvo B10M | | | | | | B44D | | 1998 |

5577	SL	PB89174	**5580**	SL	PB89267	**5584**	SL	PB89293	**5587**	SL	PC92292
5578	SL	PB89265	**5582**	SL	PB89285	**5585**	SL	PC92290	**5588**	SL	PC92294
5579	SL	PB89266	**5583**	SL	PZ92815	**5586**	SL	PC92291			

5593	SL	SN97729	Volvo B12M 13.7m	-	BC47D	2003
5594	RI	PJ97702	Volvo B10M	-	B44D	1998
5595	SL	PJ97703	Volvo B10M	-	B44D	1998
5596	SL	PJ97716	Volvo B10M	-	B44D	1998
5597	RI	SN97728	Volvo B12M	-	B44D	2003
5598	SL	SN97748	Volvo B12M	-	B44D	2003
5599	SL	SN97740	Volvo B12M	-	B44D	2003
5600	RI	PE94537	Volvo B10BLE	-	N32D	1998
5601	HA	PE94486	Volvo B10BLE	-	N32D	1998
5602	SV	PE94487	Volvo B10BLE	-	N32D	1998
5603	MA	PE94488	Volvo B10BLE	-	N32D	1998
5604	SE	PE94504	Volvo B10BLE	-	N32D	1998
5605	UD	PE94505	Volvo B10BLE	-	N32D	1998
5606	UD	XE92630	Volvo B10BLE	-	N32D	1998
5607	RI	PE94520	Volvo B10BLE	-	N32D	1998
5608	SE	OS89899	Volvo B10M	-	B45D	1996
5609	ST	SN97878	Volvo B12M 13.7m	-	B55D	2003
5610	ST	SU97087	Volvo B12M	-	B44D	2003
5611	ST	SU97085	Volvo B12M	-	B44D	2003
5612	MA	UE92316	Mercedes-Benz Vario 0614	Mercedes-Benz		1998

5613-5626 — Volvo B10M — Åbenrå — B47D — 1999

5613	ST	PL97649	**5617**	ST	PX96183	**5621**	ST	PX96186	**5624**	FS	PX96173
5614	ST	PL97648	**5618**	ST	PX96184	**5622**	ST	PX96188	**5625**	FN	PT95333
5615	ST	PX96181	**5619**	ST	PX96185	**5623**	ST	PX96191	**5626**	ST	PT95332
5616	ST	PX96182	**5620**	ST	PX96179						

5627-5630 — Volvo B10MA — Åbenrå — AB56D — 1999

5627	ST	PX96224	**5628**	ST	PX96225	**5629**	ST	PX96226	**5630**	ST	PX96227

5631	ST	PR94857	Mercedes-Benz OB6140	-	-	1999
5632	ST	PL91038	Mercedes-Benz OB6140	-	-	1999
5633	ST	OZ92464	Volvo B10M	Åbenrå	B47D	1997
5634	ST	TD96638	Mercedes-Benz Vario O614	Mercedes-Benz		2004
5635	ST	TK89525	Mercedes-Benz Vario O614	Mercedes-Benz		2004
5636	ST	PT95335	Volvo B10M	Åbenrå	B47D	1997
5637	ST	VH94955	Volvo B12M	Volvo	B47D	2007
5638	ST	XD91378	Volvo B7R	Volvo	N47D	2008
5639	ST	XD91379	Volvo B7R	Volvo	N47D	2008

5701	ND	VJ92443	DAB Citybus S11	DAB	N15D	1995
5702	ND	RN88117	Volvo B10BLE	Åbenrå	N32D	2000
5703	ND	TD95436	Volvo B10M	Åbenrå	B47D	1987
5704	ND	LP94443	Volvo B10M	Åbenrå	B47D	1988
5705	ND	NB89308	Volvo B10M	Åbenrå	B47D	1992
5706	NY	OU94973	Volvo B6LE	-		1997
5707	ND	OV91931	Volvo B10M	Åbenrå	B47D	1997

5708-5712 — Volvo B10M HLB — Åbenrå — BC42D — 1997-98

5708	ND	PB89085	**5710**	ND	OZ94283	**5711**	ND	PC89056	**5712** ND PC89105
5709	ND	OZ94278							

5713-5717 — Volvo B10M — Åbenrå — B47D — 1999

5713	ND	RC91287	**5715**	ND	RC91289	**5716**	ND	RC91290	**5717** ND RC91291
5714	ND	RC91288							

5718	ND	RD94746	DAB Citybus S11	DAB	N15D	1999
5719	ND	RD94747	DAB Citybus S11	DAB	N15D	1999
5720	ND	RD90971	Volvo B10M	Åbenrå	B47D	1999
5721	ND	RE94201	Volvo B10M	Åbenrå	B47D	1999
5722	ND	RP91142	Volvo B10BLE	Åbenrå	N32D	2000
5723	ND	RT95351	MAN 12.220 9m	-	N26D	2001

5724-5730 — Volvo B10M HLB — Åbenrå — BC44D — 2001

5724	ND	RT95270	**5726**	ND	RT95272	**5728**	ND	RT95274	**5730** ND RT95276
5725	ND	RT95271	**5727**	ND	RT95273	**5729**	ND	RT95275	

5731-5743 — Scania OmniLink — Scania — N45D — 2002-05

5731	ND	SM97681	**5735**	ND	TJ88588	**5738**	ND	TJ88591	**5741**	ND	TJ88593
5732	ND	TJ88585	**5736**	ND	TJ88589	**5739**	ND	TJ88555	**5742**	ND	TJ88594
5733	ND	TJ88586	**5737**	ND	TJ88590	**5740**	ND	TJ88592	**5743**	ND	TZ97302
5734	ND	TJ88587									

5744	ND	UU97057	Volvo B12M	Volvo	B47D	2006
5745	ND	PP94329	DAB Citybus S11	DAB	N15D	1998
5746	ND	ME94417	DAF/DAB S12	DAB	B47D	1990
5747	KI	PP89432	Volvo B10M	Åbenrå	BC44D	1998
5748	KI	RN88014	Volvo B10M	Åbenrå	BC44D	2000
5749	NY	OD92656	Volvo B10M	Åbenrå	B47D	1995
5750	NY	OD92657	Volvo B10M	Åbenrå	B47D	1995
5751	KI	OD92658	Volvo B10M	Åbenrå	B47D	1995
5752	NY	OX94453	Volvo B10M	Åbenrå	B47D	1995
5753	NK	RT95350	MAN 12.220 9m	-	N26D	2001

5754-5759 — MAN 18.310 — MAN — N47D — 2003

5754	NK	SX88391	**5756**	NK	SX88393	**5758**	NK	SX88395	**5759** NK SX88396
5755	NK	SX88392	**5757**	NK	SX88394				

5760	KI	PX96500	MAN 12.220 9m	-	B16D	1999
5761	NK	RT95499	MAN 12.220 9m	-	B16D	1999

5762-5768 — Volvo B10BLE — - — N32D — 1997

5762	RI	PP89362	5764	MA	OV91887	5766	RI	XD93063	5768	MA	OV91896
5763	HE	OV91884	5765	MA	OV91886	5767	RI	XD93037			

5769-5776 — Volvo B10M HLB — Åbenrå — BC44D — 1998-2001

5769	HO	PP89434	5771	SL	RT95266	5774	RI	PE91021	5776	RI	RD91919
5770	RI	RT95267	5772	HO	RT95265	5775	RI	RD91904			

5777	RI	MR93432	Volvo B10M	Åbenrå	B47D	1991

5778-5784 — Volvo B10M HLB — Åbenrå — BC44D — 2001

5778	RI	RT95277	5780	RI	RT95262	5782	RI	RT95264	5784	RI	RT95269
5779	RI	RT95261	5781	RI	RT95263	5783	RI	RT95268			

5785-5789 — Volvo B12M HLB — Volvo — NC44D — 2006

5785	RI	UU97026	5787	RI	UU97104	5788	RI	UU97034	5789	RI	UU97040
5786	RI	UU97025									

5792-5795 — Volvo B10M HLB — Åbenrå — BC44D — 1996

5792	SO	OL96921	5793	SO	OM97903	5794	SO	OM97918	5795	SO	OM97927

5796	ND	XD91377	Volvo B7R	Volvo	N47D	2008
6023	JK	DAW714	Peugeot Partner	Peugeot	M8	2006
6030	JK	WCE566	Peugeot Partner	Peugeot	M8	2006
6101	JG	HSO203	Scania L113CLB	Scania	B55D	1995
6102	JD	HSL103	Scania L113CLB	Scania	B55D	1995
6103	HB	PFT339	Scania L113CLB	Carrus	B52D	1994
6104	HB	OL97128	Volvo B10LA	Säffle S	N--D	1996
6105	SD	RL92.435	Volvo B10LA	Säffle S	N--D	2000
6112	JD	HLR063	Scania L113CLB	DAB	B55D	1995
6113	JD	HTS063	Scania L113CLB	DAB	B55D	1995
6114	JK	TFY796	Volvo B10B	Carrus Delta Star	B42D	1993
6117	SD	GNX410	Scania CN113ALB	Scania	AB70D	1996
6118	SD	GOE130	Scania CN113ALB	Scania	AB70D	1996

6126-6132 — Scania L113TLL — Carrus — AB56D — 1996

6126	JD	AEO541	6128	SV	AEO661	6130	SV	AES611	6132	JG	ASX602
6127	JD	AER941	6129	JG	AES661	6131	SD	ASX692			

6133	JD	RZ88061	Scania OmniLine L94UB	DAB	N47D	2002
6134	SD	SD91475	Scania OmniLine CL94UB 12m	Scania	N47D	2002
6135	SD	SL96650	Scania OmniLine CL94UB 12m	Scania	N47D	2002

6136-6142 — Scania L94UB — Vest — N--D — 2001

6136	HB	SCH145	6138	HB	SCH094	6140	HB	SCH106	6142	HB	SKL304
6137	HB	SCH103	6139	HB	SCH175	6141	HB	SKL439			

6143	JG	DLE670	Volvo B10M-70	Vest	AB55D	1997
6144	JG	DLE630	Volvo B10M-70	Vest	AB55D	1997

6146-6159 — Scania L94UB 14.8m — Lahti — N38D — 2003

6146	JD	STU583	6150	JG	TPM226	6154	JG	TPL931	6157	JG	TPL703
6147	JD	TPM241	6151	JG	TPM082	6155	JG	TPL922	6158	JG	TPL691
6148	JD	TPM238	6152	JG	TPM079	6156	JG	TPL916	6159	JG	TPL673
6149	JD	TPM232	6153	JG	TPM073						

6160-6166 — Scania L94UB 13.5m — Vest — N34D — 2003

6160	JD	TPL913	6162	JD	TPL955	6164	JD	TPL712	6166	JD	TPM811
6161	JD	TPL940	6163	JD	TPL685	6165	JD	TPM802			

6167-6171 — Scania L94UB 13.5m — Vest — N34D — 2003

6167	SV	TSX610	6169	SV	TSX601	6170	SV	TSX559	6171	SV	TSX583
6168	SV	TSX562									

6172-6176 — Scania L94 UB 14.8m — Vest — N38D — 2003

6172	SV	TSX574	6174	SV	TSX532	6175	SV	TSX538	6176	SV	TSX547
6173	SV	TSX580									

6198	SV	RGP523	Scania L94 UB	Vest	N34D	2000
6199	SD	RGP529	Scania L94 UB	Vest	N34D	2000

Malmo in Sweden is the location for this view of tri-axle Scania OmniCity 6222, XDB832. Recent awards of contracts have increased the numbers of buses which Arriva operates in the country. *Tom Johnson*

6221-6263

Scania OmniCity CL94 UB6   Scania   N48D   2005

6221	SD	XDC498	**6232**	SD	XDB747	**6243**	SD	XDB772	**6254**	SD	XDB682
6222	SD	XDB832	**6233**	SD	XDB667	**6244**	SD	XDB761	**6255**	SD	XDB762
6223	SD	XDB822	**6234**	SD	XDB652	**6245**	SD	XDB702	**6256**	SD	XDB647
6224	SD	XDB707	**6235**	SD	XDB641	**6246**	SD	XDB771	**6257**	SD	XDC037
6225	SD	XDB692	**6236**	SD	XDB632	**6247**	SD	XDB762	**6258**	SD	XDB841
6226	SD	XDB827	**6237**	SD	XDB631	**6248**	SD	XDB757	**6259**	SD	XDC053
6227	SD	XDB701	**6238**	SD	XDC453	**6249**	SD	XDB737	**6260**	SD	XDB681
6228	SD	XDB671	**6239**	SD	XDB637	**6250**	SD	XDB642	**6261**	SD	XDB651
6229	SD	XDB691	**6240**	SD	XDB831	**6251**	SD	XDB697	**6262**	SD	XDB847
6230	SD	XDB661	**6241**	SD	XDB622	**6252**	SD	XDB651	**6263**	SD	XDB862
6231	SD	XDB687	**6242**	SD	XDB617	**6253**	SD	XDB857			

6264-6272

Scania CL94 UB6   Scania   N49D   2007

6264	SD	XYD388	**6267**	SD	XYD497	**6269**	SD	XYD508	**6271**	SD	XYD397
6265	SD	XYD478	**6268**	SD	XYD387	**6270**	SD	XYD417	**6272**	SD	XYD367
6266	SD	XYD378									

6300-6309

MAN Lion's City G   MAN   AN46D   2007

6300	MO	ALR532	**6303**	MO	ALR792	**6306**	MO	FFW958	**6308**	MO	FEC737
6301	MO	ALR661	**6304**	MO	ALR642	**6307**	MO	CBJ997	**6309**	MO	ALS333
6302	MO	ALR790	**6305**	MO	ALR652						

6310-6313

MAN Lion's City G   MAN   AN46D   2008

6310	MO	BNA670	**6311**	MO	BNA886	**6312**	MO	BNH454	**6313**	MO	BNH426

6359-6362

Scania OmniCity   Scania   N32D   1999

6359	MO	MZA536	**6360**	MO	FMS873	**6361**	MO	FMS686	**6362**	MO	MNV681

6368	TG	SKK910	Neoplan Euroliner N3316/3L	Neoplan	C62D	2001
6369	TG	OAB577	Volvo B10M-65	Volvo	BC54D	1999
6370	TG	JLG734	Volvo B10M-65	Volvo	BC54D	1999
6371	TG	KCY735	Volvo B10M-65	Volvo	BC54D	1999

6391-6398 — Scania OmniLink CL94UB — Scania — N32D — 2001

6391	u	SFW553	6393	JL	SHA709	6395	JL	SHB571	6397	JL	SOB811
6392	SD	SFW559	6394	JL	SHA715	6396	JL	SOB844	6398	JL	SOB832

6399	AN	TUS256	DAB GS200 8.6m	DAB	N18D	1996
6400	AN	TWG436	DAB GS200 8.6m	DAB	N18D	1996
6424	AN	DHE670	Volvo B10BLE	Carrus City L	N36D	1997
6425	AN	DHE680	Volvo B10BLE	Carrus City L	N36D	1997

6451-6458 — Scania OmniCity CN94UB (cng) — Scania — N32D — 1998-99

6451	TG	JKF208	6453	TG	JJZ208	6455	TG	JKA158	6457	TG	DSO512
6452	AN	JKA408	6454	u	JKC218	6456	TG	JJZ138	6458	TG	DSO632

6461-6479 — Scania OmniCity CL94UB — Scania — N32D — 2001

6461	JL	SEA778	6466	JL	SFA697	6471	JL	SFA682	6476	JL	SFW565
6462	JL	SEA769	6467	JL	SFC013	6472	JL	SFA679	6477	JL	SFW595
6463	JL	SEA841	6468	JL	SFA691	6473	JL	SFW571	6478	JL	SFW538
6464	JL	SEA847	6469	JL	SFA796	6474	JL	SFW580	6479	JL	SHA700
6465	JL	SFA700	6470	JL	SFA685	6475	JL	SFX187			

6482	HB	SHU676	DAB Citybus S11	DAB	B36D	1994
6485	HB	SGW388	DAB Citybus S11	DAB	B36D	1994
6486	HB	SJK781	DAB Citybus S11	DAB	B36D	1994

6488-6493 — DAB Citybus S11 — DAB — B36D — 2000

6488	TG	RGP517	6490	TG	RGP574	6492	HB	RGP568	6493	HB	RGP610
6489	TG	RGP442	6491	TG	RGP577					

6516-6538 — Volvo B10BLE (cng) — Volvo Åbenrå — N30D — 2001

6516	MO	SHC256	6522	MO	SHC274	6528	MO	SHC727	6534	MO	SHB772
6517	MO	SHC259	6523	MO	SHC277	6529	MO	SHC736	6535	MO	SHB781
6518	MO	SHC252	6524	MO	SHC280	6530	MO	SHC739	6536	MO	SHD922
6519	MO	SHC265	6525	MO	SHC283	6531	MO	SHC748	6537	MO	SHD925
6520	MO	SHC288	6526	MO	SHC708	6532	MO	SHB760	6538	MO	SHD931
6521	MO	SHC271	6527	MO	SHC718	6533	MO	SHB763			

6539-6544 — Volvo B10LA (cng) — Säffle — AN48D — 2001

6539	MO	SHH016	6541	MO	SHH025	6543	MO	SHH031	6544	MO	SHH034
6540	MO	SHH019	6542	MO	SHH028					

6545-6550 — MAN NL313 (cng) — MAN — N30D — 2005

6545	HB	XBD788	6547	HB	XBC066	6549	HB	XBC047	6550	HB	XBC057
6546	HB	XBC056	6548	HB	XBC202					

6551-6558 — MAN NL313 (cng) — MAN — N30D — 2006-08

6551	HB	RZS673	6553	HB	ARC613	6555	HB	ARC374	6557	HB	BNA655
6552	HB	SAM553	6554	HB	DDB634	6556	HB	ARC405	6558	HB	BNA650

6559	HB	DCA850	Volvo B10L (cng)	Carrus	N32D	1999
6568	HB	DCE600	Volvo B10L (cng)	Carrus	N32D	1999
6570	HB	OOP953	Volvo B10L (cng)	Carrus	N32D	1999

6576-6626 — MAN NL313 (cng) — MAN — N30D — 2005

6576	HB	XBC076	6589	HB	XBC107	6602	HB	XBC211	6615	HB	XBC241
6577	HB	XBC091	6590	HB	XBC152	6603	HB	XBC212	6616	HB	XBC261
6578	HB	XBC071	6591	HB	XBC171	6604	HB	XBC176	6617	HB	XBC257
6579	HB	XBC077	6592	HB	XBC067	6605	HB	XBC206	6618	HB	XBC231
6580	HB	XBC092	6593	HB	XBC157	6606	HB	XBC217	6619	HB	XBC247
6581	HB	XBC117	6594	HB	XBC177	6607	HB	XBC221	6620	HB	XBC256
6582	HB	XBC136	6595	HB	XBC192	6608	HB	XBC232	6621	HB	XBC291
6583	HB	XBC132	6596	HB	XBC141	6609	HB	XBC216	6622	HB	XBC262
6584	HB	XBC082	6597	HB	XBC187	6610	HB	XBC227	6623	HB	XBC236
6585	HB	XBC131	6598	HB	XBC196	6611	HB	XBC242	6624	HB	XBC252
6586	HB	XBC142	6599	HB	XBC222	6612	HB	XBC237	6625	HB	XBC266
6587	HB	XBC156	6600	HB	XBC207	6613	HB	XBC197	6626	HB	XBC267
6588	HB	XBC166	6601	HB	XBC201	6614	HB	XBC246			

Purchased for a new contract in the city of Malmo, and wearing Skånetrafiken colours is MAN 6302, ALR790, one of fourteen MAN Lion's City G buses operating the network. *Tom Johnson*

6627-6632

MAN Lion's City NL313 (cng)   MAN   N30D   2006

6627	HB	XMW744	**6629**	HB	XMW749	**6631**	HB	XSY560	**6632**	HB	RZT325
6628	HB	XMW714	**6630**	HB	XMW748					

6633	HB	REO601	MAN Lion's City A21 12m	MAN	N29D	2009
6634	HB	REO601	MAN Lion's City A21 12m	MAN	N29D	2009

6641-6649

MAN Lion's City A21 (cng)   MAN   N30D   2008

6641	JK	BLS716	**6644**	JK	BLS550	**6646**	JK	BLS634	**6648**	JK	BLS685
6642	JK	BLS726	**6645**	JK	EEE013	**6647**	JK	BLS645	**6649**	JK	BLS681
6643	JK	CDL251								

6650-6667

MAN Lion's City NL313 (cng)   MAN   N30D   2006

6650	MO	RZT460	**6655**	MO	TRP760	**6660**	MO	XWE499	**6664**	MO	ALG553
6651	MO	SRO031	**6656**	MO	TRO121	**6661**	MO	RZT295	**6665**	MO	ALG597Skånetrafiken
6652	MO	RZT445	**6657**	MO	RZS745	**6662**	MO	RZS703	**6666**	MO	ALG643
6653	MO	RZT373	**6658**	MO	RZT262	**6663**	MO	ALG489	**6667**	MO	DDS656
6654	MO	RZT394	**6659**	MO	RZT388					

6703-6729

Scania Omnicity CN94UA 18m   Scania   AN44D   2001

6703	JK	SCH199	**6710**	JK	SDD955	**6717**	JK	SEA832	**6724**	JK	SFA721
6704	JK	SCH181	**6711**	JK	SDD952	**6718**	JK	SEA820	**6725**	JK	SFA706
6705	JK	SCH184	**6712**	JK	SDE007	**6719**	JK	SEA817	**6726**	JK	SFW556
6706	JK	SCH226	**6713**	JK	SDD958	**6720**	JK	SEA784	**6727**	JK	SFW568
6707	JK	SCH220	**6714**	JK	SDD949	**6721**	JK	SEA775	**6728**	JK	SFW547
6708	JK	SDD964	**6715**	JK	SDM460	**6722**	JK	SEA766	**6729**	JK	SFZ649
6709	JK	SDD967	**6716**	JK	SDM457	**6723**	JK	SFA859			

6730-6744

Volvo B10BLE   Åbenrå   N45D   2001-02

6730	JK	DEM368	**6734**	JK	BUL618	**6738**	JK	BRS808	**6742**	JK	CDT528
6731	JK	BOO598	**6735**	JK	BUL128	**6739**	JK	BUK508	**6743**	JK	SD88214
6732	JK	CDT568	**6736**	JK	BUL298	**6740**	JK	BUL028	**6744**	JK	SD88166
6733	JK	BRS048	**6737**	JK	BRS748	**6741**	JK	BUK818			

6745-6750 — MAN Lion's City A26 14.7m — MAN — N40D — 2008

6745	JK	TXL604	6747	JK	TXF553	6749	JK	TWE682	6750	JK	XYF875
6746	JK	BMD390	6748	JK	TXG121						

6751-6759 — MAN Lion's City A78 12m — MAN — N30D — 2008

6751	JK	BHZ262	6754	JK	BHZ323	6756	JK	BHZ903	6758	JK	BHZ952
6752	JK	EEB599	6755	JK	CXF158	6757	JK	BHZ983	6759	JK	EBC319
6753	JK	BHZ283									

6760-6770 — MAN Lion's City A78 12m — MAN — N30D — 2008

6760	JK	ECB032	6763	JK	CZG247	6766	JK	BJB079	6769	JK	BHZ328
6761	JK	CWF079	6764	JK	BHZ210	6767	JK	BHZ353	6770	JK	BHZ312
6762	JK	BHZ184	6765	JK	BHZ964	6768	JK	BHZ463			

6830-6843 — Volvo B10BLE — Åbenrå — N32D — 2000

6830	MO	RWG871	6834	MO	RMK088	6838	MO	RMK106	6841	MO	RMK607
6831	MO	RMJ985	6835	MO	RMK091	6839	MO	RMK109	6842	MO	RMK613
6832	MO	RMK004	6836	MO	RMK094	6840	MO	RMK601	6843	MO	RMK619
6833	MO	RMK082	6837	MO	RMK103						

6844	AN	FGC857	Volvo B12BLE 13.8m	Åbenrå	N25D	2004
6845	AN	ANM627	Volvo B12BLE 13.8m	Åbenrå	N52D	2004
6846	AN	ANM307	Volvo B12BLE 13.8m	Åbenrå	N52D	2004

6849-6867 — Volvo B10BLE — Åbenrå — N32D — 2000

6849	MO	RJS325	6854	MO	RJS337	6859	MO	RJS145	6864	MO	RJS265
6850	MO	RJS328	6855	MO	RJS340	6860	MO	RJS169	6865	MO	RJS268
6851	MO	RRD190	6856	MO	RJS343	6861	MO	RJS172	6866	MO	RJS271
6852	MO	RJS331	6857	MO	RJS346	6862	MO	RJS178	6867	MO	RJS277
6853	MO	RJS334	6858	MO	RJS349	6863	MO	RJS256			

6929	SD	TDD520	Scania L94 UB 13.8m	Scania	N55D	2002

7001-7025 — Scania OmniLink 18m — Scania — AN56D — 2009

7001	SX	ECD210	7008	SX	ECD207	7014	SX	CDP083	7020	SX	BOK472
7002	SX	BOK685	7009	SX	BOK364	7015	SX	BOK391	7021	SX	BOK521
7003	SX	BOK691	7010	SX	BOK443	7016	SX	DEF951	7022	SX	BOK562
7004	SX	BOK841	7011	SX	BOK426	7017	SX	BOK423	7023	SX	BOK540
7005	SX	BOK855	7012	SX	DEF943	7018	SX	BOK400	7024	SX	CDP093
7006	SX	BOK857	7013	SX	BOK382	7019	SX	BOK376	7025	SX	BOK541
7007	SX	BOK363									

7026-7043 — Scania OmniLink (Ethanol) — Scania — AN56D — 2009

7026	SM	BON617	7031	SM	DEG171	7036	SM	BON674	7040	SM	BON691
7027	SM	BON626	7032	SM	BON645	7037	SM	BON654	7041	SM	DEG180
7028	SM	BON627	7033	SM	DEG173	7038	SM	EET012	7042	SM	BON696
7029	SM	CDP325	7034	SM	CDP326	7039	SM	BON646	7043	SM	BON685
7030	SM	BON631	7035	SM	BON660						

7044-7056 — Scania CL94UA (Ethanol) — Scania — AN59D — 2006

7044	SM	XOS891	7048	SM	XOS837	7051	SM	XOS992	7054	SM	XOS842
7045	SM	XOS847	7049	SM	XOT037	7052	SM	XOS896	7055	SM	XOS907
7046	SM	XOS871	7050	SM	XOS877	7053	SM	XOS952	7056	SM	XOS946
7047	SM	XOS846									

7057-7068 — Scania CL94UB — Scania — N34D — 2006

7057	SM	XLG672	7060	SM	XLK365	7063	SM	XLK386	7066	SM	XLK326
7058	SM	XLK396	7061	SM	XLK356	7064	SM	XLK376	7067	SM	XLK316
7059	SM	XLK335	7062	SM	XLK336	7065	SM	XLK375	7068	SM	XLK505

7069	SM	KLM154	Scania CN113 ALB	Lahti	AB61D	1995
7070	SM	KKC274	Scania CN113 ALB	Lahti	AB61D	1995

7071-7075 — Volvo B10M-A — Vest — AB67D — 2000-02

7071	SX	SOR313	7073	SX	RJZ184	7074	SX	RJZ229	7075	SX	RJZ232
7072	SX	SOR322									

7076	SM	DKK950	Scania CN113 ALB	Lahti	AB61D	1997

When pictured at Arhus, these two Volvo buses had just been added to the Arriva network. Now numbered 8701, RL92430, like many of the older Volvo buses with Vest bodywork is now in the reserve fleet. *Bill Potter*

7077-7088

		Volvo B10M-A		Vest		AB67D	1999-2002		
7077	SX	OWR528	**7080**	SX	SOR181	**7083**	SX	SOR256	**7086**	SX	SOR295
7078	SX	SJJ643	**7081**	SX	SOR226	**7084**	SX	SOR262	**7087**	SX	SOR304
7079	SX	SJJ646	**7082**	SX	SOR235	**7085**	SX	SOR268	**7088**	SX	SOR307

7089-7092

		Scania CN113 CLB		Lahti		B41D	1997		
7089	SM	DAG659	**7090**	SM	DAH559	**7091**	SM	DBZ899	**7092**	SM	DBZ919

7093	SX	SSJ640	Volvo B10M-A		Vest		AB67D	2001		

7094-7107

		Scania CN113 CLB		Lahti		B41D	1995-96		
7094	SM	KGF424	**7098**	SM	JKC435	**7102**	SM	BRA914	**7105**	SM	BRB854
7095	SM	KHM134	**7099**	SM	JKF085	**7103**	SM	BRB614	**7106**	SM	DEK769
7096	SM	JLZ175	**7100**	SM	JMG345	**7104**	SM	BRB774	**7107**	SM	DEK699
7097	SM	JMZ145	**7101**	SM	BRA814					

7108	SX	DEL389	Volvo B10BLE	-	N43D	2001
7109	SM	DEL629	Volvo B10BLE	-	N43D	2001
7110	SX	EFB999	Volvo B10BLE	-	N43D	2001
7111	SM	UNS440	Scania	Berkhof	BC32D	1998
7112	SM	BPB720	Scania	Berkhof	BC32D	1998
7113	SX	BOX410	Volvo B10BLE	-	N43D	2001
7114	SM	XYL909	Scania	Berkhof	BC32D	1998
7115	SM	EUZ379	Scania	Berkhof	BC32D	1998
7116	SX	UOD900	Volvo B10BLE	-	N43D	2001
7117	SM	XTH909	Scania	Berkhof	BC32D	1998
7118	SM	BSL039	Scania	Berkhof	BC32D	1998

7119-7143

		Volvo B10BLE		Vest		N31D	1999-2002		
7119	SX	BOX440	**7126**	SM	SXT649	**7132**	SM	BSK619	**7138**	SM	XYF360
7120	SX	DEH221	**7127**	SM	BRP909	**7133**	SM	SXN079	**7139**	SX	WHL260
7121	SM	XZZ919	**7128**	SM	BOO004	**7134**	SM	XZC159	**7140**	SX	BSK689
7122	SM	UNG090	**7129**	SM	BRP919	**7135**	SM	XLN959	**7141**	SX	SYS349
7123	SM	CDU469	**7130**	SM	XZN069	**7136**	SM	BRP999	**7142**	SM	BOX370
7124	SM	BOX130	**7131**	SM	CDR840	**7137**	SM	CDW609	**7143**	SM	XXA809
7125	SM	XTB179								

7144	SM	GPR410	Scania CN113 ALB		Lahti				AN70D	1996	
7145	SM	DHD620	Volvo B10BLE		-				N36D	1997	
7146	SM	DHD690	Volvo B10BLE		-				N36D	1997	
7147	SM	DHE580	Volvo B10BLE		-				N36D	1997	
7148	SM	DFS589	Scania CN94 UB						N34D	1997	

7150-7157   Volvo B10BLE   Vest   N31D   1998-99

7150	SX	ATJ249	**7152**	SX	DEK119	**7154**	SX	BRR809	**7156**	SM	BPR750
7151	SX	ASB539	**7153**	SX	BRR759	**7155**	SM	UNG700	**7157**	SX	XJU534

7158-7162   Scania CN94 UB   -   N32D   2001

7158	SX	SFA796	**7160**	SX	SFA697	**7161**	SM	SFC013	**7162**	SM	SFA691
7159	SX	SHA736									

7163	SM	SCH208	Scania CN94 UA		Lahti			AN44D	2001
7164	SM	SCH208	Scania CN94 UA		Lahti			AN44D	2001

7165-7170   Volvo B10M   Vest   B40D   1998-99

7165	SM	JGC055	**7167**	SM	JGD015	**7169**	SM	GMW990	**7170**	SM	GMX740
7166	SM	JGC355	**7168**	SM	GMU630						

7171	SX	SC90691	Volvo B10BLE	-	N45D	2002
7198	SM	GPR410	Scania CN113 CLL	Lahti	N36D	1997
7199	SM	GPR410	Scania CN113 ALB	Lahti	AN63D	1997
7209	HG	NN91908	Volvo B10	Åbenrå	B39D	1994
7210	HG	NX90841	Volvo B10	Åbenrå	B39D	1994
7211	HG	OB94391	Volvo B10	Åbenrå	B39D	1995
8050	FY	NX93077	DAF SB220 12m	DAB	B47D	1994
8302	HV	TU97352	Volvo B10M	Åbenrå	B39D	1997
8303	HV	OV89881	DAF/DAB Citybus S12	Silkeborg	B47D	1997
8304	DJ	OV89882	DAF/DAB Citybus S12	Silkeborg	B47D	1997
8305	HV	OV89883	DAF/DAB Citybus S12	Silkeborg	B47D	1997
8308	FN	OZ91788	Scania/DAB	Silkeborg	N38D	1997
8417	OV	MS97.695	Volvo B10M	Åbenrå	B39D	1992
8419	FY	RH95.798	Volvo B10M	Vest	BC47D	2000
8421	FY	PX90675	Volvo B10M	Vest	BC47D	1999
8422	FY	PX90674	Volvo B10M	Vest	BC47D	1999
8423	FY	PX90671	Volvo B10M	Vest	BC47D	1999
8424	HA	PX90718	Volvo B10M	Vest	BC47D	1999
8425	RO	PZ96612	Volvo B10M	Vest	BC47D	1999
8426	MA	RK95274	Scania L94	DAB	BC47D	2000
8427	UD	RX97899	Scania OmniLink CL94 UB 12m	Scania	N47D	2001
8428	MA	SN97877	Volvo B12M	Åbenrå	B47D	2003
8430	HV	TN94780	Volvo B10M	Vest	B50D	2000
8431	HV	TN94781	Volvo B10M	Vest	B50D	2000
8432	HV	TN94757	Volvo B10M	Vest	B50D	2000
8433	HV	SL91898	Volvo B10M	Vest	B50D	2002
8434	MA	TU91783	Scania OmniLink CL94 UB 12m	DAB facelift	N47D	1998
8435	UD	RN88130	Volvo B10M	Åbenrå	B47D	2000
8436	AA	TB94707	Volvo B10M 12.7m	Vest	B50D	2004
8437	AA	TB94708	Volvo B10M	Vest	B50D	2004
8438	AA	PX96113	Volvo B10M	Vest	B50D	1999
8439	AA	PX96109	Volvo B10M	Vest	B50D	1999
8440	AA	RN88779	Volvo B10M	Vest	B50D	2000
8441	AA	RN88780	Volvo B10M	Vest	B50D	2000
8442	AA	RN88804	Volvo B10M	Vest	B50D	2000
8443	AA	RN88188	Volvo B10M	Vest	B50D	2000
8444	AA	RS92754	Volvo B10M	Vest	B50D	2000
8445	AA	PX96122	Volvo B10M	Vest	B50D	1999
8446	AA	PX96123	Volvo B10M	Vest	B50D	1999
8447	AA	RN88196	Volvo B10M	Vest	B50D	2000
8448	AA	PX96108	Volvo B10M	Vest	B50D	1999
8449	AA	RZ92524	Volvo B10M	Vest	B50D	2001
8450	AA	SD92720	Volvo B10M	Vest	B50D	2002
8451	AA	RS92753	Volvo B10M	Vest	B50D	2000
8452	AA	RL92253	Volvo B10M	Vest	B50D	2000
8453	AA	PX96129	Volvo B10M	Vest	B50D	1999
8454	AA	PX96130	Volvo B10M	Vest	B50D	1999
8455	AA	OD92511	Scania L113 CLB	Scania	B42D	1995
8456	AA	RD91795	Volvo B10M	Vest	B50D	1999
8457	AA	OD92510	Scania L113 CLB	Scania	B42D	1995
8458	AA	RL92426	Volvo B10M	Vest	B50D	2000
8459	AA	RN88178	Volvo B10M	Vest	B50D	2000

8460	AA	RN88179	Volvo B10M		Vest		B50D	2000		
8461	AA	RN88180	Volvo B10M		Vest		B50D	2000		
8462	AA	RL90808	Volvo B10M		Vest		B50D	2000		
8463	AA	RP90928	Volvo B10M		Vest		B50D	2000		
8464	AA	RL92413	Volvo B10M		Vest		B50D	2000		
8465	AA	RN88131	Volvo B10M		Vest		B50D	2000		
8466	AA	RN88187	Volvo B10M		Vest		B50D	2000		
8467	AA	PX96112	Volvo B10M		Vest		B50D	1999		
8468	AA	RN88129	Volvo B10M		Vest		B50D	2000		

8472-8486

Volvo B12BLE — Carrus — N47D — 2004

8472	AA	TB96602	8476	AA	TB96655	8480	AA	TB96608	8484	AA	TB95429
8473	AA	TB96603	8477	AA	TB96605	8481	AA	TB96654	8485	AA	TB96660
8474	AA	TB96657	8478	AA	TB96606	8482	AA	TB96659	8486	AA	TB96656
8475	AA	TB96604	8479	AA	TB96607	8483	AA	TB96658			

8487	AA	RL97023	Scania		Scania/DAB		N47D	2000		
8488	AA	TC90472	Scania OmniLine L94UB		DAB		N--D	2004		
8489	LO	UT95386	Scania 13.8m		Lahti Laventre		N49D	2006		
8490	HV	UT95387	Scania 13.8m		Lahti Laventre		N49D	2006		
8491	TY	UT95388	Scania 13.8m		Lahti Laventre		N49D	2006		
8492	AA	UT95389	Scania 13.8m		Lahti Laventre		N49D	2006		
8493	FJ	UT95399	Scania 13.8m		Lahti Laventre		N49D	2006		
8494	AA	UX91358	Scania OmniLink CL94UB 12m		Scania		N--D	2007		
8495	AA	UX91359	Scania OmniLink CL94UB 12m		Scania		N--D	2007		
8496	AA	UX91360	Scania OmniLink CL94UB 12m		Scania		N--D	2007		
8497	AA	UX91532	Scania 13.8m		Lahti Laventre		N49D	2007		
8498	AA	UY93706	Scania 13.8m		Lahti Laventre		N49D	2007		
8499	AA	VB94299	Scania 13.8m		Lahti Laventre		N49D	2007		
8500	AA	VH93298	Scania 13.8m		Lahti Laventre		N49D	2007		
8501	AA	UU96996	Volvo B12BLE 12m		Carrus		N44D	2006		
8502	AA	UX95679	Volvo B12BLE 12m		Carrus		N44D	2007		
8503	AA	UU96930	Volvo B12BLE 13.7m		Carrus		N46D	2006		
8504	AA	UU96931	Volvo B12BLE 13.7m		Carrus		N46D	2006		
8505	ES	OX91583	Scania OmniCity		Scania		N47D	1997		
8506	ES	RV95455	Scania OmniLink CL94 UB 12m		Scania		N31D	2001		
8507	ES	RZ88231	Scania OmniLink CL94 UB 12m		Scania		N31D	2001		
8508	SO	SD91538	Scania OmniLine		Scania			2002		

8509-8534

Scania OmniLink CL94 UB 12m — Scania — N31D — 2002

8509	ES	SH93739	8516	ES	SH93744	8523	ES	SH93749	8529	ES	SH93775
8510	ES	SH93770	8517	ES	SH93745	8524	ES	SH93750	8530	ES	SH93776
8511	ES	SH93716	8518	ES	SH93746	8525	ES	SH93771	8531	ES	SH93777
8512	ES	SH93740	8519	ES	SH93747	8526	ES	SH93772	8532	ES	SH93778
8513	ES	SH93741	8520	ES	SH93748	8527	ES	SH93773	8533	ES	SH93779
8514	ES	SH93742	8521	ES	SH93717	8528	ES	SH93774	8534	ES	SH93780
8515	ES	SH93743	8522	ES	SH93718						

8536	u	RL96987	Scania OmniLink CL94 UB 12m		DAB facelift		N47D	2000		
8537	RO	RN96159	Scania OmniLink CL94 UB 12m		DAB facelift		N47D	2000		

8543-8550

Volvo B12BLE — Carrus — N47D — 1998

8543	KO	OL99340	8545	KO	PP97785	8547	KO	PE94524	8549	KO	PE94538
8544	KO	PO97786	8546	KO	TL89709	8548	KO	PE94519	8550	KO	PE94539

8551	KO	TJ94029	Scania OmniLink CL94 UB 12m		Scania		N31D	2004		
8552	KO	TJ94030	Scania OmniLink CL94 UB 12m		Scania		N31D	2004		
8553	KO	RN88118	Volvo B12BLE 12m		Carrus		N44D	2000		
8554	KO	RZ88244	Scania OmniLink CL94 UB 12m		Scania		N31D	2001		
8555	ES	RZ88245	Scania OmniLink CL94 UB 12m		Scania		N31D	2001		
8556	KO	SC91809	Scania OmniLink CL94 UB 12m		Scania		N31D	2002		
8557	KO	SD91568	Scania OmniLink CL94 UB 12m		Scania		N31D	2002		

8558-8568

Volvo B12BLE — Volvo — N47D — 2002-03

8558	KO	SM96039	8561	KO	SM96050	8564	KO	SM96067	8566	KO	ST97883
8559	KO	SM96040	8562	KO	SM96051	8565	KO	SM96068	8568	KO	UR94549
8560	KO	SM96045	8563	KO	SM96052						

8569	HE	MS91106	Volvo B10M		Vest		B50D	2000		
8570	HE	NB89391	Volvo B10M		Vest		B50D	2000		
8571	u	RN96160	Scania OmniLink CL94 UB 12m		DAB facelift		N47D	2000		
8573	RM	NB89414	DAF/DAB S12		DAB		B47D	1990		
8574	RM	NB97560	DAF/DAB S12		DAB		B47D	1990		

A Sweden Volvo B10BLE with Åbenrå bodywork is 6836, RMK094, seen here in Malmo. *Tom Johnson*

8578	LA	NM90119	DAF/DAB S12		DAB			B47D	1990
8579	SJ	MS90996	Volvo B10M		Vest			B50D	2000
8580	SJ	MS90997	Volvo B10M		Vest			B50D	2000
8581	AR	RC89301	Mercedes-Benz Sprinter		Mercedes-Benz			M	1999
8583	RM	NX93003	Volvo B10M		Vest			B50D	1994
8584	RM	NZ93978	DAF/DAB S12		DAB			B47D	1995
8585	RM	OD92669	Volvo B10M		Vest			B50D	1995
8586	RM	OJ92348	Volvo B10M		Vest			B50D	2000
8587	u	RL92345	Volvo B10M		Vest			B50D	2000
8588	RM	NU93417	DAF/DAB S12		DAB			B47D	1995
8589	RM	NU93418	DAF/DAB S12		DAB			B47D	1995
8593	RM	OD92636	Volvo B10M		Vest			B50D	2000
8594	LA	TB92523	Mercedes-Benz Sprinter		Mercedes-Benz			M	1999
8595	RM	OL88383	Volvo B10M		Vest			B50D	2000
8599	HE	RL96986	Scania CL94 UB 12m		DAB facelift			N47D	2000
8601	RA	UR90590	Volvo B7RLE		Volvo 8700			N40D	2006
8602	RA	UR90591	Volvo B7RLE		Volvo 8700			N40D	2006

8603-8607

			Volvo B12BLE		Volvo			N39D	2003
8603	RA	ST97904	**8605**	RA	ST97910	**8606**	RA	ST97912	**8607**	RA	ST97919
8604	RA	ST97909								

8609-8617

			Volvo B10L		Volvo 8600			N39D	1996-98
8609	RA	OM97836	**8612**	RA	OX94374	**8614**	HE	PL97604	**8616**	RA	PL97617
8610	HE	OM97841	**8613**	HE	OX94379	**8615**	RA	PL97614	**8617**	RA	PL97630
8611	RA	OX94373								

8618-8623

			Volvo B10BLE		Volvo			N40D	2000
8618	RA	RP91114	**8620**	RA	RP91122	**8622**	RA	RP91115	**8623**	RA	RP91121
8619	RA	RP91123	**8621**	RA	RP91112					

8624	RA	VH95011	Volvo B7BLE		Volvo 8700			N40D	2007
8625	RA	VH95012	Volvo B7BLE		Volvo 8700			N40D	2007
8626	RA	VH95013	Volvo B7BLE		Volvo 8700			N40D	2007
8628	HE	MS91107	Volvo B10M		Vest			B50D	1994
8629	RY	NX93038	Volvo B10M		Vest			B50D	1994
8630	RY	NX93039	Volvo B10M		Vest			B50D	1994
8631	HE	NX93037	Volvo B10M		Vest			B50D	1994

8633	VI	RZ97955	Scania L94	Scania	N47D	2001
8634	HA	RE94208	Volvo B10M	Vest	B50D	1999
8635	HA	PX96187	Volvo B10M	Vest	B50D	1999
8636	RY	PJ97721	Volvo B10M	Vest	B50D	1998
8639	SJ	MC92536	Volvo B10M	Vest	B50D	1990
8641	SJ	NN89578	Volvo B10M	Vest	B50D	1993
8642	w	NN89579	Volvo B10M	Vest	B50D	1993
8643	SK	NN89598	Volvo B10M	Vest	B50D	1993
8644	u	RD91986	Volvo B10MA	Vest	AB66D	1999

8645-8649

			Scania L94	DAB facelift	N47D	2000

8645	BO	RL96988	**8647**	BO	RL96990	**8648**	BO	RL96991	**8649**	BO	RL96992
8646	BO	RL96989									

8650	u	RP90963	Volvo B10M 13.7m	Vest	B50D	2000
8651	SK	RL92470	Volvo B10BLE	Volvo	N40D	2000
8652	JE	RY88107	Scania L94	Scania	N47D	2001
8653	HO	PL97732	Volvo B10MA	Vest	AB66D	1998
8654	SK	OY97173	Volvo B10BLE	Volvo	N40D	1997
8655	SK	OY97172	Volvo B10BLE	Volvo	N40D	1997
8656	SK	TC94003	Volvo B10BLE	Volvo	N40D	1997
8657	RA	OM97804	Volvo B10M	Vest	B50D	1996
8658	LA	OS89983	Volvo B10MA	Vest	AB66D	1996
8659	RY	PJ97739	Volvo B10M	Vest	B50D	1998
8660	SO	RL96939	Scania CL94 UB 12m	DAB facelift	N47D	2000
8663	HE	OS89253	Volvo B10M	Vest	B50D	1996
8664	SK	OS89255	Volvo B10M	Vest	B50D	1996
8665	SK	OS89256	Volvo B10M	Vest	B50D	1996
8567	JE	TP93418	Scania OmniLink CL94 UB 12m	Scania	N31D	2005
8568	JE	TP93419	Scania OmniLink CL94 UB 12m	Scania	N31D	2005
8669	u	RP90962	Volvo B10M 13.7m	Vest	B50D	2000
8671	SK	OL97126	Volvo B10M	Vest	B40D	1996
8672	u	RM96995	Scania CL94 UB 12m	DAB facelift	N47D	2000
8573	AR	RY88119	Scania OmniLink CL94 UB 12m	Scania	N31D	2001
8574	AR	RZ99057	Scania OmniLink CL94 UB 12m	Scania	N31D	2001
8575	AR	TR95130	Scania OmniLink CL94 UB 12m	Scania	N31D	2005
8676	SA	RD94836	Scania CL94 UB 12m	DAB facelift	N47D	1999
8677	RA	UR90589	Volvo B7RLE	Volvo 8700	N40D	2006
8678	SK	OS89961	Volvo B10M	Vest	B50D	1996
8679	LA	UZ94742	Volvo B10M	Vest	B50D	1998
8680	RY	UX89357	Volvo B10M	Vest	B50D	1998
8681	RY	PJ97757	Volvo B10M	Vest	B50D	1998
8682	HA	UZ94743	Volvo B10M	Vest	B50D	1998
8683	HA	UZ94774	Volvo B10M	Vest	B50D	1998
8684	RY	PX96097	Volvo B10M	Vest	B50D	1999
8685	HA	RD94888	Scania CL94 UB 12m	DAB facelift	N47D	1999
8686	w	RD94889	Scania CL94 UB 12m	DAB facelift	N47D	1999
8688	LA	PE92003	Volvo B10M	Vest	B50D	1998
8691	u	UZ94845	Scania CL94 UB 12m	DAB facelift	N47D	2000
8692	UD	UZ95841	Scania CL94 UB 12m	DAB facelift	N47D	2000
8693	AR	UZ94897	Volvo B10M 13.7m	Vest	B50D	2000
8694	AR	RL92347	Volvo B10M 13.7m	Vest	B50D	2000
8695	AR	UZ94896	Volvo B10M 13.7m	Vest	B50D	2000
8596	u	RS97266	Scania OmniLine	Scania	NC47D	2000
8597	GR	RS97154	Scania OmniLine	Scania	NC47D	2000
8698	u	RM91513	Volvo B10MA	Voct	AB66D	2000
8699	RO	RL97017	Scania CL94 UB 12m	DAB facelift	N47D	2000
8700	u	RL97018	Scania CL94 UB 12m	DAB facelift	N47D	2000
8701	u	RL92430	Volvo B10M 13.7m	Vest	B50D	2000
8702	u	RP97563	Scania OmniLine	Scania	NC47D	2000

8703-8707

			Volvo B10M	Vest	B50D	2000

8703	SA	RL92431	**8705**	SA	RS90131	**8706**	u	RE92095	**8707**	u	RP97560
8704	SA	RE92067									

8708	u	RM91512	Volvo B10MA	Vest	AB66D	2000
8709	HO	RP97561	Volvo B12M	Volvo Carrus	BC44D	2000

8710-8717

			Scania OmniLine	Scania	NC47D	2000

8710	u	RP97564	**8712**	HA	RS97157	**8714**	u	RS97267	**8716**	u	RS97269
8711	HA	RS97156	**8713**	u	RS97158	**8715**	FA	RS97268	**8717**	u	RS97270

8718	u	RL97014	Scania CL94 UB 12m	DAB facelift	N47D	2000		
8719	u	RP97562	Scania OmniLine	Scania	NC47D	2000		
8720	UD	UZ94846	Scania OmniLine 13.6m	Lahti Flyer	NC55D	2001		

8721-8733   Scania OmniLine   Scania   NC47D   2000-01

8721	VI	RS97155	8725	HT	RT96586	8728	VI	RT96893	8731	RO	RT96896
8722	FA	RT96582	8726	HT	RT96587	8729	JE	RT96894	8732	u	RT96897
8723	GR	RT96583	8727	u	RT96892	8730	JE	RT96895	8733	RO	RT96584
8724	RO	RT96898									

8734	OD	RS90289	Volvo B10M	Vest	B50D	2001		
8735	OD	RT94703	Volvo B10M	Vest	B50D	2001		

8736-8739   Volvo B10MA   Vest   AB66D   2001

8736	HT	SB94592	8737	HT	SB94567	8738	OD	SB94594	8739	OD	SB94593

8740-8744   Scania OmniLine   Scania   NC47D   2001

8740	OD	RV95329	8742	FA	RT96899	8743	u	RT96585	8744	UD	RS96891
8741	OD	RV95330									

8745-8751   Volvo B12M   Volvo   BC55D   2002

8745	JE	SM95991	8747	GR	SM95992	8749	OD	SH90380	8751	OD	SM95889
8746	GR	SM96003	8748	OD	SH90381	8750	OD	SH90382			

8752	RA	SH93632	Scania OmniLine	Scania	NC47D	2002		
8753	RY	SM96002	Volvo B12M	Volvo Carrus	BC44D	2002		
8754	RA	SU97086	Volvo B12M	Volvo Carrus	BC44D	2003		
8755	OD	TT88410	Scania OmniLine	Scania	NC47D	2005		
8756	OD	TT88270	Scania OmniLine	Scania	NC47D	2005		
8757	HT	VH95033	Volvo B12M 13.7m	Volvo 8700	BC55D	2007		
8758	SA	VH94956	Volvo B12M 12m	Volvo 8700	BC47D	2007		

8759-8768   Volvo B12M 13.7m   Volvo 8700   BC55D   2007

8759	HT	VH95055	8762	HT	VH95083	8765	OR	VJ96917	8767	HT	VJ96988
8760	HT	VH95056	8763	HT	VH95097	8766	HT	VJ96980	8768	HT	VJ96987
8761	SK	VH95072	8764	HT	VJ96904						

8769	SK	VJ96981	Volvo B12M-A	Volvo 8700	AC65D	2007		
8770	SK	VJ97023	Volvo B12M 13.7m	Volvo 8700	BC55D	2007		
8771	SK	VN92492	Volvo B12M 13.7m	Volvo 8700	BC55D	2007		
8772	SK	VS88406	Volvo B12M-A	Volvo 8700	AC65D	2007		
8773	RA	VT91976	Volvo B7R	Volvo	N47D	2008		
8774	SK	VS95959	Volvo B10M	Vest	B50D	1999		
8775	u	VS96062	Volvo B10M	Vest	B50D	1999		
9022	RA	OZ94277	Volvo B10M	Vest	B50D	1999		

9026-9048   Volvo B10BLE   Volvo   N32D   1997-98

9026	HE	TV96476	9038	AA	PE94616	9042	VE	TV96500	9047	KK	PJ97559
9035	AA	PE94591	9040	KO	TV96477	9044	KK	PJ97560	9048	HT	RM91003

9060	SJ	KV94741	Volvo B10M	DAB Silkeborg	B39D	1986		
9061	SJ	KV94742	Volvo B10M	DAB Silkeborg	B39D	1986		
9066	HE	PJ89239	Neplan	Neoplan	B26D	1998		
9067	HE	PJ89235	Neplan	Neoplan	B26D	1998		

9070-9101   Volvo B10BLE   Volvo   N45D   1997-2000

9070	HE	OV91913	9083	HJ	RN90414	9090	ES	RN95489	9101	SM	BDM460
9082	HO	RN90413	9085	VE	OV92039						

9102	SM	KNH164	Scania CN113 ALB	Lahti	AN70D	1995		
9103	SM	KMU024	Scania CN113 ALB	Lahti	AN70D	1995		

Depots and codes: Copenhagen area - Amager (AM); Ejby (EJ); Falster (FA); Gladsaxe (GX); Glostrup (GL); Hillerød (HI); Holbæk (HO); Kokkedal (KK); Maribo (MA); Næstved (ND); Ringsted (RI); Roskilde (RK); Ryvange (RG); Sorø (SO).

Other depots in Denmark and Sweden: Aalborg (AA); Ängelholm-Helsingborg (AN); Århus (AR); Årslev (AS); Djursland (DJ); Ebeltoft (EB); Esbjerg (ES); Fyn North [Assens, Kerteminde, Bogense & Kildemosevej] (FN); Fyn South [Ærø, Nyborg, Ringe, Faaborg, Lohals & Svendborg] (FS); Fyn [Odense] (FY); Gislaved (GD); Gränna (JG); Haderslev (HV); Helsingborg (HG); Herning (HG); Hornslet (HT); Horsens (HN); Jönköping (JK); Jönköping Väst (JL); Løkken (LO); Malmö (MO); Nordjyllands [Hjørring, Hobro, Løkken, Sæby & Tversted] (ND); Odense (OV); Odder (OD); Ø.Kippinge (KI); Randers (RA); Rønde/Rostved (RO); Rudkøbing (RU); Ryomgård (RY); Skælskør (SK);Skanderborg (SG); Söderslätt regional (SD); Stockholm Ekerö (SX); Stockholm Märsta (SM); Svalöv (SV); Svendborg (SE); Trelleborg (TG); Vejile (VJ); Vinderup (VI); FDM Sjællandsringen (SJ)

ARRIVA NEDERLANDS

Arriva Nederlands BV, Trambaan 3, postbus 626, 8440 AP Heerenveen

53-70
Mercedes-Benz Integro O550 — Mercedes-Benz — C40F — 2003-09

53	AS	BT-TT-56	64	GS	BR-NL-82	67	GS	BR-NL-79	69	SV	BR-NL-77
54	AS	BV-PZ-37	65	GS	BR-NL-81	68	GS	BR-NL-78	70	GS	BR-NL-76
55	AS	BN-JB-91	66	GS	BR-NL-80						

152	LS	VR-10-LR	Mercedes-Benz Citaro O405G	Mercedes-Benz	AB49F	1993

158-165
Mercedes-Benz Citaro O530G — Mercedes-Benz — AB50F — 1999-2002 Seating varies

158	TS	BH-VJ-15	160	AM	BH-XX-29	162	GS	BN-HN-57	164	GS	BN-HN-60
159	GR	BH-XX-31	161	GR	BH-XX-30	163	GS	BN-HN-59	165	GS	BN-HN-61

190-199
Mercedes-Benz Tourismo O350 — Mercedes-Benz — C51F — 1995-2002

190	GR	BD-RX-88	193	GR	BL-VH-95	196	WT	BH-FH-71	198	SN	BJ-BG-36
191	DM	BF-LH-16	194	GR	BL-VH-96	197	WT	BH-HG-65	199	GR	BL-BG-74
192	GR	BG-NP-90	195	HV	BL-VH-80						

221-234
Mercedes-Benz Citaro O530 G — Mercedes-Benz — AN50D — 2004

221	GR	BP-NH-41	225	GR	BP-NH-53	229	GR	BP-NH-59	232	GR	BP-NL-71
222	GR	BP-NH-50	226	GR	BP-NH-52	230	GR	BP-NH-39	233	GR	BP-NL-74
223	GR	BP-NH-51	227	GR	BP-NH-56	231	GR	BP-NL-69	234	GR	BP-NL-75
224	GR	BP-NH-54	228	GR	BP-NH-57						

241	GR	BR-NN-60	Mercedes-Benz Citaro O530 G	Mercedes-Benz	AN49D	2005

251-260
MAN Lion's City CNG — MAN — AN46D — 2008

251	GR	BV-HH-09	254	GR	BV-HH-12	257	GR	BV-HP-22	259	GR	BV-HP-24
252	GR	BV-HH-10	255	GR	BV-HH-13	258	GR	BV-HP-23	260	GR	BV-HP-27
253	GR	BV-HH-11	256	GR	BV-HP-21						

457-464
Setra S317 HDH — Setra — C50F — 2005

457	GR	BP-LG-16	459	HN	BP-LT-05	461	SN	BL-TT-46	463	LS	BN-RR-71
458	WT	BP-LH-10	460	WT	BP-LV-72	462	SN	BL-TT-50	464	LS	BN-PT-94

465	GR	BP-JR-38	VDL Bus SBR3000 15m	Berkhof E3000HD	C62D	2004
466	GR	BV-VX-22	Mercedes-Benz Tourismo O350	Mercedes-Benz	C55F	2000

As we go to press the route network around Groningen is about to change significantly. Here, Park & Ride bus 256, BV-HP-21, illustrates the livery applied which has also been captured in a Rietze model. *Harry Laming*

Two buses being evaluated with Arriva Netherlands are a MAN Lion's City (numbered 1001) and an Optare Tempo, numbered 1002. 1001, BV-GX-98, is currently allocated to Appingedam. *Harry Laming*

521-540

Mercedes-Benz Citaro O530 · Mercedes-Benz · N34D · 2002-03

521	AS	BN-JD-07	526	AS	BN-JD-02	531	GR	BN-JB-96	536	GR	BN-TS-65
522	GR	BN-JD-06	527	GR	BN-JD-01	532	GR	BN-JB-95	537	GR	BN-TS-66
523	AS	BN-JD-05	528	GR	BN-JB-99	533	GR	BN-JB-93	538	GR	BN-TS-67
524	GR	BN-JD-04	529	GR	BN-JB-98	534	GR	BN-TS-61	539	GR	BN-TS-62
525	AS	BN-JD-03	530	GR	BN-JB-97	535	GR	BN-TS-64	540	GR	BN-TS-68

551-557

Mercedes-Benz Citaro O530 · Mercedes-Benz · N35D · 2005

551	GR	BP-NH-40	553	GR	BP-NH-44	555	GR	BP-NH-47	557	GR	BP-NH-49
552	GR	BP-NH-43	554	GR	BP-NH-46	556	GR	BP-NH-48			

721-730

Mercedes-Benz Citaro O530 · Mercedes-Benz · N28D · 2001

721	DO	BL-RF-50	724	DO	BL-RF-50	727	DO	BL-RF-50	729	DO	BL-RF-50
722	DO	BL-RF-50	725	DO	BL-RF-50	728	DO	BL-RF-50	730	DO	BL-RF-50
723	DO	BL-RF-50	726	DO	BL-RF-50						

751-784

Mercedes-Benz Citaro O530 · Mercedes-Benz · N26D · 2007

751	SH	BT-GT-95	760	SH	BT-GX-08	769	SH	BT-GX-24	777	SH	BT-HF-89
752	SH	BT-GV-99	761	SH	BT-GX-09	770	SH	BT-GX-26	778	SH	BT-HF-88
753	SH	BT-GX-01	762	SH	BT-GX-11	771	SH	BT-HG-02	779	SH	BT-HF-87
754	SH	BT-GX-02	763	SH	BT-GX-13	772	SH	BT-HG-01	780	SH	BT-HF-86
755	SH	BT-GX-03	764	SH	BT-GX-14	773	SH	BT-HF-97	781	SH	BT-HF-85
756	SH	BT-GX-04	765	SH	BT-GX-17	774	SH	BT-HF-94	782	SH	BT-HF-83
757	SH	BT-GX-05	766	SH	BT-GX-18	775	SH	BT-HF-93	783	SH	BT-HF-82
758	SH	BT-GX-06	767	SH	BT-GX-19	776	SH	BT-HF-92	784	SH	BT-HF-81
759	SH	BT-GX-07	768	SH	BT-GX-20						

1001	AP	BV-GX-98	MAN Lion's City A78	MAN	N40D	2007
1002	GR	BV-VH-41	Optare Tempo	Optare	N37D	2007
1125	GS	BD-FT-05	Mercedes-Benz O408	Mercedes-Benz	B49D	1995
1141	DR	BS-FT-05	Iveco EuroRider B89	Den Oudsten	B45D	1995
1142	LS	BD-JB-28	Iveco EuroRider B89	Den Oudsten	B45D	1995
1153	GR	BD-BS-16	Den Oudsten B91	Den Oudsten Alliance	BC47D	1996

1155	WT	BD-BS-18	Den Oudsten B91	Den Oudsten Alliance	BC47D	1996
1258	AM	BD-ZB-16	Mercedes-Benz O408	Mercedes-Benz	B49D	1996
1266	TG	BD-ZF-16	Mercedes-Benz O408	Mercedes-Benz	B49D	1990
1271	LS	BD-TS-66	Den Oudsten B95	Den Oudsten Alliance	N45D	1997
1276	WT	BD-TS-67	Den Oudsten B95	Den Oudsten Alliance	N45D	1997
1280	ZK	BF-GJ-38	Iveco EuroRider	Berkhof 2000NL	N44D	1997
1288	GR	BF-GJ-13	Iveco EuroRider	Berkhof 2000NL	N44D	1997

2194-2203
Den Oudsten B95 — Den Oudsten Alliance — B45D — 1997

2194	AL	BF-LG-44	**2197**	SK	BF-LG-87	**2199**	AS	BF-LJ-29	**2202**	GS	BF-LG-42
2195	UN	BF-LG-43	**2198**	UN	BF-LG-83	**2200**	UN	BF-LG-40	**2203**	GS	BF-LH-20
2196	AP	BF-LG-89									

2217-2220
Den Oudsten B95 — Den Oudsten Alliance — B45D — 1997

2217	AP	BF-XV-78	**2218**	AP	BF-XV-24	**2219**	AS	BF-XV 23	**2220**	AP	BF-XV-22

3154	GS	BS-HR-95	VDL Bus SB200	VDL Berkhof	N41D	2008
3158	GS	BR-ZL-75	VDL Bus SB200	VDL Berkhof	N41D	2008
4751	GR	VX-18-DL	Volvo B10M-61	Berkhof 2000NL	B45D	1993

4771-4777
Mercedes-Benz O408 — Mercedes-Benz — B49D — 1993

4771	WT	BB-DL-39	**4773**	HV	BB-DL-76	**4775**	ZE	BB-DL-35	**4777**	GR	BB-DL-37
4772	LS	BB-DL-31	**4774**	WT	BB-DL-33	**4776**	LS	BB-DL-86			

5800	GS	46-DL-RX	Mercedes-Benz Vito 208D	Mercedes-Benz	M8	1999
5803	VD	42-DL-XT	Mercedes-Benz Vito 208D	Mercedes-Benz	M8	1999

5810-5821
Mercedes-Benz Integro O550ÜL 15m — NC50D — 2000

5810	GS	BH-TD-29	**5813**	GS	BH-TN-93	**5816**	GS	BH-TN-90	**5819**	LK	BH-TR-99
5811	GS	BH-TN-95	**5814**	GS	BH-TN-92	**5817**	GS	BH-TR-97	**5820**	LK	BH-TS-01
5812	GS	BH-TN-94	**5815**	GS	BH-TN-91	**5818**	LK	BH-TR-98	**5821**	LK	BH-TS-02

5822-5834
DAF SB220 — Berkhof Excellence 2000 — N36D — 2000

5822	GS	BJ-DF-10	**5826**	TG	BJ-DF-16	**5829**	TG	BJ-DF-21	**5832**	GR	BJ-DF-25
5823	AM	BJ-DF-12	**5827**	TG	BJ-DF-18	**5830**	AM	BJ-DF-23	**5833**	GR	BJ-DF-27
5824	AM	BJ-DF-14	**5828**	TG	BJ-DF-19	**5831**	GR	BJ-DF-24	**5834**	GS	BJ-DR-02
5825	AM	BJ-DF-15									

5835-5845
DAF SB220 — Berkhof Excellence 2000 — N42D — 2000

5835	LK	BJ-DP-99	**5838**	GS	BJ-DP-93	**5841**	ML	BJ-DP-90	**5844**	LK	BJ-DP-86
5836	LK	BJ-DP-96	**5839**	GS	BJ-DP-92	**5842**	ML	BJ-DP-89	**5845**	ML	BJ-DP-83
5837	GS	BJ-DP-95	**5840**	LK	BJ-DP-91	**5843**	ML	BJ-DP-88			

5846	VD	96-JS-XX	Mercedes-Benz Vito 208D	Mercedes-Benz	M8	2002
5847	MK	97-JS-XX	Mercedes-Benz Vito 208D	Mercedes-Benz	M8	2002
5848	VG	98-JS-XX	Mercedes-Benz Vito 208D	Mercedes-Benz	M8	2002
5849	AS	99-JS-XX	Mercedes-Benz Vito 208D	Mercedes-Benz	M8	2002

5850-5899
Dennis Dart SLF — Alexander ALX200 — N39D — 2000-01

5850	GS	BL-JD-45	**5866**	GS	BJ-ZT-77	**5875**	AS	BJ-ZT-87	**5885**	AS	BJ-ZT-76
5851	WT	BJ-VB-33	**5867**	GS	BJ-ZJ-38	**5876**	AS	BJ-ZT-90	**5887**	GS	BL-BJ-82
5852	TL	BJ-XN-84	**5868**	GS	BJ-ZJ-39	**5878**	SG	BJ-ZT-91	**5889**	ML	BL-BJ-78
5853	EM	BJ-XN-81	**5869**	GS	BJ-ZJ-40	**5879**	WT	BJ-ZT-86	**5892**	EM	BL-BS-20
5854	WT	BJ-XN-82	**5870**	PU	BJ-ZJ-41	**5880**	GS	BJ-ZT-94	**5893**	GS	BL-BS-21
5856	EM	BJ-XN-80	**5871**	TL	BJ-ZJ-42	**5881**	GS	BJ-ZT-93	**5894**	GS	BL-BS-22
5857	WT	BJ-XN-77	**5872**	SH	BJ-ZJ-43	**5882**	GS	BJ-ZT-92	**5896**	GS	BL-BV-37
5858	EM	BJ-XN-83	**5873**	VG	BJ-ZJ-44	**5883**	GS	BL-DX-66	**5898**	ML	BL-GN-77
5859	SK	BL-BS-18	**5874**	EN	BJ-ZJ-45	**5884**	GS	BL-BJ-77	**5899**	SK	BJ-TB-91

5900-5919
Volvo B10BLE — Carrus — N40D — 1997 — Arriva Sverige, 2002

5900	ZK	BL-SH-07	**5906**	ZK	BN-BS-01	**5911**	GS	BN-FN-94	**5916**	GS	BN-HT-39
5902	ZK	BL-VJ-93	**5907**	ZK	BN-FJ-84	**5912**	GS	BL-TX-80	**5917**	GS	BN-FR-71
5903	ZK	BN-FH-34	**5908**	ZK	BN-PJ-70	**5913**	GS	BL-ZP-63	**5918**	GS	BN-HV-20
5904	ZK	BL-VZ-99	**5909**	ZK	BN-DG-29	**5914**	GS	BL-TN-74	**5919**	GS	BL-SH-08
5905	ZK	BL-XR-75	**5910**	GS	BL-SP-71	**5915**	GS	BL-VF-77			

5920-5927
DAF SB120 — Wrightbus Cadet — N24D — 2002

5920	GO	BL-TG-43	**5922**	WT	BL-TG-49	**5924**	WT	BL-TG-47	**5926**	WT	BL-TG-48
5921	WT	BL-TG-50	**5923**	WT	BL-TG-46	**5925**	WT	BL-TG-45	**5927**	WT	BL-TG-44

The British influence is seen in the Dutch fleet by Alexander-bodied Darts and Wrightbus-bodied SB200s. Of the former type, 5850, BL-JD-45, is seen heading for Roden in the Drenthe area of the country. *Harry Laming*

5928-5939

DAF SB200 — Wrightbus Commander — N37D — 2002

5928	AS	BL-XH-68	5931	VD	BL-XH-67	5934	LK	BL-XH-70	5937	SK	BL-XH-80
5929	AS	BL-XH-73	5932	VD	BL-XH-71	5935	LK	BL-XH-77	5938	SV	BL-XH-76
5930	GS	BL-XH-66	5933	AS	BL-XH-74	5936	SK	BL-XH-78	5939	SG	BL-XH-79

5941-5999

DAF SB200 — Wrightbus Commander — N42D — 2002 — Seating varies

5941	VD	BL-BP-84	5956	LK	BL-DS-81	5971	SK	BN-HD-75	5986	EM	BN-HS-82
5942	SV	BL-BP-85	5957	PU	BL-DS-82	5972	VD	BN-HD-76	5987	WT	BN-HS-66
5943	VD	BL-BP-86	5958	SV	BN-DS-83	5973	VD	BN-HD-78	5988	TL	BN-HS-67
5944	VD	BL-BP-87	5959	SV	BN-DS-84	5974	SK	BN-HD-79	5989	TL	BN-HS-68
5945	VD	BL-BP-89	5960	VD	BN-DS-85	5975	SK	BN-HD-81	5990	TL	BN-HS-69
5946	WT	BL-BP-90	5961	SV	BN-DS-86	5976	SK	BN-HD-87	5991	TL	BN-HS-83
5947	WT	BL-BP-91	5962	EM	BN-DS-87	5977	AS	BN-HD-73	5992	TL	BN-HS-76
5948	WT	BL-BP-92	5963	EM	BN-DS-88	5978	AS	BN-HD-77	5993	TL	BN-HS-77
5949	WT	BL-BP-94	5964	EM	BN-DS-89	5979	AS	BN-HD-88	5994	TL	BN-HS-78
5950	WT	BL-BP-96	5965	EM	BN-DS-90	5980	AS	BN-HD-89	5995	TL	BN-HS-80
5951	AP	BL-BP-97	5966	EM	BN-DS-91	5981	AS	BN-HD-90	5996	TL	BN-HS-81
5952	AP	BL-BP-82	5967	SV	BN-DS-93	5982	SV	BN-HS-70	5997	TL	BN-HS-73
5953	AP	BL-DS-78	5968	EM	BN-DS-95	5983	SV	BN-HS-71	5998	TL	BN-HS-74
5954	SV	BL-DS-79	5969	EM	BN-DS-96	5984	AP	BN-HS-72	5999	TL	BN-HS-75
5955	LK	BL-DS-80	5970	SK	BN-DS-97	5985	DO	BN-HS-79			

6022	VD	BP-PN-08	Volkswagen City Bus	Volkswagen	M8	2004

6023-6028

Mercedes-Benz Sprinter 309 CDi — Mercedes-Benz — M18 — 2006

6023	AP	BR-VF-69	6025	GR	BL-PP-65	6027	VD	BL-GR-26	6028	AP	BL-FH-61
6024	AP	BR-VF-72	6026	AP	BN-FG-26						

6030-6045

Irisbus Crossway — Irisbus — N20F — 2007

6030	GR	BS-XS-02	6034	GO	BS-ZB-48	6038	GO	BS-ZB-52	6042	DO	BS-ZB-57
6031	GR	BS-XS-04	6035	GO	BS-ZB-49	6039	GO	BS-ZB-53	6043	DO	BS-ZB-58
6032	HE	BS-ZB-46	6036	GO	BS-ZB-50	6040	GO	BS-ZB-54	6044	DO	BS-ZB-59
6033	GO	BS-ZB-47	6037	GO	BS-ZB-51	6041	DO	BS-ZB-56	6045	DO	BS-ZB-60

The interurban livery applied to the buses in the Netherlands is shown on Mercedes-Benz Integro 7172, BP-NT-51, pictured in Groningen. Q-Liner branding is applied to the interurban network. *Harry Laming*

6071	ZK	96-TB-JZ	Volkswagen City Bus	Volkswagen	M8	2006		

6073-6078 Mercedes-Benz Sprinter 207 CDi Mercedes-Benz M8 2007

6073	AL	94-TV-XF	6075	MK	41-XB-RG	6077	MK	81-TX-FK	6078	TL	11-FG-KP
6074	MK	89-TV-XF	6076	AL	55-XB-RG						

6079	TL	48-FS-DX	Volkswagen City Bus	Volkswagen	M8	2000		

6080-6086 Mercedes-Benz Sprinter 207 CDi Mercedes-Benz M8 2004

6080	TL	08-LS-LD	6082	TL	37-PT-XL	6084	TL	25-NV-HK	6086	TL	89-PG-XZ
6081	TL	84-PG-XZ	6083	TL	38-PT-XL	6085	TL	35-PT-XL			

6100-6109 DAF SB200 Berkhof Ambassador N43D 2002

6100	LK	BN-HX-03	6103	VD	BN-JB-07	6106	SG	BN-HH-82	6108	SG	BN-JT-79
6101	LK	BN-HX-04	6104	VD	BN-JB-08	6107	SG	BN-HH-85	6109	SG	BN-JT-80
6102	LK	BN-HX-06	6105	SG	BN-JB-09						

6111-6122 Mercedes-Benz Vito 208 Mercedes-Benz M8 2003

6111	PU	35-LD-NK	6114	GR	08-LF-XH	6117	TL	07-LF-XH	6121	PU	28-LK-BF
6112	VG	36-LD-NK	6115	TL	05-LF-XH	6119	AS	26-LK-BF	6122	EM	29-LK-BF
6113	AP	34-LD-NK	6116	TL	06-LF-XH	6120	GR	27-LK-BF			

6123	SK	82-PJ-KL	Volkswagen City Bus	Volkswagen	M8	2004
6124	AP	31-PR-JP	Volkswagen City Bus	Volkswagen	M8	2004
6125	AP	32-PR-JP	Volkswagen City Bus	Volkswagen	M8	2004
6126	TL	63-RG-DN	Mercedes-Benz Vito 208	Mercedes-Benz	M8	2005
6127	TL	54-RK-BR	Mercedes-Benz Vito 208	Mercedes-Benz	M8	2005
6128	LK	25-RS-JL	Volkswagen City Bus	Volkswagen	M8	2005
6129	PU	52-SF-FR	Volkswagen City Bus	Volkswagen	M8	2005

6131-6158 Mercedes-Benz O550 Mercedes-Benz Integro NC43D 2003

6131	SK	BN-NF-13	6139	GS	BN-NF-25	6146	AL	BN-NG-13	6153	AL	BN-NG-40
6132	SK	BN-NF-15	6140	GS	BN-NF-06	6147	GS	BN-NG-08	6154	GO	BN-NX-74
6133	SK	BN-NF-17	6141	GO	BN-NG-39	6148	AL	BN-NG-04	6155	GO	BN-NX-78
6134	SK	BN-NF-20	6142	GO	BN-NG-38	6149	AL	BN-NG-02	6156	AL	BN-NX-76
6136	SK	BN-NF-22	6143	GO	BN-NG-34	6150	AL	BN-NG-01	6157	GS	BN-NX-77
6137	GS	BN-NF-23	6144	GO	BN-NG-19	6151	AL	BN-NF-99	6158	GS	BN-NX-73
6138	GO	BN-NF-24	6145	AL	BN-NG-16	6152	AL	BN-NF-98			

6171-6189 — DAF SB200 — Berkhof Ambassador — NC43D — 2003

6171	LK	BN-VX-54	6176	WT	BN-VX-59	6181	EM	BN-VX-64	6186	WT	BN-VX-69
6172	SV	BN-VX-55	6177	WT	BN-VX-60	6182	AS	BN-VX-65	6187	LK	BN-VX-70
6173	WT	BN-VX-56	6178	UZ	BN-VX-61	6183	LK	BN-VX-66	6188	AS	BN-VX-71
6174	SV	BN-VX-57	6179	WT	BN-VX-62	6184	LK	BN-VX-67	6189	LK	BN-VX-72
6175	VD	BN-VX-58	6180	EM	BN-VX-63	6185	LK	BN-VX-68			

6200-6301 — DAF SB200 — Wrightbus Commander — N42D — 2003

6200	AL	BN-PN-23	6226	MK	BN-RP-10	6252	AL	BN-SG-15	6277	MK	BN-TB-25
6201	GO	BN-PN-27	6227	MK	BN-RP-11	6253	AL	BN-SG-17	6278	MK	BN-TB-26
6202	GO	BN-PN-30	6228	DO	BN-RP-12	6254	AL	BN-TB-43	6279	MK	BN-TB-27
6203	GO	BN-PN-32	6229	AL	BN-RP-13	6255	AL	BN-TB-45	6280	MK	BN-TB-28
6204	GO	BN-PN-33	6230	AL	BN-RP-15	6256	AL	BN-TB-47	6281	AL	BN-TB-29
6205	GO	BN-PN-38	6231	AL	BN-RP-17	6257	AL	BN-TB-49	6282	AL	BN-TB-30
6206	GO	BN-PN-39	6232	AL	BN-RP-18	6258	AL	BN-TB-51	6283	AL	BN-TB-31
6207	GO	BN-PN-35	6233	AL	BN-RP-20	6259	AL	BN-TB-53	6284	DO	BN-TB-32
6208	GO	BN-PN-41	6234	AL	BN-RP-21	6260	AL	BN-TR-56	6285	DO	BN-TB-33
6209	GO	BN-PN-42	6235	AL	BN-RP-23	6261	DO	BN-TR-58	6286	AL	BN-TB-35
6210	AL	BN-RD-38	6236	AL	BN-RP-26	6262	AL	BN-TR-59	6287	DO	BN-TB-37
6211	AL	BN-RD-39	6237	AL	BN-RP-28	6263	AL	BN-TR-61	6288	AL	BN-TB-38
6212	DO	BN-RD-40	6238	AL	BN-RP-30	6264	AL	BN-TR-62	6289	AL	BN-TB-41
6213	AL	BN-RD-41	6239	AL	BN-RP-32	6265	AL	BN-TR-63	6290	AL	BN-TR-69
6214	AL	BN-RD-47	6240	AL	BN-SF-92	6266	AL	BN-TR-64	6291	DO	BN-TR-70
6215	AL	BN-RD-48	6241	AL	BN-SF-94	6267	AL	BN-TR-66	6292	DO	BN-TR-71
6216	AL	BN-RD-49	6242	AL	BN-SF-96	6268	AL	BN-TR-67	6293	DO	BN-TR-76
6217	AL	BN-RD-50	6243	AL	BN-SF-98	6269	AL	BN-TR-68	6294	DO	BN-TR-77
6218	AL	BN-RD-51	6244	AL	BN-SF-99	6270	AL	BN-SG-19	6295	DO	BN-TR-78
6219	AL	BN-RD-52	6245	AL	BN-SG-01	6271	GO	BN-SG-24	6296	DO	BN-TR-79
6220	MK	BN-RN-83	6246	AL	BN-SG-03	6272	AL	BN-SG-25	6297	DO	BN-TR-81
6221	MK	BN-RN-87	6247	AL	BN-SG-04	6273	MK	BN-SG-26	6298	DO	BN-TR-83
6222	MK	BN-RN-84	6248	AL	BN-SG-06	6274	MK	BN-SG-27	6299	DO	BN-TR-84
6223	MK	BN-RN-85	6249	AL	BN-SG-08	6275	MK	BN-SG-28	6300	DO	BN-TR-85
6224	MK	BN-RN-86	6250	AL	BN-SG-11	6276	MK	BN-TB-24	6301	DO	BN-TR-86
6225	MK	BN-RP-08	6251	AL	BN-SG-13						

6387	LS	BF-XL-12	MAN 11.220	Berkhof 2000NLE	B25D	1997

7151-7174 — Mercedes-Benz Integro O550ÜL — Mercedes-Benz — NC50D — 2004

7151	GS	BP-NN-71	7158	VD	BP-NN-76	7164	AS	BP-NT-63	7170	EM	BP-NT-54
7152	GS	BP-NN-73	7159	VD	BP-NN-78	7165	AS	BP-NT-59	7171	EM	BP-NT-53
7153	VD	BP-NN-74	7160	VD	BP-NN-79	7166	EM	BP-NT-58	7172	EM	BP-NT-51
7155	VD	BP-NN-80	7161	AS	BP-NN-81	7167	EM	BP-NT-56	7173	EM	BP-NT-47
7156	VD	BP-NN-84	7162	AS	BP-NT-60	7168	EM	BP-NT-57	7174	EM	BP-NT-46
7157	VD	BP-NN-86	7163	AS	BP-NT-61	7169	EM	BP-NT-55			

7181-7184 — Mercedes-Benz O550ÜL — Mercedes-Benz Integro L — NC50D — 2005

7181	PU	BR-NL-85	7182	PU	BR-NL-84	7183	PU	BR-NL-83	7184	GS	BR-XV-91

7321	TL	37-JV-VN	Volkswagen City Bus	Volkswagen	M8	2002
7322	TL	89-JV-XP	Volkswagen City Bus	Volkswagen	M8	2002
7323	TL	TJ-DX-10	Mercedes-Benz Sprinter	Mercedes-Benz	M8	1998

7331-7354 — Mercedes-Benz Sprinter — Mercedes-Benz — M8 — 2007

7331	VG	37-XZ-RH	7337	VG	34-XZ-RH	7343	VG	33-XZ-RH	7349	VG	40-XZ-RH
7332	VG	42-XZ-RH	7338	VG	46-XZ-RH	7344	VG	47-XZ-RH	7350	VG	41-XZ-RH
7333	VG	32-XZ-RH	7339	VG	35-XZ-RH	7345	VG	54-XZ-RH	7351	VG	45-XZ-RH
7334	VG	53-XZ-RH	7340	VG	44-XZ-RH	7346	VG	36-XZ-RH	7352	VG	51-XZ-RH
7335	VG	43-XZ-RH	7341	VG	49-XZ-RH	7347	VG	33-XZ-RH	7353	VG	52-XZ-RH
7336	VG	38-XZ-RH	7342	VG	48-XZ-RH	7348	VG	39-XZ-RH	7354	OT	50-XZ-RH

7801-7823 — Mercedes-Benz Citaro O530G — Mercedes-Benz — AN58D — 2004-05

7801	ZK	BP-NZ-66	7807	GS	BP-NZ-75	7813	AP	BP-NZ-85	7819	AP	BP-NZ-91
7802	ZK	BP-NZ-68	7808	GS	BP-NZ-76	7814	GS	BP-NZ-86	7820	AP	BP-NZ-92
7803	ZK	BP-NZ-69	7809	GS	BP-NZ-77	7815	GS	BP-NZ-87	7821	AS	BP-NZ-93
7804	ZK	BP-NZ-71	7810	GS	BP-NZ-81	7816	GS	BP-NZ-88	7822	AS	BP-NZ-94
7805	ZK	BP-NZ-73	7811	GS	BP-NZ-83	7817	AP	BP-NZ-89	7823	AS	BP-NZ-95
7806	GS	BP-NZ-74	7812	GS	BP-NZ-84	7818	AP	BP-NZ-90			

7830	GS	BP-SG-53	Scania OmniCity CN94UA	Scania	AN62D	2004
7831	ML	BP-SG-52	Scania OmniCity CN94UA	Scania	AN62D	2004

Purchased for the Amsterdam network were forty-eight Mercedes-Benz Citaro articulated buses and a similar number of Scania OmniCity buses. From the former, 7903, BR-LV-65, is seen in the city. *Mark Doggett*

7851-7898   Scania OmniCity CN94UA   Scania   AN57D   2005

7851	PU	BR-NP-73	7863	PU	BR-NP-87	7875	PU	BR-NR-19	7887	PU	BR-NR-01
7852	PU	BR-NP-74	7864	PU	BR-NP-88	7876	PU	BR-NR-22	7888	PU	BR-NP-03
7853	PU	BR-NP-75	7865	PU	BR-NP-96	7877	PU	BR-NR-24	7889	PU	BR-NP-10
7854	PU	BR-NP-77	7866	PU	BR-NP-97	7878	PU	BR-NR-26	7890	PU	BR-NP-12
7855	PU	BR-NP-78	7867	PU	BR-NP-99	7879	PU	BR-NR-28	7891	PU	BR-NP-13
7856	PU	BR-NP-80	7868	PU	BR-NR-02	7880	PU	BR-NP-90	7892	PU	BR-NP-15
7857	PU	BR-NP-81	7869	PU	BR-NR-08	7881	PU	BR-NP-91	7893	PU	BR-NP-18
7858	PU	BR-NP-82	7870	PU	BR-NR-09	7882	PU	BR-NP-92	7894	PU	BR-NP-20
7859	PU	BR-NP-83	7871	PU	BR-NR-11	7883	PU	BR-NP-93	7895	PU	BR-NP-21
7860	PU	BR-NP-84	7872	PU	BR-NR-14	7884	PU	BR-NP-94	7896	PU	BR-NP-23
7861	PU	BR-NP-85	7873	PU	BR-NR-16	7885	PU	BR-NP-95	7897	PU	BR-NP-25
7862	PU	BR-NP-86	7874	PU	BR-NR-17	7886	PU	BR-NP-98	7898	PU	BR-NP-27

7901-7948   Mercedes-Benz Citaro O530G   Mercedes-Benz   AN57D   2005

7901	PU	BR-NF-15	7913	PU	BR-LZ-70	7925	PU	BR-NJ-72	7937	PU	BR-NJ-95
7902	PU	BR-LV-67	7914	PU	BR-LZ-73	7926	PU	BR-NJ-73	7938	PU	BR-NJ-96
7903	PU	BR-LV-65	7915	PU	BR-LZ-74	7927	PU	BR-NJ-74	7939	PU	BR-NJ-94
7904	PU	BR-LV-60	7916	PU	BR-LZ-68	7928	PU	BR-NJ-76	7940	PU	BR-NJ-98
7905	PU	BR-LV-57	7917	PU	BR-LZ-66	7929	PU	BR-NJ-77	7941	PU	BR-NJ-99
7906	PU	BR-LV-54	7918	PU	BR-LZ-67	7930	PU	BR-NJ-78	7942	PU	BR-NL-02
7907	PU	BR-LV-70	7919	PU	BR-NJ-65	7931	PU	BR-NJ-79	7943	PU	BR-NL-03
7908	PU	BR-LV-72	7920	PU	BR-NJ-67	7932	PU	BR-NJ-84	7944	PU	BR-NL-04
7909	PU	BR-LV-53	7921	PU	BR-NJ-68	7933	PU	BR-NJ-85	7945	PU	BR-NL-05
7910	PU	BR-LZ-75	7922	PU	BR-NJ-69	7934	PU	BR-NJ-87	7946	PU	BR-NL-06
7911	PU	BR-LV-56	7923	PU	BR-NJ-70	7935	PU	BR-NJ-92	7947	PU	BR-NN-68
7912	PU	BR-LZ-71	7924	PU	BR-NJ-71	7936	PU	BR-NJ-93	7948	PU	BR-NN-70

The Scania OmniCity buses that run in the Amsterdam area are represented by 7883, BR-NP-93, which is seen operating route 100. *Mark Doggett*

8001-8069 — VDL Bus SB200 / VDL Berkhof Ambassador — N40D — 2005

8001	AP	BP-LT-50	8019	AS	BP-LT-58	8037	PU	BP-LT-77	8054	SK	BP-NS-96
8002	AP	BP-LD-76	8020	AS	BP-LT-59	8038	PU	BP-LT-78	8055	GS	BP-NS-97
8003	AP	BP-LD-75	8021	SK	BP-LT-60	8039	PU	BP-LT-79	8056	SV	BP-LT-98
8004	WT	BP-LD-74	8022	SK	BP-LT-61	8040	PU	BP-LT-80	8057	ZE	BP-NS-22
8005	ZK	BP-LT-51	8023	SK	BP-LT-62	8041	PU	BP-LT-81	8058	ZE	BP-NS-23
8006	WT	BP-LD-72	8024	SK	BP-LT-63	8042	PU	BP-LT-82	8059	EM	BP-NS-26
8007	SK	BP-LD-71	8025	SK	BP-LT-64	8043	PU	BP-LT-84	8060	GS	BP-NS-27
8008	SK	BP-LD-71	8026	GS	BP-LT-65	8044	PU	BP-LT-85	8061	ZE	BP-NS-28
8009	ZK	BP-LT-52	8027	EM	BP-LT-66	8045	ZE	BP-LT-87	8062	ZE	BP-NS-29
8010	SK	BP-LD-70	8028	WT	BP-LT-67	8046	ZE	BP-LT-88	8063	VG	BP-NS-30
8011	WT	BP-LD-68	8029	WT	BP-LT-68	8047	AL	BP-LT-89	8064	VG	BP-NS-14
8012	WT	BP-LD-67	8030	PU	BP-LT-69	8048	AL	BP-LT-90	8065	EM	BP-NS-16
8013	GS	BP-LD-66	8031	PU	BP-LT-70	8049	AL	BP-LT-91	8066	LK	BP-NS-18
8014	SV	BP-LT-53	8032	PU	BP-LT-71	8050	AL	BP-LT-92	8067	EM	BP-NS-19
8015	SV	BP-LT-54	8033	PU	BP-LT-73	8051	AL	BP-LT-93	8068	VG	BP-NS-20
8016	AP	BP-LT-55	8034	PU	BP-LT-74	8052	AL	BP-LT-94	8069	EM	BP-NS-21
8017	AP	BP-LT-56	8035	PU	BP-LT-75	8053	EM	BP-LT-95			
8018	AP	BP-LT-57	8036	PU	BP-LT-76						

8071-8090 — Scania OmniCity C94UA 12m — Scania — N39D — 2006

8071	PU	BS-DZ-41	8076	PU	BS-DZ-45	8081	PU	BR-TD-79	8086	PU	BS-TD-70
8072	PU	BS-DZ-42	8077	PU	BS-DZ-47	8082	PU	BR-TD-78	8087	PU	BS-TD-81
8073	PU	BS-DZ-39	8078	PU	BS-DZ-50	8083	PU	BR-TD-72	8088	PU	BS-TD-80
8074	PU	BS-DZ-43	8079	PU	BS-DZ-51	8084	PU	BR-TD-74	8089	PU	BS-TD-84
8075	PU	BS-DZ-44	8080	PU	BS-DZ-79	8085	PU	BR-TD-77	8090	PU	BS-TD-83

The 2009-10 Arriva Bus Handbook

8201-8244   VDL Bus SB200   VDL Ambassador   N40D   2007

8201	SH	BT-BH-94	8212	VG	BT-BJ-12	8223	VG	BT-BX-23	8234	VG	BT-BX-43
8202	SH	BT-BH-96	8213	VG	BT-BJ-14	8224	VG	BT-BX-24	8235	VG	BT-BX-44
8203	SH	BT-BH-97	8214	VG	BT-BJ-15	8225	VG	BT-BX-28	8236	SH	BT-BX-47
8204	SH	BT-BH-99	8215	VG	BT-BJ-16	8226	VG	BT-BX-29	8237	SH	BT-DF-81
8205	SH	BT-BJ-03	8216	VG	BT-BJ-19	8227	VG	BT-BX-30	8238	SH	BT-DF-84
8206	SH	BT-BJ-05	8217	VG	BT-BJ-20	8228	VG	BT-BX-31	8239	SH	BT-DF-86
8207	SH	BT-BJ-07	8218	VG	BT-BJ-21	8229	VG	BT-BX-33	8240	SH	BT-DF-88
8208	ZE	BT-BJ-08	8219	VG	BT-BX-15	8230	VG	BT-BX-34	8241	VG	BT-DF-91
8209	ZE	BT-BJ-09	8220	VG	BT-BX-19	8231	VG	BT-BX-36	8242	VG	BT-DF-94
8210	ZE	BT-BJ-10	8221	VG	BT-BX-20	8232	VG	BT-BX-38	8243	ZE	BT-DF-95
8211	ZE	BT-BJ-11	8222	VG	BT-BX-22	8233	VG	BT-BX-40	8244	ZE	BT-DF-96

8301-8363   VDL Bus SB200   VDL Ambassador   N40D   2007

8301	HE	BT-JB-55	8317	HE	BT-LN-48	8333	HE	BT-LN-84	8349	HE	BT-LP-31
8302	HE	BT-JB-56	8318	HE	BT-LN-51	8334	HE	BT-LN-87	8350	HE	BT-LP-33
8303	HE	BT-JB-57	8319	HE	BT-LN-53	8335	HE	BT-LN-88	8351	HE	BT-LP-35
8304	HE	BT-JB-58	8320	HE	BT-LN-54	8336	HE	BT-LN-89	8352	HE	BT-LP-36
8305	HE	BT-JB-60	8321	HE	BT-LN-55	8337	HE	BT-LN-91	8353	HE	BT-LP-37
8306	HE	BT-JB-61	8322	HE	BT-LN-57	8338	HE	BT-LN-92	8354	HE	BT-LP-38
8307	HE	BT-JB-62	8323	HE	BT-LN-58	8339	HE	BT-LN-93	8355	HE	BT-LP-39
8308	HE	BT-JB-65	8324	HE	BT-LN-65	8340	HE	BT-LN-95	8356	HE	BT-LP-41
8309	HE	BT-JB-66	8325	HE	BT-LN-66	8341	HE	BT-LN-97	8357	HE	BT-LP-42
8310	HE	BT-JB-68	8326	HE	BT-LN-67	8342	HE	BT-LN-98	8358	HE	BT-LP-43
8311	HE	BT-JP-92	8327	HE	BT-LN-68	8343	HE	BT-LP-07	8359	HE	BT-LP-44
8312	HE	BT-JP-93	8328	HE	BT-LN-69	8344	HE	BT-LP-08	8360	HE	BT-LP-47
8313	HE	BT-JP-94	8329	HE	BT-LN-73	8345	HE	BT-LP-09	8361	HE	BT-ND-31
8314	HE	BT-JP-98	8330	HE	BT-LN-74	8346	HE	BT-LP-26	8362	HE	BT-ND-30
8315	HE	BT-JR-01	8331	HE	BT-LN-76	8347	HE	BT-LP-27	8363	HE	BT-ND-29
8316	HE	BT-LN-46	8332	HE	BT-LN-77	8348	HE	BT-LP-28			

8364-8382   VDL Bus SB200   VDL Ambassador   N40D   2007

8364	OT	BT-ND-28	8369	OT	BT-ND-23	8374	OT	BT-ND-06	8379	OT	BT-NB-93
8365	OT	BT-ND-27	8370	OT	BT-ND-21	8375	OT	BT-ND-02	8380	OT	BT-NB-95
8366	OT	BT-ND-26	8371	OT	BT-ND-11	8376	OT	BT-NB-99	8381	OT	BT-NP-15
8367	OT	BT-ND-25	8372	OT	BT-ND-10	8377	OT	BT-NB-98	8382	OT	BT-NP-14
8368	OT	BT-ND-24	8373	OT	BT-ND-08	8378	OT	BT-NB-96			

9042	GS	BG-ZP-25	Volvo B10MG	Den Oudsten B88	AB60D	1999
9043	GS	BG-ZP-23	Volvo B10MG	Den Oudsten B88	AB60D	1999
90	WT	BB-GF-85	DAF SB3000KS	-	C52F	1994
90	GS	BB-HB-77	Mercedes-Benz Tourismo O350	Mercedes-Benz	C51F	1994

Depots and Codes:

AL	Alblasserdam	HE	Heinenoord	SH	's Hertogenbosch
AM	Ameland	HV	Heerenveeng	SK	Stadskanaal
AP	Appingedam	LK	Leek	SN	Sneek
AS	Assen	LS	Leeuwarden Stad	SV	Surhuisterveen
DG	Dieverbrug	MK	Meerkerk	TG	Terschelling
DM	Dokkum	ML	Meppel	TL	Tiel
DO	Dordrecht	ON	Oude Tonge	UZ	Uithuizen
DR	Drachten	OO	Oosterworlde	VD	Veendam
EM	Emmen	OT	Oude Tonge	VG	Veghel
GO	Gorinchem	PU	Purmerend	WT	Winschoten
GR	Groningen Stadt	RM	Ruilbussen Meppel	ZE	Zeeland
GS	Groningen Streek	SG	Schiermonnikoo	ZK	Zoutkamp

AUTOBUS SIPPEL

Autobus-Dreischmeyer GmbH; Sippel Travel GmbH; Autobus Sippel GmbH,
Hessenstraße 16, 65719 Hofheim, Germany

1	Ke	WI-RS501	Mercedes-Benz Vario O815	-	C30F	2003
2	Ke	WI-RS402	Mercedes-Benz Sprinter 313 cdi	Mercedes-Benz	M8	2006
4	Wa	WI-RS214	Mercedes-Benz O405	Mercedes-Benz	B45D	1986
5	Wa	WI-RS805	Mercedes-Benz Travego O580	Mercedes-Benz	C46F	2005
6	Bü	WI-JP74	Mercedes-Benz O405	Mercedes-Benz	B45D	1990
7	Wa	WI-RS507	Mercedes-Benz O405	Mercedes-Benz	B45D	1991
12	St	WI-RS412	Kässbohrer S215 UL	Kässbohrer Setra	BC54D	1990
13	Wa	WI-RS913	Mercedes-Benz Citaro O530	Mercedes-Benz	B27D	2001
14	Wa	WI-RS514	Mercedes-Benz 614D	-	B18F	1992
15	Ob	WI-RS915	Mercedes-Benz O405G	Mercedes-Benz	AB60D	1987
18	Ke	WI-RS618	Mercedes-Benz Vario O815	Mercedes-Benz	C21F	2001
19	Ke	WI-RS419	Mercedes-Benz Sprinter 313	Mercedes-Benz	M8	2005
20	Wa	WI-RS728	Mercedes-Benz O405	Mercedes-Benz	B45D	1988
21	Ke	WI-RS421	Mercedes-Benz Sprinter 313	Mercedes-Benz	M8	2005
22	Wa	WI-RS722	Mercedes-Benz O405G	Mercedes-Benz	AB61D	1986
23	Wa	WI-RS923	Mercedes-Benz Citaro O530G	Mercedes-Benz	AN56D	2001
25	Ke	WI-RS625	Mercedes-Benz O405N	Mercedes-Benz	N35D	1995
26	Mz	WI-RS726	Mercedes-Benz O405N	Mercedes-Benz	N37D	1996
27	Ke	WI-RS727	Mercedes-Benz Citaro O530	Mercedes-Benz	N38D	2002
30	Wa	WI-RS930	Mercedes-Benz O405	Mercedes-Benz	B45D	1992
31	Ke	WI-RS931	Mercedes-Benz O405GN	Mercedes-Benz	AB49D	1998
32	Mz	WI-RS632	Mercedes-Benz O405N	Mercedes-Benz	B42D	1990
36	Ke	WI-RS736	Mercedes-Benz O405GN	Mercedes-Benz	AB49D	1998
38	Bü	WI-RS538	Mercedes-Benz Citaro O530G	Mercedes-Benz	AB56D	2000
40	St	WI-S4140	Mercedes-Benz O405	Mercedes-Benz	B45D	1992
42	Ke	WI-RS942	Mercedes-Benz Citaro O530	Mercedes-Benz	N35D	1998
43	Li	WI-RS743	Mercedes-Benz O405	Mercedes-Benz	B45D	1992
44	Er	WI-RS244	Mercedes-Benz O405N	Mercedes-Benz	B38D	1994
45	Wa	WI-RS245	Mercedes-Benz O405N	Mercedes-Benz	B45D	1994
48	Wa	WI-RS448	Mercedes-Benz O405	Mercedes-Benz	B45D	1992
50	Ke	WI-S2050	Mercedes-Benz Citaro O530	Mercedes-Benz	N38D	2002
51	St	WI-RS451	Kässbohrer S215 UL	Kässbohrer Setra	BC54D	1991
52	Ha	WI-RS752	Kässbohrer S215 UL	Kässbohrer Setra	BC53D	1994
53	St	WI-RS553	Kässbohrer S215 HR	Kässbohrer Setra	C59F	1989
55	Wa	WI-XR988	Mercedes-Benz O405	Mercedes-Benz	B45F	1985
56	Ke	WI-RS756	Mercedes-Benz Citaro O530	Mercedes-Benz	N37D	2002
57	Ke	WI-RS257	Mercedes-Benz Citaro O530	Mercedes-Benz	N38D	2002
58	Mz	WI-RS758	Mercedes-Benz O405N	Mercedes-Benz	B35D	1996
60	Wa	WI-RS836	Mercedes-Benz O405	Mercedes-Benz	B45D	1991
62	Wa	WI-RS762	Mercedes-Benz Citaro O530	Mercedes-Benz	N37D	2003
67	Bü	WI-RS767	Mercedes-Benz Citaro O530	Mercedes-Benz	N40D	2001
69	Ke	WI-RS269	Mercedes-Benz Citaro O530	Mercedes-Benz	N35D	1998
73	Ha	WI-RS173	Kässbohrer S213 RL	Kässbohrer Setra	BC50D	1986
74	Er	Wi-RS774	Mercedes-Benz O405N	Mercedes-Benz	B37D	1995
75	Wa	WI-RS775	Mercedes-Benz O405N	Mercedes-Benz	B37D	1995
77	Bü	WI-RS977	Mercedes-Benz Citaro O530 Ü	Mercedes-Benz	NC43D	2000
78	Wa	WI-RS878	Mercedes-Benz O350 RHD	Mercedes-Benz	C49F	2005
79	Rü	WI-RS879	Mercedes-Benz O405G	Mercedes-Benz	AB61D	1987
80	Us	WI-RS780	Mercedes-Benz O405	Mercedes-Benz	B45D	1992
81	Wa	WI-RS781	Mercedes-Benz O405GN	Mercedes-Benz	AB53D	1995
82	Ke	WI-RS782	Mercedes-Benz Citaro O530	Mercedes-Benz	N37D	2001
83	Ke	WI-RS683	Mercedes-Benz O405N	Mercedes-Benz	N37D	2000
84	Wa	WI-RS284	Mercedes-Benz O405	Mercedes-Benz	B45D	1992
85	Ke	WI-RS785	Mercedes-Benz Citaro O530	Mercedes-Benz	N37D	2001
86	Wa	WI-AS586	Mercedes-Benz O405	Mercedes-Benz	B45D	1990
87	Wa	WI-AS687	Mercedes-Benz O405	Mercedes-Benz	B45D	1990
88	Wa	WI-RS688	Mercedes-Benz O405	Mercedes-Benz	B45D	1990
89	Wa	WI-RS489	Mercedes-Benz O405	Mercedes-Benz	B45D	1990
90	St	WI-S1090	Mercedes-Benz O405	Mercedes-Benz	B45D	1991
92	Ha	WI-RS792	MAN 893	MAN	C54F	1994
93	Kn	WI-RS993	Mercedes-Benz O404 15RH	Mercedes-Benz	C53F	1993
94	Wa	WI-RS894	Mercedes-Benz O405	Mercedes-Benz	B45D	1993
95	Wa	WI-RS595	Mercedes-Benz O405	Mercedes-Benz	B45D	1993

Since the last edition of this Arriva Handbook Arriva has increased the number of operations which it has in Germany along with several rail franchises. Located near Frankfurt, Sippel continues to expands. Seen at the airport is 57, WI-RS257 a Mercedes-Benz Citaro. *Harry Laming*

96	Wa	WI-RS596	Mercedes-Benz O405	Mercedes-Benz	B45D	1993
97	Wa	WI-RS597	Mercedes-Benz O405	Mercedes-Benz	B45D	1993
99	Wa	WI-RS799	Mercedes-Benz O405G	Mercedes-Benz	AB52D	1990
101	GG	WI-S1101	Mercedes-Benz Cito O520	Mercedes-Benz	N16D	2001
102	Ke	WI-S1102	Mercedes-Benz Cito O520	Mercedes-Benz	N16D	2001
103	Ho	WI-RS803	Mercedes-Benz Citaro O530G	Mercedes-Benz	N50D	2002
104	Bü	WI-RS704	Mercedes-Benz Citaro O530 Ü	Mercedes-Benz	NC49D	2001
105	Li	WI-RS105	Mercedes-Benz Citaro O530	Mercedes-Benz	N37D	2001
107	Ke	WI-RS907	Mercedes-Benz Citaro O530	Mercedes-Benz	N31D	2002
108	Al	WI-RS908	Mercedes-Benz Citaro O530	Mercedes-Benz	N41D	2001
109	Er	WI-S1313	Mercedes-Benz Citaro O530	Mercedes-Benz	N37D	2001
110	Ke	WI-S1323	Mercedes-Benz Citaro O530	Mercedes-Benz	N37D	2001
111	Ke	WI-S1324	Mercedes-Benz Citaro O530	Mercedes-Benz	N37D	2001
112	Ke	WI-RS712	Mercedes-Benz Vario O815	-	B15D	2001
113	Ke	WI-RS613	Mercedes-Benz Vario O815	-	B15D	2001
114	Mz	WI-RS914	Mercedes-Benz Citaro O530G	Mercedes-Benz	AN51D	2001
115	Mz	WI-S2305	Mercedes-Benz Citaro O530G	Mercedes-Benz	AN51D	2001
117	Wa	WI-RS317	Mercedes-Benz O405	Mercedes-Benz	B45D	1988
118	Wa	WI-RS318	Mercedes-Benz O405	Mercedes-Benz	B45D	1988
121	Bü	WI-RS721	Mercedes-Benz Citaro O530	Mercedes-Benz	N38D	2001
122	Wa	F-ST2122	Mercedes-Benz O350	Mercedes-Benz	BC44D	2002
123	Wa	WI-RS163	Mercedes-Benz O405GN	Mercedes-Benz	AB52D	1992
124	Mz	WI-GS224	Mercedes-Benz O405GN	Mercedes-Benz	AB53D	1996
125	Mz	WI-GS225	Mercedes-Benz O405N	Mercedes-Benz	B37D	1996
126	Wa	WI-RS926	Mercedes-Benz Citaro O530	Mercedes-Benz	N37D	2002
132	Mz	WI-RS832	Mercedes-Benz O405N	Mercedes-Benz	B37D	1996
133	Wa	WI-RS433	Kässbohrer S215 UL	Kässbohrer Setra	BC54D	1990
134	Ke	WI-RS134	Mercedes-Benz Citaro O530	Mercedes-Benz	N43D	2002
136	Bü	WI-RS936	Mercedes-Benz Citaro O530	Mercedes-Benz	N38D	2001
138	Wa	WI-RS638	Kässbohrer S215 HR	Kässbohrer Setra	C59F	1989
139	Wa	WI-RS339	Mercedes-Benz O405N	Mercedes-Benz	B35D	1997
144	Wa	WI-XW509	Mercedes-Benz O405	Mercedes-Benz	B43D	1986
145	Wa	WI-XW515	Mercedes-Benz O405	Mercedes-Benz	B43D	1986
146	Wa	WI-XW506	Mercedes-Benz O405	Mercedes-Benz	B43D	1986
147	Wa	WI-RS147	Mercedes-Benz Tourismo O350	Mercedes-Benz	C50F	2004
148	Ke	WI-RS748	Mercedes-Benz O405GN	Mercedes-Benz	AN49D	1997

149	Ez	WI-RS749	Mercedes-Benz Citaro O530	Mercedes-Benz	N37D	2003	
150	Mz	WI-RS157	Mercedes-Benz O405N	Mercedes-Benz	B35D	1997	
151	Mz	WI-RS851	Mercedes-Benz Citaro O530G	Mercedes-Benz	AN50D	2001	
154	Wa	WI-RS854	Mercedes-Benz O405	Mercedes-Benz	B43D	1988	
155	Wa	WI-RS955	Mercedes-Benz O404 RH	Mercedes-Benz	C47F	2001	
157	Li	WI-RS247	Mercedes-Benz O405GN	Mercedes-Benz	AB46D	1997	
158	Er	WI-RS258	Mercedes-Benz O405N	Mercedes-Benz	B35D	1997	
160	St	WI-RS160	Kässbohrer S215 UL	Kässbohrer Setra	BC54D	1990	
164	Ke	WI-RS464	Mercedes-Benz Citaro O530	Mercedes-Benz	N33D	2001	
165	St	F-ST1165	Mercedes-Benz Tourismo O350	Mercedes-Benz	C51F	2007	
169	Mz	WI-RS569	Mercedes-Benz Citaro O530	Mercedes-Benz	N37D	1998	
170	Mz	WI-S1170	Mercedes-Benz Citaro O530	Mercedes-Benz	N39D	2001	
171	Wa	F-ST4171	Mercedes-Benz Tourismo O350	Mercedes-Benz	C51F	1999	
172	Wa	F-ST4172	Mercedes-Benz Tourismo O350	Mercedes-Benz	C51F	1999	
174	Ke	WI-RS674	Mercedes-Benz Citaro O530 G	Mercedes-Benz	AN61D	1999	
175	Rü	WI-RS275	Mercedes-Benz Citaro O530 G	Mercedes-Benz	AN61D	1999	
176	Rü	WI-RS976	Mercedes-Benz Citaro O530 G	Mercedes-Benz	AN61D	1999	
178	Ke	WI-RS378	Mercedes-Benz Citaro O530 G	Mercedes-Benz	AN61D	1999	
179	St	F-ST2179	Mercedes-Benz Tourismo O350	Mercedes-Benz	C51F	2007	
180	St	F-ST2180	Mercedes-Benz Tourismo O350	Mercedes-Benz	C51F	2007	
181	Wa	WI-RS981	Mercedes-Benz Citaro O405N	Mercedes-Benz	B35D	1999	
182	Wa	WI-RS982	Mercedes-Benz Citaro O405N	Mercedes-Benz	B35D	1999	
183	Ke	WI-RS983	Mercedes-Benz Citaro O530	Mercedes-Benz	N37D	1999	
184	Mz	WI-RS684	Mercedes-Benz Citaro O530 G	Mercedes-Benz	AN48D	1999	
185	Ke	WI-RS675	Mercedes-Benz Citaro O530 G	Mercedes-Benz	AN48D	1999	
186	Wa	WI-RS886	Mercedes-Benz Citaro O530	Mercedes-Benz	N32D	1999	
180	St	F-ST2189	Mercedes-Benz Tourismo O350	Mercedes-Benz	C51F	2007	
190	Wa	F-ST1190	Mercedes-Benz Tourismo O350	Mercedes-Benz	C51F	2000	
191	Wa	WI-RS791	Mercedes-Benz O405N	Mercedes-Benz	B37D	2000	
192	Wa	WI-RS592	Mercedes-Benz Cito O520	Mercedes-Benz	N20D	1999	
193	Wa	WI-RS193	Mercedes-Benz Citaro O530N	Mercedes-Benz	N37D	1998	
194	Ke	WI-RS794	Mercedes-Benz Citaro O530G	Mercedes-Benz	AB55D	2000	
195	Mz	GG-PL195	Mercedes-Benz Citaro O530G	Mercedes-Benz	AB50D	2002	
196	Mz	WI-RS856	Mercedes-Benz Citaro O530G	Mercedes-Benz	AB53D	2001	
197	Wa	F-ST2197	Mercedes-Benz Citaro O350	Mercedes-Benz	C45F	2002	
198	Er	WI-RS698	Mercedes-Benz Citaro O530	Mercedes-Benz	N36D	2001	
199	Wa	F-ST2199	Mercedes-Benz Citaro O350	Mercedes-Benz	C45F	2002	
200	Wa	F-EF1000	Mercedes-Benz Travego O580	Mercedes-Benz	C32F	2000	

201-228

		Volvo B7700		Volvo	N36D	2008	

201	-	WI-RS201	208	-	WI-RS408	215	-	WI-RS315	222	-	WI-RS522
202	-	WI-RS302	209	-	WI-RS409	216	-	WI-RS216	223	-	WI-RS423
203	-	WI-RS203	210	-	WI-RS450	217	-	WI-RS217	224	-	WI-RS224
204	-	WI-RS304	211	-	WI-RS411	218	-	WI-RS418	225	-	WI-RS925
205	-	WI-RS195	212	-	WI-RS312	219	-	WI-RS319	226	-	WI-RS226
206	-	WI-RS906	213	-	WI-RS293	220	-	WI-RS480	227	-	WI-RS327
207	-	WI-RS277	214	-	WI-RS614	221	-	WI-RS621	228	-	WI-RS328

229-248

		Volvo B7700A		Volvo	AN--D	2008	

229	-	WI-RS328	234	-	WI-RS733	239	-	WI-RS438	244	-	WI-RS843
230	-	WI-RS329	235	-	WI-RS734	240	-	WI-RS439	245	-	WI-RS844
231	-	WI-RS730	236	-	WI-RS935	241	-	WI-RS110	246	-	WI-RS845
232	-	WI-RS431	237	-	WI-RS436	242	-	WI-RS241	247	-	WI-RS846
233	-	WI-RS732	238	-	WI-RS437	243	-	WI-RS842	248	-	WI-RS847

301	Wa	WI-AP401	Mercedes-Benz Citaro O530	Mercedes-Benz	N37D	1999	
302	Mz	WI-TK447	Mercedes-Benz O405	Mercedes-Benz	B45D	1993	
303	Wa	WI-TK433	Mercedes-Benz O405	Mercedes-Benz	B45D	1991	
304	Wa	WI-TK434	Mercedes-Benz O405	Mercedes-Benz	B45D	1992	
306	Kt	WI-TK436	Mercedes-Benz O405	Mercedes-Benz	B45D	1989	
307	Kt	WI-TK437	Mercedes-Benz O405	Mercedes-Benz	B45D	1996	
308	Mz	WI-JP108	Mercedes-Benz Citaro O530	Mercedes-Benz	N37D	1998	
309	Mz	WI-JP109	Mercedes-Benz Citaro O530	Mercedes-Benz	N37D	1998	
310	Mz	WI-TK440	Mercedes-Benz O405	Mercedes-Benz	B45D	1986	
311	Mz	WI-TK441	Mercedes-Benz O405G	Mercedes-Benz	AB64D	1993	
314	Er	WI-JP114	Mercedes-Benz Citaro O530	Mercedes-Benz	N27D	1998	
315	Wa	WI-JP105	Mercedes-Benz Citaro O530	Mercedes-Benz	N27D	1998	
317	Wa	WI-TK417	Mercedes-Benz O405	Mercedes-Benz	B45D	1992	
318	Mz	WI-TK448	Mercedes-Benz O405	Mercedes-Benz	B45D	1995	
325	Mz	WI-S1325	Mercedes-Benz Citaro O530	Mercedes-Benz	N30D	2001	
326	Ke	WI-S1326	Mercedes-Benz Citaro O530	Mercedes-Benz	N35D	2003	

Another Mercedes-Benz Citaro, this time showing the latest frontal style is number 376, WI-RS376, which was added to the fleet in 2007. The model now being the main type operated, although a few Volvo integral 7700s have now been added. *Harry Laming*

327	Mz	WI-S1327	Mercedes-Benz Citaro O530	Mercedes-Benz	N35D	2003
331	-	WI-RS831	Volvo 7700A	Volvo	AN--D	2008
333	-	WI-RS833	Volvo 7700A	Volvo	AN--D	2008
336	Mz	WI-RS336	Mercedes-Benz Citaro O530 K	Mercedes-Benz	N35D	2003
342	Mz	WI-RS342	Mercedes-Benz Citaro O530 K	Mercedes-Benz	N35D	2003
344	Mz	WI-RS344	Mercedes-Benz Citaro O530	Mercedes-Benz	N35D	2003
346	Mz	WI-RS346	Mercedes-Benz Citaro O530	Mercedes-Benz	N35D	2003
351	Mz	WI-RS351	Mercedes-Benz Citaro O530	Mercedes-Benz	N35D	2003
352	Mz	WI-RS352	Mercedes-Benz Citaro O530	Mercedes-Benz	N35D	2003
357	Mz	WI-RS357	Mercedes-Benz Citaro O530	Mercedes-Benz	N35D	2003
358	-	WI-RS358	Volvo 7700A	Volvo	AN--D	2008
361	Ke	WI-RS361	Mercedes-Benz Citaro O530	Mercedes-Benz	N35D	2007
367	Ke	WI-RS367	Mercedes-Benz Citaro O530	Mercedes-Benz	N35D	2007
371	Ke	WI-RS371	Mercedes-Benz Citaro O530	Mercedes-Benz	N35D	2007
373	Ke	WI-RS373	Mercedes-Benz Citaro O530	Mercedes-Benz	N35D	2007
376	Ke	WI-RS376	Mercedes-Benz Citaro O530	Mercedes-Benz	N35D	2007
377	Ke	WI-RS377	Mercedes-Benz Citaro O530	Mercedes-Benz	N35D	2007
379	Ke	WI-RS379	Mercedes-Benz Citaro O530	Mercedes-Benz	N35D	2007
381	Ke	WI-RS381	Mercedes-Benz Citaro O530G	Mercedes-Benz	AN--D	2008
383	Ke	WI-RS383	Mercedes-Benz Citaro O530G	Mercedes-Benz	AN--D	2008
389	Ke	WI-RS389	Mercedes-Benz Citaro O530G	Mercedes-Benz	AN--D	2008
391	Ke	WI-RS391	Mercedes-Benz Citaro O530	Mercedes-Benz	N35D	2007
392	Ke	WI-RS392	Mercedes-Benz Citaro O530	Mercedes-Benz	N35D	2007
393	Ke	WI-RS393	Mercedes-Benz Citaro O530	Mercedes-Benz	N35D	2007
398	-	WI-RS390	Volvo 7700	Volvo	N--D	2008
400	Li	GI-YC400	Setra S315 UL	Setra	NC56D	2000
401	Li	GI-YC401	Setra S315 UL	Setra	NC56D	2000
402	Wa	GI-YI402	Mercedes-Benz Integro O550	Mercedes-Benz	C56F	2001
403	Li	GI-YI403	Mercedes-Benz Integro O550	Mercedes-Benz	C56F	2001
404	Us	GI-ED404	Mercedes-Benz O408	Mercedes-Benz	BC50D	1992
405	Us	GI-ED405	Mercedes-Benz O408	Mercedes-Benz	BC50D	1991
407	Li	GI-AD407	Mercedes-Benz Vario O814	Mercedes-Benz	C20F	1995
416	Ho	GI-DM416	Kässbohrer S215 HR	Kässbohrer Setra	C54F	1988
417	Us	GI-DM417	Mercedes-Benz O405	Mercedes-Benz	B44D	1997
418	Us	GI-DM418	Mercedes-Benz O405	Mercedes-Benz	B38D	1986
419	Li	GI-DM419	Mercedes-Benz O405	Mercedes-Benz	B44D	1992
420	Li	GI-AD420	Mercedes-Benz Tourismo O350	Mercedes-Benz	C50F	2001
241	Ke	WI-RS813	Mercedes-Benz Sprinter 313 cdi	Mercedes-Benz	M8	2006

Depots: Altenstadt (Al); Büdingen (Bü); Düsseldorf (Dü); Echzell (Ez); Erlangen (Er); Groß-Gerau (GG); Hanau (Ha); Hofheim (Hm); Kaiserslautern (Kn); Kastel (Kt); Kelsterbach (Ke); Lich (Li); Mainz (Mz); Oberursel (Ob); Rüsselsheim (Rü); Stuttgart (St); Usingen (Us); Wallau (Wa).

BB TOURISTIK

BB-Touristik GmbH, Augustastr 33, 17235 Neustrelitz, Germany

01	R	NZ T923	Mercedes-Benz O405N	Mercedes-Benz	B45D	1992
02	R	NZ Y274	Mercedes-Benz O405	Mercedes-Benz	B39D	1993
03	R	NZ AC131	Mercedes-Benz O405	Mercedes-Benz	B39D	1993
04	R	MST E99	Mercedes-Benz O405	Mercedes-Benz	B37D	1994
06	R	MST DY91	MAN NL222	MAN	N35D	1996
07	R	MST FK63	Mercedes-Benz O405	Mercedes-Benz	B36D	1999
08	R	MST MD85	MAN SG242	MAN	AB61D	1989
10	R	MST UJ69	MAN NL263	MAN	N32D	1998
12	R	MST V835	MAN NG323	MAN	AN32D	2001
13	R	MST P808	MAN NL263	MAN	N32D	2002
03	T	MST W404	Mercedes-Benz O404	Mercedes-Benz	C46D	1998
05	T	MST WZ42	Mercedes-Benz O404	Mercedes-Benz	C46D	1999
02	T	MST RF61	Mercedes-Benz O404	Mercedes-Benz	C46D	2000
04	T	MST W350	Mercedes-Benz Tourismo O350	Mercedes-Benz	C47F	2000
06	T	MST A350	Mercedes-Benz Tourismo O350	Mercedes-Benz	C51F	2001
07	T	MST OP100	Temsa Opalin 9	Temsa	C30F	2005

Depot: Neustrelitz, Mecklenburg-West Pomerania BBR - BB Reisen; BBT - BB Touristik.

Arriva Sippel has gained services within Frankfurt Airport. Here another Citaro, 371, WI-RS371 is seen with the main airport car park in the background. *Harry Laming*

BILS

Verkehrsbetriebe Bils GmbH, Haberkamp 2-6, 48324 Sendenhorst-Albersloh, Germany

WAF PA420	Mercedes-Benz O405G	Mercedes-Benz	AB62D	1989
WAF PA519	Mercedes-Benz O405	Mercedes-Benz	B44D	1991
WAF PA429	Mercedes-Benz O405	Mercedes-Benz	B44D	1992
WAF PA105	Mercedes-Benz O405	Mercedes-Benz	B44D	1992
WAF PA851	Mercedes-Benz O407	Mercedes-Benz	BC49D	1992
WAF PA435	Mercedes-Benz O407N	Mercedes-Benz	BC49D	1992
WAF PA986	Mercedes-Benz O405N	Mercedes-Benz	B44D	1992
WAF PA424	Mercedes-Benz O405N	Mercedes-Benz	B44D	1993
WAF PA135	Mercedes-Benz O405	Mercedes-Benz	B44D	1993
WAF PA472	Mercedes-Benz O405	Mercedes-Benz	B44D	1993
WAF PA421	Mercedes-Benz O405	Mercedes-Benz	B44D	1993
WAF PA412	Mercedes-Benz O405G	Mercedes-Benz	AB62D	1993
WAF PA223	Mercedes-Benz O407	Mercedes-Benz	BC49D	1993
WAF PA376	Mercedes-Benz O407	Mercedes-Benz	BC49D	1993
WAF PA329	Mercedes-Benz O407	Mercedes-Benz	B53D	1994
WAF PA647	Mercedes-Benz O405GN	Mercedes-Benz	AB50D	1994
WAF PA159	Setra S215 UL	Setra	BC49D	1994
WAF PA166	Setra S215 UL	Setra	BC49D	1995
WAF PA178	Setra S215 UL	Setra	BC49D	1995
WAF PA137	Mercedes-Benz O614	Frenzle	M7	1995
WAF PA747	Mercedes-Benz O405GN	Mercedes-Benz	AB52D	1996
WAF PA31	Setra SG321UL	Setra	AB79D	1996
WAF PA370	Setra SG321UL	Setra	AB79D	1996
WAF PA368	Mercedes-Benz Vario O814	Auwärter Teamstar	C24F	1996
WAF PA273	Mercedes-Benz Integro O550	Mercedes-Benz	C53F	1997
WAF PA205	Mercedes-Benz O405N	Mercedes-Benz	B36D	1997
WAF PA226	Mercedes-Benz O405N	Mercedes-Benz	B36D	1997
WAF PA364	Mercedes-Benz O405N	Mercedes-Benz	B36D	1997
WAF PA754	Mercedes-Benz O405N	Mercedes-Benz	B36D	1997
WAF PA908	Mercedes-Benz O405N	Mercedes-Benz	B43D	1997
WAF PA160	Mercedes-Benz O405GN	Mercedes-Benz	AB58D	1998
WAF PA168	Mercedes-Benz O405GN	Mercedes-Benz	AB58D	1998
WAF PA230	Mercedes-Benz O405GN	Mercedes-Benz	AB58D	1998
WAF PA498	Setra S315NF	Setra	N45D	1998
WAF PA512	Setra S315NF	Setra	N45D	1998
WAF PA559	Setra S315NF	Setra	N45D	1998
WAF PA562	Setra S315NF	Setra	N45D	1998
WAF PA215	Mercedes-Benz Integro O550	Mercedes-Benz	C53F	1998
WAF PA306	Mercedes-Benz Citaro O530	Mercedes-Benz	N37D	1998
WAF PA318	Mercedes-Benz Citaro O530	Mercedes-Benz	N37D	1998
WAF PA438	Mercedes-Benz Citaro O530	Mercedes-Benz	N37D	1998
WAF PA263	Mercedes-Benz Sprinter 412	Auwärter Sprinter	C16F	1999
WAF PA358	Mercedes-Benz Sprinter 412	Koch Sprinter	M7	1999
WAF PA275	Mercedes-Benz O405GN	Mercedes-Benz	AB52D	1999
WAF PA992	Mercedes-Benz 316 cdi	Mercedes-Benz	M6	2000
WAF PA147	Setra S315NF	Setra	N43D	2000
WAF PA316	Setra S315NF	Setra	N43D	2000
WAF PA354	Setra S315NF	Setra	N43D	2000
WAF PA373	Setra S315NF	Setra	N43D	2000
WAF PA258	Mercedes-Benz Citaro O530	Mercedes-Benz	N37D	2001
WAF PA282	Mercedes-Benz Citaro O530	Mercedes-Benz	N37D	2001
WAF PA289	Mercedes-Benz Citaro O530	Mercedes-Benz	N37D	2001
WAF PA297	Mercedes-Benz Citaro O530	Mercedes-Benz	N37D	2001
WAF PA304	Mercedes-Benz Citaro O530	Mercedes-Benz	N37D	2001
WAF PA324	Mercedes-Benz Citaro O530	Mercedes-Benz	N37D	2001
WAF PA739	Mercedes-Benz Citaro O530G	Mercedes-Benz	AN52D	2000
WAF PA960	Mercedes-Benz Vario O815	Auwärter Teamstar	C25F	2001
WAF PA515	Mercedes-Benz Sprinter 211	Mercedes-Benz	M8	2002
WAF PA261	Setra S315 GT-HD	Setra	C49F	2002
WAF PA74	Setra S315NF	Setra	N42D	2002
WAF PA488	Setra S315NF	Setra	N42D	2002
WAF PA387	Setra S315NF	Setra	N42D	2002
WAF PA384	Mercedes-Benz O815	Koch 815	M7	2002
WAF PA436	Mercedes-Benz O412	Koch Sprinter	M7	2002
WAF PA189	Mercedes-Benz Citaro O530	Mercedes-Benz	N37D	2002

WAF PA167	Mercedes-Benz Citaro 0530	Mercedes-Benz	N37D	2003
WAF PA298	Mercedes-Benz Citaro 0530	Mercedes-Benz	N37D	2003
WAF PA393	Mercedes-Benz Citaro 0530	Mercedes-Benz	N37D	2003
WAF PA394	Mercedes-Benz Citaro 0530	Mercedes-Benz	N37D	2003
WAF PA310	Setra S415 HD	Setra	C44F	2003
WAF PA291	Mercedes-Benz 412D	Koch Sprinter	M7	2003
WAF PA332	Mercedes-Benz Vario 0815	Auwärter Teamstar	C25F	2004
WAF PA143	Mercedes-Benz Citaro 0530	Mercedes-Benz	N37D	2004
WAF PA144	Mercedes-Benz Citaro 0530	Mercedes-Benz	N37D	2004
WAF PA187	Mercedes-Benz Citaro 0530	Mercedes-Benz	N37D	2004
WAF PA173	Setra S315 NF	Setra	N42D	2004
WAF PA164	Setra S315 NF	Setra	N42D	2004
WAF PA311	Setra S317 GT-HD	Setra	C52F	2004
WAF PA233	Setra SG321 UL	Setra	AN75D	2004
WAF PA183	Setra S411 HD	Setra	C32F	2004
WAF PA202	Setra S411 HD	Setra	C36F	2004
WAF PA184	Setra S415 GT-HD	Setra	C44F	2004
WAF PA410	Setra S416 GT-HD	Setra	C48F	2004
WAF PA190	Setra S417 H-HD	Setra	C59F	2004
WAF PA208	Setra S415 GT	Setra	C49F	2005
WAF PA126	Setra S415 HD	Setra	C49F	2005
WAF PA213	Setra SG321 UL	Setra	BC77F	2005
WAF PA402	Mercedes-Benz Citaro 0530	Mercedes-Benz	N37D	2006
WAF PA403	Mercedes-Benz Citaro 0530	Mercedes-Benz	N35D	2006
WAF PA405	Mercedes-Benz Citaro 0530	Mercedes-Benz	N35D	2006
WAF PA406	Mercedes-Benz Citaro 0530G	Mercedes-Benz	AN50D	2006
WAF PA206	Mercedes-Benz Citaro 0530G	Mercedes-Benz	AN50D	2006
WAF PA237	Mercedes-Benz Citaro 0530	Mercedes-Benz	N37D	2007
WAF PA241	Mercedes-Benz Citaro 0530	Mercedes-Benz	N37D	2007
WAF PA656	Mercedes-Benz Citaro 0530	Mercedes-Benz	N37D	2007
WAF PA702	Mercedes-Benz Citaro 0530	Mercedes-Benz	N37D	2007
WAF PA246	Setra S415NF	Setra	N40D	2007
WAF PA248	Setra S415NF	Setra	N40D	2007
WAF PA257	Setra S415NF	Setra	N40D	2007
WAF PA267	Setra S415NF	Setra	N43D	2007
WAF PA221	Setra S417 GT-HD	Setra	C50F	2007
WAF PA180	Mercedes-Benz 308	Mercedes-Benz	M8	200
WAF PA120	Mercedes-Benz 415 cdi	Mercedes-Benz	N8	200
WAF PA122	Mercedes-Benz Citaro 0530	Mercedes-Benz	N37D	200
WAF PA127	Mercedes-Benz Citaro 0530	Mercedes-Benz	N37D	200
WAF PA129	Mercedes-Benz Citaro 0530	Mercedes-Benz	N37D	200
WAF PA145	Mercedes-Benz Citaro 0530	Mercedes-Benz	N37D	200
WAF PA146	Mercedes-Benz Citaro 0530G	Mercedes-Benz	AN50D	200
WAF PA149	Mercedes-Benz Citaro 0530	Mercedes-Benz	N37D	200
WAF PA172	Mercedes-Benz Citaro 0530 Ü	Mercedes-Benz	N37D	200
WAF PA227	Mercedes-Benz Tourino	Mercedes-Benz	C	200
WAF PA325	Mercedes-Benz Citaro 0530	Mercedes-Benz	N37D	200
WAF PA326	Mercedes-Benz Citaro 0530	Mercedes-Benz	N37D	200
WAF PA327	Mercedes-Benz Citaro 0530	Mercedes-Benz	N37D	200
WAF PA275	Mercedes-Benz Citaro 0530	Mercedes-Benz	N37D	200

Depots: Abersloh [Haberkamp 2-6, 48324 Sendenhorst]; Ahlen [Am Neuen Baum 20, 59229 Ahlen] and Niederlassung Warendorf [Dreibrückenstrasse. 28, 48231 Warendorf] in North Rhine-Westphalia.

CeBus

Nienburger Strasse 50, 29225 Celle, Lower Saxony, Germany

CE-KC20	Mercedes-Benz O405N	Mercedes-Benz	B--D	-
CE-KC22	Mercedes-Benz O405GN	Mercedes-Benz	AB--D	-
CE-KC23	Mercedes-Benz O405GN	Mercedes-Benz	AB--D	-
CE-KC27	Mercedes-Benz O405GN	Mercedes-Benz	AB--D	-
CE-KC28	Mercedes-Benz O405N	Mercedes-Benz	B--D	-
CE-KC29	Mercedes-Benz O405N	Mercedes-Benz	B--D	-
CE-KC30	Mercedes-Benz O405N	Mercedes-Benz	B--D	-
CE-KC31	Mercedes-Benz O405G	Mercedes-Benz	AB--D	-
CE-KC32	Mercedes-Benz O405GN	Mercedes-Benz	AB--D	-
CE-KC34	Mercedes-Benz O405GN	Mercedes-Benz	AB--D	-
CE-KC35	Mercedes-Benz O405GN	Mercedes-Benz	AB--D	-
CE-KC36	Mercedes-Benz O405GN	Mercedes-Benz	AB--D	-
CE-KC37	Mercedes-Benz O405GN	Mercedes-Benz	AB--D	-
CE-KC38	Mercedes-Benz O405GN	Mercedes-Benz	AB--D	-
CE-KC39	Mercedes-Benz O405GN	Mercedes-Benz	AB--D	-
CE-KC67	Mercedes-Benz Citaro O530G	Mercedes-Benz	AN50D	-
CE-KC68	Mercedes-Benz Citaro O530G	Mercedes-Benz	AN50D	-
CE-KC69	Mercedes-Benz Citaro O530G	Mercedes-Benz	AN50D	-
CE-KC71	Mercedes-Benz Citaro O530G	Mercedes-Benz	AN50D	-
CE-KC74	Mercedes-Benz Citaro O530G	Mercedes-Benz	AN50D	-
CE-KC75	Mercedes-Benz Citaro O530G	Mercedes-Benz	AN50D	-
CE-KC76	Mercedes-Benz Citaro O530G	Mercedes-Benz	AN50D	-
CE-KC78	Mercedes-Benz Citaro O530G	Mercedes-Benz	AN50D	-
CE-KC80	Mercedes-Benz O405GN	Mercedes-Benz	AB--D	-
CE-KC81	Mercedes-Benz O405GN	Mercedes-Benz	AB--D	-
CE-KC82	Mercedes-Benz O405GN	Mercedes-Benz	AB--D	-
CE-KC83	Mercedes-Benz O405GN	Mercedes-Benz	AB--D	-
CE-KC84	Setra S315 NF	Setra	N42D	-
CE-KC85	Setra S315 NF	Setra	N42D	-
CE-KC86	Mercedes-Benz Citaro O530	Mercedes-Benz	N37D	-
CE-KC87	Setra SG321 ÜL	Setra	AC--D	-
CE-KC220	Mercedes-Benz Citaro O530G	Mercedes-Benz	AN50D	-
CE-KC221	Mercedes-Benz Citaro O530G	Mercedes-Benz	AN50D	-
CE-KC222	Mercedes-Benz Citaro O530G	Mercedes-Benz	AN50D	-
CE-KC239	Setra S213 ÜL	Setra	AC--D	-
CE-KC259	Setra SG321 ÜL	Setra	AC--D	-
CE-KC260	Mercedes-Benz O405G	Mercedes-Benz	AB--D	-
CE-KC275	Setra SG180	Setra	AC--D	-
CE-KC269	Mercedes-Benz O405G	Mercedes-Benz	AB--D	-
CE-KC282	Mercedes-Benz O405	Mercedes-Benz	B--D	-
CE-KC283	Mercedes-Benz O405G	Mercedes-Benz	AB--D	-
CE-KC284	Mercedes-Benz O405G	Mercedes-Benz	AB--D	-
CE-KC287	Setra SG219 SL	Setra	AC--D	-
CE-KC288	Setra SG219 SL	Setra	AC--D	-
CE-KC291	Mercedes-Benz O405G	Mercedes-Benz	AB--D	-
CE-KC293	Mercedes-Benz O405G	Mercedes-Benz	AB--D	-
CE-KC294	Mercedes-Benz Citaro O530G	Mercedes-Benz	AN50D	-
CE-KC295	Mercedes-Benz O405G	Mercedes-Benz	AB--D	-
CE-KC296	Mercedes-Benz O405G	Mercedes-Benz	AB--D	-
CE-KC298	Mercedes-Benz Citaro O530	Mercedes-Benz	N37D	-
CE-KC299	Mercedes-Benz Citaro O530	Mercedes-Benz	N37D	-
CE-DP812	Mercedes-Benz Integro O550-19	Mercedes-Benz	C53F	-
CE-DP813	Mercedes-Benz Integro O550	Mercedes-Benz	C53F	-
CE-DP814	Mercedes-Benz Integro O550	Mercedes-Benz	C53F	-
CE-DP816	Mercedes-Benz O407	Mercedes-Benz	BC--D	-
CE-DP843	Mercedes-Benz O407	Mercedes-Benz	BC--D	-
CE-DP848	Setra SG221 ÜL	Setra	AC--D	-
CE-DP849	Mercedes-Benz O408	Mercedes-Benz	BC--D	-
CE-DP852	Setra SG221 ÜL	Setra	AC--D	-
CE-DP853	Setra SG221 ÜL	Setra	AC--D	-
CE-DP854	Mercedes-Benz O408	Mercedes-Benz	BC--D	-
CE-DP855	Mercedes-Benz O408	Mercedes-Benz	BC--D	-
CE-DP856	Setra SG221 ÜL	Setra	AC--D	-
CE-DP857	Setra SG221 ÜL	Setra	AC--D	-

CE-DP858	Setra SG321 ÜL	Setra		AC--D	-
CE-DP861	Setra SG221 ÜL	Setra		AC--D	-
CE-DP876	Mercedes-Benz Integro O550	Mercedes-Benz		C53F	-
CE-DP877	Mercedes-Benz Integro O550	Mercedes-Benz		C53F	-
CE-DP878	Mercedes-Benz Integro O550	Mercedes-Benz		C53F	-
CE-DP879	Mercedes-Benz Integro O550-79	Mercedes-Benz		C53F	-
CE-DP880	Mercedes-Benz Integro O550-80	Mercedes-Benz		C53F	-

Depot: Celle

SBN - Südbrandenburger Nahverkehr

Südbrandenburger Nahverkehrs GmbH, Spremberger Str 23, 01968 Senftenberg, Germany

35010	OSL-KV10	MAN 313 ÜL	MAN	N	-
35014	OSL-KV14	MAN 313 ÜL	MAN	N	-
35015	OSL-KV15	Mercedes-Benz Integro O550	Mercedes-Benz	NC	
35016	OSL-KV16	Mercedes-Benz Sprinter 616cdi	Mercedes-Benz	M	
35017	OSL-KV17	Mercedes-Benz Citaro O530 Ü	Mercedes-Benz	N	-
35018	OSL-KV18	Mercedes-Benz Citaro O530 GÜ	Mercedes-Benz	AN	-
35019	OSL-KV19	Mercedes-Benz Citaro O530 GÜ	Mercedes-Benz	AN	-
35020	OSL-KV20	Mercedes-Benz Integro O550	Mercedes-Benz	NC	-
35021	OSL-KV21	Mercedes-Benz Integro O550	Mercedes-Benz	NC	-
35024	OSL-KV24	Mercedes-Benz Integro O550	Mercedes-Benz	NC	-
35025	OSL-KV25	Mercedes-Benz Integro O550	Mercedes-Benz	NC	-
35026	OSL-KV26	Mercedes-Benz Integro O550	Mercedes-Benz	NC	-
35027	OSL-KV27	Mercedes-Benz Integro O550	Mercedes-Benz	NC	-
35028	OSL-KV28	Mercedes-Benz Integro O550	Mercedes-Benz	NC	-
35029	OSL-KV29	Mercedes-Benz Integro O550	Mercedes-Benz	NC	-
35030	OSL-KV30	Mercedes-Benz Integro O550	Mercedes-Benz	NC	-
35031	OSL-KV31	Mercedes-Benz Integro O550	Mercedes-Benz	NC	-
35032	OSL-KV32	Mercedes-Benz Integro O550-19	Mercedes-Benz	NC	-
35033	OSL-KV33	Mercedes-Benz Integro O550-19	Mercedes-Benz	NC	-
35034	OSL-KV34	Mercedes-Benz Integro O550	Mercedes-Benz	NC	-
35035	OSL-KV35	Mercedes-Benz Integro O550	Mercedes-Benz	NC	-
35036	OSL-KV36	Mercedes-Benz Integro O550	Mercedes-Benz	NC	-
35037	OSL-KV37	Mercedes-Benz Integro O550	Mercedes-Benz	NC	-
35038	OSL-KV38	Mercedes-Benz Integro O550	Mercedes-Benz	NC	-
35039	OSL-KV39	Mercedes-Benz Integro O550-19	Mercedes-Benz	NC	-
35040	OSL-KV40	Mercedes-Benz Integro O550-19	Mercedes-Benz	NC	-
35042	OSL-KV42	Mercedes-Benz Integro O550	Mercedes-Benz	NC	-
35043	OSL-KV43	Mercedes-Benz Integro O550	Mercedes-Benz	NC	-
35044	OSL-KV44	MAN 313 ÜL	MAN	N	-
35045	OSL-KV45	MAN 313 ÜL	MAN	N	-
35046	OSL-KV46	Mercedes-Benz Integro O550	Mercedes-Benz	NC	-
35047	OSL-KV47	Mercedes-Benz Citaro O530 Ü	Mercedes-Benz	N	-
35048	OSL-KV48	Mercedes-Benz Integro O550	Mercedes-Benz	NC	-
35049	OSL-KV49	Mercedes-Benz Integro O550	Mercedes-Benz	NC	-
35050	OSL-KV50	Mercedes-Benz Integro O550	Mercedes-Benz	NC	-
35051	OSL-KV51	Mercedes-Benz Integro O550	Mercedes-Benz	NC	-
35054	OSL-KV54	MAN Lion's City Ü	MAN	NC	-
35055	OSL-KV55	MAN Lion's City Ü	MAN	NC	-
35056	OSL-KV56	MAN Lion's City Ü	MAN	NC	-
35057	OSL-KV57	Mercedes-Benz Sprinter City 65	Mercedes-Benz	N	
35062	OSL-KV62	Mercedes-Benz 414	Mercedes-Benz	M	
35063	OSL-KV63	Mercedes-Benz Citaro O530 Ü	Mercedes-Benz	N	-
35065	OSL-KV65	Mercedes-Benz Citaro O530 Ü	Mercedes-Benz	N	-
35074	OSL-KV74	Mercedes-Benz 414	Mercedes-Benz	M	
35075	OSL-KV75	Mercedes-Benz Citaro O530 Ü	Mercedes-Benz	N	-
35077	OSL-KV77	Mercedes-Benz Integro O550-19	Mercedes-Benz	NC	-
35078	OSL-KV78	Mercedes-Benz Integro O550-19	Mercedes-Benz	NC	-
35079	OSL-KV79	Mercedes-Benz Integro O550-19	Mercedes-Benz	NC	-
35080	OSL-KV80	Mercedes-Benz Integro O550-19	Mercedes-Benz	NC	-

35081	OSL-KV81	Mercedes-Benz Citaro O530 Ü	Mercedes-Benz	N	-
35085	OSL-KV85	Mercedes-Benz Integro O550	Mercedes-Benz	NC	-
35086	OSL-KV86	Mercedes-Benz Integro O550	Mercedes-Benz	NC	-
35087	OSL-KV87	Mercedes-Benz Integro O550	Mercedes-Benz	NC	-
35088	OSL-KV88	Mercedes-Benz Integro O550	Mercedes-Benz	NC	-
35089	OSL-KV89	MAN 313 ÜL	MAN	N	-
35090	OSL-KV90	MAN 313 ÜL	MAN	N	-
35091	OSL-KV91	MAN 313 ÜL	MAN	N	-
35092	OSL-KV92	MAN 313 ÜL	MAN	N	-
35093	OSL-KV93	Mercedes-Benz Citaro O530 Ü	Mercedes-Benz	N	-
35094	OSL-KV94	Mercedes-Benz O345	Mercedes-Benz	BC	-
35095	OSL-KV95	Mercedes-Benz O345	Mercedes-Benz	BC	-
35096	OSL-KV96	Mercedes-Benz Sprinter 616cdi	Mercedes-Benz	M	-
35097	OSL-KV97	Mercedes-Benz Sprinter 616cdi	Mercedes-Benz	M	-
35101	OSL-KV101	Mercedes-Benz Citaro O530 Ü	Mercedes-Benz	N	-
35102	OSL-KV102	Mercedes-Benz Citaro O530 Ü	Mercedes-Benz	N	-
35103	OSL-KV103	Mercedes-Benz O345	Mercedes-Benz	BC	-
35104	OSL-KV104	Mercedes-Benz O345	Mercedes-Benz	BC	-
35105	OSL-KV105	Mercedes-Benz O345	Mercedes-Benz	BC	-
35106	OSL-KV106	Mercedes-Benz O345	Mercedes-Benz	BC	-
35107	OSL-KV107	Mercedes-Benz O345	Mercedes-Benz	BC	-
35108	OSL-KV108	Mercedes-Benz O345	Mercedes-Benz	BC	-

Arriva Holland has established a touring company. Here, 457, BP-LG-16, is one of eight Setra S317 HDHs operated. The Setra coaches are produced at Neu Ulm on the Bavaria/Baden-Württemberg border. *Harry Laming*

NV - Neißeverkehr

Neißeverkehr GmbH, Dubrauweg 47, 03172 Guben, Germany

210	Gu	SPN-N942	MAN NL202	MAN	N D	-
240	Gu	SPN-N984	Volvo B10M-A	Säffle SG18	AB D	-
242	Gu	SPN-N963	Setra S319 ÜL	Setra	NC D	-
244	Gu	SPN-NV271	Setra S319 ÜL	Setra	NC D	-
264	Gu	SPN-NV235	MAN Lion Regio	MAN	NC D	-
265	Gu	SPN-NV236	MAN Lion Regio	MAN	NC D	-
266	Gu	SPN-NV237	MAN Lion Regio	MAN	NC D	-
267	Gu	SPN-NV238	Volvo 8700	Volvo	NC D	-
268	Gu	SPN-NV239	Volvo 8700	Volvo	NC D	-
269	Gu	SPN-NV240	Volvo 8700	Volvo	NC D	-
270	Gu	SPN-NV241	Volvo 8700	Volvo	NC D	-
271	Gu	SPN-NV242	Volvo 8700	Volvo	NC D	-
272	Gu	SPN-NV243	Volvo 8700	Volvo	NC D	-
273	Gu	SPN-NV244	Volvo 8700	Volvo	NC D	-
281	Gu	SPN-N970	Volvo B10B	Volvo SL12	NC D	-
282	Gu	SPN-N971	Volvo B10B	Volvo SL12	NC D	-
312	Gu	SPN-NV201	MAN A37	MAN	N D	-
315	Gu	SPN-N943	MAN NL202	MAN	N D	-
318	Gu	SPN-N940	MAN NL202	MAN	N D	-
342	Gu	SPN-N964	Setra S319 ÜL	Setra	NC D	-
344	Gu	SPN-NV272	Setra S319 ÜL	Setra	NC D	-
361	Gu	SPN-N953	MAN ÜL313	MAN	NC D	-
362	Gu	SPN-N954	MAN ÜL313	MAN	NC D	-
363	Gu	SPN-N955	MAN ÜL313	MAN	NC D	-
364	Gu	SPN-N956	MAN ÜL313	MAN	NC D	-
365	Gu	SPN-N965	MAN ÜL313	MAN	NC D	-
366	Gu	SPN-N966	MAN ÜL313	MAN	NC D	-
367	Gu	SPN-NV230	MAN ÜL313	MAN	NC D	-
368	Gu	SPN-N959	MAN ÜL313	MAN	NC D	-
369	Gu	SPN-NV233	MAN Lion Regio	MAN	NC D	-
385	Gu	SPN-N951	Volvo B10B	Volvo SL12	NC D	-
386	Gu	SPN-N952	Volvo B10B	Volvo SL12	NC D	-
410	Gu	SPN-NV200	MAN NL303	MAN	N D	-
416	Gu	SPN-N944	MAN NL202	MAN	N D	-
417	Gu	SPN-N941	MAN NL202	MAN	N D	-
440	Gu	SPN-N960	Volvo B10M-A	Säffle SG18	AB D	-
441	Gu	SPN-N961	Volvo B10M-A	Säffle SG18	AB D	-
442	Gu	SPN-N962	Setra S319 ÜL	Setra	NC D	-
444	Gu	SPN-NV270	Setra S319 ÜL	Setra	NC D	-
461	Gu	SPN-NV231	MAN ÜL313	MAN	NC D	-
462	Gu	SPN-N969	MAN ÜL313	MAN	NC D	-
463	Gu	SPN-N967	MAN ÜL313	MAN	NC D	-
464	Gu	SPN-NV234	MAN Lion Regio	MAN	NC D	-
465	Gu	SPN-N968	MAN ÜL313	MAN	NC D	-
468	Gu	SPB-S170	Mercedes-Benz O407	Mercedes-Benz	BC49D	1993
473	Gu	SPB-EA97	Mercedes-Benz O407	Mercedes-Benz	BC49D	1993
475	Gu	SPN-N957	MAN ÜL313	MAN	NC D	-
476	Gu	SPN-E962	Mercedes-Benz O407	Mercedes-Benz	BC49D	1993
477	Gu	SPN-NV232	MAN ÜL313	MAN	NC D	-
478	Gu	SPN-N958	MAN ÜL313	MAN	NC D	-
480	Gu	SPN-E960	Mercedes-Benz O408	Mercedes-Benz	B	-
481	Gu	SPN-E961	Mercedes-Benz O408	Mercedes-Benz	B	-

Depot: Guben (Gu)

KVG

KVG Stade GmbH & Co KG, Harburger Straße 96, 21680 Stade, Germany

579	HD	STD-S4026	Kässbohrer S215 UL		Kässbohrer Setra		BC54D	1989				
582	HD	STD-S4029	Kässbohrer S215 UL		Kässbohrer Setra		BC54D	1989				
595	ST	STD-S 8	Kässbohrer S8 9m		Kässbohrer Setra		C25C	1953				
597	BU	STD-S 4050	Mercedes-Benz O407		Mercedes-Benz		B54D	1991				

598-604

			Mercedes-Benz O405		Mercedes-Benz		B54D	1990-91				
598	Lü	STD-S 4046	**600**	HD	WL-KD 716	**602**	HD	WL-KD 580	**604**	Lü	LG-ER 216	

606-636

			Kässbohrer S215 UL		Kässbohrer Setra		BC54D	1990-91				
606	BU	LG-ER 214	**627**	ST	STD-S 4054	**632**	HD	WL-HY 382	**634**	HD	WL-HY 384	
617	HD	WL-KD 713	**628**	BU	STD-S 4055	**633**	HD	WL-HY 383	**636**	Lü	LG-ER 105	

637	CX	STD-S 4058	Kässbohrer SG 219 SL		Kässbohrer Setra		BC57D	1991				
638	CX	CUX-AT638	Kässbohrer SG 219 SL		Kässbohrer Setra		BC57D	1991				
639	ST	STD-S 4059	Kässbohrer SG 221 UL		Kässbohrer Setra		BC70D	1991				
640	ST	STD-S 4061	Kässbohrer SG 219 SL		Kässbohrer Setra		BC70D	1991				
641	ST	STD-S 3649	Kässbohrer SG 219 SL		Kässbohrer Setra		BC70D	1991				
642	BU	STD-S4062	Mercedes-Benz O405		Mercedes-Benz		B45D	1991				
644	BU	STD-S4063	Mercedes-Benz O405		Mercedes-Benz		B45D	1991				

649-654

			Mercedes-Benz O405N		Mercedes-Benz		B42D	1991				
649	Lü	LG-ER 218	**652**	Lü	LG-ER 114	**653**	CX	CUX-AN 653	**654**	CX	CUX-AK 654	
650	Lü	LG-ER 112										

656-669

			Kässbohrer S215 UL		Kässbohrer Setra		BC54D	1992				
656	Lü	LG-ER 121	**660**	HD	WL-DW 748	**668**	HD	STD-L 1009	**669**	BU	STD-L 1008	
658	HD	WL-DW 739	**667**	SO	STD-L 1011							

670-676

			Kässbohrer S215 RL		Kässbohrer Setra		BC54D*	1992		*Seating varies		
670	HD	CUX-CN 670	**672**	ST	STD-L 1004	**674**	ST	STD-L 1006	**676**	HD	CUX-DR 676	
671	BU	STD-L 1003	**673**	Lü	STD-L 1005	**675**	BU	STD-L 1007				

677-683

			Mercedes-Benz O405N		Mercedes-Benz		B42D	1992				
677	BU	STD-L 1000	**679**	Lü	STD-L 1001	**681**	Lü	LG-ER 220	**683**	BU	LG-ER 126	
678	ST	STD-L 1002	**680**	Lü	LG-ER 219	**682**	Lü	LG-ER 161				

687	Lü	LG-ER 687	Kässbohrer SG221 UL		Kässbohrer Setra		ABC70D	1993				

691-699

			Kässbohrer S215 UL		Kässbohrer Setra		BC54D	1993				
691	HD	WL-KV 103	**694**	BU	STD-L 1028	**696**	HD	CUX-DC 696	**698**	BU	STD-L 1025	
692	HD	WL-KV 104	**695**	BU	CUX-DC 695	**697**	HD	STD-L 1026	**699**	HD	LG-ER 699	
693	BU	STD-L 1027										

701	HD	LG-ER701	Kässbohrer S215 RL		Kässbohrer Setra		BC54D	1993				
702	HD	WL-KV101	Kässbohrer S215 RL		Kässbohrer Setra		BC54D	1993				
703	HD	WL-KV102	Kässbohrer S215 RL		Kässbohrer Setra		BC54D	1993				
704	HD	WL-DW307	Mercedes-Benz O405		Mercedes-Benz		B45D	1993				

705-718

			Mercedes-Benz O405N		Mercedes-Benz		B42D*	1993		*Seating varies		
705	Lü	STD-L1021	**709**	BU	STD-L1019	**713**	Lü	LG-ER713	**716**	CX	CUX-LT716	
706	BU	STD-L1022	**710**	Lü	LG-ER710	**714**	CX	CUX-CX714	**717**	CX	CUX-LT717	
707	Lü	STD-L1023	**711**	Lü	LG-ER711	**715**	CX	CUX-EH715	**718**	CX	CUX-LT718	
708	BU	STD-L1024	**712**	Lü	LG-ER712							

723-726

			Mercedes-Benz O405N		Mercedes-Benz		B42D	1994				
723	HD	WL-KV 144	**724**	Lü	LG-ER 724	**725**	Lü	LG-ER 725	**726**	Lü	LG-ER 726	

727-739 Mercedes-Benz O408 — Mercedes-Benz — BC54D — 1994

727	Lü	LG-ER 727	731	HD	LG-ER 731	734	ST	LG-ER 734	737	Lü	LG-ER 737
728	Lü	LG-ER 728	732	ST	LG-ER 732	735	HD	LG-ER 735	738	Lü	STD-L 1043
729	Lü	LG-ER 729	733	ST	LG-ER 733	736	HD	LG-ER 736	739	HD	WL-KV 145
730	Lü	STD-L 1042									

740-746 Kässbohrer S215 UL — Kässbohrer Setra — BC54D — 1994

740	HD	STD-L 1031	742	ST	STD-L 1111	744	ST	STD-L 1035	746	ST	STD-L 1037
741	HD	STD-L 1032	743	ST	STD-L 1034	745	BU	STD-L 1036			

747	ST	STD-L 1038	Kässbohrer SG219 SL	Kässbohrer Setra	BC70D	1994
748	ST	STD-L 1039	Kässbohrer SG219 SL	Kässbohrer Setra	BC70D	1994
749	ST	STD-L 1041	Kässbohrer SG219 SL	Kässbohrer Setra	BC70D	1994
750	BU	STD-L 1044	Mercedes-Benz O405N	Mercedes-Benz	BC54D	1994
751	BU	STD-L 1045	Mercedes-Benz O405N	Mercedes-Benz	BC54D	1994
752	BU	STD-L 1046	Mercedes-Benz O405N	Mercedes-Benz	BC54D	1994
757	ST	STD-L 1051	Kässbohrer SG219 SL	Kässbohrer Setra	BC70D	1995

759-768 Mercedes-Benz O405N — Mercedes-Benz — B42D — 1996

759	Lü	LG-ER 759	762	Lü	LG-ER 762	765	Lü	LG-ER 765	767	Lü	LG-ER 767
760	Lü	LG-ER 760	763	Lü	LG-ER 763	766	Lü	LG-ER 766	768	Lü	LG-ER 768
761	Lü	LG-ER 761	764	Lü	LG-ER 764						

773	ST	WL-BU601	MAN 469L	MAN	N25D	1997
778	ST	WL-BU606	MAN 469L	MAN	N25D	1997
781	ST	STD-L 1058	Mercedes-Benz O405N	Mercedes-Benz	B42D	1997
782	BU	STD-L 1059	Mercedes-Benz O405N	Mercedes-Benz	B42D	1997
783	Lü	STD-L 1056	Setra S315NF	Setra	N46D	1997
784	HD	STD-L 1057	Setra S315NF	Setra	N46D	1997

786-789 Mercedes-Benz O405N — Mercedes-Benz — B42D — 1998

786	Lü	LG-ER 786	787	Lü	LG-ER 787	788	ST	STD-L 1061	789	BU	STD-L 1062

790	ST	STD-L 1063	Setra S315UL	MAN	BC54	1998
791	HD	STD-L 1064	MAN 263 NÜ	MAN	N45D	1998
792	HD	WL-KV 84	MAN 903	MAN	N44D	1997

798-805 Mercedes-Benz O405N — Mercedes-Benz — B43D — 1999

798	ST	STD-L1066	800	ST	STD-L1068	802	BU	STD-L1071	804	BU	STD-L1073
799	ST	STD-L1067	801	BU	STD-L1069	803	BU	STD-L1072	805	Lü	LG-ER805

806-813 MAN NL 263 — MAN — N35D — 2000

806	BU	STD-L1082	808	BU	STD-L1084	810	ST	STD-L1086	812	ST	STD-L1088
807	BU	STD-L1083	809	ST	STD-L1085	811	ST	STD-L1087	813	ST	STD-L1089

814-817 Setra S315 UL — Setra — BC50D — 2000

814	ST	STD-L 1077	815	ST	STD-L 1076	816	BU	STD-L 1074	817	ST	STD-L 1075

818	HD	WL-KV 162	Mercedes-Benz O405N	Mercedes-Benz	B45D	2000
819	HD	WL-KV 163	Mercedes-Benz O405N	Mercedes-Benz	B45D	2000

820-824 Mercedes-Benz O405N — Mercedes-Benz — B42D — 2000

820	ST	STD-L 1078	822	Lü	LG-ER 820	823	Lü	LG-ER 821	824	Lü	LG-ER 824
821	ST	STD-L 1079									

825	BU	STD-L 1081	Mercedes-Benz O405N	Mercedes-Benz	B45D	2000

827-831 MAN NL 263 — MAN — N35D — 2001

827	CX	CUX-CB 827	829	ST	STD-L 1094	830	ST	STD-L 1095	831	ST	STD-L 1096
828	CX	CUX-CB 828									

832	CX	CUX-CB 832	Mercedes-Benz Citaro O530	Mercedes-Benz	N44D	2001
833	Lü	LG-ER 833	Mercedes-Benz Citaro O530 Ü	Mercedes-Benz	NC42D	2001

834-837 Setra S315 UL — Setra — BC49D — 2001

834	BU	STD-L 1092	835	SO	WL-KV 216	836	HD	WL-KV 218	837	ST	STD-L 1093

838	HD	WL-KV 217	Setra S319	Setra	C66D	2001
839	HD	WL-KV 13	MAN A03	MAN	C45F	2000

840-845 MAN NL 263 MAN N37D 2002

| 840 | BU | STD-L 1097 | 842 | BU | STD-L 1099 | 844 | Lü | LG-ER 844 | 845 | Lü | LG-KV 845 |
| 841 | BU | STD-L 1098 | 843 | Lü | LG-KV 843 | | | | | | |

846	ST	STD-L 1101 Mercedes-Benz Citaro O530	Mercedes-Benz	N38D	2002
847	ST	STD-L 1102 Setra S319UL	Setra	C68D	2002
849	BU	STD-L 1104 Setra S319UL	Setra	C66D	2002

850-853 Setra S315 Setra BC54D 2002

| 850 | HD | WL-KV 350 | 851 | HD | WL-KV 251 | 852 | SO | WL-KV 452 | 853 | ST | STD-L 1105 |

| 854 | HD | WL-KV 854 Mercedes-Benz Citaro O530 | Mercedes-Benz | N38D | 2002 |
| 855 | HD | WL-KV 855 Mercedes-Benz Citaro O530 | Mercedes-Benz | N38D | 2002 |

856-859 Mercedes-Benz Citaro O530 Ü Mercedes-Benz NC46D 2002

| 856 | ST | STD-L 1106 | 857 | ST | STD-L 1107 | 858 | ST | STD-L 1108 | 859 | ST | STD-L 1149 |

| 860 | HD | WL-BU 360 MAN NL 313 (CNG) | MAN | 31 | 2002 |
| 861 | HD | WL-BU 361 MAN NL 313 (CNG) | MAN | 31 | 2002 |

862-866 MAN NL 263 MAN N37D 2003

| 862 | Lü | LG-KV 862 | 864 | Lü | LG-KV 864 | 865 | Lü | LG-KV 865 | 866 | Lü | LG-KV 866 |
| 863 | Lü | LG-KV 863 | | | | | | | | | |

867-873 Mercedes-Benz Citaro O530 Ü Mercedes-Benz NC46D 2003

| 867 | Lü | LG-KV 867 | 869 | Lü | LG-KV 869 | 871 | ST | STD-L 1113 | 873 | CX | CUX-CB 873 |
| 868 | HD | WL-KV 107 | 870 | ST | STD-L 1112 | 872 | ST | STD-L 1114 | | | |

874	Lü	LG-KV 874 Mercedes-Benz Integro O550	Mercedes-Benz	C66F	2003
875	Lü	LG-KV 875 Mercedes-Benz Integro O550	Mercedes-Benz	C66F	2003
876	HD	WL-KV 86 Setra S319UL	Setra	BC68D	2003
877	BU	STD-L 1115 Setra S319UL	Setra	BC68D	2003
878	BU	STD-L 1116 Setra S319UL	Setra	BC68D	2003
879	ST	STD-L 1117 Setra S315UL	Setra	BC52D	2003
880	BU	STD-L 1118 Setra S315UL	Setra	BC52D	2003
881	BU	STD-L 1119 Setra S315UL	Setra	BC52D	2003
882	ST	STD-L 1122 Setra S319UL	Setra	BC68D	2003
883	CX	CUX-CB 883 Mercedes-Benz Citaro O530 Ü	Mercedes-Benz	NC46D	2003
884	Lü	CUX-CB 884 Mercedes-Benz Integro O550 L	Mercedes-Benz	C66F	2003

885-889 MAN NL 313 CNG MAN N31D 2004

| 885 | Lü | LG-ER 885 | 887 | Lü | LG-ER 887 | 888 | Lü | LG-AI 888 | 889 | Lü | LG-KV 889 |
| 886 | Lü | LG-KV 886 | | | | | | | | | |

| 890 | ST | STD-L 1121 Setra S319 UL | Setra | BC68D | 2004 |

891-896 Mercedes-Benz Citaro O530 Ü Mercedes-Benz NC46D 2004

| 891 | HD | WL-KV 63 | 893 | HD | WL-KV 64 | 895 | ST | STD-L 1123 | 896 | ST | STD-L 1125 |
| 892 | HD | WL-KV 58 | | | | | | | | | |

| 897 | Lü | CUX-CB 897 Mercedes-Benz Integro O550 L | Mercedes-Benz | C66F | 2003 |

898-901 Setra S319 UL Setra BC68D 2004

| 898 | HD | WL-KV 23 | 899 | HD | WL-KV 29 | 900 | SO | WL-KV 47 | 901 | ST | STD-L 1126 |

902-905 MAN NL 313 CNG MAN N35D 2005

| 902 | HD | WL-BU 607 | 903 | HD | WL-BU 608 | 904 | HD | WL-BU 609 | 905 | HD | WL-BU 610 |

| 906 | Lü | LG-KV 906 Mercedes-Benz Citaro O530 MÜ | Mercedes-Benz | NC50D | 2004 |

907-917 Mercedes-Benz Citaro O530 Ü Mercedes-Benz NC46D* 2005 *914-7 are NC44D

907	ST	STD-L 1127	910	ST	STD-L 1131	913	Lü	LG-ER 913	916	HD	WL-KV 816
908	ST	STD-L 1128	911	ST	STD-L 1132	914	HD	WL-KV 914	917	HD	WL-KV 917
909	ST	STD-L 1129	912	CX	CUX-BC 912	915	HD	WL-KV 915			

918-921 — Mercedes-Benz Citaro O530 MÜ — Mercedes-Benz — NC50D — 2005

918	HD	WL-KV 918	919	HD	WL-KV 919	920	ST	STD-L 1134	921	CX	CUX-CB 921

922-926 — Setra S319 NF — Setra — N56D — 2005

922	HD	WL-KV 422	924	HD	WL-KV 924	925	ST	STD-L 1135	926	Lü	LG-ER 926
923	HD	WL-KV 923									

927	HD	WL-KV 33	Mercedes-Benz Vario O905	Mercedes-Benz	B13F	2005
928	Lü	LG-KV 928	Mercedes-Benz Citaro O530 G	Mercedes-Benz	AN50D	2006

929-932 — Mercedes-Benz Citaro O530 MÜ — Mercedes-Benz — NC50D — 2005

929	Lü	LG-KV 929	930	Lü	LG-KV 930	931	Lü	LG-KV 931	932	Lü	LG-KV 932

933-938 — Mercedes-Benz Citaro O530 Ü — Mercedes-Benz — N46D — 2006

933	HD	WL-KV 533	935	HD	WL-KV 535	937	ST	STD-L 1138	938	ST	STD-L 1139
934	HD	WL-KV 934	936	ST	STD-L 1137						

939	ST	STD-L 1141	Volvo B12BLE 12m	Volvo 8700	NC44F	2006
940	BU	STD-L 1142	Volvo B12BLE 12m	Volvo 8700	NC44F	2006
941	HD	WL-KV 941	Mercedes-Benz Citaro O530 LÜ	Mercedes-Benz	N66D	2006
942	HD	WL-KV 942	Mercedes-Benz Citaro O530 LÜ	Mercedes-Benz	N66D	2006
943	ST	STD-L 1143	Mercedes-Benz Citaro O530 LÜ	Mercedes-Benz	N66D	2006
944	ST	STD-L 1144	Volvo B12BLE 14.5m	Volvo 8700	NC60F	2006
945	BU	STD-L 1145	Volvo B12BLE 14.5m	Volvo 8700	NC60F	2006
946	ST	STD-L 1146	Mercedes-Benz Integro O550	Mercedes-Benz	C54F	2006
947	ST	STD-L 1147	Mercedes-Benz Integro O550	Mercedes-Benz	C54F	2006
948	ST	STD-L 1136	Setra S317 UL	Setra	N62D	2006
950	Lü	STD-U1000	MAN 5D200F	MAN	C49D	-
951	Lü	STD-L 1148	MAN NG 272	MAN	AN58D	1993
952	Lü	LG-ER952	Volvo 7700A	Volvo	AN--D	2008
953	Lü	LG-ER953	Volvo 7700A	Volvo	AN--D	2008
954	Lü	LG-ER954	Volvo 7700A	Volvo	AN--D	2008

955-960 — Volvo 7700 — Volvo — N44D — 2008

955	Lü	LG-ER955	957	Lü	LG-ER957	959	Lü	LG-ER959	960	Lü	LG-ER960
956	Lü	LG-ER956	958	Lü	LG-ER958						

961-976 — Volvo 8700 — Volvo — NC44D — 2008

961	ST	ST-L1152	965	ST	ST-L1156	969	ST	ST-L1164	973	ST	ST-L1161
962	ST	ST-L1153	966	ST	ST-L1157	970	ST	ST-L1163	974	ST	ST-L1162
963	ST	ST-L1154	967	ST	ST-L1158	971	ST	ST-L1160	975	ST	ST-L1167
964	ST	ST-L1155	968	ST	ST-L1159	972	ST	ST-L1166	976	ST	ST-L1168

977-980 — Mercedes-Benz Citaro O530 G — Mercedes-Benz — AN50D — 2009

977	Lü	LG-KV977	978	Lü	LG-KV978	979	Lü	LG-KV970	980	Lü	LG-KV980

Depots: Buxtehude (BU); Cuxhaven (CX) Hittfeld (HI); Lüneburg (Lü); Soltau (SO); Stade (ST) and Winsen (Wi))

VOG

Verkehrsbetrieb Osthannover GmbH, Biermannstr. 33, 29221 Celle, Niedersachsen

26	Wi	LG-KS126	Setra SG219 SL			Setra		ABC59D	2002	

28-38			Mercedes-Benz 0407			Mercedes-Benz		B52D	1992-95	
28	SO	LG-KS128	34	Wi	LG-KS134	36	Wi	WL-AY36	38	Wi	LG-KS 138
30	Wi	LG-KS130	35	SO	SFA-DR981	37	Wi	WL-JA 37			

39-49			Mercedes-Benz 0408			Mercedes-Benz		B54D	1993-96	*seating varies
39	Wi	WL-ZA 39	42	SO	SFA-JS 442	46	Wi	LG-KS 146	48	Wi	WL-ZK 48
40	Wi	WL-ZA 40	44	Wi	LG-KS 144	47	Wi	LG-KS 147	49	Wi	LG-KS 149
41	SO	SFA-JS 441	45	SO	LG-KS 145						

50	Wi	WL-YK 50	Mercedes-Benz 0404			Mercedes-Benz		50	1996	

51-58			Mercedes-Benz 0408			Mercedes-Benz		B54D	1993-96	*seating varies
51	Wi	WL-GK 51	54	Wi	WL-UX 54	56	Wi	LG-KS 156	58	Wi	LG-KS 158
53	Wi	WL-AL 53	55	Wi	LG-KS 155	57	Wi	LG-KS 157			

72-81			Mercedes-Benz Integro 0550			Mercedes-Benz		C54F	2000	
72	SO	LG-KS 172	75	SO	LG-KS 175	78	Lü	LG-KS 178	80	Lü	LG-KS 180
73	Lü	LG-KS 173	76	Lü	LG-KS 176	79	Lü	LG-KS 179	81	Lü	LG-KS 181
74	Lü	LG-KS 174	77	Lü	LG-KS 177						

82	Wi	LG-KS 182	Mercedes-Benz Integro 0550 L	Mercedes-Benz	C66F	2002
83	Lü	LG-KS 183	Mercedes-Benz Citaro 0530 Ü	Mercedes-Benz	NC50D	2005
84	Lü	LG-KS 184	Mercedes-Benz Citaro 0530 Ü	Mercedes-Benz	NC50D	2005
85	SO	LG-KS 185	Mercedes-Benz Integro 0550 L	Mercedes-Benz	C62F	2003

86-93			Mercedes-Benz Citaro 0530 Ü			Mercedes-Benz		NC50D	2005	*90-3 are type MÜ
86	Lü	LG-KS 586	88	Lü	LG-KS 588	90	Lü	LG-KS 590	92	Lü	LG-KS 592
87	Lü	LG-KS 587	89	Lü	LG-KS 589	91	Lü	LG-KS 591	93	Lü	LG-KS 593

94	Lü	LG-KS 594	Mercedes-Benz Integro 0550 L	Mercedes-Benz	C62F	2005
95	Lü	LG-KS 595	Mercedes-Benz Integro 0550 L	Mercedes-Benz	C62F	2006
99	Lü	LG-KS 183	Mercedes-Benz Citaro 0530 G	Mercedes-Benz	AN50D	2008

Depots: Buxtehude (BU); Cuxhaven (CX) Hittfeld (HI); Lüneburg (Lü); Soltau (SO); Stade (ST) and Winsen (Wi))

In the absence of any pictures from the VOG operation, here is the Optare Tempo currently on evaluation with Arriva Netherlands. 1002, BV-VH-41, is operating from Groningen depot.
Harry Laming

ARRIVA PORTUGAL

Arriva Portugal, Edificio Guimarães, Rua Eduardo de Almeida, No 162, 2°Sala-C,
4810-440 Guimarães, Portugal

75	FA	CJ-48-40	Scania K112S	Alfredo Caetano	C51D	1984	Belos Transportes, Setúbal, 1999
85	GU	JS-97-53	Scania K112S	Irmãos Mota	C49D	1986	
130	FA	99-70-EH	Mercedes-Benz O303	Mercedes-Benz	BC55D	1986	Germany, 1994
147	FA	16-81-HT	Mercedes-Benz O303	Mercedes-Benz	C51D	1984	Germany, 1997
148	FA	13-06-HJ	Mercedes-Benz O303	Mercedes-Benz	C51D	1988	Germany, 1996
159	PA	46-36-ML	Mercedes-Benz O405	Mercedes-Benz	B44D	1986	Germany, 1998
160	FA	46-37-ML	Mercedes-Benz O303	Mercedes-Benz	BC53D	1991	Germany, 1998

163-179   Mercedes-Benz O405   Mercedes-Benz   B44D   1987-91   Germany, 2003

163	PA	98-34-VU	168	PA	25-06-VU	172	PA	45-10-VX	176	PA	98-35-VU
164	PA	25-10-VU	169	GU	45-08-VX	173	PA	98-36-VU	177	GU	98-37-VU
165	GU	25-08-VU	170	GU	25-11-VU	174	PA	45-09-VX	178	PA	98-33-VU
166	PA	25-09-VU	171	PA	98-32-VU	175	PA	25-12-VU	179	PA	25-13-VU
167	PA	25-07-VU									

180-196   Mercedes-Benz O405   Mercedes-Benz   B44D   1987-91   Arriva Skandianvian, 2006

180	GU	76-70-NM	185	PA	75-95-ZN	189	PA	76-19-ZQ	193	PA	90-59-ZZ
181	PA	76-20-ZQ	186	PA	76-21-ZQ	190	PA	89-90-ZQ	194	PA	70-12-MM
182	PA	76-18-ZQ	187	PA	76-00-ZN	191	PA	26-72-ZU	195	PA	62-53-OE
183	PA	75-93-ZN	188	PA	75-97-ZN	192	PA	90-55-ZZ	196	PA	28-48-PQ
184	PA	76-22-ZQ									

261	FA	72-54-HT	Mercedes-Benz O303/15R	Mercedes-Benz	C49D	1987	
269	FA	76-65-NM	Mercedes-Benz O303/15R	Mercedes-Benz	C49D	1987	
270	FA	76-66-NM	Mercedes-Benz O303/15R	Mercedes-Benz	C49DT	1984	
271	GU	SR-74-98	Volvo B58-60P	Irmãos Mota (1998)	C53D	1978	
272	GU	85-FQ-46	Mercedes-Benz O405	Mercedes-Benz	B53D	1997	
273	GU	22-FR-91	Mercedes-Benz O405	Mercedes-Benz	B53D	1997	
300	FA	RS-59-62	Pegaso 5036	Salvador Caetano	C47D	1983	
302	GU	QN-06-19	Volvo B10M-55G	CAMO	AB49D	1987	
308	GU	QQ-79-80	Volvo B10M-55G	CAMO	AB52D	1987	
309	GU	81-DH-99	Volvo B10M-55G	Sunsundegui	AB50D	1997	
310	GU	81-DH-98	Volvo B10M-55G	Sunsundegui	AB50D	1997	
311	GU	86-FV-97	Mercedes-Benz O405 GN	Mercedes-Benz	AB53D	1995	
312	GU	40-GC-89	Mercedes-Benz O405 GN	Mercedes-Benz	AB53D	1995	
313	GU	57-HE-00	Volvo B10M-55G	Castrosua	AB67D	1997	
331	GU	24-40-BT	Volvo B10B	CAMO	B37D	1993	
376	GA	70-16-TI	Volvo B10M-60	DAB	BC51D	1988	Arriva Danmark, 2002

380-390   Volvo B10M-60   DAB   BC51D   1989-90   Arriva Danmark, 2001-03

380	GA	70-17-TI	384	PA	97-08-TD	387	PA	81-16-TH	389	GA	85-58-TL
381	PA	89-21-SR	385	PA	07-29-XC	388	PA	81-17-TH	390	GA	35-65-UP
382	FA	07-28-XC									

| 391 | PA | 89-73-UP | Volvo B10R | Åbenrå | BC49D | 1987 | Arriva Danmark, 2003 |

392-400   Volvo B10M-60   Åbenrå   BC51D*   1987-91   Arriva Denmark, 2003

| 392 | FA | 59-25-UP | 395 | PA | 35-61-UP | 397 | GA | 35-59-UP | 400 | FA | 97-43-UZ |
| 393 | PA | 35-64-UP | 396 | GA | 35-63-UP | 398 | PA | 35-62-UP | | | |

401-420   Mercedes-Benz OH1634L   Irmãos Mota   C51D   1994

401	FA	23-30-EH	406	GA	90-45-EI	411	FA	79-16-EJ	416	GA	42-79-EL
402	FA	49-07-EH	407	GA	90-46-EI	412	FA	79-37-EJ	417	GA	42-80-EL
403	FA	49-08-EH	408	GA	90-47-EI	413	FA	79-38-EJ	418	GA	42-81-EL
404	GA	49-09-EH	409	GA	90-48-EI	414	GA	79-39-EJ	419	GA	42-82-EL
405	GA	90-35-EI	410	FA	79-15-EJ	415	GA	42-78-EL	420	FA	42-88-EL

No.	Op.	Reg.	Chassis	Body	Type	Year
421	FA	VI-16-66	Mercedes-Benz O303	Irmãos Mota	C51D	1990
422	FA	17-59-ZL	Mercedes-Benz O303	Irizar	C52D	1904
423	FA	17-58-ZL	Mercedes-Benz O303	Irizar	C52D	1984

424-434

Mercedes-Benz — Irmãos Mota — NC55D — 2006-07

No.	Op.	Reg.	No.	Op.	Reg.	No.	Op.	Reg.	No.	Op.	Reg.
424	FA	48-CD-67	427	GU	95-DQ-63	430	GU	95-DQ-66	433	FA	95-DQ-69
425	GU	72-DO-73	428	GU	95-DQ-64	431	FA	95-DQ-67	434	FA	95-DQ-70
426	GU	72-DO-74	429	GU	95-DQ-65	432	FA	95-DQ-68			

435-438

Mercedes-Benz OC500 RF — Marcopolo — NC55D — 2008

No.	Op.	Reg.	No.	Op.	Reg.	No.	Op.	Reg.	No.	Op.	Reg.
435	GA	97-FZ-28	436	GA	97-FZ-29	437	GA	97-FZ-30	438	GA	97-FZ-31

441-487

Mercedes-Benz O408 — Mercedes-Benz — BC49D — 1991

No.	Op.	Reg.	No.	Op.	Reg.	No.	Op.	Reg.	No.	Op.	Reg.
441	GU	90-63-ZZ	453	FA	56-71-ZU	465	FA	56-79-ZU	477	GA	23-AI-87
442	GU	56-72-ZU	454	FA	26-73-ZU	466	FA	75-91-ZN	478	GU	13-AS-28
443	GA	23-AT-62	455	GA	84-AC-56	467	FA	56-70-ZU	479	GA	23-AI-88
444	GU	81-AL-31	456	PA	90-58-ZZ	468	FA	75-96-ZN	480	GU	12-AS-29
445	GA	84-AC-58	457	PA	90-60-ZZ	469	GA	90-61-ZZ	481	GA	81-AL-29
446	FA	75-94-ZN	458	GU	26-71-ZU	470	GU	13-AS-30	482	GU	86-AH-12
447	GU	56-68-ZU	459	FA	75-98-ZN	471	FA	81-AL-28	483	GU	86-AH-13
448	GU	56-69-ZU	460	FA	23-AI-84	472	FA	86-AH-10	484	GA	23-AT-63
449	FA	56-67-ZU	461	GA	90-57-ZZ	473	FA	23-AI-86	485	GA	84-AC-57
450	GU	90-62-ZZ	462	FA	56-81-ZU	474	GU	81-AL-27	486	GU	86-AH-11
451	GU	56-80-ZU	463	GU	84-AC-55	475	GA	23-AT-64	487	GU	81-AL-30
452	FA	75-99-ZN	464	FA	75-92-ZN	476	GA	23-AI-85			

No.	Op.	Reg.	Chassis	Body	Type	Year	Notes
501	GU	75-37-DJ	Scania K113CLB	Irmãos Mota	C51D	1994	
502	GU	66-89-FI	Scania K113CLB	Irmãos Mota	C51D	1995	
503	GU	61-34-ND	Scania K124IB4	Irmãos Mota	C51D	1999	
504	GU	70-12-NT	Scania K124IB4	Irmãos Mota	C55D	1999	

521-527

Scania K114IB4 — Caetano Bus — C59D — 2002

No.	Op.	Reg.	No.	Op.	Reg.	No.	Op.	Reg.	No.	Op.	Reg.
521	GA	10-70 TT	523	FA	10-68-TT	525	GU	10-66-TT	527	GU	10-64-TT
522	FA	10-69-TT	524	FA	10-67-TT	526	GU	10-65-TT			

No.	Op.	Reg.	Chassis	Body	Type	Year	Notes
589	PA	RD-57-84	Scania K113CLB	UTIC (Lisboa)	C49D	1988	
592	PA	UA-48-68	Scania K113CLB	UTIC (Lisboa)	C49D	1989	Cruz e Neves, Ilhavo, 2002
594	PA	VC-38-25	Scania K113CLB	UTIC (Lisboa)	C49D	1990	EVA Transportes, Faro, 2002
601	FA	68-35-SD	Scania L94IB4	CAMO	B47D	2001	
602	FA	68-34-SD	Scania L94IB4	CAMO	B47D	2001	
603	FA	62-12-SG	Scania L94IB4	CAMO	B47D	2001	
604	GU	62-06-SG	Scania L94IB4	CAMO	B47D	2001	
651	GU	32-05-GU	Scania K113CLL	Irmãos Mota	B47D	1996	
652	GU	32-06-GU	Scania K113CLL	Irmãos Mota	B47D	1996	
701	FA	QX-01-42	Scania K113CLB	Irmãos Mota	C(63)DT	1991	
702	FA	QX-01-43	Scania K113CLB	Irmãos Mota	C(63)DT	1991	
703	PA	29-69-FB	Scania K113CLB	Irmãos Mota	C(61)DT	1995	
704	PA	29-70-FB	Scania K113CLB	Irmãos Mota	C(61)DT	1995	
705	PA	06-01-IG	Scania K113CLB	Irmãos Mota	C(59)DT	1997	
801	GU	05-56-ZM	Mercedes-Benz Sprinter 616	Mercedes-Benz	M8	2005	
802	GU	69-72-ZP	Mercedes-Benz Sprinter 616	Mercedes-Benz	M8	2005	

Portugal 313, an articulated Volvo B10M-55G with Castrosua bodywork is seen in full livery.

One of four Mercedes-Benz OC500s with Marcopolo bodywork supplied in 2008 is 437, 97-FZ-30, seen here shortly after delivery.

4044	GU	97-41-ER	Volvo B10B	CAMO		N35D	1995
4045	GU	97-42-ER	Volvo B10B	CAMO		N35D	1995
4046	GU	43-71-GR	Volvo B10B	Irmãos Mota		N35D	1996
4047	GU	43-72-GR	Volvo B10B	Irmãos Mota		N35D	1996
4048	GU	58-51-HU	Volvo B10B	CAMO		N35D	1997
4049	GU	58-52-HU	Volvo B10B	CAMO		N35D	1997
4050	GU	99-34-LA	Volvo B10B	CAMO		N34D	1998
4051	GU	99-35-LA	Volvo B10B	CAMO		N34D	1998
4052	GU	99-24-QR	Volvo B10BLE	CAMO		N34D	2000
4053	GU	96-89-SS	Volvo B7L	CAMO		N34D	2002
4054	GU	96-90-SS	Volvo B7L	CAMO		N34D	2002
4055	GU	19-22-ST	Volvo B7L	CAMO		N34D	2002
4056	GU	27-57-SX	Volvo B7L	CAMO		N34D	2002
4057	GU	24-40-BT	Volvo B7L	CAMO		N34D	2002
4058	GU	35-33-UX	Mercedes-Benz Citaro O530	Mercedes-Benz		N33D	2003
4059	GU	35-34-UX	Mercedes-Benz Citaro O530	Mercedes-Benz		N33D	2003
4060	GU	70-09-XH	Mercedes-Benz Citaro O530	Mercedes-Benz		N33D	2004

4061-4075

	Scania L113 CLB		Marcopolo	N37D	1996

4061	GU	20-09-HF	4065	GU	20-13-HF	4069	GU	20-17-HF	4073	GU	49-68-HG
4062	GU	20-10-HF	4066	GU	20-14-HF	4070	GU	49-66-HG	4074	GU	49-70-HG
4063	GU	20-11-HF	4067	GU	20-15-HF	4071	GU	49-67-HG	4075	GU	49-71-HG
4064	GU	20-12-HF	4068	GU	20-16-HF	4072	GU	49-69-HG			

5017	FA	52-76-ZR	Mercedes-Benz OC500	Irmãos Mota		C39D	2005

5018-5028

	Mercedes-Benz O405	Mercedes-Benz	B44D*	1989-90	*Seating varies

5018	FA	46-40-ML	5021	FA	62-52-OE	5024	FA	83-81-QC	5027	FA	67-03-QT
5019	FA	46-41-ML	5022	FA	99-69-OI	5025	FA	83-82-QC	5028	FA	71-99-QV
5020	FA	70-11-MM	5023	FA	28-47-PQ	5026	FA	67-02-QT			

5030	FA	58-EQ-34	Iveco TurboDaily 65C18	Marcopolo		N17D	2007
5031	FA	68-BP-89	Mercedes-Benz OC500 EL	Marcopolo		C43D	2006

Depots: Famalicão (FA); Garfe (GA); Guimarães (GU) and Parada (PA)

TST

Transportes Sul do Tejo S.A, Rua Marcos de Portugal, nº 10 – 2810 Laranjeiro, Portugal

31	AL	87-37-EI	DAF FA45	URB	B19	1994
32	AL	87-35-EI	DAF FA45	URB	B19	1994
33	AL	39-47-FZ	DAF FA45	URB	B19	1995
34	SX	08-GN-34	Mercedes-Benz Sprinter 518cdi	Mercedes-Benz	N15C	2008
35	SX	16-GQ-67	Mercedes-Benz Sprinter 518cdi	Mercedes-Benz	N15C	2008
36	SX	58-GT-88	Mercedes-Benz Sprinter 518cdi	Mercedes-Benz	N15C	2008
41	AL	61-20-XR	Mercedes-Benz O404	Mercedes-Benz	C49	1997
42	AL	61-23-XR	Mercedes-Benz O404	Mercedes-Benz	C49	1997
43	AL	51-29-XR	Mercedes-Benz O404	Mercedes-Benz	C49	1997
44	AL	51-31-XR	Mercedes-Benz O404	Mercedes-Benz	C49	1997
45	AL	51-31-XR	Mercedes-Benz O404	Mercedes-Benz	C55	1999
46	VA	QO-44-22	MAN 16-290	TUR	C49	1988
47	SE	QO-44-70	MAN 16-290	TUR	C49	1988
51	VA	QQ-88-16	MAN 16-290	TUR	C49	1989
61	SE	OQ-55-82	MAN 16-290	TUR	C49	1991
62	SE	OQ-55-84	MAN 16-290	TUR	C49	1991
68	AZ	QR-31-58	MAN 24-360	TUR	C73	1989
69	AZ	QR-31-57	MAN 24-360	TUR	C73	1989
71	SE	QS-00-30	MAN 16-290	TUR	C51	1989
72	SE	QN-95-82	MAN 16-290	-	C49	1988

90-96

MAN 18-320 — C49 1995

90	AL	46-93-FV	92	AL	80-79-GA	94	AZ	80-83-GA	96	AZ	80-85-GA
91	AZ	46-94-FV	93	AL	80-80-GA						

100-104

MAN 18-360 — C59 2003-04

100	MO	49-40-VA	102	MO	49-42-VA	103	MO	49-43-VA	104	BA	17-18-XM
101	MO	49-41-VA									

111	MO	35-49-SO	Neoplan N4011	Neoplan	28	1994
112	MO	35-52-SO	Neoplan N4011	Neoplan	28	1994
113	MO	35-54-SO	Neoplan N4011	Neoplan	28	1994
150	MO	51-GN-76	Mercedes-Benz Integro O550	Mercedes-Benz	C69F	2002
151	MO	09-GJ-67	Mercedes-Benz Integro O550	Mercedes-Benz	C69F	2002
152	MO	50-GJ-40	Mercedes-Benz Integro O550	Mercedes-Benz	C69F	2002
197	AL	93-GH-61	Mercedes-Benz Integro O550	Mercedes-Benz	C49F	2000
198	AL	93-GH-62	Mercedes-Benz Integro O550	Mercedes-Benz	C49F	1999
199	SE	80-FN-68	Mercedes-Benz Integro O550	Mercedes-Benz	C53F	1997

200-217

Mercedes-Benz Integro O550 Mercedes-Benz NC55D 2001

200	MO	05-03-SU	205	SE	05-08-SU	210	AL	05-56-SU	214	AL	05-60-SU
201	MO	05-04-SU	206	MO	05-09-SU	211	AL	05-58-SU	215	AL	05-61-SU
202	MO	05-05-SU	207	AZ	05-10-SU	212	AL	05-62-SU	216	AL	05-62-SU
203	MO	05-06-SU	208	AL	05-11-SU	213	AL	05-59-SU	217	AL	05-63-SU
204	SE	05-07-SU	209	AL	05-55-SU						

218	AL	85-FB-63	Mercedes-Benz Integro O550	Mercedes-Benz	C55F	1998
219	SE	62-FN-55	Mercedes-Benz Integro O550	Mercedes-Benz	C49F	1999
220	SE	51-BC-25	Mercedes-Benz O814	-	15	1998
221	SE	92-BC-86	Mercedes-Benz O814	-	15	1998
222	SE	35-BF-90	Mercedes-Benz O814	-	15	1998

228-240

Mercedes-Benz O407 Mercedes-Benz NC49D 1992-96

228	MO	36-94-ZG	232	MO	51-32-XR	235	MO	91-72-XH	238	MO	67-42-XH
229	MO	60-39-XP	233	MO	40-71-XL	236	MO	91-71-XH	239	MO	30-63-XH
230	MO	02-17-XN	234	MO	34-50-XJ	237	MO	67-43-XH	240	MO	90-40-XG
231	MO	02-16-XN									

289	SX	05-89-FA	Mercedes-Benz O405	Mercedes-Benz	NC41D	1995
290	SX	05-91-FA	Mercedes-Benz O405	Mercedes-Benz	NC41D	1995
291	SX	05-92-FA	Mercedes-Benz O405	Mercedes-Benz	NC41D	1995
292	SX	05-98-FA	Mercedes-Benz O405	Mercedes-Benz	NC41D	1995
293	SX	18-10-FA	Mercedes-Benz O405	Mercedes-Benz	NC41D	1995

The TST fleet is dominated by Mercedes-Benz products with eighteen Mercedes-Benz Integro O550s added in 2001. Representing the batch is 200, 05-03-SU, which is based at Moita. *Ken MacKenzie*

304-320

			Mercedes-Benz O303	Mercedes-Benz	044D	1983-85
304	VA	67-55-NE	**306**	VA	67-58-NE	**313**	AL	22-70-NL	**320**	AL	74-94-NL

321	MO	05-BL-39	Mercedes-Benz O405N	Mercedes-Benz	37	1994
322	MO	05-BL-40	Mercedes-Benz O405N	Mercedes-Benz	37	1994
323	VA	95-GJ-24	Mercedes-Benz O405NU	Mercedes-Benz	42	1995
324	MO	35-BL-22	Mercedes-Benz O405N	Mercedes-Benz	37	1994
325	VA	18-GI-94	Mercedes-Benz O405NU	Mercedes-Benz	42	1995
326	VA	18-GI-95	Mercedes-Benz O405NU	Mercedes-Benz	42	1996
327	VA	90-GI-33	Mercedes-Benz O405NU	Mercedes-Benz	42	1995
328	MO	92-BO-10	Mercedes-Benz O408	Mercedes-Benz	49	1992
329	VA	12-EJ-04	Mercedes-Benz O407	Mercedes-Benz	49	1995
330	VA	12-EJ-06	Mercedes-Benz O407	Mercedes-Benz	49	1995
331	SE	77-EJ-88	Mercedes-Benz O407	Mercedes-Benz	49	1995
332	MO	57-BP-60	Mercedes-Benz O408	Mercedes-Benz	49	1992
336	SE	09-EM-64	Mercedes-Benz O407	Mercedes-Benz	49	1995
337	SE	10-FI-90	Mercedes-Benz O408	Mercedes-Benz	49	1996
338	SE	50-GJ-38	Mercedes-Benz O407	Mercedes-Benz	53	1996
339	MO	25-BS-81	Mercedes-Benz O407	Mercedes-Benz	44	1997
340	VA	75-GI-06	Mercedes-Benz O407	Mercedes-Benz	49	1995
341	MO	36-BT-13	Mercedes-Benz O407	Mercedes-Benz	49	1993
343	MO	36-BT-14	Mercedes-Benz O405N	Mercedes-Benz	37	1993
344	MO	17-DU-91	Mercedes-Benz O408	Mercedes-Benz	49	1996
345	MO	34-DX-05	Mercedes-Benz O408	Mercedes-Benz	49	1996
346	MO	35-EH-19	Mercedes-Benz O408	Mercedes-Benz	49	1996
348	AL	31-BI-30	Mercedes-Benz O405	Mercedes-Benz	44	1995
349	AL	31-BI-29	Mercedes-Benz O405N	Mercedes-Benz	37	1993
350	AL	35-BL-23	Mercedes-Benz O405N	Mercedes-Benz	37	1993
351	AL	26-BM-52	Mercedes-Benz O405N	Mercedes-Benz	37	1993
352	AL	35-BL-20	Mercedes-Benz O405N	Mercedes-Benz	37	1993
353	AL	87-BJ-03	Mercedes-Benz O405N	Mercedes-Benz	37	1993
354	AL	00-BM-36	Mercedes-Benz O405N	Mercedes-Benz	37	1993
355	AL	26-BM-51	Mercedes-Benz O405N	Mercedes-Benz	37	1993
356	VA	26-BM-50	Mercedes-Benz O405	Mercedes-Benz	44	1994

357	MO	63-BM-15	Mercedes-Benz O405N	Mercedes-Benz	37	1993
358	AL	17-BQ-71	Mercedes-Benz O405N	Mercedes-Benz	44	1995
359	VA	24-DI-16	Mercedes-Benz O405	Mercedes-Benz	44	1995
360	AL	68-BQ-34	Mercedes-Benz O405N	Mercedes-Benz	44	1995
361	AL	68-BQ-31	Mercedes-Benz O405N	Mercedes-Benz	44	1994
362	VA	60-BI-86	Mercedes-Benz O405	Mercedes-Benz	44	1994
363	VA	44-BI-75	Mercedes-Benz O405	Mercedes-Benz	44	1994
364	VA	44-BI-76	Mercedes-Benz O405	Mercedes-Benz	44	1994
365	VA	35-BL-42	Mercedes-Benz O405	Mercedes-Benz	44	1994
366	VA	24-DI-15	Mercedes-Benz O405	Mercedes-Benz	44	1995
367	MO	24-BV-70	Mercedes-Benz O407	Mercedes-Benz	53	1994
368	VA	57-BF-85	Mercedes-Benz O408	Mercedes-Benz	49	1993
369	VA	83-BF-43	Mercedes-Benz O408	Mercedes-Benz	49	1993
370	VA	83-BF-49	Mercedes-Benz O405	Mercedes-Benz	44	1994
371	VA	83-BF-48	Mercedes-Benz O405	Mercedes-Benz	44	1994
372	VA	44-BI-73	Mercedes-Benz O405	Mercedes-Benz	44	1994
373	MO	30-BX-07	Mercedes-Benz O407	Mercedes-Benz	49	1995
374	SE	07-CE-89	Mercedes-Benz O408	Mercedes-Benz	53	1996
375	SE	30-BX-08	Mercedes-Benz O407	Mercedes-Benz	49	1995
376	SE	72-BH-26	Mercedes-Benz O408	Mercedes-Benz	53	1993
377	SX	51-FV-92	Mercedes-Benz O408	Mercedes-Benz	53	1995
378	MO	07-CE-87	Mercedes-Benz O408	Mercedes-Benz	53	1994
379	SX	68-FV-78	Mercedes-Benz O408	Mercedes-Benz	53	1995
380	SX	54-49-KB	Mercedes-Benz O307	Mercedes-Benz	49	1986
381	SX	05-FZ-99	Mercedes-Benz O408	Mercedes-Benz	49	1996
383	VA	85-FS-95	Mercedes-Benz O407	Mercedes-Benz	53	1997
384	VA	07-DA-85	Mercedes-Benz O407	Mercedes-Benz	49	1995
385	VA	07-DA-84	Mercedes-Benz O407	Mercedes-Benz	49	1995
386	VA	84-DC-23	Mercedes-Benz O407	Mercedes-Benz	49	1995
387	SX	05-FZ-98	Mercedes-Benz O408	Mercedes-Benz	49	1996
388	VA	10-DH-58	Mercedes-Benz O405	Mercedes-Benz	44	1995
389	VA	10-DH-52	Mercedes-Benz O405	Mercedes-Benz	44	1995
390	SX	23-94-JM	Mercedes-Benz O305	Mercedes-Benz	36	1987
391	MO	07-DA-86	Mercedes-Benz O408	Mercedes-Benz	49	1993
394	MO	65-DH-08	Mercedes-Benz O405	Mercedes-Benz	44	1995
395	VA	65-DH-07	Mercedes-Benz O405	Mercedes-Benz	44	1995
396	SX	49-BH-35	Mercedes-Benz O405	Mercedes-Benz	44	1994
397	SX	49-BH-37	Mercedes-Benz O405	Mercedes-Benz	44	1995
398	SX	49-BH-34	Mercedes-Benz O405	Mercedes-Benz	44	1994
399	SX	49-BH-36	Mercedes-Benz O405	Mercedes-Benz	44	1996
401	SX	02-BQ-00	Mercedes-Benz O405N	Mercedes-Benz	44	1996
402	SX	21-BG-92	Mercedes-Benz O405N	Mercedes-Benz	39	1996
408	SX	02-BQ-01	Mercedes-Benz O405N	Mercedes-Benz	44	1996
409	SX	17-DU-90	Mercedes-Benz O405	Mercedes-Benz	44	1995
410	SX	01-BQ-97	Mercedes-Benz O405N	Mercedes-Benz	44	1996
411	SX	68-BQ-32	Mercedes-Benz O405N	Mercedes-Benz	44	1996
412	MO	47-79-HG	Setra S215 SL	Kässbohrer Setra	46	1985
413	VA	22-38-QG	Setra S215 UL	Kässbohrer Setra	49	1990
414	VA	22-39-QG	Setra S215 UL	Kässbohrer Setra	49	1990
415	SX	43-BF-89	Mercedes-Benz O 405	Mercedes-Benz	44	1995
416	SX	43-BF-90	Mercedes-Benz O 405	Mercedes-Benz	44	1995
417	VA	22-41-QG	Kässbohrer S215 UL	Kässbohrer Setra	49	1990
418	VA	33-61-QL	Kässbohrer S215 UL	Kässbohrer Setra	53	1985
419	VA	33-62-QL	Kässbohrer S215 UL	Kässbohrer Setra	49	1985
420	VA	33-63-QL	Kässbohrer S215 UL	Kässbohrer Setra	49	1989
421	VA	33-64-QL	Kässbohrer S215 UL	Kässbohrer Setra	49	1989
422	VA	16-54-MX	Kässbohrer S215 SL	Kässbohrer Setra	53	1987
423	SX	25-BS-78	Mercedes-Benz O405	Mercedes-Benz	44	1995
424	MO	93-51-MV	Kässbohrer S215 SL	Kässbohrer Setra	46	1987
425	SX	25-BS-77	Mercedes-Benz O405	Mercedes-Benz	44	1995
426	MO	93-53-MV	Kässbohrer S215 SL	Kässbohrer Setra	46	1987
427	SX	68-BQ-36	Mercedes-Benz O405N	Mercedes-Benz	44	1995
428	VA	93-55-MV	Kässbohrer S215 SL	Kässbohrer Setra	46	1987
429	VA	24-92-PN	Kässbohrer S215 UL	Kässbohrer Setra	53	1987
433	VA	73-12-OU	Kässbohrer S215 UL	Kässbohrer Setra	53	1986
435	MO	05-FZ-91	Setra S319 ÜL	Setra	NC71D	1999
437	VA	06-16-PP	Kässbohrer S215 UL	Kässbohrer Setra	53	1987
438	MO	01-FN-50	Setra S319 ÜL	Setra	NC71D	1999
440	AZ	79-EF-91	Setra S319 ÜL	Setra	NC71D	2000
441	AZ	39-EN-10	Setra S319 ÜL	Setra	NC71D	2000
442	MO	26-EM-10	Setra S319 ÜL	Setra	NC71D	1999
443	MO	26-EM-09	Setra S319 ÜL	Setra	NC71D	2000
444	MO	07-FR-03	Setra S319 ÜL	Setra	NC71D	1999

445	MO	26-EM-11	Setra S319 ÜL	Setra	NC71D	1999	
446	AL	14-DT-71	Setra S315 ÜL	Setra	NC53D	1995	
447	MO	79-EF-89	Setra S315 ÜL	Setra	NC47D	1999	
448	AL	50-DT-63	Setra S315 ÜL	Setra	NC53D	1998	
459	SE	62-FN-58	Setra S315 ÜL	Setra	NC53D	1995	
460	AL	14-DT-70	Setra S315 ÜL	Setra	NC53D	1995	
461	AL	14-DT-72	Setra S315 NF	Setra	NC45D	1996	
463	AL	62-FN-59	Setra S315 NF	Setra	NC47D	1997	
464	AL	58-GL-16	Setra S315 ÜL	Setra	NC53D	1997	
473	SE	45-AZ-05	Mercedes-Benz O408	Mercedes-Benz	49	1992	
475	MO	62-FN-56	Mercedes-Benz O405N	Mercedes-Benz	36	1995	
477	SX	89-FJ-17	Mercedes-Benz O405N	Mercedes-Benz	34	1995	
478	SX	98-FJ-20	Mercedes-Benz O405N	Mercedes-Benz	34	1995	
479	SX	78-FL-14	Mercedes-Benz O405N	Mercedes-Benz	34	1995	
480	SX	78-FL-15	Mercedes-Benz O405N	Mercedes-Benz	34	1995	
481	SX	17-FQ-37	Mercedes-Benz O405N	Mercedes-Benz	34	1995	
482	SX	17-FQ-38	Mercedes-Benz O405N	Mercedes-Benz	34	1995	
483	MO	24-AS-35	Mercedes-Benz O405	Mercedes-Benz	37	1994	
484	SX	95-FB-61	Mercedes-Benz O405	Mercedes-Benz	44	1995	
485	VA	00-AV-75	Mercedes-Benz O405N	Mercedes-Benz	44	1994	
486	SX	95-FB-64	Mercedes-Benz O405	Mercedes-Benz	44	1995	
487	VA	00-AV-77	Mercedes-Benz O405N	Mercedes-Benz	44	1993	
489	MO	59-AU-45	Mercedes-Benz O405N	Mercedes-Benz	41	1993	
490	SX	01-ER-39	Mercedes-Benz O405N	Mercedes-Benz	44	1996	
491	SX	04-ER-66	Mercedes-Benz O405N	Mercedes-Benz	34	1997	
492	SX	04-ER-67	Mercedes-Benz O405N	Mercedes-Benz	36	1997	
493	SX	22-ET-50	Mercedes-Benz O405N	Mercedes-Benz	35	1997	
494	AL	21-AO-37	Mercedes-Benz O405N	Mercedes-Benz	45	1995	

495-499

Mercedes-Benz O 405 — Mercedes-Benz — 44 — 1993-94

495	AL	21-AO-39	497	AL	80-AQ-85	498	AL	90-AQ-03	499	MO	90-AQ-04
496	AL	80-AQ-84								

500-504

MAN 12-220 — - — 31 — 2003

500	AL	65-00-VC	502	AL 65-02-VC	503	AL 52-25-VC	504	AL 65-04-VC
501	AL	65-01-VC						

506-516

MAN 18-310 — - — 41 — 2005-06

506	AL 32-BO-23	509	AL 32-BO-26	512	AL 04-AB-43	515	AL 04-AB-46
507	AL 32-BO-24	510	AL 04-AB-41	513	AL 04-AB-44	516	AL 04-AB-47
508	AL 32-BO-25	511	AL 04-AB-42	514	AL 04-AB-45		

518	VA	51-AJ-65	Mercedes-Benz O405N	Mercedes-Benz	36	1993
519	VA	51-AJ-64	Mercedes-Benz O405N	Mercedes-Benz	36	1993
520	VA	83-AH-87	Mercedes-Benz O405N	Mercedes-Benz	41	1993
521	MO	47-AI-19	Mercedes-Benz O405	Mercedes-Benz	44	1994
522	MO	47-AI-20	Mercedes-Benz O405N	Mercedes-Benz	41	1994
523	MO	47-AI-21	Mercedes-Benz O405N	Mercedes-Benz	37	1994
524	MO	60-AI-86	Mercedes-Benz O405N	Mercedes-Benz	36	1995
525	MO	60-AI-85	Mercedes-Benz O405N	Mercedes-Benz	41	1993
526	VA	60-AI-69	Mercedes-Benz O405N	Mercedes-Benz	42	1994
527	MO	60-AI-83	Mercedes-Benz O405N	Mercedes-Benz	37	1994
528	VA	77-AI-20	Mercedes-Benz O405	Mercedes-Benz	36	1995
529	VA	44-AL-72	Mercedes-Benz O407	Mercedes-Benz	49	1995
530	VA	47-AI-18	Mercedes-Benz O405	Mercedes-Benz	36	1995
531	AL	69-AI-52	Mercedes-Benz O405N	Mercedes-Benz	37	1994
532	AL	69-AI-48	Mercedes-Benz O405N	Mercedes-Benz	37	1994
533	AL	77-AI-19	Mercedes-Benz O405	Mercedes-Benz	44	1994
536	VA	86-95-PL	Kässbohrer S215 HR	Kässbohrer Setra	51	1987

537-543

Kässbohrer S215 H — Kässbohrer Setra — 55 — 1990

537	VA	65-07-OC	539	VA	34-38-OH	542	VA	81-17-OH	543	VA	34-39-OH
538	VA	34-37-OH	540	VA	81-16-OH				

544	VA	29-06-PT	Kässbohrer S215 HR	Kässbohrer Setra	51	1987
545	VA	44-AL-71	Mercedes-Benz O407	Mercedes-Benz	49	1995
546	VA	44-AL-73	Mercedes-Benz O407	Mercedes-Benz	49	1996
547	VA	33-66-QL	Kässbohrer S215 H	Kässbohrer Setra	51	1987
548	VA	55-44-NU	Kässbohrer S215 H	Kässbohrer Setra	53	1987
549	VA	75-AL-61	Mercedes-Benz O407	Mercedes-Benz	49	1995

550-552			Scania L113 CLB		-		45	1994	
550	AL	09-40-DZ	551	AL	09-41-DZ	552	AL	09-42-DZ	
554	VA	66-AO-92	Mercedes-Benz O407		Mercedes-Benz		49	1995	

556-564			Scania L113 CLB		-		45	1994	
556	AL	71-71-ED	558	AL	74-82-EE	561	AL	57-81-EC	563	AL	57-79-EC
557	AL	71-72-ED	560	AL	56-42-EE	562	AL	57-78-EC	564	AL	57-80-EC

565-582			Mercedes-Benz O 408		Mercedes-Benz		49*	1991-94	Seating varies
565	SE	23-77-RP	570	SE	90-66-SF	575	SE	88-21-UV	579	SE	21-71-XR
566	AL	23-75-RP	571	SE	90-67-SF	576	SE	88-23-UV	580	AL	36-82-ZE
567	SE	23-76-RP	572	SE	84-27-UR	577	SE	90-39-XG	581	SE	13-24-ZF
568	SE	90-69-SF	573	SE	88-24-UV	578	SE	91-70-XH	582	SE	49-AB-90
569	SE	90-68-SF	574	SE	88-20-UV						

583-588			Mercedes-Benz O 407		Mercedes-Benz		49	1992-94	Seating varies
583	SE	00-02-ZE	585	SE	33-04-XE	587	SE	00-03-ZE	588	SE	76-30-ZE
584	SE	27-02-ZF	586	SE	67-44-XH						

589	SX	60-47-VU	Mercedes-Benz O405N	Mercedes-Benz	37	1991
590	SX	66-50-VT	Mercedes-Benz O405N	Mercedes-Benz	37	1991

591-599			Mercedes-Benz O405N		Mercedes-Benz		41	1992	
591	SX	65-20-VR	594	SX	18-96-VQ	596	SX	18-98-VQ	598	SX	22-45-VP
592	SX	55-38-VQ	595	SX	66-73-VP	597	SX	22-46-VP	599	SX	04-64-VP
593	SX	18-97-VQ									

600-650			Mercedes-Benz O 405		Mercedes-Benz		B44D*	1985-89	*Seating varies
600	AL	77-47-LM	612	AL	45-84-NN	626	SX	24-46-PZ	638	AL	86-12-QP
601	AL	77-49-LM	613	AL	49-68-OP	627	SX	24-47-PZ	639	AL	86-13-QP
602	AL	77-45-LM	614	AL	49-69-OP	628	SX	59-50-QB	640	AL	86-14-QP
603	AL	77-46-LM	615	SX	49-72-OP	629	SX	22-42-QG	641	AL	01-93-RD
604	AL	77-51-LM	617	SX	49-75-OP	630	SX	22-43-QG	642	AL	14-85-RG
605	AL	77-48-LM	618	SX	49-76-OP	631	SX	22-45-QG	644	AL	18-23-RI
606	SX	85-95-NB	619	SX	49-77-OP	632	SX	22-46-QG	645	AL	18-26-RI
607	SX	86-01-NB	620	SX	49-78-OP	633	SX	22-47-QG	646	SX	18-24-RI
608	SX	86-02-NB	622	SX	81-66-PP	634	SX	69-64-QG	647	SX	18-25-RI
609	SX	86-04-NB	623	SX	24-43-PZ	635	AL	69-49-QP	648	SX	31-48-RL
610	SX	86-05-NB	624	SX	24-44-PZ	636	AL	69-50-QP	649	SX	31-46-RL
611	SX	81-69-PP	625	SX	24-45-PZ	637	AL	69-51-QP	650	VA	77-50-LM

616	SX	23-GM-87	Mercedes-Benz O405N	Mercedes-Benz	B37D	1997
621	SX	23-GM-73	Mercedes-Benz O405N	Mercedes-Benz	B37D	1997

651-699			Mercedes-Benz O405N		Mercedes-Benz		B44D*	1985-93	*Seating varies
651	MO	78-33-MS	663	SX	63-81-NI	677	w	63-86-NI	689	SX	45-87-NN
652	MO	86-03-NB	664	VA	78-34-MS	678	VA	63-87-NI	690	SX	78-31-MS
653	MO	67-57-NE	665	w	78-35-MS	679	w	63-88-NI	691	w	81-65-PP
654	AL	31-49-RL	666	SX	78-36-MS	680	VA	45-77-NN	692	SX	81-67-PP
655	AL	31-47-RL	667	SX	78-37-MS	681	w	45-78-NN	693	SX	01-92-RD
656	SX	31-77-RL	669	w	42-95-MZ	682	VA	45-79-NN	694	SX	84-49-SM
657	SX	46-61-SJ	671	SX	42-97-MZ	683	SX	45-80-NN	695	SX	84-48-SM
658	SX	99-58-SJ	672	SX	85-94-NB	684	SX	45-81-NN	696	SX	41-98-SS
659	SX	60-49-SJ	673	SX	63-82-NI	685	SX	45-82-NN	697	SX	41-99-SS
660	w	57-68-MS	674	w	63-83-NI	686	VA	45-83-NN	698	SX	87-AF-45
661	SX	57-69-MS	675	SX	63-84-NI	687	VA	45-85-NN	699	SX	42-00-SS
662	VA	57-71-MS									

704	SE	30-01-BO	Renault B110-50	-	M12	1993

705-730			Mercedes-Benz O405		Mercedes-Benz		B44D*	1986-91	*Seating varies
705	SX	23-80-RP	712	SX	23-84-RP	718	SX	25-46-RZ	725	SX	84-25-UR
706	SX	23-78-RP	713	SX	25-44-RZ	720	SX	25-48-RZ	726	SX	84-28-UR
707	SX	23-79-RP	714	SX	21-30-RQ	721	SX	73-32-SZ	727	SX	84-29-UR
708	SX	23-81-RP	715	SX	21-31-RQ	722	SX	73-33-SZ	728	SX	88-22-UV
709	SX	21-29-RQ	716	SX	21-27-RQ	723	SX	73-34-SZ	729	SX	97-16-UX
710	SX	21-28-RQ	717	SX	25-45-RZ	724	SX	84-23-UR	730	SX	19-69-VA
711	SX	23-83-RP	718	SX	25-46-RZ						

731	SX	26-57-XL	Mercedes-Benz O405N		Mercedes-Benz		B37D	1991			
732	SX	55-66-XN	Mercedes-Benz O405		Mercedes-Benz		B36D	1991			
733	SX	55-65-XN	Mercedes-Benz O405		Mercedes-Benz		B44D	1991			
734	SX	88-70-XO	Mercedes-Benz O405		Mercedes-Benz		B44D	1994			

735-744
Mercedes-Benz O405N — Mercedes-Benz — B37D* — 1991-94 — *Seating varies

735	SX	87-AF-47	738	AL	01-67-ZC	741	AL	60-38-XP	743	AL	66-49-VT
736	SX	87-AF-48	739	AL	01-68-ZC	742	AL	05-69-XE	744	AL	44-18-VR
737	SX	87-AF-46	740	AL	61-21-XR						

745	AL	03-65-VM	Mercedes-Benz O405		Mercedes-Benz		B36D	1987
746	AL	14-20-VL	Mercedes-Benz O405N		Mercedes-Benz		B34D	1990
747	AL	14-21-VL	Mercedes-Benz O405N		Mercedes-Benz		B34D	1990

748-762
Mercedes-Benz O405 — Mercedes-Benz — B44D* — 1986-93 — *Seating varies

748	AL	29-06-VD	752	AL	46-59-SJ	756	AL	68-56-SM	760	AL	84-47-SM
749	AL	36-03-VC	753	SX	23-85-RP	757	AL	06-37-SO	761	AL	41-97-SS
750	AL	23-82-RP	754	AL	46-60-SJ	758	AL	68-57-SM	762	AL	42-01-SS
751	AL	60-41-SJ	755	AL	46-62-SJ	759	AL	68-55-SM			

763	SX	66-AG-48	Mercedes-Benz O405N		Mercedes-Benz		B37D	1994
764	SX	66-AG-47	Mercedes-Benz O405N		Mercedes-Benz		B37D	1994

765-775
Mercedes-Benz O405 — Mercedes-Benz — B43D — 1988-89

765	VA	07-14-QL	768	SX	07-17-QL	771	VA	07-26-QL	774	MO	84-26-UR
766	VA	07-15-QL	769	VA	07-18-QL	772	MO	84-22-UR	775	VA	48-26-ZR
767	SX	07-16-QL	770	VA	07-19-QL	773	MO	84-24-UR			

776	SX	60-AH-60	Mercedes-Benz O407		Mercedes-Benz		N49D	1994
777	SX	13-BC-32	Mercedes-Benz O405		Mercedes-Benz		B44D	1993
778	SX	04-BE-86	Mercedes-Benz O405		Mercedes-Benz		B44D	1993
779	SX	57-BF-86	Mercedes-Benz O407		Mercedes-Benz		N49D	1996
781	VA	59-49-QB	MAN SG242		-		AB54D	1988
782	VA	59-51-QB	MAN SG242		-		AB54D	1988
786	AZ	RQ-54-68	MAN SG292F		-		AB69D	1991
793	MO	61-22-XR	Volvo B10M-55		-		AB77D	1995
797	MO	51-30-XR	Volvo B10M-55		-		AB77D	1995

805-822
Volvo B10M-55 — - — AB52D — 1987-91

805	SX	RP-51-85	813	MO	QT-18-26	817	SX	QT-85-33	820	AL	PQ-38-82
807	SX	RP-69-33	815	SX	QT-33-70	819	AL	OQ-61-71	822	AL	PQ-38-80
812	MO	QT-18-27	816	SX	QT-71-27						

824-831
Mercedes-Benz O305G — Mercedes-Benz — AB66D — 1983-85

824	VA	07-12-JP	826	VA	55-73-KD	829	SX	46-70-LI	831	SX	46-73-LI
825	SX	07-14-JP	828	SX	46-71-LI			

834	MO	66-CE-43	Mercedes-Benz O405G		Mercedes-Benz		AB52D	1996
835	MO	07-CE-88	Mercedes-Benz O405G		Mercedes-Benz		AB52D	1996

841-849
Volvo B6-50 — - — B33D — 1994

841	SX	43-42-DT	843	SX	43-44-DT	846	MO	21-12-EA	848	MO	21-14-EA
842	SX	43-43-DT	845	MO	21-11-EA	847	MO	21-13-EA	849	MO	21-15-EA

850	VA	47-FC-62	Van Hool A318		Van Hool		B36D	1996
851	MO	47-FC-63	Van Hool A318		Van Hool		B36D	1996
852	VA	09-AV-99	Van Hool A318		Van Hool		B40D	1995
853	VA	10-AV-00	Van Hool A318		Van Hool		B40D	1995

854-858
Van Hool A300 — Van Hool — B39D — 1999

854	SX	23-EG-30	856	SX	23-EG-32	857	SX	35-EH-18	858	SX	97-EX-90
855	SX	23-EG-31									

859	VA	78-FL-13	Van Hool A318		Van Hool		B40D	1995

860-868
Volvo B10M-55 — - — AB77D — 1995

860	MO	00-01-ZE	863	AZ	58-95-ZG	865	MO	95-40-ZG	867	MO	58-96-ZG
861	MO	27-38-ZD	864	MO	36-83-ZE	866	MO	95-41-ZG	868	MO	02-98-ZI
862	AZ	27-37-ZD									

869	VA	TB FL 10	Van Hool A318	Van Hool		B36D	1996	
876	VA	45-49-PP	MAN UEL292	MAN		R40D	1990	
878	VA	29-08-PT	MAN SR292	MAN		B53D	1987	

881-899

DAF SB3000 — BC59D 1993-96

881	AZ	00-98-EE	887	AZ	00-99-EE	892	AZ	54-22-CQ	896 AZ 36-98-CQ
882	AZ	94-97-CQ	888	AZ	01-01-EE	893	AZ	54-27-CQ	897 AZ 75-99-HH
883	AZ	94-96-CQ	889	AZ	01-02-EE	894	AZ	36-97-CQ	898 AZ 76-01-HH
885	AZ	94-95-CQ	890	AZ	98-88-EF	895	AZ	74-13-CQ	899 AZ 15-89-HS
886	AZ	94-98-CQ	891	AZ	54-26-CQ				

900	MO	83-69-SC	Hyundai	Huundai	M8	2001	
901	MO	01 AZ 66	Hyundai	Huundai	M8	2001	

903-913

Mercedes-Benz O303 — Mercedes-Benz — BC49D* — 1980-91 — *seating varies

903	SE	42-26-ME	908	VA	33-65-QL	910	SE	09-21-LU	913 VA 09-24-LU
906	VA	77-65-PE	909	VA	09-20-LU				

920	AL	29-05-GT	Toyota 43P850		C23F	1994
921	AL	29-06-GT	Toyota 43P850		C23F	1994
922	AL	77-75-IG	Toyota Coaster HB31R	Caetano Optimo	C21F	1997
923	AL	72-87-JX	Toyota Coaster HB31R	Caetano Optimo	C21F	1997
924	AL	74-25-HH	Toyota 43P850		C23F	1994
929	SE	13-56-HM	Scania K113 CLB	-	C54F	1988
930	AL	VX-75-24	Scania K113 CLB	-	C49F	1991
931	SE	VX-75-25	Scania K113 CLB	-	C49F	1991
932	AL	71-21-IH	Scania L113 CLB	-	C51F	1997
933	AL	95-13-IO	Scania L113 CLB	-	C51F	1997
934	AL	89-99-IP	Scania K113 CLB	-	C51F	1997
935	AL	90-00-IP	Scania K113 CLB	-	C51F	1997
936	AL	43 01-JL	Scania K113 CLB	-	C51F	1998
937	AL	05-10-JO	Scania K113 CLB	-	C51F	1998
938	AL	05-16-IX	Volvo B10B	-	C51F	1997
939	AL	05-19-IX	Volvo B10B	-	C51F	1997
940	AL	78-13-LM	MAN 18370 A	-	C51F	1998
941	AL	93-26-PA	MAN 18310 A	-	C51F	2000
942	AL	93-31-PA	MAN 18310 A	-	C51F	2000
943	AL	93-28-PA	MAN 18310 A	-	C51F	2000

944-948

MAN 18-350 — C55F 2001

944	AZ	29-48-SE	946	AL	29-52-SE	947	AZ	29-53-SE	948 AZ 29-54-SE
945	AL	29-51-SE							

956	AL	04-DC-40	Mercedes-Benz 396.493	-	C55F	1998
957	AL	63-DF-94	Mercedes-Benz 396.493	-	C55F	1999

960-967

Mercedes-Benz OH1627L — C51F 1996

960	AZ	24-28-HH	962	AZ	24-32-HH	964	AZ	24-34-HH	966 AZ 24-36-HH
961	AZ	24-29-HH	963	AL	24-33-HH	965	AZ	24-35-HH	967 AZ 24-37-HH

968-982

Mercedes-Benz O1829 — C52F 1997

968	AZ	19-51-IU	972	AZ	43-67-JA	976	AZ	43-65-JA	980 AZ 43-78-JA
969	AZ	19-83-IU	973	AZ	43-77-JA	977	AZ	43-66-JA	981 AZ 54-17-JA
970	AZ	43-62-JA	974	AZ	43-76-JA	978	AZ	43-72-JA	982 AZ 43-68-JA
971	AZ	43-63-JA	975	AZ	43-75-JA	979	AZ	43-74-JA	

983	AZ	93-25-PA	MAN 18350A	-	C51F	2000
984	AZ	93-30-PA	MAN 18310A	-	C51F	2000
985	AZ	89-55-RD	MAN 18310	-	C55F	2001

990-994

Mercedes-Benz O408 — Mercedes-Benz — BC49D 1995

990	MO	05-68-XE	992	AZ	05-71-XE	993 AZ 33-02-XE	994 AL 33-03-XE
991	MO	05-70-XE				

Depots: Almada (AL); Moita (MO); Sesimbra (SE); Seixal (SX) and Varzinha (VA).

ARRIVA NOROESTE

Arriva (Iasa-Finisterre), Poligono de Sabon, Parcela 31-32, 15142 Arteixo, La Coruña, España

231	FE	C-6734-A	Kässbohrer S215H	Kässbohrer Setra	C55D	1991
232	LU	C-6735-A	Kässbohrer S215H	Kässbohrer Setra	C55D	1991
243	CO	C-1910-AY	DAF SB3000	Castrosua	C55F	1991
245	FE	C-1912-AY	DAF SB3000	Castrosua	C55F	1991
246	LU	C-1913-AY	DAF SB3000	Castrosua	C55F	1991
248	CO	C-0746-AZ	Mercedes-Benz O303/15	Irizar	C55F	1992
250	CO	C-2011-BB	Mercedes-Benz O303/15	Irizar	C55F	1992
251	SA	C-2012-BB	Mercedes-Benz O303/15	Irizar	C55F	1992
255	FE	C-7264-BB	Pegaso 5226	Castrosua	C55F	1992
256	LU	C-7265-BB	Pegaso 5226	Castrosua	C55F	1992
257	FE	C-7266-BB	Pegaso 5226	Castrosua	C55F	1992
258	FE	C-7267-BB	Pegaso 5226	Castrosua	C55F	1992
264	FE	C-9126-BD	Pegaso 5226	Castrosua	C55F	1993
265	FE	C-9127-BD	Pegaso 5226	Castrosua	C55F	1993
266	FE	C-9128-BD	Pegaso 5226	Castrosua	C55F	1993
267	FE	C-9129-BD	Pegaso 5226	Castrosua	C55F	1993
268	LU	LU-5964-O	Kässbohrer S215HD	Kässbohrer Setra	C55F	1993
269	CO	LU-5965-O	Kässbohrer S215HD	Kässbohrer Setra	C55F	1993
270	LU	C-9130-BD	Volvo B12	Irizar	C55D	1993
271	LU	C-9131-BD	Volvo B12	Irizar	C55D	1993
286	FE	C-1998-BH	Scania K113TLA	Irizar Dragon	C(81)D	1994
298	FE	C-9928-AX	Kässbohrer S215HD	Kässbohrer Setra	C55D	1991
299	LU	C-0041-BD	Kässbohrer S215HD	Kässbohrer Setra	C55D	1993
300	CO	C-4724-BJ	Mercedes-Benz O404 RH	Irizar	C55D	1994
301	SA	C-4725-BJ	Mercedes-Benz O404 RH	Irizar	C55D	1994
303	FE	LU-4913-O	Pegaso 5226	Castrosua	C56D	1993
304	CO	C-2809-BK	Pegaso 5226	Hispano Carrocera	C55D	1994
307	LU	C-9810-BK	Pegaso CC95.9.E18	Unvi	C36D	1995
309	CO	C-3742-BL	Mercedes-Benz O1117	Ferqui	C36C	1995
315	FE	C-6101-BN	Mercedes-Benz O1117	Ferqui	C36C	1996
316	FE	C-6424-BN	Mercedes-Benz O1117	Ferqui	C36C	1996
317	FE	C-6425-BN	Mercedes-Benz O1117	Ferqui	C36C	1996
322	LU	C-5538-BT	Setra Seida 412MH	Setra	C55D	1997
323	CO	C-5539-BT	Setra Seida 412MH	Setra	C55D	1997
324	LU	C-5748-BT	Setra Seida 412MH	Setra	C55D	1997
325	LU	C-5749-BT	Setra Seida 412MH	Setra	C55D	1997
326	LU	C-0297-BU	MAN 10.220	Ferqui	C38C	1997
327	FE	C-0358-BV	MAN 10.220	Ferqui	C38C	1997
328	LU	C-0359-BV	MAN 10.220	Ferqui	C38C	1997
329	FE	C-0360-BV	MAN 10.220	Ferqui	C38C	1997
330	CO	C-3683-BX	MAN 13.220	Ugarte	C43D	1998
331	CO	C-3684-BX	MAN 13.220	Ugarte	C43D	1998
332	LU	C-3685-BX	MAN 13.220	Ugarte	C43D	1998
333	LU	C-3686-BX	MAN 13.220	Ugarte	C43D	1998
334	LU	C-7698-BY	MAN 13.220	Ugarte	C43D	1998
336	FE	C-7700-BY	MAN 13.220	Ugarte	C43D	1998
337	CO	C-7701-BY	MAN 13.220	Ugarte	C43D	1998
338	LU	C-0336-CB	Scania K94IB	OVI	C47D	1999
339	FE	C-0337-CB	Scania K94IB	OVI	C47D	1999
341	FE	C-0339-CB	Scania K94IB	OVI	C47D	1999
400	LU	C-8493-BZ	Iveco Mago 59.12	Indcar	C27D	1999
401	LU	C-8494-BZ	Iveco Mago 59.12	Indcar	C27D	1999
402	LU	C-8495-BZ	Iveco Mago 59.12	Indcar	C27D	1999
403	FE	C-8496-BZ	Iveco Mago 59.12	Indcar	C27D	1999
464	SA	C-9239-BX	MAN 16.360	Irizar	C64D	1991
474	FE	C-9544-BF	Pegaso 5226	Castrosua	C56D	1993
479	FE	C-4402-BM	MAN 18.310	Castrosua	C56D	1995
482	FE	C-9561-BP	MAN 18.310	Castrosua	C56D	1996
486	LU	C-5304-BX	MAN 18.350	Irizar	C57D	1999
487	SA	C-5305-BX	MAN 18.350	Irizar	C57D	1999
491	SA	C-5395-CF	Volvo B12	Irizar Century	C50D	1999
492	FE	C-5396-CF	Volvo B12	Irizar Century	C50D	1999
493	FE	C-5397-CF	Volvo B12	Irizar Century	C50D	1999
494	FE	C-5398-CF	Volvo B12	Irizar Century	C50D	1999
495	CO	C-5399-CF	Volvo B12	Irizar Century	C50D	1999

496	CO	C-5400-CF	Volvo B12	Irizar Century	C50D	1999
497	FE	C-5401-CF	Volvo B12	Irizar Century	C50D	1999
498	FE	C-5402-CF	Volvo B12	Irizar Century	C50D	1999
499	LU	C-8930-CF	Volvo B12	Irizar Century	C50D	1999
500	CO	C-8928-CF	Volvo B12	Irizar Century	C50D	1999
501	LU	C-8929-CF	Volvo B12	Irizar Century	C50D	1999
502	CO	C-9814-CF	Volvo B12	Irizar Century	C50D	1999
504	SA	C-9816-CF	Volvo B12	Irizar Century	C50F	1999
505	LU	C-9817-CF	Volvo B12	Irizar Century	C50F	1999
507	CO	C-0638-CG	Volvo B12	Irizar Century	C50F	1999
508	FE	C-0639-CG	Volvo B10M	Irizar InterCentury	C54F	1999
510	CO	C-0641-CG	Volvo B10M	Irizar InterCentury	C54F	1999
511	LU	C-0642-CG	Volvo B10M	Irizar InterCentury	C54F	1999
512	CO	C-0643-CG	Volvo B10M	Irizar InterCentury	C54F	1999
513	FE	C-0644-CG	Volvo B10M	Irizar InterCentury	C54F	1999
514	FE	C-0645-CG	Volvo B10M	Irizar InterCentury	C54F	1999
515	FE	C-0646-CG	Volvo B10M	Irizar InterCentury	C54F	1999
517	LU	C-0648-CG	Volvo B12	Irizar Century	C50F	1999
518	LU	C-0649-CG	Volvo B12	Irizar Century	C50F	1999
519	SA	C-0650-CG	Volvo B10M	Irizar InterCentury	C54F	1999
520	FE	C-0651-CG	Volvo B10M	Irizar InterCentury	C54F	1999
521	CO	C-0625-CG	Volvo B10M	Irizar InterCentury	C54F	1999
522	CO	C-0653-CG	Volvo B10M	Irizar InterCentury	C54F	1999
523	CO	C-0654-CG	Volvo B10M	Irizar InterCentury	C54F	1999
524	LU	C-0655-CG	Volvo B12	Irizar Century	C50F	1999
525	FE	C-0656-CG	Volvo B10M	Irizar InterCentury	C54F	1999
526	FE	C-0657-CG	Volvo B10M	Irizar InterCentury	C54F	1999
527	CO	C-0658-CG	Volvo B10M	Irizar InterCentury	C54F	1999
528	SA	C-0659-CG	Volvo B10M	Irizar InterCentury	C54F	1999
529	FE	C-0660-CG	Volvo D10M	Irizar InterCentury	C54F	1999
530	CO	C-0661-CG	Volvo B10M	Irizar InterCentury	C54F	1999
532	LU	4278-BSW	Volvo B12	Irizar Century	C55F	1996
533	CO	4276-BSW	Volvo B12	Sunsundegui	C56F	1996
534	CO	3838-BWJ	Volvo B12	Irizar Century	C55F	1995
535	CO	3839-BWJ	Volvo B12	Irizar Century	C55F	1996
536	CO	4016-BWJ	Volvo B12	Irizar Century	C55F	1997
537	FE	0471-BZN	Scania 420	Irizar Century	C71F	2002
538	FE	0547-BZN	Scania 420	Irizar Century	C71F	2002
539	SA	0343-BZN	Scania K114	Irizar Century	C59F	2002
540	LU	0517-BZN	Scania K114	Irizar Century	C59F	2002
541	FE	7452-BZN	Scania K114	Irizar Century	C59F	2002
542	LU	0324-BZN	Scania K114	Irizar Century	C59F	2002
543	LU	9571-BZN	Scania K114	Irizar Century	C59F	2002
544	CO	0359-BZN	Scania K114	Irizar Century	C59F	2002
545	LU	0402-BZN	Scania K114	Irizar Century	C59F	2002
546	FE	5007-CZN	Volvo B10B	Noge	C57F	2000
547	CO	5051-CZN	Volvo B10B	Noge	C57F	2000
548	FE	5084-CZN	Volvo B10B	Noge	C57F	2000
549	SA	5104-CZN	Volvo B10B	Noge	C57F	2000
550	LU	C-7703-BU	Scania L113 CLA	Irizar Century	C55F	1997
551	CO	C-0806-BT	Scania L113 CLA	Irizar Century	C55F	1997
552	LU	C-7704-BU	Scania L113 CLA	Irizar Century	C55F	1997
553	FE	8173-CZR	Scania L113 CLB	Irizar InterCentury	C55F	1996
554	SA	7192-CZV	Volvo B10B	Noge	C57F	2000
555	FE	7218-CZV	Scania L113 CLB	Irizar InterCentury	C55F	1996
556	FE	9516-CZV	Volvo B10B	Noge	C57F	2000
557	CO	5671-CZY	Volvo B10B	Noge	C57F	2000
558	CO	9882-DPK	Irisbus Eurorider C35	Irizar InterCentury	C59F	2005
559	CO	9891-DPK	Irisbus Eurorider C35	Irizar InterCentury	C59F	2005
560	CO	9881-DPK	Irisbus Eurorider C35	Irizar InterCentury	C59F	2005
561	LU	9885-DPK	Irisbus Eurorider C35	Irizar InterCentury	C59F	2005
562	SA	9895-DPK	Irisbus Eurorider C35	Irizar InterCentury	C59F	2005
563	LU	9921-DPK	Irisbus Eurorider C35	Irizar InterCentury	C59F	2005
564	CO	9913-DPK	Irisbus Eurorider C35	Irizar InterCentury	C59F	2005
565	FE	9916-DPK	Irisbus Eurorider C35	Irizar InterCentury	C59F	2005
566	FE	9899-DPK	Irisbus Eurorider C35	Irizar InterCentury	C59F	2005
567	CO	9902-DPK	Irisbus Eurorider C35	Irizar InterCentury	C59F	2005
569	LU	2314-DWL	MAN 18-350	Noge	C57F	2000
570	LU	2023-FFM	Irisbus Eurorider C35	Irizar InterCentury	C59F	2006
571	FE	1993-FFM	Irisbus Eurorider C35	Irizar InterCentury	C59F	2006
572	CO	1997-FFM	Irisbus Eurorider C35	Irizar InterCentury	C59F	2006
573	FE	5043-FFX	Irisbus Eurorider C35	Irizar InterCentury	C59F	2006
574	FE	7070-FFV	Irisbus Eurorider C35	Irizar InterCentury	C59F	2006

Since the last edition of this Handbook Autocares Mallorca have repainted all the buses into the yellow and red livery. Illustrating the change is 46, IB-3531-CN, an Iveco 391E with Castrosua bodywork. It is seen in Puerto Pollensa. *Colin Martin*

575	FE	7084-FFV	Irisbus Eurorider C35	Irizar InterCentury	C59F	2006
578	FE	6457-FFW	Irisbus Eurorider C35	Irizar InterCentury	C59F	2006
579	FE	M-0287-SY	Volvo B10B	Castrosua	B40D	2000
580	LU	3481-FFP	MAN 18-420	Noge	C55F	2008
581	LU	3457-FFP	MAN 18-420	Noge	C55F	2008

Depots: La Coruña (CO); Ferrol (FE); Lugo (LU) and Santiago de Compostela (SA)

Autocares Mallorca

Autocares Mallorca, Camino Vell Mal Pas, Alcudia
Bus Nort Balear, Gremi Fusters, Poligono Son Castello, Palma

35	AM	PM-4846-BZ	Iveco 391E	Castrosua CS40	B39D	1995
36	AM	PM-4851-BZ	Iveco 391E	Castrosua CS40	B39D	1995
37	AM	PM-0457-CB	Iveco 391E	Ugarte CX-Elite	C55D	1995
38	AM	PM-1248-CB	Iveco 391E	Ugarte CX-Elite	C55D	1995
39	AM	PM-0304-CH	Iveco 391E	Castrosua CS40	B39D	1996
40	AM	PM-0305-CH	Iveco 391E	Castrosua CS40	B39D	1995
41	AM	PM-4402-CG	Ford Transit	Ford	C14D	1996
42	AM	IB-1249-CM	Iveco 80E18	Indcar Mago	C30F	1997
43	AM	PM-9734-CM	Iveco 391E	Irizar InterCentury	BC55D	1997
44	AM	PM-9735-CM	Iveco 391E	Irizar InterCentury	BC55D	1997
45	AM	PM-3530-CN	Iveco 391E	Castrosua CS40	B39D	1997
46	AM	IB-3531-CN	Iveco 391E	Castrosua CS40	B39D	1997
47	AM	IB-8194-CV	Iveco 391E	Irizar Century	C55D	1998
48	AM	IB-7901-CW	Iveco 391E	Unvi Cidade II	B26D	1998
51	AM	IB-2184-CZ	Mercedes-Benz O405	Mercedes-Benz	B55D	1998

Also in Puerto Pollensa is 65, IB-5686-CY, an Iveco 391E with Irizar InterCentury bodywork. Irizar have bodied a variety of chassis for their home market and from 2009 will be supplying an integral model to the remainder of Europe. *Colin Martin*

53	AM	IB-3540-DG	Iveco 391E		Ayats Atlas	C55D	1999		
56	AM	IB-5157-DG	Iveco Daily 35-10		Iveco	C13D	1999		
57	AM	IB-5158-DG	Iveco Daily 35-10		Iveco	C13D	1999		
58	AM	IB-3046-DN	Iveco 391E		Noge Touring	C55D	2000		
60	AM	IB-9413-DN	Iveco 391E		Unvi Cidade II	B44D	2000		
61	AM	IB-9414-DN	Iveco 391E		Unvi Cidade II	B44D	2000		
62	AM	IB-4796-DP	Iveco 391E		Ugarte Nobus	C55D	2000		
63	AM	IB-4737-CY	MAN 1190		Arabus	C35D	1988		
64	AM	IB-5685-CY	Iveco 391E		Irizar InterCentury	BC53D	1998		
65	AM	IB-5686-CY	Iveco 391E		Irizar InterCentury	BC53D	1998		

66-70
Iveco EuroRider 397E.12.35 — Irizar InterCentury — BC55D — 2002-03

66	AM	8867BWL	68	AM	8784BWL	69	AM	7280CHJ	70	AM	7321CHJ
67	AM	8678BWL									

71-76
Iveco EuroRider 397E — Unvi Cidade II — B44D — 2003

71	AM	2543CHR	73	AM	2671CHR	75	AM	6244CJD	76	AM	6293CJD
72	AM	2630CHR	74	AM	2717CHR						

77	AM	IB-0120-DF	Iveco EuroRider	OVI Radial	BC45D	1999
80	AM	0359-DYW	Irisbus EuroRider	Irizar Century	C55D	2006
81	AM	0388-DYW	Irisbus EuroRider	Irizar Century	C55D	2006
82	AM	3129-CHG	Iveco EuroRider	Irizar InterCentury	BC55D	1996
83	AM	4735-FSL	Irisbus EuroRider	Irizar InterCentury	BC55D	2007
84	AM	IB-4069-BZ	Iveco	UNVI Mago	C25D	1995
85	AM	7430-GFL	Irisbus EuroRider 397E	Andecar Viana	C55D	2008

59	BN	IB-2124-DB	MAN 24.420	Obradors ST400	C(75)D	1995
75	BN	IB-7717-CL	Ford Transit	Ford	M14	1997
76	BN	IB-9451-DD	MAN 18.310	Sunsundegui Stylo	B55D	1999
78	BN	IB-7573-CN	Iveco EuroRider 391E	Ugarte CX-Elite	C55D	1997
81	BN	9830-BWT	Iveco EuroRider 391E.12.35	Irizar InterCentury	C55D	2002
86	BN	3283-CHJ	Iveco EuroRider 397E.12.35	Irizar InterCentury	C55D	1996
88	BN	3061-CHG	Iveco EuroRider 397E.12.35	Irizar InterCentury	C55D	1996

89	BN	3290-CHG	Iveco EuroRider 397E.12.35	Irizar InterCentury	C55D	1996
91	BN	C-2524-BT	Mercedes-Benz 01829	Irizar	C55D	1989
93	BN	8953-FRG	Irisbus EuroRider 397E.12.35	Irizar InterCentury	C55D	2007
94	BN	5666-GDM	Irisbus C65	Indcar	C18F	2008
95	BN	IB-4666-CT	MAN	Ayats	C80D	1997
96	BN	7884-BHR	Iveco Daily 80E18	Indcar Mago	C16F	2001

ESFERA BUS

Pol Industrial Fin de Semana. Avda, Gumersindo Llorente 54, 28022 Madrid

7	M-8290-NT	Renault FR1	Irizar Century	C55F	1992
10	M-2880-UL	Renault FRH	Irizar Century	C55F	1997
13	M-7603-YZ	Renault FRX	Andecar Viana	C55F	2000
15	96-4-BHC	Mercedes-Benz O404	Ugarte Nobus	C55F	2001
17	M-7230-TT	Ford Transit	Ferqui	M13	1996
18	8124-CHX	MAN 10.225	Andecar Sensca	C30F	2003
19	1739-CJF	MAN 10.225	Andecar Sensca	C28F	2003
21	M-3721-XH	Iveco EuroRider 35	Noge Touring	C55F	1999
22	M-7048-TX	MAN 11.190	Ugarte Corvo CX	C37F	2002
23	7997-BRW	Mercedes-Benz Sprinter 413	Noge Sprinter	C18F	2003
24	3911-CMN	MAN 18.463	Andecar Viana	C55F	2003
25	4389-CNB	MAN 18.463	Noge Touring	C50F	2003
26	2426-BHX	Iveco EuroRider 38	Andecar Viana	C55F	2001
27	9334-DHN	MAN 18.463	Andecar Viana	C55F	2005
28	4716-DJW	MAN 18.463	Noge Touring	C50F	2005
29	2250-CHP	MAN 10.225	Andecar Seneca	C28F	2003
30	4929-DPK	Volkswagen	Andecar 111.1	C18F	2005
31	4966-DPK	Volkswagen	Andecar 111.1	C18F	2005
32	9459-DPR	Volkswagen	Andecar 111.1	C18F	2005
33	8232-CJJ	Mercedes-Benz Sprinter 413	Noge Sprinter	C16F	2003
36	3457-FFP	MAN 18.920	Noge Touring	C55F	1997
37	3481-FFP	MAN 18.920	Noge Touring	C55F	1997
38	1653-FFT	MAN 18.920	Noge Touring	C55F	1997
39	6104-CKW	Volvo B12B	Irizar PB	C53F	2003
40	0650-BHY	Volvo B12B	Irizar PB	C54F	2001
41	0651-BHY	Volvo B12B	Irizar PB	C54F	2001
42	-	MAN 18.463	Noge Touring	C55F	2002
43	-	Iveco EuroRider 397E	Noge Titanium	C50F	2008
61	M-2261-ZC	Volvo B12B	Noge Touring	C55F	2000
72	M-4072-SK	Scania K113 CLB	Irizar Century	C55F	1995
73	8124-CHX	MAN 10.225	Ferqui	C35F	2000
74	-	Mercedes-Benz Sprinter 416	Noge Sprinter	C18F	2001
75	-	Scania K124 EB	Noge Titanium	C55F	2001
76	-	Toyota	Caetano	C25F	2002
77	-	Mercedes-Benz Sprinter 413	Noge Sprinter	C19F	2002
78	-	Ford Transit	Ferqui Sunset	M16	2002
79	- -	Mercedes-Benz OC500	Irizar Century	C55D	2003
80	-	Volvo B7R	Irizar Century	C50D	2003
81	- -	Mercedes-Benz OC500	Irizar Century	C55D	2004
82	-	MAN 10.225	Ferqui Solera	C33F	2004
83	- -	Mercedes-Benz OC500	Andecar	C55D	2004
84	-	Mercedes-Benz 923	Beulas Gianino	C30F	2005
85	-	Volvo B7R	Irizar Century	C55F	2005

ARRIVA ITALY - SAB

SAB Autoservizi srl, Piazza Marconi 4, 24122 Bergamo

35-38		Iveco 370.97.24		Portesi		Regional bus		1985-89	
35	BG922939	36	BG922940	37	BG930888			38	BG940393

39	BGA52588	Iveco 370.97.S24	Portesi	Regional bus	1991
40	BGA50162	Iveco 370.97.S24	Portesi	Regional bus	1991
42	BGA50161	Iveco 370.97.024	Portesi	Regional bus	1991
44	BGA50160	Iveco 370.97.S24	Portesi	Regional bus	1991
45	BGA34200	Iveco 370.97.S24	Portesi	Regional bus	1990
76	AW749PY	Mercedes-Benz O303/9R	Mercedes	Regional bus	1990
80	BGD12750	Mercedes-Benz O303/10R	Mercedes	Coach	1994
81	BG918869	Mercedes-Benz O303/10R	Bianchi	Regional bus	1990
83	BG918870	Mercedes-Benz O303/10R	Bianchi	Regional bus	1990
84	BGA40904	Mercedes-Benz O303/10R	Bianchi	Regional bus	1990
86	AH911KM	Iveco 380.10.29 EuroClass	Orlandi	Regional bus	1996
87	AH915KM	Iveco 380.10.29 EuroClass	Orlandi	Regional bus	1996
107	BG952197	Iveco 370.10S.24	Ivecoi	Regional bus	1989

108-113		Iveco 370.10S.24		Portesi		Regional bus		1989	
108	BG952200	110	BG952198	112	BG960676			113	BG960675
109	BG952199	111	BG953420						

114	AD666TP	Iveco 370.10S.24	Desimon	Regional bus	1990
115	AH401KF	Iveco 380.10.29 EuroClass	Orlandi	Regional bus	1995
116	AH878KF	Iveco 380.10.29 EuroClass	Orlandi	Regional bus	1995
120	AH875KF	Iveco 380.10.29 EuroClass	Orlandi	Regional bus	1995
138	BGA65439	Kässbohrer SG221 UL	Kässbohrer Setra	Regional bus	1991
139	BGA74947	Kässbohrer SG221 UL	Kässbohrer Setra	Regional bus	1991
140	BGB18812	Kässbohrer SG221 UL	Kässbohrer Setra	Regional bus	1992

142-145		Setra S 215 UL		Kässbohrer		Regional bus		1991-92	
142	BGB51904	143	BGB51903	144	BGB01838			145	BGB01837

In addition to its bus services, Arriva operates a cable car between Albino and Selvino, where midi-buses continue into the mountain villages. Working the connecting service at the Albino terminus is 45, BGA34200, an Iveco 370 with Portesi bodywork. *Bill Potter*

Parked outside the upper cable car station at Selvino is 81, BG918869, a Mercedes-Benz O303 with Bianchi bodywork. *Bill Potter*

159	BS458WY	Cacciamali TCI 970 Sigma 2	Cacciamali	Regional bus	2001
160	BS460WY	Cacciamali TCI 970 Sigma 2	Cacciamali	Regional bus	2001
161	BG886418	Iveco 315.8.17	Iveco	Regional bus	1988
162	BG886413	Iveco 315.8.17	Iveco	Regional bus	1988
164	AN015PP	Iveco 315.8.18	Orlandi	Regional bus	1996
165	AN014PP	Iveco 315.8.18	Orlandi	Regional bus	1996
166	AN013PP	Iveco 315.8.18	Orlandi	Regional bus	1996

168-186

				Cacciamali TCI 970 Sigma 2	Cacciamali	Interurban	2008-09

168	BS456WY	173	BS791WZ	178		183	
169	BS459WY	174	BS895WZ	179	BV335VS	184	
170	BS457WY	175	BS896WZ	180		185	BX177BE
171	BS560WT	176	BS777WT	181	BV382VR	186	CE896TB
172	BS790WZ	177	BV355VS	182			

187	CW712PC	Irisbus EuroRider 397.10.31	10.8m	2004	

261-264

		Iveco 370.12.S30	Bianchi	Regional bus	1991

261	BGA48629	262	BGA52589	263	BGA48630	264	BGA52590

308	BGA55605	Mercedes-Benz O303/15R	Bianchi	Regional bus	1988-91
309	BGA55606	Mercedes-Benz O303/15R	Bianchi	Regional bus	1988-91
310	BGB54150	Kässbohrer S215 HRI	Kässbohrer Setra	Regional bus	1992

321-329

		Iveco 380.12.35 EuroClass	Orlandi	Regional bus	1995-96

321	AH880KF	324	AH882KF	326	AH912KM	328	AH914KM
322	AH403KF	325	AH400KF	327	AH910KM	329	AH909KM
323	AH881KF						

332-346

		Iveco 393.12.35 My Way	Irisbus-Orlandi	Regional bus	2000

332	BG305PC	336	BG301PC	340	BG297PC	344	BG293PC
333	BG304PC	337	BG300PC	341	BG296PC	345	BG292PC
334	BG303PC	338	BG299PC	342	BG295PC	346	BG291PC
335	BG302PC	339	BG298PC	343	BG294PC		

A recent addition to the SAB fleet is BredaMenarini M 221 404, AN974PP, a less-common model in the fleet. It is seen at the main depot in Bergamo during 2009, carrying the orange colour used on city services. *Bill Potter*

347	CP755CX	Mercedes-Benz O405 NU	Mercedes	Regional bus	1992
348	AW278TP	Mercedes-Benz O404	Mercedes	Regional bus	1998
349	AW279TP	Mercedes-Benz O404	Mercedes	Regional bus	1998

350-364

Mercedes-Benz O408 — Mercedes — Regional bus — 1997

350	AT083SN	353	AT084SN	356	AT080SN	358	AT088SN
351	AT086SN	354	AT087SN	357	AT079SN	364	AT085SN
352	AT081SN	355	AT082SN				

365-387

Mercedes-Benz O405 NU — Mercedes-Benz — Regional bus — 1998

365	AW205TM	371	AW386TM	377	AW383TM	383	AW396TM
366	AW220TM	372	AW387TM	378	AW395TM	384	AW392TM
367	AW203TM	373	AW388TM	379	AW384TM	385	AW394TM
368	AW206TM	374	AW382TM	380	AW391TM	386	BE193NY
369	AW204TM	375	AW389TM	381	AW393TM	387	BE194NY
370	AW219TM	376	AW390TM	382	AW397TM		

388-398

Mercedes-Benz Citaro O530 NU — Mercedes-Benz — Regional bus — 2001

388	BN683RW	391	BN689RW	394	BN681RW	397	BN690RW
389	BN688RW	392	BP287ZA	395	BN687RW	398	BP288ZA
390	BN677RW	393	BN693RW	396	BN694RW		

400-403

Kässbohrer S300 NC — Kässbohrer Setra — City bus — 1991-92

| 400 | AT158SM | 401 | AT159SM | 402 | AT889SM | 403 | AT989SM |

| 404 | AN974PP | BredaMenarini M 221.1 | Menarini | City bus | 1997 |

405-418

Mercedes-Benz Citaro O530 NU — Mercedes-Benz — Regional bus — 2000-01

405	BN678RW	411	BN692RW	414	BN682RW	417	BM871EW
406	BP289ZA	412	BN691RW	415	BN679RW	418	BN680RW
407	BN684RW	413	BP292ZA	416	BN686RW		

419-422	Mercedes-Benz O407		Mercedes		Regional bus		1992
419 BZ825WZ	**420**	BZ826WZ	**421**	BZ827WZ		**422**	BZ828WZ

423-429	Mercedes-Benz O407		Mercedes		Regional bus		1992
423 CP756CZ	**425**	CP757CZ	**427**	CP754CZ		**429**	CP496CZ
424 CP752DA	**426**	CP495CZ	**428**	CP753CZ			

430-447	Iveco 393.12.35 My Way		Iveco		Regional bus		2000
430 BM961EV	**435**	BM262EW	**440**	BM259EW		**444**	BM874EW
431 BM962EV	**436**	BM263EW	**441**	BM257EW		**445**	BM875EW
432 BM963EV	**437**	BM260EW	**442**	BM872EW		**446**	BM876EW
433 BM964EV	**438**	BM264EW	**443**	BM873EW		**447**	BM877EW
434 BM261EW	**439**	BM258EW					

448-451	Iveco 393.12.35 My Way		Iveco-Orlandi		Regional bus		2000
448 BM569EY	**449**	BM571EY	**450**	BM570EY		**451**	BM568EY

452	CE897TB	Irisbus 393.12.35 My Way	Irisbus-Orlandi	Regional bus	2003
453	CH278PD	Kässbohrer S300 NC	Kässbohrer Setra	City bus	1993
454	CH633NZ	Kässbohrer S300 NC	Kässbohrer Setra	City bus	1992
455	CR546AH	Irisbus Domino 2001 HD	Irisbus Orlandi	Regional bus	2004
456	CR545AH	Irisbus Domino 2001 HD	Irisbus Orlandi	Regional bus	2004
457	CT986EC	Mercedes-Benz Citaro O530 NU	Mercedes-Benz	Regional bus	2005
458	CT987EC	Mercedes-Benz Citaro O530 NU	Mercedes-Benz	Regional bus	2005
459	CT988EC	Mercedes-Benz Integro O550 UL	Mercedes-Benz	Regional bus	2005

484-489	Mercedes-Benz O303/15R		Bianchi		Coach	1986-88	
484 BG498PC	**487**	BG497PC	**488**	BG907343		**489**	BG912768
486 BG499PC							

492	BGA65274	Iveco 370.12.S30	Bianchi	Coach	1991
493	CB768MJ	Iveco 370.12.S30	Orlandi	Regional bus	1992
494	BGB29553	Iveco 370.12.S30	Orlandi	Coach	1992
495	BGB29554	Iveco 370.12.S30	Orlandi	Coach	1992
496	BGB31233	Iveco 370.12.S30	Orlandi	Coach	1992
498	AT356TE	Mercedes-Benz O350 Tourismo	Mercedes	Coach	1997
499	AT537TE	Mercedes-Benz O350 Tourismo	Mercedes	Coach	1997

501-511	Mercedes-Benz O404		Mercedes		Coach		1992-93
501 AW352RV	**505**	AW145PZ	**507**	AW888PZ		**510**	BE213NY
503 AW144PZ	**506**	AW146PZ	**508**	AW488RA		**511**	BF264NR

512	CD086FF	Mercedes-Benz O350 Tourismo	Mercedes	Coach	2002
513	CD743FE	Mercedes-Benz O350 Tourismo	Mercedes	Coach	2002

514-517	Irisbus Domino 2001 HD		Irisbus Orlandi		Regional bus		2004
514 CR045AH	**515**	CR046AH	**516**	CR991AH		**517**	CR992AH

550-555	Mercedes-Benz O407		Mercedes		Regional bus		1994-95 Arriva Danmark, 2005
550 CW217AL	**552**	CW219AL	**554**	CW221AL		**555**	CW222AL
551 CW218AL	**553**	CW220AL					

556	CT698ZF	Mercedes-Benz Integro O550 UL	Mercedes-Benz	Interurban	2001
557	CW898AM	Mercedes-Benz Citaro O530 LU	Mercedes-Benz	Regional bus	2003
558	DD561RG	MAN NL263 11.95m	MAN	Regional bus	2002
559	DE933XJ	Mercedes-Benz O405 NU	Mercedes-Benz	Regional bus	1996

560-564	Mercedes-Benz Integro O550 UL		Mercedes-Benz		Interurban		2008
560 DK366WP	**562**	DP678KW	**563**	DR090CR		**564**	DR220CR
561 DN633DG							

600-603	Mercedes-Benz Citaro O530NU		Mercedes-Benz		Regional bus		2001
600 CT412ZF	**601**	BP695ZY	**602**	BP541ZY		**603**	CT104ZF

613-617	Irisbus Turbocity CityClass U491		Irisbus		Regional bus		2006-07
613 DC797TX	**615**	DE473TG	**616**	DE607TG		**617**	DE673TG
614 DE513TG							

Two Irisbus models are shown here. Citybus 625, DN235DG, is an integral Irisbus CityClass 491 from the 2008 delivery while 341, BG296PC is a Myway regional bus. *Bill Potter*

618-623
BredaMenarini M 240 10.5m   Bredabus   Urban   2008

618	DL054SN	620	DL655SN	622	DK779WP	623	DK895WP
619	DK777WP	621	DK778WP				

624	DN234DG	Irisbus CityClass 491	Irisbus	Urban	2008
625	DN235DG	Irisbus CityClass 491	Irisbus	Urban	2008
626	DN236DG	Irisbus CityClass 491	Irisbus	Urban	2008

631-642
Irisbus Ares N 10.6m   Irisbus   Regional bus   2005

631	CW708AL	634	CW710AL	637	CW829AL	640	CW078AM
632	CW707AL	635	CW709AL	638	CW830AL	641	CW353AM
633	CW706AL	636	CW758AL	639	CW957AL	642	CW354AM

643-658
Irisbus Crossway 10.5m   Irisbus   Regional bus   2007

643	DH076GP	647	DH073GP	651	DH071GP	655	DH070GP
644	DH075GP	648	DH424GP	652	DH662GP	656	DH462GP
645	DH074GP	649	DH663GP	653	DH664GP	657	DH069GP
646	DH423GP	650	DH072GP	654	DH425GP	658	DH068GP

659-674
Irisbus Crossway 10.5m*   Irisbus   Regional bus   2008   *668-70 are 12.8m

659	DN425DG	663	DN287DG	667	DP211KW	671	DR783CR
660	DN431DG	664	DN280DG	668	DP212KW	672	DR784CR
661	DN278DG	665	DN288DG	669	DP213KW	673	DT364EE
662	DN279DG	666	DN281DG	670	DP214KW	674	DT365EE

701-724
MAN Lion's City U 12m   MAN   Regional bus   2007

701	DE474TG	707	DE515TG	713	DE665TG	719	DE518TG
702	DE475TG	708	DE519TG	714	DE666TG	720	DE670TG
703	DE476TG	709	DE602TG	715	DE667TG	721	DE606TG
704	DE477TG	710	DE603TG	716	DE668TG	722	DE671TG
705	DE478TG	711	DE604TG	717	DE669TG	723	DE880TG
706	DE514TG	712	DE605TG	718	DE517TG	724	DE672TG

Proving a popular vehicle is the MAN Lion's City U of which ten were added to SAB fleet in 2007. First numerically, 725, DN246DG, is seen arriving at Bergamo for fuel and cleaning before heading back to the villages. *Bill Potter*

725-734

MAN Lion's City U 12m — MAN — Regional bus — 2007

725	DN246DG	728	DN249DG	731	DN252DG	733	DN254DG
726	DN247DG	729	DN250DG	732	DN253DG	734	DN255DG
727	DN248DG	730	DN251DG				

963	vintage	Menarini M231	Irisbus	Suburban	19xx	
964	DR063CR	Probus 215 SCB	Probus	Regional	2009	

965-968

Cacciamali TCI 840 — Cacciamali — Interurban — 2008

965	DR525CR	966	DR447CR	967	DR448CR	968	DR449CR

969-977

Mercedes-Benz Sprinter 515 — Sprinter 65 — Minibus — 2007

969	DE709TG	972	DG873MV	974	DG875MV	976	DG877MV
970	DE710TG	973	DG874MV	975	DG876MV	977	DG878MV
971	DG872MV						

978	BX755BF	Iveco Daily F45.12	Iveco	Regional bus	2001
979	BS105XA	Iveco EuroPolis 9.15	Iveco	City bus	2001
980	BS224XA	Iveco EuroPolis 9.15	Iveco	City bus	2001
981	AW170TP	Mercedes-Benz Vario O814	Beluga	Coach	1987 ex Zambetti
984	CE543TA	Fiat Ducato L2.8JTD	Fiat	Coach	2003
985	BX274BG	Iveco 65.C15 THESI	Cacciamali	Regional bus	2001
986	BY309EF	Iveco 65.C15 THESI	Cacciamali	Regional bus	2002
987	BX275BG	Iveco 65.C15 THESI	Cacciamali	Regional bus	2001
988	BY308EF	Iveco 65.C15 THESI	Cacciamali	Regional bus	2002
991	AT949TT	Iveco Daily F45.12	Iveco	Regional bus	1997
992	AT948TT	Iveco Daily F45.12	Iveco	Regional bus	1997
993	AT947TT	Iveco Daily F45.12	Iveco	Regional bus	1997
994	AT950TT	Iveco Daily F45.12	Iveco	Regional bus	1997
995	BE248NZ	Iveco Daily F45.12	Iveco	Regional bus	1999
996	BE249NZ	Iveco Daily F45.12	Iveco	Regional bus	1999
999	AW385TM	Fiat Ducato	Fiat	Regional bus	1996
1004	BG963500	Iveco 370.97.S24	Portesi	Regional bus	1989
1011	BGB28561	Iveco 370.12.S30	Orlandi	Regional bus	1992

Ten Volvo articulated B7LAs operate from Bergamo depot, the batch carrying examples of both city and urban liveries. Illustrating the model is 1209, CH635NZ, about to re-enter service following its visit to the depot in Bergamo. *Bill Potter*

1013	BGB28562	Iveco 370.12.S30	Orlandi	Regional bus	1992
1051	BG883403	Iveco 370.97.S24	Portesi	Regional bus	1988
1058	BG963501	Iveco 370.97.S24	Portesi	Regional bus	1989
1061	AT219SM	Iveco 370.12.L25	Portesi	Regional bus	1989
1062	BG969683	Iveco 370.12.L25	Portesi	Regional bus	1989
1063	BG970248	Iveco 370.12.L25	Portesi	Regional bus	1989
1067	BGA57919	Iveco 370.10S.24	Iveco	Regional bus	1991
1069	BGB11432	Kässbohrer SG 221 UL	Kässbohrer Setra	Regional bus	1992

1081-1086

		Mercedes-Benz O530GNU	Mercedes-Benz Citaro	Regional bus	2001
1081	BS585XB	1083	BS844XB	1085	BS988WY	1086	BS101WY
1082	BS455WY	1084	BS578WY				

1087	BB553TB	BredaMenarini M 321	BredaMenarini	City bus	1999
1111	BG922858	Mercedes-Benz O303/10R	Bianchi	Regional bus	1988
1112	BG940225	Mercedes-Benz O303/10R	Menarini	Regional bus	1989
1113	BGA68889	Volvo B10M	Portesi	Regional bus	1991
1114	BGA68890	Volvo B10M	Portesi	Regional bus	1991

1201-1210

		Volvo B7LA	-	City/Regional bus	2000-01
1201	CH280PD	1204	CH483PD	1207	CH096NZ	1209	CH635NZ
1202	CH281PD	1205	CH563PD	1208	CH140NZ	1210	CH636NZ
1203	CH484PD	1206	CH139NZ				

1211-1216

		Mercedes-Benz O405GN	Mercedes-Benz	AN57D	1994	Arriva Denmark, 2005-07
1211	CW968VS	1213	CW969VS	1215	-	1216	-
1212	CW969VS	1214	CW969VS				

1217	DK894WP	BredaMenarini M 321 18m	BredaMenarini	City bus	1999
1219	DR668ZB	BredaMenarini M 321 18m	BredaMenarini	City bus	1999
1220	u-airport	BredaMenarini M 321 18m	BredaMenarini	City bus	1999

SIA

Società Italiana Autoservizi SpA, Via Cassala 3/a, 25126 Brescia

22	BSA41550	Iveco 370.12.30	Orlandi	Coach	-
23	BSA41551	Iveco 370.12.30	Orlandi	Coach	-
24	BSA83566	Iveco 370.12.S30	Orlandi	Coach	-
25	BSA83567	Iveco 370.12.S30	Orlandi	Coach	-
28	BSD42532	Iveco 370.12.S30	Orlandi	Coach	-
32	BSE05236	Iveco 370.12.S30	Orlandi	Coach	1992
33	BSE05235	Iveco 370.12.S30	Orlandi	Coach	1992
34	BSE72407	Iveco 315.8.18	Orlandi	Coach	1993
35	BSF08559	Iveco 370.12.S30	Dallavia	Coach	1994
36	AF 080XT	Iveco 370.12.SE35	Dallavia	Coach	1996
37	AF 330XW	Iveco 370.12.SE35	Dallavia	Coach	1996
38	AZ 085MC	Iveco 380.12.38 EuroClass HD	Orlandi	Coach	1998
39	AZ 423MC	Mercedes-Benz O404/15R HD	Mercedes-Benz	Coach	1998
40	AZ 431MC	Mercedes-Benz O404/15R HD	Mercedes-Benz	Coach	1998
41	BE 792ZJ	Iveco 380.12.38 EuroClass HD	Orlandi	Coach	1999
42	BT062GC	Iveco 391.12.35 EuroRider	Orlandi	Coach	2001
45	CD209BW	Mercedes-Benz O350	Mercedes-Benz Tourismo	Coach	2002
46	CD124BW	Mercedes-Benz O350	Mercedes-Benz Tourismo	Coach	2002
47	CD125BW	Mercedes-Benz O350	Mercedes-Benz Tourismo	Coach	2002
48	DK751HY	Irisbus Domino HD	Irisbus	Coach	2008
67	DE034SB	Irisbus 370.12.S30	?	Regional bus	?

84-87		Iveco 370.12.SE35 12m	Orlandi	Regional bus	1998
84	AZ 264MD	**85**	AZ 888MC	**86**	AZ 783MD	**87**	AZ 784MD

92	BSB55781	BredaMenarini M 3001.12L	Bredabus	City bus	1989
94	BSB55779	BredaMenarini M 3001.12L	Bredabus	City bus	1989
95	BSB90063	BredaMenarini M 2001.10L	Bredabus	City bus	1990
96	CY322CX	BredaMenarini 220-LU	Bredabus	City bus	19
97	CY323CX	BredaMenarini 220-LU	Bredabus	City bus	19
98	DK670HY	BredaMenarini M240 NU	Bredabus	City bus	19
99	DK653HY	BredaMenarini M240 NU	Bredabus	City bus	19

116	BSE19611	Inbus AID 280 FT	Inbus	Regional bus	1992
117	BSE62370	MAN SG 292	Inbus	Regional bus	1993
118	AF 079XT	Inbus AID 280 FT	Inbus	Regional bus	1985
121	AP 164NW	Inbus AID 280 FT	Inbus	Regional bus	1991

122-126		Mercedes-Benz O530 GNU 18m	Mercedes-Benz Citaro	Regional bus	2001
122	BV 951 DY	**124**	BV 579 DW	**125**	BV 580 DW	**126**	BV 585 DW
123	BV 581 DW						

127	CP730TT	Mercedes-Benz Citaro O530 GNU	Mercedes-Benz	Regional bus	1998
128	CN584HF	Mercedes-Benz Citaro O530 GNU	Mercedes-Benz	Regional bus	1998
129	CN334HF	Mercedes-Benz Citaro O530 GNU	Mercedes-Benz	Regional bus	1998
130	DB387GE	Mercedes-Benz O405 G	Mercedes-Benz	Regional bus	19
131	DB388GE	Mercedes-Benz O405 G	Mercedes-Benz	Regional bus	19
132	DK654HY	BredaMenarini M321 U	Bredabus	City bus	19

133-142		Mercedes-Benz Citaro O530 GNU	Mercedes-Benz	Regional bus	2008
133	DN323MS	**136**	DN380MS	**139**	DN379MS	**141**	DN320MS
134	DN325MS	**137**	DN322MS	**140**	DN319MS	**142**	DN318MS
135	DN324MS	**138**	DN321MS				

162	BSB29130	Iveco 315.8.17 7.5m	Iveco	Regional bus	1983-87

163-167		Iveco 315.8.18 7.6m	Orlandi	Regional bus	1993
163	BSE62369	**165**	BSE62367	**166**	BSE62366	**167**	BSE62365
164	BSE62368						

New Cacciamali TCI 840s have been added to several of the Italian fleets in recent months, displacing older midi-coaches that provide services to the villages in the hills of the Lombardy region. SIA received ten of this model, including 178, DP021HG, seen here at the Brescia headquarters. *Bill Potter*

168-172

Iveco 315.8.S18 7.6m   Orlandi   Regional bus   1996-98

168	AF870XR	170	AF869XR	171	AZ350NF	172	AZ351NF
169	AF868XR						

173-183

Cacciamali TCI 840   Cacciamali   Interurban   2008

173	DN862MS	176	DP018HG	179	DP125HG	182	DP128HG
174	DP020HG	177	DP019HG	180	DP126HG	183	DP129HG
175	DP017HG	178	DP021HG	181	DP127HG		

200	DL831NA		Mercedes-Benz O550 Integro L	Mercedes-Benz	Interurban		2008

250-290

Mercedes-Benz O408   Mercedes-Benz   Regional bus   1996-97

250	AF436XX	261	AF440XX	272	AP634NW	282	AP636NW
251	AF701XX	262	AF441XX	273	AP481NW	283	AP632NW
252	AF437XX	263	AF787XX	274	AP490NW	284	AP635NW
253	AF710XX	265	AF705XX	275	AP683NW	285	AP489NW
254	AF708XX	266	AF709XX	276	AP491NW	286	AP488NW
256	AF707XX	267	AP487NW	277	AP492NW	287	AP486NW
257	AF702XX	268	AP482NW	278	AP493NW	288	AP631NW
258	AF703XX	269	AP495NW	279	AP494NW	289	AP633NW
259	AF704XX	270	AP485NW	280	AP483NW	290	AP682NW
260	AF439XX	271	AP496NW	281	AP484NW		

291-295

Mercedes-Benz O405 NU   Mercedes-Benz   Regional bus   1998

291	BH071WK	293	BH073WK	294	BH074WK	295	BH075WK
292	BH072WK						

Enhancements to the routes 201 and 202 that links Brescia with Lake Garda have seen an increase in the numbers of vehicles required on this route, including some with route lettering. Running as a duplicate, Irisbus Ares 446, CY863CZ, is seen returning to the city. *Bill Potter*

296-322
Mercedes-Benz O530 NU   Mercedes-Benz Citaro   Regional bus   2001

296	BT242GD	303	BT108GD	310	BT763GD	317	BV403DV
297	BT100GD	304	BT109GD	311	BT107GD	318	BV582DW
298	BT101GD	305	BT110GD	312	BT244GD	319	BV402DV
299	BT102GD	306	BT112GD	313	BT764GD	320	BV583DW
300	BT103GD	307	BT111GD	314	BT761GD	321	BV401DV
301	BT104GD	308	BT106GD	315	BT243GD	322	BV584DW
302	BT105GD	309	BV404DV	316	BT762GD		

323	CG989SP	Mercedes-Benz O530 NU	Mercedes-Benz Citaro	Regional bus	2006
324	CV103XV	Mercedes-Benz O530 NU	Mercedes-Benz Citaro	Regional bus	2006

350-362
Mercedes-Benz O345   Mercedes-Benz Conecto   Regional bus   2003-04

350	CL043AY	354	CL982AY	357	CL190AY	360	CL979AY
351	CL042AY	355	CL983AY	358	CL191AY	361	CL985AY
352	CL980AY	356	CL981AY	359	CL984AY	362	CL099AY
353	CL189AY						

428	BSD33607	Iveco 370.97.S24	Portesi	Regional bus	1991
429	BSD33608	Iveco 370.97.S24	Portesi	Regional bus	1991
430	BSD33604	Iveco 370.97.S24	Portesi	Regional bus	1991
432	BSD39302	Iveco 370.10.S24	Iveco	Regional bus	1991
438	AD 459WC	Iveco 370.97.S24	Portesi	Regional bus	1995
439	AD 458WC	Iveco 370.97.S24	Portesi	Regional bus	1995

440-450
Irisbus Ares 10.6m   Irisbus   Interurban   2007

440	CY857CZ	443	CY860CZ	446	CY863CZ	449	CY866CZ
441	CY858CZ	444	CY861CZ	447	CY864CZ	450	CY867CZ
442	CY859CZ	445	CY862CZ	448	CY865CZ		

451-470
Irisbus Crossway 10.66m   Irisbus   Interurban   2008

451	DE784SB	456	DE785SB	461	DN005MS	466	DN012MS
452	DE951SB	457	DE948SB	462	DN006MS	467	DN641MS
453	DE950SB	458	DN002MS	463	DN007MS	468	DN642MS
454	DH572CS	459	DN003MS	464	DN008MS	469	DN639MS
455	DE949SB	460	DN004MS	465	DN889MS	470	DN640MS

Illustrating the route branding, 142, DN318MS, is one of ten Mercedes-Benz Citaro buses delivered for the service in 2008. As can be seen from the background, the SIA and SAIA operations now share the same depot in Brescia. *Bill Potter*

504	BSE88082	Iveco 370.12.L25	Portesi		Regional bus	1993	
505	BSE97138	Iveco 370.12.L25	Portesi		Regional bus	1993	
506	AF649XH	Iveco 370.12.SE35	Orlandi		Regional bus	1994	
507	AD126VP	Iveco 370.12.SE35	Orlandi		Regional bus	1995	
508	AF784XJ	Iveco 370.12.SE35	Orlandi		Regional bus	1995	
509	AP382NT	Mercedes-Benz O408	Mercedes-Benz		Regional bus	1997	
510	AZ605ND	Mercedes-Benz O408	Mercedes-Benz		Regional bus	1998	
511	AZ736NF	Mercedes-Benz O405 N2	Mercedes-Benz		City bus	1998	
512	BE684ZJ	Iveco EuroRider 391.12.29	Orlandi		Regional bus	1999	
513	BR642FF	Iveco EuroRider 391.12.29	Orlandi		Regional bus	2000	
514	BR112FG	Iveco EuroRider 391.12.29	Orlandi		Regional bus	2000	
644	BSD39303	Iveco 370.12.L25	Iveco		Regional bus	1991	
645	BSD39308	Iveco 370.12.L25	Iveco		Regional bus	1991	

646-658		Iveco 370.12.SE35		Iveco		Regional bus	1995
646	AF 402XJ	650	AF 405XJ	653	AF 395XJ	656	AF 392XJ
647	AF 404XJ	651	AF 396XJ	654	AF 393XJ	657	AF 401XJ
648	AF 417XL	652	AF 406XJ	655	AF 403XJ	658	AF 391XJ
649	AF 394XJ						

659-690		Iveco MyWay 393.12.35		Iveco		Regional bus	2000
659	BM 314 FV	676	BM 417 FV	681	BM 689 FV	686	BM 692 FV
666	BM 685 FV	677	BM 418 FV	682	BM 690 FV	687	BM 555 FF
667	BM 321 FV	678	BM 687 FV	683	BM 420 FV	688	BM 554 FF
668	BM 686 FV	679	BM 419 FV	684	BM 691 FV	689	BM 553 FF
674	BM 416 FV	680	BM 688 FV	685	BM 326 FV	690	BM 552 FF
675	BM 325 FV						

691-698		Iveco MyWay 393.12.35		Irisbus		Interurban	2008
691	DE787SB	693	DE786SB	695	DE720SB	697	DE721SB
692	DE947SB	694	DE719SB	696	DE788SB	698	DE789SB

Newly delivered to SIA when your editor visited was the first of a batch of MAN Lion's City buses for the operations. 713, DV714RG, is seen preparing to depart for Salò in Lake Garda where an outstation is located.
Bill Potter

699-712   Iveco Crossway 12m   Irisbus   Interurban   2008

699	DN056MS	703	DN339MS	707	DN697MS	710	DN696MS
700	DN013MS	704	DN057MS	708	DN801MS	711	DN863MS
701	DN011MS	705	DN010MS	709	DN865MS	712	DN695MS
702	DN014MS	706	DN800MS				

713	DV714RG	MAN Lion's City U 12m	MAN	Regional bus	2009

802-810   Kässbohrer S300 NC   Kässbohrer Setra   City bus   1992-94

802	AP 002NZ	805	AP 005NZ	807	AP 007NZ	809	AP 009NZ
803	AP 003NZ	806	AP 006NZ	808	AP 008NZ	810	AP 010NZ
804	AP 004NZ						

811-831   Mercedes-Benz O405N   Mercedes-Benz   City bus   1998-99

811	AZ 933NE	817	AZ 112NF	822	AZ 935NE	827	AZ 119NF
812	AZ 932NE	818	AZ 114NF	823	AZ 934NE	828	AZ 937NE
813	AZ 108NF	819	AZ 115NF	824	AZ 938NE	829	AZ 107NF
814	AZ 110NF	820	AZ 936NE	825	AZ 117NF	830	BE 794ZJ
815	AZ 109NF	821	AZ 116NF	826	AZ 118NF	831	BE 793ZJ
816	AZ 113NF						

832	CN335HF	Mercedes-Benz O530 NU 12m	Mercedes-Benz Citaro	Regional bus	1997

ASF

ASF Autolinee SRL - 22100 Como - Via Asiago, 16/18

1003	AX556YS	Iveco 315.8.18 Poker	Iveco	ALI	19
1004	BB758KB	Iveco 315.8.18 Orlandi	Iveco	ALI	19
1005	BE009MY	Cacciamali A59	Cacciamali	ALI	19
1006	AX556YS	Iveco 315.8.18 Orlandi	Iveco	ALI	19

1007-1012		Iveco Daily 65C	Cacciamali	ALI	19				
1007	BV211SX	**1009**	BZ619TB	**1011**	CJ091XJ			**1012**	CJ336XJ
1008	BV382SX	**1010**	CJ754XH						

1013	DC049JP	Toyota Optimo BB50L	Caetano Optimo	ALI	19
1014	DC048JP	Toyota Optimo BB50L	Caetano Optimo	ALI	19
1015	DF658ZA	Iveco Daily 65C	Cacciamali	ALI	19
1016	DM119KV	Iveco A50	Tourys	ALI	19
1017	DR546KK	Volkswagen Crafter	Volkswagen	ALI	19
1105	DK018JP	Iveco 370.10.24	Portesi	ALI	20
1106	DK019JP	Iveco 370.10.24	Portesi	ALI	20
1130	AG409TE	Iveco 370E.9.27	Dalla Via	ALI	19

1131-1135		MAN 11.220	De Simon	ALI	19				
1131	AN759EX	**1133**	AX548YS	**1134**	AX549YS			**1135**	AX550YS
1132	AN758EX								

1136	BE011MY	Iveco 370.10.24	Portesi	ALI	20
1137	BE650MY	Mercedes-Benz O303	Mercedes-Benz	ALI	19
1138	BE154MY	Mercedes-Benz O303	Mercedes-Benz	ALI	19
1140	BJ809NR	Cacciamali TCI970	Cacciamali	ALI	19
1141	BJ708NR	Cacciamali TCI970	Cacciamali	ALI	19
1142	BJ593NR	Cacciamali TCI970	Cacciamali	ALI	19

1143-1146		MAN 11.220	De Simon	ALI	19				
1143	BP587PP	**1144**	BP588PP	**1145**	BP841PP			**1145**	BP840PP

1147	BP737PR	Mauri SAS	?	ALI	19
1148	CF472HJ	Cacciamali Engin TCI 800	Cacciamali	ALI	19
1149	CL515JL	Cacciamali Engin TCI 800	Cacciamali	ALI	19

1150-1157		Cacciamali Engin TCI 800	Cacciamali	ALI	19				
1150	CZ353MH	**1152**	DE764YY	**1154**	DF092YZ			**1156**	DF075YZ
1151	CZ363MH	**1153**	DF073YZ	**1155**	DF091YZ			**1157**	DF076YZ

1158	DF129YZ	Cacciamali TCI 972	Cacciamali	ALI	19
1159	DM142KV	Cacciamali Engin TCI 800	Cacciamali	ALI	19
1160	DM204KV	Cacciamali Engin TCI 800	Cacciamali	ALI	19

1161-1166		Cacciamali TCI 972	Cacciamali	ALI	20				
1161	DM348KV	**1163**	DM153KV	**1165**	DK989JP			**1166**	DM079KV
1162	DK988JP	**1164**	DM103KV						

1263	CO947141	Iveco 370.10.24 10.6M	Iveco	Regional	19
1266	CO896829	Sitcar 161.24	Sitcar	Regional	19
1267	CO896828	Iveco 370.10.24 10.6m	Iveco	Regional	19
1268	CO050966	Iveco 370.10.24 10.6m	Iveco	Regional	19
1270	AG319TM	BredaMenariniBus M120/1E	Breda	Regional	19
1272	BP079ZY	Mauri SAS	?	ALI	19
1273	CF221HH	Volvo B12B	Volvo 8700	Coach	19
1274	CR764PP	Irisbus EuroRider 397.10.31	?	Coach	20
1275	CX613DW	De Simon Starbus LN2801	?	Coach	20
1276	CX243ZG	Scania IN3	de Simon	Coach	20
1277	CX242ZG	Scania IN3	de Simon	Coach	20
1278	CZ078MH	Scania IN3	de Simon	Coach	20
1279	CZ079MH	Scania IN3	de Simon	Coach	20
1324	COA04084	Iveco 370.12.25	Iveco	Regional	19
1325	COA45564	Siccar 166.30	?	Regional	19
1331	COA92635	Iveco 370.12.25	Iveco	Regional	19

Further expansion in Italy has seen Arriva take over the management of ASF which is based at Como on the southern edge of Lake Como. Introducing a new model to the group, the Irisbus Europolis, is 1001, DT627FX, seen operating on the narrow town streets shortly after entering service. *Bill Potter*

1332	COB18826	Iveco 370.12.25	Iveco	Regional	19
1333	COB18823	Iveco 370.12.25	Iveco	Regional	19
1337	COD46831	Iveco 370.12.30	Iveco	Regional	19
1338	COD44785	Iveco 370.12.30	Iveco	Regional	19
1340	AE301HL	Iveco 370.12.30	Iveco	Regional	19
1341	AG738TG	BredaMenariniBus M120/1E	Breda	Regional	19
1343	AP467VW	Iveco 391E.12.29	Iveco	Regional	19
1344	AP909VY	Iveco 391E.12.29	Iveco	Regional	19
1346	BX353BR	Irisbus 391E.12.29	Irisbus	Regional	20
1347	BZ748TD	Mercedes-Benz O550 Integro	Mercedes-Benz	Regional	20
1348	DP481CA	Ayats Bravo 1 391E	Ayats	Coach	20

1349-1354   Irisbus 399 EL75   Irisbus   Regional   20

1349	CL436JJ	1351	CL765JJ	1353	CL894JJ	1354	CL435JJ
1350	CL457JJ	1352	CL895JJ				

1355-1359   de Simon IL3   de Simon   Regional

1355	CL476JK	1357	CX240ZG	1358	CX239ZG	1359	CX238ZG
1356	CX241ZG						

1360-1364   Irisbus Karosa C956   Irisbus   Regional

1360	DC275JP	1362	CZ620MH	1363	CZ621MH	1364	CZ622MH
1361	CZ619MH						

1365	DE444YY	Scania OmnuLink	Scania	Regional	
1366	DE787YY	Scania OmnuLink	Scania	Regional	
1367	DE765YY	Scania OmnuLink	Scania	Regional	
1504	CL640JL	Irisbus 380.12.35	-	Regional	
1601	BZ596SZ	Irisbus 380.10.35	-	Regional	
2001	DT627FX	Irisbus Europolis TCC760	Cacciamali	ALS	2009
2002	DS960PK	Irisbus Europolis TCC760	Cacciamali	ALS	2009
2127	CO949209	Inbus U150	Sicca 181 CU		

2132-2136   Irisbus 200E 9.23m   Irisbus   ALS

2132	BK746MB	2134	BK983MB	2135	BK686MB	2136	BK984MB
2133	BK867MB						

Heading away from Como towards the Saginoi shopping centre is BredaMenariniBus M240, 3299, DE423YY. Based at Bologna in Italy, BredaMenarinibus is one of the main Italian bus suppliers and normally uses Deutz engines. *Bill Potter*

2201	DK306JP	Irisbus 491E 10.8m		Irisbus		ALS		
2202	DK307JP	Irisbus 491E 10.8m		Irisbus		ALS		
2203	DK308JP	Irisbus 491E 10.8m		Irisbus		ALS		

2257-2262
Iveco 571.10.20 — Iveco — ALS

2257	CO949206	**2259**	CO949207	**2261**	CO961281	**2262**	CO974967
2258	CO949208	**2260**	CO949205				

2265	COA04085	BredaMenariniBus M201/NS		Breda		Regional		

2266-2275
BredaMenariniBus M220/NS — Breda — Regional

2266	COB29650	**2269**	COD46236	**2272**	AA627ZZ	**2274**	AE138HH
2267	COB29651	**2270**	COD46235	**2273**	AA628ZZ	**2275**	AE139HH
2268	COB29652	**2271**	COD46234				

2277-2292
Irisbus 491E 10.8m — Irisbus — ALS

2277	BJ632NR	**2281**	BJ594NR	**2285**	BJ123NR	**2289**	BJ407NR
2278	BJ034NR	**2282**	BJ633NR	**2286**	BJ032NR	**2290**	BJ595NR
2279	BJ356NR	**2283**	BJ357NR	**2287**	BJ406NH	**2291**	BJ300NR
2280	BJ592NR	**2284**	BJ707NR	**2288**	BJ301NR	**2292**	BJ825NR

2293-2299
Irisbus 491E 10.8m — Irisbus — ALS

2293	DC616JP	**2295**	DC618JP	**2297**	DK310JP	**2299**	DK305JP
2294	DC617JP	**2296**	DK311JP	**2298**	DK309JP		

2301-2304
Irisbus 491E 10.8m — Irisbus — ALS

2301	AP587WA	**2302**	AP594WA	**2303**	AP589WA	**2304**	AP588WA

2305	AX007XR	Mercedes-Benz O405 N	Mercedes-Benz	B--D	19
2306	AP310WA	BredaMenariniBus M221	Menarini	B--D	
2307	BE589MZ	Irisbus 491E 12m	Irisbus	B--D	
2308	BE007MY	Mercedes-Benz O405 N	Mercedes-Benz	B--D	19

2309-2313 · Setra S300 NC · Setra · N D

2309	BE004MY	2311	BE015MY	2312	BE018MY	2313	BE017MY
2310	BE014MY						

2314	BJ124MP	BredaMenariniBus M240	Menarini	B--D	

2315-2337 · Mercedes-Benz Citaro O530 · Mercedes-Benz · N--D

2315	BJ631NR	2321	BJ469NR	2327	BJ527NR	2333	BJ401NR
2316	BJ840NP	2322	BJ528NR	2328	BJ467NR	2334	BJ673NR
2317	BJ247NR	2323	BJ248NR	2329	BJ465NR	2335	BJ400NR
2318	BJ529NR	2324	BJ630NR	2330	BJ705NR	2336	BJ675NR
2319	BJ808NR	2325	BJ399NR	2331	BJ674NR	2337	BJ466NR
2320	BJ468NR	2326	BJ122NR	2332	BJ402NR		

2338	BP450PP	Irisbus 491E 12m	Irisbus	N--D
2339	CR869PN	Irisbus 491E 12m	Irisbus	N--D
2340	CC697FB	Scania OmniCity CN94 UB	Scania	N--D
2341	CL979JK	Solaris Ubino 12m	Solaris	N--D
2342	CX551DW	MAN NL202	MAN	N--D
2346	DC408JP	Solaris Ubino 12m	Solaris	N--D

2343-2351 · BredaMenariniBus M240 11.9m · Breda · Regional

2343	CX336ZG	2345	CX338ZG	2348	DF174YZ	2350	DF173YZ
2344	CX337ZG	2347	DE445YY	2349	DF191YZ	2351	DF190YZ

2371	COB18825	BredaMenariniBus 12m	Siccar 286	Regional
2372	COB18824	BredaMenariniBus 12m	Siccar 286	Regional

2373-2379 · Irisbus 590E 12m · Irisbus · ALS

2373	AG312TM	2375	AG314TM	2377	AG316TM	2379	AG318TM
2374	AG313TM	2376	AG315TM	2378	BC671FV		

2380-2386 · Setra S300 NC · Setra · N D

2380	AN821EW	2382	AN466EW	2384	AN461EW	2386	AN901EW
2381	AN463EW	2383	AN464EW	2385	AN462EW		

2388-2399 · Irisbus 591E 12m · Irisbus · ALS

2388	AP709VW	2391	AP465VW	2394	AP590WA	2397	AP586WA
2389	AP710VW	2392	AP464VW	2395	AP593WA	2398	AP591WA
2390	AP466VW	2393	AX027XR	2396	AP592WA	2399	AP595WA

2401	COB18827	BredaMenariniBus 17.5m	Siccar 386	AB--D	
2402	COB19901	BredaMenariniBus 17.5m	Siccar 386	AB--D	
2403	COB53310	BredaMenariniBus 17.5m	Siccar 386	AB--D	
2404	AG486TT	Irisbus 590E 17.5m	Irisbus	AB--D	
2405	AG487TT	Irisbus 590E 17.5m	Irisbus	AB--D	
2406	BC378FV	MAN NG272 17.9m	MAN	AB--D	
2407	DS959PK	BredaMenariniBus M321 18m	Menarini	AB--D	
2408	CD964PW	BredaMenariniBus M321 18m	Menarini	AB--D	
2409	u	BredaMenariniBus M321 18m	Menarini	AB--D	
2501	DH553VP	Scania OmniLink 13.6m	Scania	N--D	
3001	CR744PP	Iveco A50 7.8m	Iveco	M	
3003	CR506PR	Mercedes-Benz Sprinter 616	Mercedes-Benz	M	
3004	CX354ZG	BredaMenariniBus M231/V 7.8m	Menarini	N--D	
3005	CX339ZG	BredaMenariniBus M231/V 7.8m	Menarini	N--D	
3006	CX340ZG	BredaMenariniBus M231/V 7.8m	Menarini	N--D	
3007	DS609PK	Iveco A50 7.8m	Iveco	M	
3013	M18S4895	Iveco Dailybus 49	Iveco	M	
3101	AX547YS	Cacciamali TCM920	Cacciamali		
3143	COB96830	Inbus U150	Sicca 181 C	B	
3144	COB96831	Inbus U150	Sicca 181 C	B	
3145	AB648AD	MAN 11.190	-	B	
3146	AG586TT	Cacciamali TCM890 9m	Cacciamali		

3147-3153 · Irisbus 200E 9.23m · Irisbus · N--D

3147	BJ017NP	3149	BJ018NP	3151	BJ020NP	3153	BJ526NR
3148	BJ019NP	3150	BJ016NP	3152	BK745MB		

3154	BP159ZZ	Mercedes-Benz O520 Cito	Mercedes-Benz	N20D	19
3155	BP160ZZ	Mercedes-Benz O520 Cito	Mercedes-Benz	N20D	19
3156	CF220HH	Mercedes-Benz O520 Cito	Mercedes-Benz	N20D	19

3201-3208 — BredaMenariniBus M240 — Menarini — N--D

3201	DE447YY	3203	DF147YZ	3205	DF127YZ	3207	DF077YZ
3202	DE424YY	3204	DF128YZ	3206	DF149YZ	3208	DF074YZ

Fleet	Reg	Model	Make	Notes
3210	DM969KV	Cacciamali TCN105 10.5m	Cacciamali	N--D
3211	DP074CA	Cacciamali TCN105 10.5m	Cacciamali	N--D
3212	DP669CA	Cacciamali TCN105 10.5m	Cacciamali	N--D

3276-3290 — Fiat 471 — Fiat — B--D

3276	MI7L6079	3287	MI5F8167	3289	MI5F8166	3290	MI7L6077
3277	MI7L6079	3288	MI5F8165				

Fleet	Reg	Model	Make	Notes
3291	AG400TN	Fiat 490	Fiat	B--D
3292	AP596WA	Cacciamali TCN105 10.5m	Cacciamali	N--D

3293-3299 — BredaMenariniBus M240 — Menarini — N--D

3293	CR932PR	3295	CX341ZG	3297	CX353ZG	3299	DE423YY
3294	CT198VY	3296	CX352ZG	3298	DE446YY		

Fleet	Reg	Model	Make	Notes
3336	M16T4450	Fiat 480	Fiat	B--D
3337	M16V7772	de Simon UL55 12m	de Simon	

3338-3342 — BredaMenariniBus M220 — Menarini — B--D

3338	AG377TH	3340	AG378TH	3341	AG374TH	3342	AG375TH
3339	AG375TH						

Fleet	Reg	Model	Make	Notes
3343	AN465EW	Setra S300 NC	Setra	N D
3344	AP625VZ	Setra S300 NC	Setra	N D

3345-3349 — Iveco 491E 12m — Iveco — N--D

3345	AP584WA	3347	AP583WA	3348	BE613MY	3349	BE612MY
3346	AP585WA						

Fleet	Reg	Model	Make	Notes
3350	BJ035NR	Irisbus 491E 11.9m	Irisbus	N--D
3351	BJ706NR	Irisbus 491E 11.9m	Irisbus	N--D
3352	BJ033NR	Irisbus 491E 11.9m	Irisbus	N--D
3403	M13G0815	Inbus 17.5m	Sicca 383 C	B
3404	M13G0814	Inbus 17.5m	Sicca 383 C	B
3405	M11L7207	Inbus 17.5m	Sicca 383 C	B
3407	M11L7209	Inbus 17.5m	Sicca 383 C	B
3408	AP009VV	Mercedes-Benz O405 GN	Mercedes-Benz	AB--D
3409	AP010VV	Mercedes-Benz O405 GN	Mercedes-Benz	AB--D
3410	AN645EY	Mercedes-Benz O405 GN	Mercedes-Benz	AB--D
3411	DT527FX	BredaMenariniBus M321 18m	Menarini	AB--D

Waiting time at Como bus terminal before a journey to Lecco where the SAL operation is based is Iveco CityClass 2396, AP592WA.
Bill Potter

SAIA

SAIA Trasporti, Via Foro Boario 4/b, 25124 Brescia.

Additional depots are located at Palazzolo sull'Oglio, Orzinuovi, Fiesse, Pralboino and Desenzano del Garda.

1	BSB72612	Menarini M101/1 12m	Menarini	Coach	1990
11	AF 737 XL	Scania	Ikarus	Coach	1996
15	AP 032 NY	Iveco 370E.12.35	Iveco	Coach	1997
17	AN 528 JX	Iveco 380.12.38.	Irisbus-Orlandi	Coach	1997
19	AP 139 NW	Renault Iliade GTX	Renault	Coach	1997
21	AP 209 NZ	Renault Iliade GTX	Renault	Coach	1998
23	AP 289 NZ	Iveco 380.12.38	Irisbus-Orlandi	Coach	1998
25	BR 459 ZT	Renault Iliade GTX	Renault	Coach	2001
29	BS D23781	Iveco 70	Cacciamali	Coach	1990
31	BS B97340	Irisbus 389E.12.43	Iveco	Coach	2003
33	CY103CX	Iveco Daily	Iveco	Coach	1996
35	DB054GE	Mercedes-Benz Tourismo O350	Mercedes-Benz	Coach	2006
37	DE537SB	Iveco Daily	Iveco	Coach	2003
39	DE722SB	Iveco Domino	Iveco	Coach	2007
60	BS 897397	Inbus AID 280.FT	Inbus	Regional bus	1985
62	BS 897399	Inbus AID 280.FT	Inbus	Regional bus	1985
68	AZ 734 ND	Mercedes-Benz O402	Mercedes-Benz	Regional bus	1986
110	BS B31560	Iveco 315.8.17	Iveco	Regional bus	1989
124	BS D23647	Inbus AID 280 FT	Inbus	Regional bus	1990
128	AP 354 NZ	Mercedes-Benz O405N	Mercedes-Benz	Regional bus	1991
130	MI 7T2510	Mercedes-Benz O303	Bianchi	Regional bus	1991
132	AF 503 XX	Iveco 370.12.30.	Orlandi	Regional bus	1991
134	BS D65556	Iveco 370.12.L.25	Iveco	Regional bus	1991
138	AP 163 NW	Inbus AID. 280.FT	Inbus	Regional bus	1991
140	BS D98773	Kässbohrer 215 UL	Kässbohrer Setra	Regional bus	1992
142	BS D99326	Volvo B10M-60	Barbi	Regional bus	1992
144	BS D99327	Volvo B10M-60	Barbi	Regional bus	1992
146	BH 433WC	Iveco GTS	Irisbus-Orlandi Domino	Regional bus	1992
150	AP 425NW	MAN NG272	MAN	Regional bus	1993
152	BC 217VM	Iveco GTS	Irisbus-Orlandi Domino	Regional bus	1993
154	BF 021VJ	MAN NG272	MAN	Regional bus	1993
156	AF 284XK	Volvo B10B	Barbi	Regional bus	1995

158-178   Mercedes-Benz O408   Mercedes-Benz   Regional bus   1996-97

158	AF 957XX	164	AF 706XX	170	AP 033NX	176	AP 036NX
160	AF 958XX	166	AF 438XX	172	AP 034NX	178	AP 037NX
162	AF 959XX	168	AP 032NX	174	AP 035NX		

180	AZ 944NF	Mercedes-Benz O405 NU	Mercedes-Benz	Regional bus	1998
182	AZ 710NF	Mercedes-Benz O405 NU	Mercedes-Benz	Regional bus	1998
184	AZ 927NF	Mercedes-Benz O405 NU	Mercedes-Benz	Regional bus	1998
190	AY 420CV	De Simon UL Scania	Desimon	Regional bus	1998
192	BA 087SM	Iveco 391E.12.35/M	Padane	Regional bus	1999
194	BE 218DN	Mauri 18EP30-1	Mauri	Regional bus	1999
196	BN 300SX	Mercedes-Benz Integro O550	Mercedes-Benz	Regional bus	2000
198	BN 518SX	Mercedes-Benz Integro O550	Mercedes-Benz	Regional bus	2000
200	BN 695SX	Mercedes-Benz Integro O550	Mercedes-Benz	Regional bus	2000
202	BN 694SX	Ayats Bravo I	Ayats	Regional bus	2000
204	BR 094FE	Ayats Bravo I	Ayats	Regional bus	2000
206	BR 096FE	Mercedes-Benz Integro O550	Mercedes-Benz	Regional bus	2000
208	BR 095FE	Mercedes-Benz Integro O550	Mercedes-Benz	Regional bus	2000

210-236   Irisbus 393.12.35 My Way   Irisbus-Orlandi   Regional bus   2000

210	BR 928FE	218	BM 324FV	226	BM 320FV	232	BM 315FV
212	BR 964FE	220	BM 316FV	228	BM 323FV	234	BM 415FV
214	BR 772FE	222	BM 317FV	230	BM 313FV	236	BM 318FV
216	BM 322FV	224	BM 319FV				

238	BP 180BG	MAN NL263 F	Autodromo	Regional bus	2000
240	BP 177BG	MAN NL263 F	Autodromo	Regional bus	2000

SAIA's operations are entirely rural heading into the countryside around Brescia. Irisbus was formed by the amalgamation of Renault and Iveco's bus interests. One of the last with a Renault badge is 268, BV874DZ. This is an Agora Moovy model which was built at the former Renault facility between 1999 and 2002, since when the Irisbus names was used. *Bill Potter*

242-254

							Mercedes-Benz Citaro O530 NU	Mercedes-Benz		Regional bus		2001

242	BT 351GC	246	BT 348GC	250	BT 537GC	254	BT 204GD
244	BT 350GC	248	BT 349GC	252	BT 538GC		

256	BT203GD	Ayats Bravo I	Ayats	Regional bus	2000
258	BT941GD	Mercedes-Benz Citaro O530 NU	Mercedes-Benz	Regional bus	2001
260	BT940GD	Mercedes-Benz Citaro O530 NU	Mercedes-Benz	Regional bus	2001
262	BV410DV	Mercedes-Benz Citaro O530 NU	Mercedes-Benz	Regional bus	2001
266	BV873DZ	Renault Agora Moovy	Renault	Regional bus	2001
268	BV874DZ	Renault Agora Moovy	Renault	Regional bus	2001
270	BZ174XN	Kässbohrer S215 UL	Kässbohrer Setra	Regional bus	1991
272	BZ169XN	MAN SU 313	MAN	Regional bus	2002
274	BZ170XN	MAN SU 313	MAN	Regional bus	2002
276	CF122JN	Irisbus 399E My Way	Irisbus Orlandi	Regional bus	2003
278	CJ391BX	Iveco 380.12.35.	Orlandi	Regional bus	1995
280	CJ309BY	MAN SG292 18m	MAN	Regional bus	1993

282-300

							Mercedes-Benz Conecto O345	Mercedes-Benz		Regional bus		2001

282	CL542AX	288	CL544AX	294	CL052AY	298	CM023EW
284	CL543AX	290	CL220AX	296	CL5624Y	300	CL561AY
286	CL541AX	292	CL560AY				

304	CP892TT	MAN NU313	MAN	Regional bus	1996
306	CP893TT	MAN NU313	MAN	Regional bus	1996
308	CP894TT	MAN NU313	MAN	Regional bus	1996
310	CP895TT	MAN NU313	MAN	Regional bus	1996
312	CP891TT	MAN NU313	MAN	Regional bus	1996
322	CR330HK	Mercedes-Benz Integro O550 UL	Mercedes-Benz	Regional bus	2004
324	CT379CZ	Setra S215 UL	Setra	Regional bus	1997
326	CT896XV	Mercedes-Benz O407	Mercedes-Benz	Regional bus	1994
328	CT896XV	Mercedes-Benz O407	Mercedes-Benz	Regional bus	1994

330-342

							Irisbus MyWay 399E.12.35	Irisbus		Regional bus		2005

330	CV667XV	334	CV671XV	338	CV669XV	342	CV673XV
332	CV672XV	336	CV668XV	340	CV670XV		

Currently out of service following the arrival of new vehicles, Mercedes-Benz O303 number 130, MI7T2510, is one of several older vehicles now awaiting re-allocation. Bodywork is by Bianchi. *Bill Potter*

344	CY157CZ	Mercedes-Benz Integro O550 UL	Mercedes-Benz	Regional bus	2006
346	BR335FF	Mercedes-Benz Integro O550 UL	Mercedes-Benz	Regional bus	2003
348	CJ565BY	Mercedes-Benz Integro O550 UL	Mercedes-Benz	Regional bus	2003
350	DB510GE	Setra S215 UL	Setra	Regional bus	1998
354	DB865WS	Mercedes-Benz Citaro O530 NU	Mercedes-Benz	Regional bus	2006
356	DE260SB	Iveco 680.12.30	Iveco	Regional bus	1989
358	DE361SB	MAN Lion's City U (A20)	MAN		2007
360	DE362SB	MAN Lion's City U (A20)	MAN		2007

362-392

		Irisbus Arway	Irisbus	Regional bus	2005-08
362	DE360SB	370	DE724SB	378	DH331CS	386	DH335CS
364	DE792SB	372	DE791SB	380	DH332CS	388	DH244CS
366	DH851CS	374	DE725SB	382	DH333CS	390	DH245CS
368	DE723SB	376	DH330CS	384	DH334CS	392	DF336CS

394	LD687NA	MAN Lion's City U	MAN	N--D	2008

396-406

		Irisbus Crossway	Irisbus	Regional bus	2008
396	DN058MS	400	DN067MS	404	DN062MS	406	DN059MS
398	DN060MS	402	DN061MS				

408	DN364MS	Mercedes-Benz Citaro O530 GNU	Mercedes-Benz	AN--D	2008
410	DN363MS	Mercedes-Benz Citaro O530 GNU	Mercedes-Benz	AN--D	2008
412	DN864MS	Cacciamali 840 TCI 8.48m	Cacciamali	N--D	2009

414-442

		Irisbus Crossway	Irisbus	Regional bus	2009
414	DN373MS	422	DN366MS	430	DN377MS	438	DN375MS
416	DN367MS	424	DN365MS	432	DN370MS	440	DN378MS
418	DN441MS	426	DN374MS	434	DN372MS	442	DN371MS
420	DN440MS	428	DN376MS	436	DN369MS		

641	BM868RA	Cacciamali Tema 331	Cacciamali	Schoolbus	19

SAL

SAL srl, Via della Pergola 2, 23900 Lecco

3002	AN848EV	MAN 11.190	Macchi	City bus	1996
3005	AN942EW	MAN 11.190	Macchi	City bus	1997
3013	COD59447	MAN 11.190	Macchi	City bus	1994
3015	AN438EW	Kässbohrer S300 NC	Kässbohrer Setra	City bus	1992
3016	AN439EW	Setra S 300 NC	Setra	City bus	1997
3017	AN440EW	Kässbohrer S300 NC	Kässbohrer Setra	City bus	1993
3010	AN441EW	Kässbohrer S300 NC	Kässbohrer Setra	City bus	1993
3026	AG533TR	BredaMenarini M 3001.12L	Bredamenarinbus	City bus	1996
3027	AG534TR	BredaMenarini M 3001.12L	Bredamenarinbus	City bus	1996
3029	AN731EW	BredaMenarini M 221	Bredamenarinbus	City bus	1997
3032	C0B03782	Kässbohrer S210 H	Kässbohrer Setra	City bus	1990
3034	AN988EW	Kässbohrer S300 NC	Kässbohrer Setra	Regional bus	1991
3035	COB83012	BredaMenarini M120/1	Bredamenarinbus	City bus	1992
3036	COA95307	Iveco 370.97.24	Portesi	Regional bus	1990
3048	COA95306	Mercedes-Benz O303/15R	Bianchi	Regional bus	1990
3057	COD59602	Setra S212 H	Setra	Regional bus	1994
3058	COD61582	Iveco 370.12.L25	Iveco	Regional bus	1990
3059	COD61581	Iveco 370.12.L25	Iveco	Regional bus	1990
3063	AN080EZ	Mercedes-Benz O303/15R	Bianchi	Regional bus	1987
3064	AP574VZ	Mercedes-Benz O303/15R	Bianchi	Regional bus	1986
3065	AN939EY	Mercedes-Benz O303/15R	Bianchi	Coach	1986
3077	MI0U7810	Iveco 580.12.24	Iveco	Regional bus	1991
3098	AN943EW	Mercedes-Benz O303/10R	Menarini	City bus	1987
3106	BB199KC	Mercedes-Benz O303/14R	Mercedes	Regional bus	1999
3109	BE328MY	Iveco Daily F45.12	Iveco	City bus	1999
3114	BF634TD	Mercedes-Benz O408	Mercedes	Regional bus	1997

Lecco is located at the southern tip on the eastern leg of Lake Como, and is where the SAL operation is based. Many of the vehicles serve the surrounding villages. The official fleet numbers are 3000 higher than the number displayed, so 3077, MI0U7810, the only Iveco 580 in the fleet, displays 77 as it arrives at the rail station. *Bill Potter*

Recent arrivals with SAL are eight of the Irisbus Crossway model. These feature high-back seating as shown by 190, DM091JM. *Bill Potter*

3115	BF635TD	Mercedes-Benz O408	Mercedes	Regional bus	1997
3116	BF636TD	Mercedes-Benz O408	Mercedes	Regional bus	1997
3117	BF637TD	Mercedes-Benz O408	Mercedes	Regional bus	1997
3118	BF638TD	Mercedes-Benz O408	Mercedes	Regional bus	1997
3119	BF830TD	Iveco Daily F45.12	Iveco	Regional bus	1998
3120	BF829TD	Iveco Daily F45.12	Iveco	Regional bus	2000
3121	BF828TD	Iveco Daily F45.12	Iveco	School bus	2000
3122	BY901MN	Iveco 393.12.35 My Way	Iveco	School bus	2000
3123	BK965MA	Mercedes-Benz O404	Mercedes	School bus	1993
3124	BJ017NR	Iveco 393.12.35 My Way	Iveco	Regional bus	2000
3125	BJ015NR	Iveco 393.12.35 My Way	Iveco	Coach	2000
3126	BJ018NR	Iveco 393.12.35 My Way	Iveco	Regional bus	2000
3127	BJ016NR	Iveco 393.12.35 My Way	Iveco	Regional bus	2000
3128	BJ014NR	Iveco 393.12.35 My Way	Iveco	Regional bus	2000
3129	BJ012NR	Iveco 393.12.35 My Way	Iveco	Regional bus	2000
3130	BJ013NR	Iveco 393.12.35 My Way	Iveco	Regional bus	2000
3131	BP600PP	DeSimon Starline 55.12	Desimon	Regional bus	2001
3132	BP722PP	DeSimon Starline 55.12	Desimon	Regional bus	2001
3133	BP262PR	Iveco EuroPolis 9.15	Iveco	City bus	2001
3134	BP261PR	Iveco EuroPolis 9.15	Iveco	Regional bus	2001
3135	BP260PR	Iveco EuroPolis 9.15	Iveco	City bus	2001
3136	BP259PR	Iveco EuroPolis 9.15	Iveco	City bus	2001
3137	BP694ZY	Iveco EuroPolis 10.50	Iveco	City bus	2001
3138	BP539ZY	Iveco EuroPolis 10.50	Iveco	Regional bus	2001
3143	BP538ZY	Mercedes-Benz Citaro O530 NU	Mercedes	City bus	2001
3144	BV031SY	Inbus 181	De Simon	City bus	1992
3145	BV004 SZ	Inbus 181	De Simon	Regional bus	1987
3146	BV937SX	Iveco Daily F45.12	Iveco	City bus	2001
3147	BZ 262 SZ	Mercedes-Benz O350 TURISMO	Mercedes	Coach	2002
3148	CD147SA	DeSimon Starline 55.12	Desimon	School bus	2002
3149	BZ646TB	Cacciamali TCI 970 Sigma 2	Cacciamali	Coach	2002
3150	CL586JJ	Mercedes-Benz O340	Mercedes	Regional bus	2001
3151	CL587JJ	Mercedes-Benz O340	Mercedes	Regional bus	2001
3152	CL478JL	MAN 272 UL 12m	MAN	Regional bus	1993

3153	CL479JL	MAN 272 UL 12m	MAN	Regional bus	1993
3154	CL501JL	MAN 272 UL 12m	MAN	Regional bus	1993
3155	CL502JL	MAN 272 UL 12m	MAN	Regional bus	1995
3156	CL477JL	MAN 313 UL 12m	MAN	Regional bus	1997
3157	CL503JL	MAN 313 UL 10.4m	MAN	Regional bus	1997
3158	CR428PP	Mercedes-Benz O404	Noleggio	Coach	1992
3159	CR005PR	MAN 313 UL 10.4m	MAN	Regional bus	1999
3160	CR670PR	MAN 272 UL 12m	MAN	Regional bus	1994
3161	CR669PR	MAN 272 UL 12m	MAN	Regional bus	1994
3162	CR668PR	MAN 292 UL 12m	MAN	Regional bus	1993
3163	CT017VY	Mercedes-Benz Citaro O530	Mercedes	Regional bus	2001
3164	CR189PR	Mercedes-Benz Citaro O530	Mercedes	Regional bus	2001
3165	CR190PR	Mercedes-Benz Citaro O530	Mercedes	City bus	2001
3166	CR191PR	Mercedes-Benz Citaro O530	Mercedes	City bus	2001
3167	CR804PN	Irisbus MyWay 393.12.35	Irisbus	Regional bus	2005
3168	CR803PN	Irisbus MyWay 393.12.35	Irisbus	Regional bus	2005
3169	CR856PN	Irisbus MyWay 393.12.35	Irisbus	Regional bus	2005
3170	CR855PN	Irisbus MyWay 393.12.35	Irisbus	Regional bus	2005
3171	CX050YW	Toyota Coaster BB50L	Caetano Optimo V		2005
3172	CX294YW	Iveco TurboDaily 59	Cacciamali		1999
3173	CX884ZG	Mercedes-Benz Integro O550 UL	Mercedes-Benz	Coach	2006
3174	DC413JP	Mercedes-Benz O405G	Mercedes-Benz	City bus	1995
3175	DE651YY	Irisbus TurboDaily 59	Irisbus		2006
3176	DE954YY	Irisbus TurboDaily 59	Irisbus		2007
3177	DF616YZ	Iveco TurboDaily 59	Iveco		1999
3179	DF317ZA	Irisbus Crossway 491.12.29	Irisbus		2007
3180	DF315ZA	Irisbus Crossway 491.12.29	Irisbus		2007
3181	DF014YZ	MAN Lion's City U (A20)	MAN		2007
3182	DF353YZ	MAN Lion's City U (A20)	MAN		2007
3183	DF011YZ	MAN Lion's City U (A20)	MAN		2007
3184	DF013YZ	MAN Lion's City U (A20)	MAN		2007
3185	DF012YZ	MAN Lion's City U (A20)	MAN		2007
3186	DF157YZ	MAN Lion's City U (A20)	MAN		2007
3187	DF158YZ	MAN Lion's City U (A20)	MAN		2007
3188	AH841CY	Cacciamali Tema 207	Cacciamali	Schoolbus	19
3189	BZ283SY	Bredamenarini M321	Bredamenarinbus	Suburban	19
3190	DM091JM	Irisbus Crossway 491.10.29	Irisbus	Interurban	2008
3191	DM092JM	Irisbus Crossway 491.10.29	Irisbus	Interurban	2008
3192	DM093JM	Irisbus Crossway 491.10.29	Irisbus	Interurban	2008
3193	DM094JM	Irisbus Crossway 491.10.29	Irisbus	Interurban	2008
3194	DM306JM	Iveco TurboDaily 59	Iveco	Interurban	2008
3203	DM305JM	MAN Lion's City U (A20)	MAN	Interurban	2008
3204	DR843GP	Irisbus Crossway 491.12.29	Irisbus	Interurban	2009
3205	DR844GP	Irisbus Crossway 491.12.29	Irisbus	Interurban	2009
3206	DN859NX	Cacciamali TCI 840	Cacciamali	Interurban	2009
3207	DN861NX	Cacciamali TCI 840	Cacciamali	Interurban	2009
3208	DN860NX	Cacciamali TCI 840	Cacciamali	Interurban	2009
3209	DR930GP	Iveco A45.10	Iveco	Schoolbus	2009

SAL received three Cacciamali TCI 840 earlier in 2009. Representing the model is 3206, DN859NX.
Bill Potter

KM

KM SpA, Via Postumia 102, 26100 Cremona

1003	CR389359	Iveco 280 RA7		Cacciamali		School bus		1988
1004	BG451PH	Iveco 45.10		Iveco		School bus		1995
1005	BH060WY	Iveco CC80E18M/86		Cacciamali		School bus		2000
1006	BV937GP	Iveco Scuolabus Turbo Daily		Cacciamali		School bus		1989
1007	DD102YG	Iveco Tema 207		Cacciamali		School bus		1996
1008	BV088ZB	Iveco 100E		Cacciamali		Schoolbus		20
1009	BV073ZB	Iveco 100E		Cacciamali		Schoolbus		20

1030-1036

		Irisbus Cityclass 491.12.29		Irisbus		Citybus		2007-08
1030	DD354YG	1032	DD355YG	1034	DF237PJ		1036	DM301JY
1031	DD352YG	1033	DD353YG	1035	DM300JY			

1037-1040

		Irisbus 200E 10.48m		Irisbus		Urban		2009
1037	DN634JT	1038	DN716JT	1039	DN316JT		1040	DS028EJ

1062	CR384195	Inbus U210/FT	Breda	City bus	1987
1064	CR390912	Menarini M 201/2NU	Menarini	City bus	1988
1066	CR410068	Inbus U210/FTN	Breda	City bus	1989
1067	CR410069	Inbus U210/FTN	Breda	City bus	1989
1068	CR410070	Inbus U210/FTN	Breda	City bus	1989
1070	CR413848	Menarini M 201/2NU	Menarini	City bus	1989
1071	CR413849	Menarini M 201/2NU	Menarini	City bus	1989
1072	CR454939	Menarini M 220/NU	Menarini	City bus	1991
1073	CR454940	Menarini M 220/NU	Menarini	City bus	1991
1074	AL597GD	CAM Bussotto NL202 FU	Autiromo	City bus	1996
1075	AP257XK	Breda Menarini M 230/1E2	Bredamenarinbus	City bus	1998
1076	AP258XK	Breda Menarini M 230/1E2	Bredamenarinbus	City bus	1998

KM was established in January 2001 to merge SAIA bus Cremona suburban and urban services with AEM, part of the Sat Group, with SAB of Bergamo as the parent. In 2002 the group was acquired by Arriva who were tasked with replacing the trolleybus system. Similar to SAL, the official fleet number is 1000 higher than that displayed with 1162, BS695LE, an Iveco 491 Cityclass, illustrated, *Ken McKenzie*

An interesting model from 2001 is the Kronos 10KV23-U, of which just five are operated. Illustrating the type is **1081, BV826GK.** *Ken MacKenzie*

1078-1082   Kronos 10KV23-U   Mauri   City bus   2001

1078	BV620GK	**1080**	BV621GK	**1081**	BV826GK	**1082**	BV827GK
1079	BV680GK						

1083-1086   Mercedes-Benz Cito O520   Mercedes-Benz   City bus   2002

1083	BV916GS	**1084**	BV014GT	**1085**	BV917GS	**1086**	BV936GP

1087	CL922MC	Irisbus Cityclass 491.12.29	Irisbus	Citybus	2004
1088	CR503TW	Irisbus Cityclass 491.12.29	Irisbus	Citybus	2005
1091	CV595BG	Iveco TurboCity 491.10.24	Irisbus	Citybus	1992
1093	DC575NY	Mercedes-Benz Sprinter 416	Mercedes-Benz	Regional bus	2006
1094	DF336PJ	Breda Menarini M 230/1E2	Bredamenarinbus	City bus	1998
1095	DF335PJ	Breda Menarini M 230/1E2	Bredamenarinbus	City bus	1998

1096-1099   Irisbus 200E 10.48m   Irisbus   Urban   2008

1096	DM445JY	**1097**	DM212JY	**1098**	DM223JY	**1099**	DM315JY

1103	BSD05103	Iveco 370.12.25L	Iveco	Regional bus	1990
1105	AF778XJ	Cam Busotto 2LS-SR	Autiromo	City bus	1995
1108	AW765EJ	Breda Menarini M 230/01 MS	Bredamenarinbus	City bus	1998
1109	BE301DN	Mercedes-Benz O405NU	Mercedes-Benz	Regional bus	1999
1110	AZ465NF	Mercedes-Benz O405NU	Mercedes-Benz	Regional bus	1998
1111	BM984FW	Mercedes-Benz O405NU	Mercedes-Benz	Regional bus	2000
1113	AF164XD	Iveco 370.12.35	Dalla Via	Regional bus	1995
1114	AZ469NF	Mercedes-Benz O405NU	Mercedes-Benz	Regional bus	1998
1128	BSB69662	Iveco 370.12.25L	Iveco	Regional bus	1990
1129	BSD05102	Iveco 370.12.25L	Iveco	Regional bus	1990
1130	BH343WC	Iveco EuroRider 391E.12.29	Iveco	Regional bus	1998
1131	BH342WC	Iveco EuroRider 391E.12.29	Iveco	Regional bus	1998
1132	AZ 468NF	Mercedes-Benz O405 NU	Iveco	Regional bus	1998

1137-1142   Ayats 2 Piani   Ayats   Regional bus   2001

1137	BM281DX	**1139**	BM 283DX	**1141**	BR 776FE	**1142**	BR 905FE
1138	BM282DX	**1140**	BR 775FE				

1143	AF 163 XD	Iveco 370.12.35	Dalla Via	Regional bus	1995	
1144	BM 280DX	Iveco 393E.12.35 My Way	Iveco	Regional bus	2001	
1145	BM 285DX	Iveco 393E.12.35 My Way	Iveco	Regional bus	2001	
1146	BM 290DX	Iveco 393E.12.35 My Way	Iveco	Regional bus	2001	
1147	AF 747XF	Iveco 370.12.35	Dalla Via	Regional bus	1995	

1148-1155 Iveco 393E.12.35 My Way — Iveco — Regional bus — 2001

1148	BM 291DX	1150	BM 284DX	1152	BM 288DX	1154 BM 292DX
1149	BM 286DX	1151	BM 287DX	1153	BM 289DX	1155 CD523YR

1156	AD940JF	Cacciamali Tema 100	Cacciamali	Interurban	19
1157	AZ 470NF	Mercedes-Benz O405NU	Mercedes-Benz	Regional bus	1998
1158	DF 183PJ	Irisbus Crossway	Irisbus	City bus	2007
1159	AZ 467NF	Mercedes-Benz O405NU	Mercedes-Benz	Regional bus	1998
1160	BS 693LE	Iveco 491.12.27 - Cityclass	Iveco	City bus	2001
1161	BS 694LE	Iveco 491.12.27 - Cityclass	Iveco	City bus	2001
1162	BS 695LE	Iveco 491.12.27 - Cityclass	Iveco	City bus	2001

1163-1166 Irisbus MyWay 399E — Irisbus — Regional bus — 2004

1163	CR 563TW	1164	CL 684ME	1165	CL 685ME	1166 CR 504TW

1167	DA 573RA	Mercedes-Benz Integro O550	Mercedes-Benz	C40F	2006
1168	DA 574RA	Mercedes-Benz Integro O550	Mercedes-Benz	C40F	2006
1169	DC 004WA	Irisbus Daily A50	Irisbus	Regional bus	2006
1170	DC 005WA	Irisbus Daily A50	Irisbus	Regional bus	2006

1171-1184 Irisbus Arway — Irisbus — Regional bus — 2007

1171	DF 025PJ	1175	DD 456YG	1178	DF 028PJ	1182 DF 030PJ
1172	DD 458YG	1176	DF 032PJ	1179	DF 029PJ	1183 DF 027PJ
1173	DD 459YG	1177	DF 026PJ	1180	DF 024PJ	1184 DF 031PJ
1174	DD 457YG					

1185	DM344JY	Irisbus Crossway	Irisbus	Interurban	2008
1186	DM343JY	Irisbus Crossway	Irisbus	Interurban	2008
1189	BS D09800	Iveco 370.12.30S	Portesi	Regional bus	1990
1190	DH499TY	Toyota Coaster BB50R	Caetano Optimo IV	Interurban	2008
1191	DM302JY	Toyota Coaster BB50R	Caetano Optimo IV	Interurban	2008
1195	AZ 466NF	Mercedes-Benz O405NU	Mercedes-Benz	Regional bus	1998
2405	AP 785XF	Iveco 370E.12.35 - Domino	Iveco-Orlandi	Coach	1998
2407	BG 322PH	Mercedes-Benz O404	Dalla Via Palladio	Coach	2000
2408	BG 323PH	Mercedes-Benz O404	Dalla Via Palladio	Coach	2000
2410	DF130PJ	Mercedes-Benz Tourismo	Mercedes-Benz	Coach	2003

RTL001	BZ 417CB	Iveco A50 C 15	Iveco	Regional bus	2002
RTL002	BZ 980CB	Mercedes-Benz O303/10R	Bianchi	Regional bus	1981

SAF

Società Autoservizi FVG SpA, Via Baldasseria Bassa 75, 33100 Udine, Italy

3	-	Kässbohrer S140 ES	Kässbohrer Setra	Vintage	C	
47	UD517390	Iveco 370.12.30 12m	Iveco	Regional	B55D	1986
52	UD662611	Iveco 370.12.25 12m	Iveco	Regional	B55D	1991
53	UD684276	Iveco 370.12.25 12m	Iveco	Regional	B55D	1991
54	UD684277	Iveco 370.12.25 12m	Iveco	Regional	B55D	1991
55	UD684278	Iveco 370.12.25 12m	Iveco	Regional	B55D	1991
56	UD744862	Iveco 370.12.30 12m	Iveco	Regional	B55D	1993
68	AG170CR	Iveco 380.12.35 12m	Iveco	Regional	B55D	1996
101	AM920SR	Scania L94IB	Irizar InterCentury 12m	Regional	B55D	1997
104	BD181LG	Scania L94IB	Irizar InterCentury 12m	Regional	B55D	1999
109	BD439LG	Scania L94IB	Irizar InterCentury 12m	Regional	B55D	1999
110	BV193MF	Mercedes-Benz O350 12m	Mercedes-Benz	Coach	C51F	2001
112	BV651MF	Mercedes-Benz O350 12m	Mercedes-Benz	Coach	C51F	2001
118	BV238MF	Mercedes-Benz O350 12m	Mercedes-Benz	Coach	C51F	2001
119	BV194MF	Mercedes-Benz Travego O580 12m		Coach	C51F	2001
120	BV195MF	Mercedes-Benz Travego O580 12m		Coach	C51F	2001
123	BV196MF	Mercedes-Benz Travego O580 12m		Coach	B51D	2001
129	BW000WE	Neoplan N4426/3	Piani 12m	Regional	B86D	2001

134	AM880SS	De Simon IL3 260 Scania	12m		Regional	53+28+1	1997
135	AM870SS	De Simon IL3 260 Scania	12m		Regional	B53D	1997
136	AM879SS	De Simon IL3 260 Scania	12m		Regional	B53D	1997
137	BV289MG	Neoplan N4426/3	Piani 12m		Regional	B86D	2001
139	BV423MG	Neoplan N4426/3	Piani 12m		Regional	B86D	2001
144	BV497MG	Neoplan N4426/3	Piani 12m		Regional	B86D	2001
145	BV740MG	Neoplan N4426/3	Piani 12m		Regional	B86D	2001
146	GO232620	Volvo B10B	12m		Regional	B55D	1994
171	PN290200	Starbus LL30 12m	Regional bus		Regional	B55D	1988
174	AG150CR	Iveco 380.12.35	Iveco		Regional	B55D	1996
175	AG160CR	Iveco 380.12.35	Iveco		Regional	B55D	1996
176	AG180CR	Iveco 380.12.35	Iveco		Regional	B55D	1996
177	AG510CS	Iveco 370E.12.35	Iveco		Regional	B55D	1996
178	AG520CS	Iveco 370E.12.35	Iveco		Regional	B53D	1996
188	UD023360	Fiat 370.12.25	Fiat		Regional	B55D	1989
189	UD663432	Iveco 370.12.30	Iveco		Regional	B55D	1991
193	DD878HP	Setra S431 DT	Setra		Regional	C(83)D	2007
194	UD684510	Fiat 315.8.17	Fiat		Regional	BC28F	1991
214	UD729328	Mercedes-Benz O303/9R	8.69		Regional	BC37F	1987
215	UD567020	Mercedes-Benz O303/15R	12m		Regional	B53D	1988
220	UD574050	Mercedes-Benz O303/15R	12m		School bus	B53D	1988
221	UD590656	Mercedes-Benz O303/15R	12m		School bus	B53D	1988

230-255		Mercedes-Benz O303/15R	12m		Regional	B53D	1988-90

230	UD620350	**242**	UD607008	**246**	UD639700	**252**	UD641030
235	UD620370	**243**	UD607005	**247**	UD639734	**253**	UD641740
237	UD616564	**244**	UD607004	**248**	UD640292	**254**	UD641040
240	UD607006	**245**	UD655850	**250**	UD656413	**255**	UD642450
241	UD607007						

258	UD682829	Mercedes-Benz O303/10R	9.23m		Regional	B39D	1991
259	UD683358	Mercedes-Benz O303/13R	10.55m		Regional	B47D	1991
262	UD709540	Mercedes-Benz O303/15R	12m		Regional	B55D	1992
263	UD711714	Mercedes-Benz O408	12m		Regional	B53D	1992
265	UD732805	Mercedes-Benz O408	12m		Regional	B53	1993
267	AA307WH	Mercedes-Benz O340	12m		Regional	C53D	1994
275	AE458RW	Mercedes-Benz O408	12m		Regional	B53D	1994
276	AE459RW	Mercedes-Benz O408	12m		Regional	B53D	1994

281-287		Mercedes-Benz O350	12m		Coach	C53F	1995-96

281	AG351CC	**283**	AG353CC	**286**	AK847RC
282	CR122TD	**284**	AG354CC	**287**	AG240CM

290	CF317HL	Mercedes-Benz O350 RHD	12m		Coach	B51D	2003

291-305		Mercedes-Benz O408	12m		Regional	B53D	1996

291	AG347CC	**295**	AG355CC	**299**	AG359CC	**303**	AG363CC
292	AG348CC	**296**	AG356CC	**300**	AG360CC	**304**	AG364CC
293	AG349CC	**297**	AG357CC	**301**	AG361CC	**305**	AG365CC
294	AG350CC	**298**	AG358CC	**302**	AG362CC		

306	BB861RY	Mercedes-Benz DF 412 40	6.94m		Regional	B18F	1999
307	BB953RY	Mercedes-Benz DF 412 40	6.94m		Regional	B18F	1999
308	BV362MF	Mercedes-Benz DF 416 40	7.m		Regional	B18F	2001
309	BE837FE	Mercedes-Benz DF 412 40	6.94m		Regional	B18F	1999
310	BE838FE	Mercedes-Benz DF 412 40	6.94m		Regional	B18F	1999
311	BF907DX	Mercedes-Benz DF 412 40	6.94m		Regional	B18F	1999
312	AG190CY	Iveco 380.12.35	12m		Regional	B55D	1996
313	AG910CZ	Iveco 370E.12.35	12m		Regional	B55D	1996
314	AG900CZ	Iveco 370E.12.35	12m		Regional	B55D	1996
316	AT960HF	Iveco Daily	6.86m		Regional	B19F	1997
319	AZ300WB	Neoplan N 4026/3	12m		Regional	C88F	1998
320	AZ488VF	Iveco 315.8.18	7.58m		Regional	B30F	1998
321	AZ489VF	Iveco 391E.12.29	12m		Regional	B53D	1998
322	AZ490VF	Iveco 391E.12.29	12m		Regional	B53D	1998
324	AZ088WB	Iveco 391E.12.29	12m		Regional	B53D	1998
325	AZ089WB	Iveco 391E.12.29	12m		Regional	B53D	1998
326	AZ099WB	Iveco 391E.12.29	12m		Regional	B53D	1998
327	AZ100WB	Iveco 380.12.35	12m		Regional	B55D	1998
328	BD310FZ	Iveco 380.12.35	12m		Regional	B55D	1999
329	BF242DX	Neoplan N 4026/3	12m		Regional	C88F	1999
330	BH578XR	Neoplan N 4026/3	12m		Regional	C88F	2000

331-342   Volvo B10B 12m   -   Regional   B53D   2000

331	BH855XR	334	BH114XS	337	BH117XS	340	BH120XS
332	BV048MG	335	BH115XS	338	BH118XS	341	BH121XS
333	BH857XR	336	BH116XS	339	BH119XS	342	BH122XS

359	AG490CL	Ncoplan N4026/3	12m	Regional	C88F	1996
360	AG500CL	Neoplan N4026/3	12m	Regional	C88F	1996

361-365   Neoplan N316 SHD   12m   Coach   C53F   1998

361	AT191JA	363	BE331FE	364	AT390JA	365	AT590JA
362	AT192JA						

369-373   Neoplan N4026/3   12m   Regional   C88F   1998

369	AZ389VF	371	AZ840VF	372	AZ047VG	373	AZ411VG
370	AZ046VG						

374	AZ899VF	Neoplan N122/3	12m	Coach	C75F	1998
375	BH568XR	Neoplan N4026/3	12m	Regional	C88F	2000
376	BH569XR	Neoplan N4026/3	12m	Regional	C88F	2000
377	BH674XT	Volvo B10B 10.8m	Volvo	Regional	B45D	2000
378	BH673XT	Volvo B10B 10.8m	Volvo	Regional	B45D	2000

379-400   Volvo B10B 12m   Volvo   Regional   B53D   2000

379	BH229XV	385	BH235XV	391	BJ051RB	396	BJ056RB
380	BH230XV	386	BH236XV	392	BJ052RB	397	BJ057RB
381	BH231XV	387	BH237XV	393	BJ053RB	398	BJ058RB
382	BH232XV	388	BH238XV	394	BJ054RB	399	BJ059RB
383	BH233XV	389	BJ049RB	395	BJ055RB	400	BJ060RB
384	BH234XV	390	BJ050RB				

446-449   De Simon Intercity IL3   12m   Regional   B53D   2001

446	BS311RN	447	BS312RN	448	BS318RN	449	BS319RN

450	CF318HL	Mercedes-Benz O350	12m	Coach	B51D	2003
461	AZ090WB	Irizar Century 12.37A	12m	Coach	B38D	1998

462-476   De Simon Intercity IL3   12m   Regional   B53D   1999

462	AZ807WB	466	AZ046WC	470	AZ812WB	474	AZ816WB
463	AZ808WB	467	AZ047WC	471	AZ813WB	475	AZ858WB
464	AZ809WB	468	AZ048WC	472	AZ814WB	476	AZ859WB
465	AZ810WB	469	AZ811WB	473	AZ815WB		

477	BJ031RB	MAN 11.22 8.82m	De Simon	Regional	B34F	2000
478	BJ028RB	Iveco A45E12	6.86m	Regional	B19F	2000
479	BJ029RB	Iveco A45E12	6.86m	Regional	B19F	2000
480	BJ030RB	Iveco A45E12	6.86m	Regional	B19F	2000
481	BJ677RB	Volvo B10B	12m	Regional	B53F	2000

482-486   Volvo B10B 10.8m   Regional   B45D   2000

482	BJ493RB	484	BJ674RB	485	BJ675RB	486	BJ676RB
483	BJ673RB						

488-493   Neoplan Euroliner N316 SHD   Neoplan   Coach   C51F   2000

488	BJ450RC	490	BH780XY	492	BH782XY	493	BH783XY
489	BJ369RC	491	BH781XY				

494	BJ244RC	Iveco A45E12	6.86m	Regional	C19F	2000
495	BJ889RC	Mercedes-Benz 416 CDI T46	6.89m	Coach	B18D	2000
497	BJ473RD	De Simon Intercity IN3	10.67m	Regional	B47D	2000
498	BJ474RD	De Simon Intercity IN3	10.67m	Regional	B47D	2000
499	BS884RN	Beulas Ministar N MAN	8.67m	Coach	C35F	2001
500	BS441RN	Volvo B12B	12m	Coach	C46F	2001
505	BE330FE	Volvo B12B	12m	Coach	C51F	1999
506	BM877SA	Iveco EuroRider 391E.12.35	12m	Coach	C48F	2001
507	BM837SA	Iveco EuroRider 391E.12.35	12m	Coach	C48F	2001
508	CN237SF	Iveco EuroRider 391E.12.35	12m	Coach	C48F	2001
509	BM809SA	Iveco EuroRider 391E.12.35	12m	Coach	C48F	2001
510	BM810SA	Iveco EuroRider 391E.12.35	12m	Regional	B53D	2001
511	BM832SA	Iveco EuroRider 391E.12.35	12m	Regional	B53D	2001
512	BM831SA	Iveco EuroRider 391E.12.35	12m	Regional	B53D	2001
513	BS487RN	Iveco EuroRider 391E.12.35	12m	Regional	B53D	2001
514	BM972SA	Iveco EuroRider 391E.12.35	12m	Regional	B53D	2001

515	BW410WE	Iveco EuroRider 391.10.35	10.8m	Regional	B45D	2002
516	R7715FZ	Mercedes Benz O350	12m	Coach	BC46F	1997
517	BW854WF	Mercedes-Benz O350	12m	Coach	BC46F	1997
518	CN236SF	Mercedes-Benz O350	12m	Coach	BC46F	1997
519	BZ548EZ	Mercedes-Benz O350	12m	Coach	BC46F	1997
520	CB526YE	Mercedes-Benz O350 RHD	12m	Coach	C53F	2003
521	CY519VV	Mercedes-Benz Integro O550	Mercedes-Benz		C40F	2006
522	CY520VV	Mercedes-Benz Integro O550	Mercedes-Benz		C40F	2006
523	BZ046FA	Iveco 370E.12.35	12m	Coach	C50F	2002
524	BZ047FA	Iveco 370E.12.35	12m	Coach	C50F	2002
525	BZ048FA	Iveco 370E.12.35	12m	Coach	C50F	2002
526	CF530HL	Neoplan N4426/3	12m	Regional	B86D	2003
527	CF531HL	Neoplan N4426/3	12m	Regional	B86D	2003
528	CF532HL	Neoplan N4426/3	12m	Regional	B86D	2003
529	CF533HL	Neoplan N4426/3	12m	Regional	B86D	2003
530	CF569HL	Mercedes-Benz O350 RHD	12m	Coach	C51F	2003
531	DC970GD	Caccimali Tema 206	9.3m	Schoolbus	B16F	2006
534	CW419WG	Mercedes-Benz DF 412	7.58M	Schoolbus	B16F	1994
535	CW420WG	Mercedes-Benz DF 412	7.08m	Schoolbus	B16F	2000
536	CW421WG	Mercedes-Benz DF 412	7.1m	Schoolbus	B16F	2005
537	CY521VV	Mercedes-Benz Integro O550	Mercedes-Benz		C40F	2006
538	CY522VV	Mercedes-Benz Integro O550	Mercedes-Benz		C40F	2006
539	CY523VV	Mercedes-Benz Integro O550	Mercedes-Benz		C40F	2006
540	CF986HL	Mercedes-Benz O350 SHD	12m	Coach	C46F	2003
541	CY524VV	Mercedes-Benz Integro O550	Mercedes-Benz		C40F	2006
542	CY525VV	Mercedes-Benz Integro O550	Mercedes-Benz		C40F	2006
543	CY526VV	Mercedes-Benz Integro O550	Mercedes-Benz		C40F	2006
544	CY527VV	Mercedes-Benz Integro O550	Mercedes-Benz		C40F	2006
545	CY528VV	Mercedes-Benz Integro O550	Mercedes-Benz		C40F	2006
546	CY529VV	Mercedes-Benz Integro O550	Mercedes-Benz		C40F	2006
547	CY530VV	Mercedes-Benz Integro O550	Mercedes-Benz		C40F	2006
548	DB816WD	Scania L124UB6 13.7m	Beulas Aura N	Coach		2006
549	DA386XH	MAN - 9.9m	Beulas Cygnus	Coach		2006
550	CW300WG	Scania L124UB6 13.7m	Beulas Aura N	Coach		2006
557	CR405TD	Scania L124UB6 13.7m	Beulas Aura	Coach	C43F	2005
558	CR406TD	Scania L124UB6 13.7m	Beulas Aura	Coach	C43F	2005

568-573

		Iveco 391E.12.29	Orlandi EuroRider	Regional	B53D	2001

568	BM937SA	570	BM939SA	572	BM878SA	573	BM879SA
569	BM938SA	571	BS486RN				

574-600

		De Simon Intercity IL3.	12m	Regional	B53D	2001

574	BS017RN	581	BS098RN	588	BS062RN	595	BS291RN
575	BS003RN	582	BS001RN	589	BS061RN	596	BS290RN
576	BS018RN	583	BS126RN	590	BS139RN	597	BS176RN
577	BS019RN	584	BS138RN	591	BS161RN	598	BS288RN
578	BS063RN	585	BS002RN	592	BS162RN	599	BS310RN
579	BS099RN	586	BS137RN	593	BS163RN	600	BS265RN
580	BS097RN	587	BS125RN	594	BS175RN		

602-608

		De Simon Intercity IL3.	12m	Regional	B53D	2001

602	BV841MF	605	BV842MF	607	BV871MF	608	BW007WE

614	CR404TD	Beulas Ministar N MAN	9.8m	Coach	C39F	2005
620	CR407TD	Beulas Ministar N MAN	13.7m	Coach	C63F	2005

622-642

		Mercedes-Benz Integro O550 U	Mercedes-Benz	Regional	B53D	2003

622	CD882SA	628	CD888SA	633	CD893SA	638	CJ747PJ
623	CH505AJ	629	CD889SA	634	CD894SA	639	CJ748PJ
624	CD884SA	630	CD890SA	635	CD895SA	640	CN331SF
625	CD885SA	631	CD891SA	636	CD896SA	641	CN332SF
626	CD886SA	632	CD892SA	637	CD897SA	642	CN333SF
627	CD887SA						

643	CJ495PJ	De Simon Intercity IN3.	10.67m	Regional	B47D	2004
644	CJ496PJ	De Simon Intercity IN3.	10.67m	Regional	B47D	2004

645-649

		Mercedes-Benz Integro O550 U	Mercedes-Benz	Regional	B53D	2005

645	CR895TB	647	CR896TB	648	CR899TB	649	CR898TB
646	CR897TB						

650	CY531VV	De Simon Intercity IN3.	10.67m			Regional	B47D	2005

655-658		MAN 11.220 8.88m		Beulas Midistar 1		Coach	C-F	2005
655	CR901TB	**656**	CR900TB	**657**	CR903TB		**658**	CR902TB

668	CR906TB	MAN 11.220 7m		Noge		Interurban	C-F	2005
669	CR905TB	MAN 11.220 7m		Noge		Interurban	C-Г	2005
673	CR904TB	MAN 11.220 10mm		Dalla Itziano		Interurban	C-F	2005

680-688		Neoplan N4426/3		Neoplan		Interurban		2006
680	CY532VV	**682**	CY533VV	**683**	CY534VV		**688**	CY535VV

692	DK726LP	TVM Marbus B4		Marbus Viveo		C30D	2008	
693	DK728LP	TVM Marbus B4		Marbus Viveo		C30D	2008	

694-698		De Simon Millle Miglia		de Simon		Interurban		2008
694	DK203GD	**696**	DK205GD	**697**	DK206GD		**698**	DK207GD
695	DK204GD							

699	DF612TL	MAN 11.220		Beaulas Midistar N		C30D		2008
701	DD719HP	Mercedes-Benz Travego O580		Mercedes-Benz		Regional	C--F	2007
707	DD720HP	Mercedes-Benz Travego O580		Mercedes-Benz		Regional	C--F	2007

711-718		MAN 11.22 8.82m		De Simon		Regional	B34D	1998
711	AT195HT	**713**	AT243JA	**715**	AT245JA		**717**	AT247JA
712	AT196HT	**714**	AT244JA	**716**	AT246JA		**718**	AT248JA

719-726		Mercedes-Benz O404 10RH		Tiziano 9.22m		Regional	B39D	2001-02
719	BS354RP	**721**	BS632RP	**723**	BS634RP		**725**	BZ915FA
720	BS631RP	**722**	BS633RP	**724**	BS635RP		**726**	BZ916FA

727-730		De Simon Intercity IN3.		10.67m		Regional	B47D	2003
727	CD795SA	**728**	CD796SA	**729**	CD797SA		**730**	CD798SA

738-750		Irisbus Arway SFR160		Irisbus		Interurban	NC--D	2007
738	DF455TL	**742**	DF459TL	**745**	DF461TL		**748**	DF458TL
739	DF466TL	**743**	DF463TL	**746**	DF462TL		**749**	DF457TL
740	DF465TL	**744**	DF460TL	**747**	DD093HP		**750**	DF456TL
741	DF464TL							

751	DK450GD	Mercedes-Benz Travego O580		Mercedes-Benz		Coach	C--F	2008
752	DK451GD	Mercedes-Benz Travego O580		Mercedes-Benz		Coach	C--F	2008
753	DP021AY	Mercedes-Benz Sprinter 515		Mercedes-Benz		School bus	N--F	2008
754	DP557AY	Iveco Dailybus		Iveco		Coach	M	2008
755	DP344AY	Neoplan Skyliner N1122/3		Neoplan		Interurban	C--D	2008

756-760		Scania OmniExpress		Scania		Interurban	NC--D	2008
756	DP540AY	**758**	DP542AY	**759**	DP543AY		**760**	DP544AY
757	DP541AY							

| 761 | DP555AY | TVM Marbus B4 | | Marbus Viveo | | C30D | 2008 | |

762-770		Irisbus Arway SFR160		Irisbus		Interurban	NC--D	2008
762	DP492AY	**764**	DP494AY	**766**	DP496AY		**769**	DP498AY
763	DP493AY	**765**	DP495AY	**768**	DP497AY		**770**	DP499AY

| 767 | BF288DX | Volvo B10B | | 12m | | Regional | B53D | 1999 |

771-775		MAN Lion's City A21		MAN		Urban	N44D	2008
771	DP545AY	**773**	DP547AY	**774**	DP548AY		**775**	DP549AY
772	DP546AY							

776-781		Bredamenarinibus Avancity L CNG 12m				City Bus	N35D	2005
776	CR907TB	**778**	CR909TB	**780**	CR910TB		**781**	CR912TB
777	CR908TB	**779**	CR911TB					

782	DP600AY	Irisbus Europolis 200E. 7.96m		Cacciamali				2008
783	DP601AY	Irisbus Europolis 200E. 7.96m		Cacciamali				2008
784	DF454TL	Irisbus Cityclass 491E.12.87		Irisbus		Urban	N--D	2007

785-799   Bredamenarinibus M240 LU3   11.96m   City Duo   N22D   2003-04

785	CN334SF	789	CF108HM	793	CH061AH	797	CH250AH
786	CN335SF	790	CF109HM	794	CH062AH	798	CH251AH
787	CN336SF	791	CF110HM	795	CH063AH	799	CH252AH
788	CF107HM	792	CH060AH	796	CH064AH		

No.	Reg	Type	Length	Class	Code	Year
800	CD435SA	De Simon 15 MT	14.80m	Regional	B65D	2002
801	DP602AY	Scania OmniCity CN94UB	Scania		N30D	2008
802	DP631AY	Scania OmniCity CN94UB	Scania		N30D	2008
803	DP603AY	Scania OmniCity CN94UB	Scania		N30D	2008
819	UD665332	Fiat 480.12.21 P	12m	City Bus	B20D	1991
822	BH018XV	Irisbus 101E.12.22 (cng)	12m	City Bus	B28D	2000
824	UD621733	Inbus U240	12m	City Duo	D20D	1989
825	AT029HT	MAN Bassotto Metano	12m	City Bus	B25D	1997
833	BH019XV	Irisbus 491E.12.22 (cng)	12m	City Bus	B28D	2000
835	AT718HV	MAN Bassotto Metano	12m	City Bus	B25D	1997
838	AT031HT	MAN Bassotto Metano	12m	City Bus	B25D	1997
839	AT719HV	MAN Bassotto Metano	12m	City Bus	B25D	1997
840	BH020XV	Irisbus 491E.12.22 (cng)	12m	City Bus	B28D	2000
848	UD665335	Fiat 480.12.21 P	12m	City Bus	B20D	1991
849	UD665336	Fiat 480.12.21 P	12m	City Bus	B20D	1991
850	BV903MF	Irisbus 491E.12.22 (cng)	12m	City Bus	B28D	2000
851	AT720HV	MAN Bassotto Metano	12m	City Bus	B25D	1997
852	UD672465	Fiat 480.12.21 P	12m	City Bus	B20D	1991
853	UD672464	Fiat 480.12.21 P	12m	City Bus	B20D	1991
854	BH022XV	Irisbus 491E.12.22 (cng)	12m	City Bus	B28D	2000
855	BH023XV	Irisbus 491E.12.22 CNG	12m	City Bus	B28D	2000
856	BH024XV	Irisbus 491E.12.22 CNG	12m	City Bus	B28D	2000
857	AT721HV	MAN Bassotto Metano	12m	City Bus	B25D	1997
858	BH025XV	Irisbus 491E.12.22 CNG	12m	City Bus	B28D	2000
859	BH026XV	Irisbus 491E.12.22 CNG	12m	City Bus	B20D	2000
860	BH027XV	Irisbus 491E.12.22 CNG	12m	City Bus	B28D	2000
861	BH028XV	Irisbus 491E.12.22 CNG	12m	City Bus	B28D	2000
862	BH029XV	Irisbus 491E.12.22 CNG	12m	City Bus	B28D	2000
863	BH030XV	Irisbus 491E.12.22 CNG	12m	City Bus	B28D	2000
864	BH031XV	Irisbus 491E.12.22 CNG	12m	City Bus	B28D	2000
874	AT030HT	MAN Bassotto Metano	12m	City Bus	B25D	1997
875	AT032HT	MAN Bassotto Metano	12m	City Bus	B25D	1997
876	AT033HT	MAN Bassotto Metano	12m	City Bus	B25D	1997
877	AT034HT	MAN Bassotto Metano	12m	City Bus	B25D	1997
878	BM940SA	Iveco 491E.12.22 (methanol)	12m	City Bus	B28D	2001
879	BM941SA	Iveco 491E.12.22 METANO	12m	City Bus	B28D	2001
880	BM847SA	Iveco 491E.12.22 METANO	12m	City Bus	B28D	2001
881	BM848SA	Iveco 491E.12.22 METANO	12m	City Bus	B28D	2001
882	BM849SA	Iveco 491E.12.22 METANO	12m	City Bus	B28D	2001
883	BM880SA	Iveco 491E.12.22 METANO	12m	City Bus	B28D	2001
884	CD013SA	Mercedes-Benz Cito 0520	9.59m	City Bus	B17D	2002
885	CD014SA	Mercedes-Benz Cito 0520	9.59m	City Bus	B17D	2002
886	CD015SA	Mercedes-Benz Cito 0520	9.59m	City Bus	B17D	2002
887	BZ627FA	Mercedes-Benz Cito 0520	8.09m	City Bus	B11D	2002
888	BZ628FA	Mercedes-Benz Cito 0520	8.09m	City Bus	B11D	2002
889	CD799SA	BMB M240GNC NU EXOBUS	10.79m	City Bus	B18D	2003
890	CD800SA	BMB M240GNC NU EXOBUS	10.79m	City Bus	B18D	2003
891	CD801SA	BMB M240GNC NU EXOBUS	10.79m	City Bus	B18D	2003
892	CD802SA	BMB M240GNC NU EXOBUS	10.79m	City Bus	B18D	2003
893	CD803SA	BMB M240GNC NU EXOBUS	10.79m	City Bus	B18D	2003
894	CH253AH	BMB M240GNC NU EXOBUS	10.79m	City Bus	B18D	2003
895	CH254AH	BMB M240GNC NU EXOBUS	10.79m	City Bus	B18D	2003
896	CH255AH	BMB M240GNC NU EXOBUS	10.79m	City Bus	B18D	2003
897	CH256AH	BMB M240GNC NU EXOBUS	10.79m	City Bus	B18D	2003
898	DF467TL	Mercedes-Benz Sprinter	7m	Urban	B11D	2007
900	CF714HL	Iveco 49.10.1/N-3,6 POLL	6.32m	City Bus	B9D	1992

TRIESTE TRASPORTI

Trieste Trasporti SpA, Via dei Lavoratori, 2 - 34144 Trieste

Trieste lies on the Italian side of the border with Slovenia, and the local fleet has seen a major intake of modern buses for use in the city.

565	CP877FT	Irisbus Europolis TCC685	Cacciamali	City bus		2004
566	CP870FT	Irisbus Europolis TCC685	Cacciamali	City bus		2004
591	CB258YE	Irisbus Europolis 924	Cacciamali	City bus		2002
592	CB350YE	Irisbus Europolis 924	Cacciamali	City bus		2002
593	CB351YE	Irisbus Europolis 924	Cacciamali	City bus		2002

594-598		Irisbus Urbano 203E.9.26		Irisbus	City bus		2007
594	DF677SY	596	DF679SY	597	DF676SY	598	DF678SY
595	DF698SY						

601	BM667RA	Setra S315 HD	Setra	C--F	-
603	BM762RA	Mercedes-Benz O404	Mercedes-Benz	C--F	-
604	BM761RA	Irisbus Dalla via 370E	Paliadio	C--F	-
605	BV038ZE	Neoplan Staliner N516	Neoplan	C--F	-
606	BV678ZC	Irisbus Domino 2001 HDH	Irisbus	C--F	-
607	CB743YE	Irisbus Orlandi	Irisbus	C--F	-
615	BM666RA	Iveco 380.12.38	Iveco	C--F	-
616	BM668RA	Iveco 380.12.38	Iveco	C--F	-
641	BM868RA	Iveco CC80E18M/86	Cacciamali	School bus	2000
642	BV039ZB	Iveco CC80E18M/86	Cacciamali	School bus	2000
643	BV074ZB	Iveco CC80E18M/86	Cacciamali	School bus	2000
646	BV072ZB	Iveco CC80E18M/86	Cacciamali	School bus	2000
647	BM605RA	Iveco 100	Cacciamali	School bus	2000
648	CB849YE	Iveco 100	Cacciamali	School bus	2000

791-795		Bredamenarini M340		Bredamenarinbus	City bus		2004
791	CP804BT	793	CP803BT	794	CP802BT	795	CP801BT
792	CP794BT						

Trieste lies on the eastern Italian Mediterranean border near Slovenia. The modern fleet includes Irisbus CityClass 491 number 810, BM595RA, seen on the coastal road. *Mark Lyons*

801-832 — Irisbus CityClass 491.12.29 — Irisbus — City bus — 2001

801	BM827RA	809	BM594RA	817	BM522RA	825	BM598RA
802	BM399RA	810	BM595RA	818	BM681RA	826	BM599RA
803	BM398RA	811	BM518RA	819	BM597RA	827	BM601RA
804	BM397RA	812	BM455RA	820	BM519RA	828	BM520RA
805	BM396RA	813	BM596RA	821	BM453RA	829	BM678RA
806	BM521RA	814	BM454RA	822	BM456RA	830	BM602RA
807	BM680RA	815	BM395RA	823	BM517RA	831	BM603RA
808	BM457RA	816	BM452RA	824	BM677RA	832	BM604RA

901-914 — Mercedes-Benz Citaro O530GN — Mercedes-Benz — AN50D — 2008

901	DJ143MF	905	DJ095MF	909	DP813GW	912	DP815GW
902	DJ101MF	906	DJ145MF	910	DP814GW	913	DP824GW
903	DJ146MF	907	DJ148MF	911	DP841GW	914	DP825GW
904	DJ100MF	908	DJ144MF				

1001-1005 — Autodrome ALE Zerai 7,7/3P E3 — Autiromo — City bus — 2001-03

1001	BV785ZB	1003	BV787ZB	1004	BV786ZB	1005	CB450YE
1002	BV788ZB						

1006-1009 — Rampini ALE Zerai 7,7/3P E3 — Rampini — City bus — 2008

1006	DP842GW	1007	DP843GW	1008	DP844GW	1009	DP845GW

1011-1020 — Breda Menarini M231/E3 — Bredamenarinbus — City bus — 2001

1011	BV501ZB	1014	BV562ZB	1017	BV561ZB	1019	BV656ZB
1012	BV502ZB	1015	BV503ZB	1018	BV637ZB	1020	BV655ZB
1013	BV635ZB	1016	BV636ZB				

1021 — Breda Menarini M231/E3 — Bredamenarinbus — City bus — 2005

1021	CP854BT

1031-1049 — Breda Menarini M240/E3 NU — Bredamenarinbus — City bus — 2001

1031	BV504ZB	1036	BV507ZB	1041	BV567ZB	1046	BV559ZB
1032	BV505ZB	1037	BV508ZB	1042	BV512ZB	1047	BV560ZB
1033	BV506ZB	1038	BV509ZB	1043	BV513ZB	1048	BV570ZB
1034	BV565ZB	1039	BV510ZB	1044	BV568ZB	1049	BV571ZB
1035	BV566ZB	1040	BV511ZB	1045	BV569ZB		

1050-1080 — Breda Menarini M240/E3 NU — Bredamenarinbus — City bus — 2002-03

1050	CB352YE	1058	CB410YE	1066	CB413YE	1074	CB127YE
1051	CB180YE	1059	CB411YE	1067	CB414YE	1075	CB152YE
1052	CB181YE	1060	CB259YE	1068	CB415YE	1076	CB128YE
1053	CB409YE	1061	CB260YE	1069	CB416YE	1077	CB129YE
1054	CB182YE	1062	CB261YE	1070	CB417YE	1078	CB151YE
1055	CB183YE	1063	CB262YE	1071	CB418YE	1079	CB150YE
1056	CB195YE	1064	CB263YE	1072	CB125YE	1080	CB149YE
1057	CB196YE	1065	CB412YE	1073	CB126YE		

Another Irisbus Cityclass 491, this time from 2007. 1273, DF738SY, is seen operating route 25 which links Cattinara with the Piazza della Borsa in Mazzini.
Mark Lyons

233

1101-1110 Breda Menarini M240/E3 LU 3P Bredamenarinbus City bus 2002

1101	CB264YE	1104	CB238YE	1107	DA977MK	1109	CB241YE
1102	CB197YE	1105	CB239YE	1108	CB198YE	1110	CB266YE
1103	CB23?YE	1106	CB265YE				

1151-1170 Breda Menarini M240/E3 LU 18m Bredamenarinbus City bus 2002-03

1151	CB419YE	1156	CB355YE	1161	CB358YE	1166	CB425YE
1152	CB420YE	1157	CB430YE	1162	CB422YE	1167	CB426YE
1153	CB267YE	1158	CB421YE	1163	CB423YE	1168	CB427YE
1154	CB353YE	1159	CB356YE	1164	CB424YE	1169	CB359YE
1155	CB354YE	1160	CB357YE	1165	CB429YE	1170	CB428YE

1201	CB835YE	Scania OmniCity CN94UB	Scania	City bus	2003
1202	DJ024MF	Irisbus Citelis C12B	Irisbus	City bus	2007
1203	DB870YH	Mercedes-Benz Cito O520	Mercedes-Benz	Citybus	2006
1204	DB998YH	Mercedes-Benz Cito O520	Mercedes-Benz	Citybus	2006
1205	DS995CZ	Irisbus Cityclass 491E.10.29	Irisbus	Citybus	2009

1211-1261 Irisbus Cityclass 491E.10.29 Irisbus Citybus 2004-05

1211	CP716FT	1224	CP864FT	1237	CT628SW	1250	CT973SW
1212	CP713FT	1225	CP865FT	1238	CT625SW	1251	CT977SW
1213	CP708FT	1226	CP866FT	1239	CT626SW	1252	CT971SW
1214	CP705FT	1227	CP867FT	1240	CT630SW	1253	CT981SW
1215	CP715FT	1228	CP868FT	1241	CT624SW	1254	CT969SW
1216	CP714FT	1229	CP869FT	1242	CT623SW	1255	CT980SW
1217	CP712FT	1230	CP871FT	1243	CT972SW	1256	CT979SW
1218	CP706FT	1231	CP872FT	1244	CT622SW	1257	CT975SW
1219	CP717FT	1232	CP873FT	1245	CT620SW	1258	CT976SW
1220	CP707FT	1233	CP874FT	1246	CT621SW	1259	CT982SW
1221	CP710FT	1234	CP875FT	1247	CT983SW	1260	CT970SW
1222	CP711FT	1235	CP876FT	1248	CT974SW	1261	CT978SW
1223	CP709FT	1236	CT627SW	1249	CT629SW		

1262-1278 Irisbus Cityclass 491E.10.29 Irisbus Citybus 2007

1262	DF695SY	1267	DF714SY	1271	DF718SY	1275	DF759SY
1263	DF697SY	1268	DF736SY	1272	DF748SY	1276	DF773SY
1264	DF717SY	1269	DF775SY	1273	DF738SY	1277	DF774SY
1265	DF715SY	1270	DF737SY	1274	DF751SY	1278	DF750SY
1266	DF716SY						

1301-1306 Mercedes-Benz Citaro O530 Mercedes-Benz Citybus 2006-07

1301	DF640SY	1303	DF623SY	1305	DF610SY	1306	DF611SY
1302	DF608SY	1304	DF609SY				

1401-1425 Irisbus Cityclass 491E.10.29 Irisbus Citybus 2007

1401	DF979SY	1408	DJ015MF	1414	DJ058MF	1420	DJ062MF
1402	DF981SY	1409	DJ018MF	1415	DJ021MF	1421	DJ064MF
1403	DF984SY	1410	DJ017MF	1416	DJ016MF	1422	DJ061MF
1404	DF983SY	1411	DJ069MF	1417	DJ065MF	1423	DJ054MF
1405	DF982SY	1412	DJ063MF	1418	DJ057MF	1424	DJ055MF
1406	DJ019MF	1413	DJ056MF	1419	DJ066MF	1425	DJ060MF
1407	DJ020MF						

1426-1441 Irisbus Citalis 12m Irisbus Citybus 2008-09

1426	DP921GW	1430	DP918GW	1434	DP890GW	1438	DP887GW
1427	DP919GW	1431	DP920GW	1435	DP884GW	1439	DA487MJ
1428	DP922GW	1432	DP886GW	1436	DA468MJ	1440	DP888GW
1429	DP883GW	1433	DP889GW	1437	DA486MJ	1441	DP885GW

1601-1607 Breda Menarini 231/5 Breda Menarini Citybus 2008

1601	DP896GW	1603	DP893GW	1605	DP891GW	1607	DP892GW
1602	DP894GW	1604	DP897GW	1606	DP895GW		

SADEM

SADEM SpA, Via della Repubblica 14, 10095 Grugliasco, Italy

116	BY642KX	Iveco 65C15/70					ALI	7.7m	2002	
159	AH664WA	Iveco 590.E12.22					ALS	12m	1995	
160	AH663WA	Iveco 590.E12.22					ALS	12m	1995	
161-165		Iveco 370E.12.35					ALSI	12m	1999	
161	BB831NA	**163**	BA840NC	**164**	BB833NA				**165**	BA841NC
162	BB832NA									
166	BS281BZ	Breda Menarini M240LS					ALS	12m	2001	
167	BS475BZ	Breda Menarini M240LS					ALS	12m	2001	
169	DM331PG	Irisbus Europolis 200E. 7.96m		Cacciamali					2008	
170	DP624ZY	Irisbus Europolis 200E. 7.96m		Cacciamali					2008	
173	DM798PG	Irisbus Europolis 200E. 7.96m		Cacciamali					2008	
209-215		Iveco Orlandi 380.12.35					ALI	12m	1999	
209	AH885VZ	**212**	AK559DN	**214**	CM568JF				**215**	AK078DP
211	AH884VZ	**213**	AK560DN							
216-224		Iveco Orlandi Sicca 391E.12.29					ALI	12m	1998-99	
216	AT109DG	**219**	AT110DG	**221**	BB375NB				**223**	BB376NB
217	AT585DG	**220**	BB347MY	**222**	BB348MY				**224**	BB349MY
218	AT296DG									
225-230		Irisbus Myway 393E.12.35					ALI	12m	2001	
225	BV767HW	**227**	BV769HW	**228**	BX110GH				**230**	BX756GH
226	BV768HW									
233-239		Irisbus Orlando Domino 397E.12.35					ALI	12m	2003	
233	CK695HV	**235**	CK355HW	**237**	CK083HW				**239**	CK951HV
234	CK818HV	**236**	CK354HW	**238**	CK625HV					
240	CN755CA	Irisbus Myway 399 EL75					ALI	12m	2004	
241	CR205NB	Irisbus Myway 399 EL75					ALI	12m	2004	
242-246		Irisbus Orlando Domino HD 397E.12.35					ALI	12m	2005-06	
242	CW652HB	**244**	CW654HB	**245**	CW107HC				**246**	CZ265AL
243	CW653HB									
247	CZ836AL	Irisbus Myway 399 EL75					ALI	12m	2006	
248	DD938GY	Irisbus Orlando Domino HD 397E.12.35					ALI	12m	2006	
249	DD819GY	Irisbus Orlando Domino HD 397E.12.35					ALI	12m	2006	
250	DD632GY	Irisbus Orlando Domino HD 397E.12.35					ALI	12m	2006	
336	BD523CG	Iveco Orlando Sicca-Poker 370E.9.27					ALI	9.7m	1999	
337	BD549TC	iveco Orlando Sicca-Poker 370E.9.27					ALI	9.7m	1999	
338-341		Irisbus Euroclass 380.12.35					ALI	12m	2001	
338	BS459PM	**339**	BS460PM	**340**	BS461PM				**341**	BS462PM
342-350		Irisbus Euroclass 380.10.35					ALI	10m	2001	
342	BS580PM	**345**	BS670PM	**347**	BS671PM				**349**	BS673PM
343	BS581PM	**346**	BS582PM	**348**	BS672PM				**350**	BS674PM
344	BS669PM									
356	TO81344R	Iveco Valle Ufita 370.12.30					ALI	12m	1991	

As mentioned under the SAB fleet, Arriva operates the cable car that links Albino with Selvino. Here is shown the car waiting time at the base station. *Bill Potter*

370-376

Irisbus Euroclass 389E.12.35 · ALI · 12m · 2003-05

370	CK537ST	372	CK962SV	374	CS262PP	376	CS264PP
371	CK594ST	373	CK963SV	375	CS263PP		

377	DB849BL	Irisbus Arway SFR160	Irisbus	N--D	2008	
378	DB850BL	Irisbus Arway SFR160	Irisbus	N--D	2008	
379	DB851BL	Irisbus Arway SFR160	Irisbus	N--D	2008	
400	TO43350W	Iveco Orlando Domino 370.12.30		ALG	12m	1994

401-404

Iveco Euroclass HD 380E.12.38 · ALI · 12m · 1996-2001

401	BC645JG	402	BC813JF	403	BS998BX	404	BS999BX

405	DM063PG	Irisbus Orlando Domino 397E.12.35	ALI	12m	-
457	TO61104N	Iveco Orlando Domino 370.12.30	ALI	12m	1990
458	TO61103N	Iveco Orlando Domino 370.12.30	ALI	12m	1990
461	AB195SY	Iveco Orlando Domino 370.12.30	ALI	12m	1991
462	AK847DS	Iveco Orlando Top Class 380.12.38	ANR	12m	1996
463	AW573NR	Iveco Euroclass HD 380E.12.38	ANR	12m	1998
464	AW574NR	Iveco Euroclass HD 380E.12.38	ANR	12m	1998
465	AW307NS	Iveco Eurocloss HD 380E.12.38	ANR	12m	1998
466	BN898EA	Iveco Euroclass HD 315E.8.18	ANR	7.58m	2000
467	BN895EA	Iveco Orlandi Sicca-Poker 370E.9.27	ANR	9.7m	2000
469	BS050BL	Irisbus Euroclass HD 380.12.38	ANR	12m	2001
470	DF307CW	Irisbus Euroclass HD 380.12.38	ANR	12m	2001
471	CV424KR	Irisbus Orlandi Domino 380.12.38	ANR	12m	2005
472	CV425KR	Irisbus Orlandi Domino 380.12.38	ANR	12m	2005
473	DM062PG	Irisbus Orlando Domino 397E.12.35	ALI	12m	-
1168	DS599SL	Irisbus Europolis 200E.8.13	ALS	10.48m	200
1171	DS729SL	Irisbus Europolis 200E.8.13	ALS	10.48m	200
1172	DS556SL	Irisbus Europolis 200E.8.13	ALS	10.48m	200

Note: This company was established in 1941, as Società Autotrasporti della Dalmazia e Montenegro.

Type codes: ALC Schoolbus; ALG Tourist coach; ALI Interurban; ALS Suburban bus; ALU Urban bus and ANR Private Hire.

SAPAV

SAPAV SpA, Corso Torino 396, 10064 Pinerolo, Italy

298	TO79338H	Iveco 370.12.30/T	ANR	12m	1988
299	TO83964H	Iveco 370.12.30/T	ANR	12m	1988
307	TO36202L	Iveco 370.12.30/T	ALI	12m	1988
308	TO40743L	Iveco 370.12.30/T	ALI	12m	1988
309	TO23593M	Iveco 370.12.30/T	ALI	12m	1989
310	TO23594M	Iveco 370.12.00/T	ALI	12m	1989
313	TO71298N	Iveco 370.12.30/T	ANR	12m	1990
317	TO93555M	Iveco 315.8.17	ANR	7.57m	1989
318	TO29942V	Iveco 370E.12.30	ANR	12m	1991

320-325

		Iveco 370E.12.35		ALI	12m	1994

320	AB033WH	322	AX219YV	324	AB879WH	325	AB877WH
321	AB035WH	323	AB878WH				

326-332

		Iveco 380.12.35		ALI	12m	1995

326	AH209VN	328	AH486VN	330	AH485VN	332	AH484VN
327	AH208VN	329	AZ210NK	331	AH483VN		

333	AK639CK	Iveco 49.10P.35P	ALU	6.32m	1996

334-338

		Iveco 380.12.35		ALI	12m	1996

334	AK017DP	336	AK019DP	337	AK020DP	338	AK021DP
335	AK018DP						

339	AK890DV	Iveco 380E.12.38HD	ANR	12m	1996
340	AK889DV	Iveco 380E.12.38HD	ANR	12m	1996
341	AK609GB	Iveco 49.12/–3.6	ALU	6.47m	1996
342	AK684GF	Iveco 49.12/N-3.6	ALU	6.47m	1996

343-347

		Iveco 370E.12.35		ALI	12m	1998

343	AW895 LA	345	AW894 LA	346	AW893 LA	347	AW897 LA
344	AW896 LA						

348	AW599 LD	Iveco 370E.12.27	ALI	9.70m	1998
349	BA557CM	Iveco 315S.8.18	ALI	7.58m	1998
350	BA571CM	Iveco 315S.8.18	ALI	7.58m	1998
351	BA558CM	Iveco 380E.12.38HD	ANR	12m	1998
352	BE206GS	Iveco 370E.12.27	ALI	9.70m	1999
353	BE209GS	Iveco 370E.12.27	ALI	9.70m	1999
354	BE207GS	Iveco 370E.12.27	ALI	9.70m	1999
355	BE210GS	Iveco 380.12.35	ALI	12m	1999
356	BE208GS	Iveco 315S.8.18	ANR	7.58	1999

357-365

		Iveco 380.12.35		ALI	12m	1999

357	BE820GS	360	BE522GT	362	BE523GT	364	BE520GT
358	BE819GS	361	BE524GT	363	BE521GT	365	BF947MA
359	BE821GS						

366-370

		Iveco 370E.9.27		ALI	9.7m	1999

366	BG356 KB	368	BG358 KB	369	BG359 KB	370	BG361 KB
367	BG357 KB						

371	BG727 KB	Iveco A45E.12	ALI	6.86m	1999

372-376

		Iveco 370E.9.27		ALI	9.7m	1999

372	BG478 KC	374	BG480 KC	375	BG481 KC	376	BG482 KC
373	BG479 KC						

377-381

Irisbus 380E.12.38HD — ANR 12m 2000

377	BK426 DA	379	BL117 WH	380	BL758 WH	381	BL757 WH
378	BL886 WF						

382-387

Irisbus Myway 393E.12.35 — ANR 12m 2000

382	BS075BJ	383	BS076BJ	386	BS079BJ	387	BS000BJ
384	BS077BJ	385	BS078BJ				

388	BS629BT	Irisbus 380.10.35	ALG	10.62m	2001
389	BS456BT	Fiat Ducato Panorama	ANR	5.03m	2001
390	BS442BV	Irisbus 380E.12.38H	ANR	11.99m	2001
391	BS441BV	Irisbus 380E.12.38HD	ANR	11.99m	2001
392	BS979BW	Iveco Happy 65.15	ALI	7.70m	2001
393	BS894BZ	Iveco A45E.12	ALI	6.86m	1999
394	BS895BZ	Mercedes-Benz Vario O815	ANR	7.64m	2001
395	BX855GH	Irisbus Europolis 200E.8.13	ALS	7.65m	2001

396-399

Irisbus Myway 393E.12.35 — ALI 12m 2002

396	BY056KY	397	BY057KY	398	BY390KY	399	BY392KY

400	BA960CK	Scudo	ALI	4.44m	1998
401	CD357AA	Mercedes-Benz Vario O815	ALI	7.64m	2002

405-411

Irisbus Myway 399.12.35 — ALI 12m 2003

405	CF155EN	407	CF157EN	409	CF159EN	411	CF161EN
406	CF156EN	408	CF158EN	410	CF160EN		

412	CF517EP	Mercedes-Benz Vario O815	ALI	7.64m	2003
413	CG735EF	Irisbus Myway 399.12.35	ALI	12m	2003
414	CG736EF	Irisbus 389E.12.43H	ALG	12m	2003
415	CG737EF	Irisbus 389E.12.43HD	ANR	11.99m	2003
416	CH347KN	Irisbus 389E.12.43HD	ANR	11.99m	2003
417	CJ175DG	Mercedes-Benz Vario O815	ALI	8.42m	2003
418	CK216HV	Breda Menarini M240	ALS	11.96m	2003
419	CK217HV	Breda Menarini M240	ALS	11.96m	2003
420	CK218HV	Breda Menarini M240	ALS	11.96m	2003

421-430

Irisbus Orlandi Euroclass 389.12.35 — ALI 12m 2003-04

421	CK478HX	424	CK479HX	427	CM704JD	429	CM051JE
422	CK832HX	425	CK833HX	428	CM705JD	430	CM052JE
423	CK587HX	426	CK834HX				

431	CN181JE	Irisbus Orlandi Euroclass 389.10.35	ALI	10.62m	2004

432-436

Irisbus Myway 399.12.35 — ALI 12m 2004

432	CN182JE	434	CN438CB	435	CN528CB	436	CN529CB
433	CN437CB						

437	CR221NA	Irisbus Domino 389E.12.43HD	ALG	12m	2004
438	CR222NA	Irisbus Domino 389E.12.43HD	ALG	12m	2004
439	CR634NB	Mercedes-Benz Vario O815	ALI	8.42m	2005
440	CS548 WL	Iveco A45.10	ALC	6.86m	2000

441-448

Irisbus Orlandi Euroclass 389.12.35 — ALI 12m 2005

441	CV568KL	443	CV570KL	445	CV751KL	447	CV753KL
442	CV569KL	444	CV750KL	446	CV752KL	448	CV754KL

449	CW113 HA	Iveco 50C15PRB	ALC	7.75m	2005

450-455

Irisbus Orlandi Euroclass 389.12.35 — ALI 12m 2006

450	CZ515AL	452	CZ517AL	454	CZ519AL	455	CZ520AL
451	CZ516AL	453	CZ518AL				

456	CZ702AL	Cacciamali TCI 8.40	ALI	8.40m	2006
457	CZ703AL	Cacciamali TCI 8.40	ALI	8.40m	2006
458	DM061PG	Iveco Orlando Sicca-Poker 370E.9.27	ALI	9.7m	1999
459	DM566PF	Iveco Orlando Sicca-Poker 370E.9.27	ALI	9.7m	1999
460	DM565PF	Iveco Orlando Sicca-Poker 370E.9.27	ALI	9.7m	1999

Type codes: ALC Schoolbus; ALG Tourist coach; ALI Interurban; ALS Suburban bus; ALU Urban bus and ANR Private Hire.

TRANCENTRUM BUS

Trancentrum Bus sro, Kancelár, Mladáá Boleslav, Boleslavská 98, Czech Republic

1S54100	Ford Transit 410L	Ford	16	2002
1S55743	Karosa C954.1360		50	2002
1S56089	Karosa C954.1360		50	2002
3S10190	Ford Transit 410L	Ford	16	2002
3S21769	Karosa C956.1074		51	2004
3S21779	Karosa C956.1074		51	2004
3S22156	Karosa C954.1360		51	2004
3S22157	Karosa C954.1360		51	2004
3S32024	Karosa C956.1074		51	2003
3S44223	Karosa C956.1074		51	2004
3S45386	Citroen Jumper 2.2		8	2004
3S88827	Karosa C954.1360		51	2003
3S89104	SOR Libchavy C10.5		47	2003
3S89105	SOR Libchavy C10.5		47	2003
3S89107	Karosa C954.1360		50	2003
3S89671	Karosa C954.1360		54	2002
3S89742	SOR Libchavy C10.5		45	2003
3S89743	SOR Libchavy C10.5		47	2003
3S89745	Karosa C954.1360		50	2002
3S89746	SOR Libchavy C10.5		47	2003
3S89748	Karosa C954.1360		50	2002
3S89749	Karosa C955.1073		52	2002
3S89750	Karosa LC735.40		45	1989
3S93729	Karosa C734.1340		45	1990
4S53670	Irisbus Daily S2000		20	2004
4S53702	Irisbus Daily S2000		20	2004
4S54621	Karosa C954.1360		50	2005
4S54717	VDL Bova Futura FHD 12-380	VDL Bova	51	2005
4S90029	Citroen Jumper 2.2		8	2005
4S90174	Karosa C954.1360		50	2005
4S90175	Karosa C954.1360		50	2005
5S04694	Karosa C954.1360		50	2005
5S04695	Karosa C954.1360		50	2005
5S19553	Bova Magiq XHD 120.D380		51	2006
5S20226	Bova Futura FHD 12-370		51	1997
5S36244	Karosa C954.1360		50	2006
5S36245	Karosa C954.1360		50	2006
5S36488	SOR Libchavy C10.5		47	2003
5S83083	Karosa C954.1360		50	2006
5S83464	SOR Libchavy C10.5		47	2003
7S15851	Karosa C934.1351		46	1999
7S15899	Bova Magiq XHD 139.D430		59	2006
7S51162	Renault Master NDD		16	2007
7S51163	Renault Master NDD		16	2007
MB8088	Karosa C 734.20		45	1988
MB8175	Karosa LC 735.20		45	1988
MB8439	Karosa C 734.40		45	1989
MB8654	Karosa C 734.40		45	1989
MB8667	Karosa C 734.40		45	1989
MB8686	Karosa LC 735.40		45	1989
MB8752	Karosa C 734.40		45	1989
MB8755	Karosa C 734.40		45	1989
MB8777	Karosa C 734.40		45	1989
MB8812	Karosa C 734.40		45	1990
MB8814	Karosa C 734.40		45	1990
MB8936	Karosa LC 735.1011		45	1990
MB8943	Karosa LC 735.1011		45	1990
MB8953	Karosa C 734.1340		45	1990
MB8989	Karosa C 734.1340		45	1990
MB9048	Karosa LC 735.1011		45	1990
MB9126	Karosa C 734.1340		45	1990
MB9150	Karosa C 734.1340		45	1990
MB9174	Karosa C 734.1340		45	1990
MB9176	Karosa C 734.1340		45	1990

MB9288	Karosa LC736.1014		46	1990
MB9424	Karosa LC736.1014		46	1991
MB9478	Karosa C734.1340		45	1991
MB9578	Karosa C734.1340		45	1991
MB9579	Karosa C734.1340		45	1991
MB9595	Karosa LC736.1014		46	1990
MB9779	Bova Futura FHD 12-290		51	1990
MB9869	Karosa C734.1340		45	1992
MB9870	Karosa C734.1340		45	1992
MBA0241	Karosa C734.1340		45	1993
MBA0243	Karosa C734.1340		45	1993
MBA0263	Karosa LC735.1011		45	1990
MBA0283	Karosa LC735.1011		45	1990
MBA0584	Karosa LC735.1011		45	1989
MBA0740	Karosa C734.1340		45	1993
MBA0741	Karosa C734.1340		45	1993
MBA0760	Karosa C734.20		45	1987
MBA0847	Karosa C734.1340		45	1993
MBA0849	Karosa C734.1340		45	1993
MBA0851	Karosa C734.1340		46	1994
MBA1045	Karosa LC735.1011		45	1994
MBA1721	Karosa LC735.1011		45	1990
MBA1835	Karosa C734.20		45	1986
MBA1836	Karosa C734.20		45	1986
MBA1841	Karosa C734.20		45	1986
MBA1842	Karosa C734.20		45	1986
MBA2308	Karosa C734.1345		46	1994
MBA3329	SOR Libchavy	SOR C 7.5	26	1997
MBA3412	Karosa C934.1351		46	1997
MBA3413	Karosa C934.1351		46	1997
MBA4067	Karosa C734.40		45	1990
MBA4769	Karosa C734.40		45	1990
MBA4784	Karosa C734.20		45	1988
MBM8601	Karosa C934.1351		46	1998
MBM8674	Renault Master L3H2		16	1999
MBM8726	Karosa C934.1351		46	1999
MBM8727	Karosa C934.1351		46	1999
MBM8937	Karosa C934.1351		46	1999
MBM8985	Mercedes-Benz Vario O814		24	2000
MBM9119	Bova Futura FHD 12-370		51	1999
MBM9204	Renault Master L3H2		16	2000
MBM9384	Karosa C934.1351		46	2000
MBM9385	Karosa C934.1351		46	2000
MBM9394	Bova Futura FHD 12-370		53	1996
MBM9496	Bova Futura FHD 12-340		50	1997
MBN2772	Bova Magiq HD 120.340		51	2000
MBN2791	Karosa C934.1351		46	2000
MBN2825	Karosa C934.1351		46	2000
MBN2963	Karosa C934.1351		46	2001
MBN2964	Karosa C934.1351		46	2001
MBN3012	Karosa C934.1351		46	2001
MBN3029	Karosa C934.1351		46	2001
MBO0397	Karosa LC735.1011		45	1990
MBO0418	Karosa C955		52	2001
MBO0442	Bova Futura FHD 12-370		49	1999
MBO9460	Karosa C954.1360		50	2002
MBO9488	Karosa C734.20		45	1987
MBO9498	Karosa C954.1360		50	2002

BOSÁK BUS

Bosák BUS spol sro, Pøíbramská 964, 263 01 Dobøíš, Èeská republika (Czech Republic)

2S70457	Karosa C734.20	45	1986
2S70464	Karosa C734.20	45	1988
2S70468	Karosa C734.1340	45	1991
2S70477	Karosa C734.40	45	1989
2S70479	Karosa C734.20	45	1987
3S08691	Karosa C734.20	45	1989
3S08692	Karosa C734.20	45	1989
PB96699	Karosa C734.20	45	1987
PBA0669	Karosa C734.40	45	1990
PBA2209	Karosa C734.40	45	1993
3S08649	Karosa C734.40	45	1990
PBA5255	Karosa C734.20	45	1986
PBA5536	Karosa C734.20	45	1987
PBA7125	Karosa C935.1039	46	2001
PBA7141	Karosa C935.1039	46	2001
PBK3857	Karosa C955.1073	51	2001
PBA52-38	Karosa C734.20	45	1986
PBA52-57	Karosa C734.20	45	1987
PB92-09	Karosa C734.20	45	1987
PB97-09	Karosa C734.20	45	1988
3S08693	Karosa C734.20	45	1987
PBL18-16	Karosa C955.1073	50	2001
PBA57-78	Karosa C734.20	45	1988
PBA57-79	Karosa C734.20	45	1988
PBA71-56	Karosa C935.1039	46	2001
2S70463	Karosa C734.20	45	1988
PBA70-72	Volvo B10-400	56	1998
PBL4385	Karosa C934.1351	49	2000
PB9119	Karosa C734.20	45	1986
5S00429	Karosa C954.1360	49	2006
2S70455	Karosa C734.20	45	1988
3S08309	Karosa C936.1038	46	2000
PBA6769	Karosa C934.1351	45	2000
4S04299	Karosa C955.1073	49	2005
4S04749	Karosa C954.1360	49	2005
4S04759	Karosa C954.1360	49	2005
3S08700	Karosa B732.1652	31	1994
3S08701	Karosa B732.1652	31	1994
3S80139	Karosa C955.1073	49	2005
3S08279	Karosa C955.1073	49	2004
3S08689	Karosa C734.20	45	1986
2S70470	Karosa C734.20	45	1986
PBL1938	Bova Futura FHD 12	51	2000
PBL1939	Bova Futura FHD 12	51	2001
5S00829	Karosa C954.1360	50	2006
PBL4282	Bova Futura FHD 12	51	1999
5S30469	Karosa C954.1360	49	2006
PBL4283	Bova Futura FHD 12	51	1999
6S22019	Irisbus SFR160	49	2007

Index to UK vehicles

Operated on behalf of Shropshire council by Arriva Midlands, Optare Solo 6004, BU03HPZ, is seen leaving Shrewsbury bus station for Pulverbatch. *Chris Clegg*

Reg	Area	Reg	Area	Reg	Area	Reg	Area
BV58URP	The Shires	BX04NDG	London	BX55FWW	London	CX06BHL	NW & Wales
BV58URR	The Shires	BX04NDJ	London	BX55FWY	London	CX06BHN	NW & Wales
BV58URS	The Shires	BX04NDK	London	BX55FWZ	London	CX06BHO	NW & Wales
BX02CLO	Tellings group	BX04NDL	London	BX55FXB	London	CX06BHP	NW & Wales
BX04MWW	London	BX04NDN	London	BX55FXC	London	CX06BHU	NW & Wales
BX04MWY	London	BX04NDU	London	BX55FXE	London	CX06BHV	NW & Wales
BX04MWZ	London	BX04NDV	London	BX55FXF	London	CX06BHW	NW & Wales
BX04MXA	London	BX04NDY	London	BX55FXG	London	CX06BHY	NW & Wales
BX04MXB	London	BX04NDZ	London	BX55FXH	London	CX06BHZ	NW & Wales
BX04MXC	London	BX04NEF	London	BX55FXJ	London	CX06BJE	NW & Wales
BX04MXD	London	BX04NEJ	London	BX55FXK	London	CX06BJF	NW & Wales
BX04MXE	London	BX04NEN	London	BX55FXL	London	CX06BJI	NW & Wales
BX04MXG	London	BX05UWV	London	BX55FXM	London	CX06BJK	NW & Wales
BX04MXH	London	BX05UWW	London	BX55FXO	London	CX06BJO	NW & Wales
BX04MXJ	London	BX05UWY	London	BX55FXP	London	CX06BJU	NW & Wales
BX04MXK	London	BX05UWZ	London	BX55FXR	London	CX06BJV	NW & Wales
BX04MXL	London	BX05UXC	London	BX55FXS	London	CX06BJY	NW & Wales
BX04MXM	London	BX05UXD	London	BX55FXU	London	CX06BJZ	NW & Wales
BX04MXN	London	BX07MXT	London	BX55FXV	London	CX06BKA	NW & Wales
BX04MXP	London	BX54EBU	Tellings group	BX55FXW	London	CX06BKD	NW & Wales
BX04MXR	London	BX54EBV	Tellings group	BX55FXY	London	CX06BKE	NW & Wales
BX04MXS	London	BX55FUH	London	BX55FXYT	London	CX06BKF	NW & Wales
BX04MXU	London	BX55FUJ	London	BX56VTU	Southern Counties	CX06BKG	NW & Wales
BX04MXV	London	BX55FUO	London	BX56VTV	Southern Counties	CX06BKJ	NW & Wales
BX04MXW	London	BX55FUP	London	BX56VTW	Southern Counties	CX06BKK	NW & Wales
BX04MXY	London	BX55FUT	London	C212GTU		CX06BKL	NW & Wales
BX04MXZ	London	BX55FUU	London	CC03HOL	Tellings group	CX06BKN	NW & Wales
BX04MYA	London	BX55FUW	London	CC04MAL	Tellings group	CX06BKO	NW & Wales
BX04MYB	London	BX55FUY	London	CC53HOL	Tellings group	CX06EAK	NW & Wales
BX04MYC	London	BX55FVA	London	CCE993	Tellings group	CX06EAM	NW & Wales
BX04MYD	London	BX55FVB	London	CE52UWW	Tellings group	CX06EAO	NW & Wales
BX04MYF	London	BX55FVC	London	CUV217C	London	CX06EAP	NW & Wales
BX04MYG	London	BX55FVD	London	CUV335C	London	CX06EAW	NW & Wales
BX04MYH	London	BX55FVF	London	CX04AXW	NW & Wales	CX06EAY	NW & Wales
BX04MYJ	London	BX55FVG	London	CX04AXY	NW & Wales	CX06EBA	NW & Wales
BX04MYK	London	BX55FVH	London	CX04AXZ	NW & Wales	CX06EBC	NW & Wales
BX04MYL	London	BX55FVJ	London	CX04AYA	NW & Wales	CX06EBD	NW & Wales
BX04MYM	London	BX55FVK	London	CX04AYB	NW & Wales	CX06EBF	NW & Wales
BX04MYN	London	BX55FVL	London	CX04AYC	NW & Wales	CX06EBG	NW & Wales
BX04MYR	London	BX55FVM	London	CX04EHV	Midlands	CX06EBJ	NW & Wales
BX04MYS	London	BX55FVN	London	CX04EHW	Midlands	CX06EBK	NW & Wales
BX04MYT	London	BX55FVP	London	CX04EHY	Midlands	CX06EBL	NW & Wales
BX04MYU	London	BX55FVQ	London	CX04EHZ	Midlands	CX06EBM	NW & Wales
BX04MYV	London	BX55FVR	London	CX04HRN	NW & Wales	CX07COJ	NW & Wales
BX04MYW	London	BX55FVS	London	CX04HRP	NW & Wales	CX07COU	NW & Wales
BX04MYY	London	BX55FVT	London	CX04HRR	NW & Wales	CX07CPE	NW & Wales
BX04MYZ	London	BX55FVU	London	CX05AAE	NW & Wales	CX07CPF	NW & Wales
BX04MYZ	London	BX55FVW	London	CX05AAF	NW & Wales	CX07CPK	NW & Wales
BX04MZD	London	BX55FVY	London	CX05AAJ	NW & Wales	CX07CPN	NW & Wales
BX04MZE	London	BX55FVZ	London	CX05AAK	NW & Wales	CX07CPO	NW & Wales
BX04MZG	London	BX55FWA	London	CX05AAN	NW & Wales	CX07CPU	NW & Wales
BX04MZJ	London	BX55FWB	London	CX05EOV	NW & Wales	CX07CPV	NW & Wales
BX04MZL	London	BX55FWG	London	CX05EOW	NW & Wales	CX07CPY	NW & Wales
BX04MZN	London	BX55FWH	London	CX05EOY	NW & Wales	CX07CPZ	NW & Wales
BX04NBK	London	BX55FWJ	London	CX05JVD	NW & Wales	CX07CRF	NW & Wales
BX04NCF	London	BX55FWK	London	CX06BGU	NW & Wales	CX07CRJ	NW & Wales
BX04NCJ	London	BX55FWL	London	CX06BGV	NW & Wales	CX07CRK	NW & Wales
BX04NCN	London	BX55FWM	London	CX06BGY	NW & Wales	CX07CRU	NW & Wales
BX04NCU	London	BX55FWN	London	CX06BGZ	NW & Wales	CX07CRV	NW & Wales
BX04NCV	London	BX55FWP	London	CX06BHA	NW & Wales	CX07CRZ	NW & Wales
BX04NCY	London	BX55FWR	London	CX06BHD	NW & Wales	CX07CSF	NW & Wales
BX04NCZ	London	BX55FWS	London	CX06BHE	NW & Wales	CX07CSO	NW & Wales
BX04NDC	London	BX55FWT	London	CX06BHF	NW & Wales	CX07CSU	NW & Wales
BX04NDD	London	BX55FWU	London	CX06BHJ	NW & Wales	CX07CSV	NW & Wales
BX04NDF	London	BX55FWV	London	CX06BHK	NW & Wales	CX07CSY	NW & Wales

While the Manchester Metrolink was being upgraded bus replacements were required both north and south of the city. Arriva North West's 2745, CX58EVU, is seen in Whitworth Street. *Richard Godfrey*

CX07CSZ	NW & Wales	CX09BFV	NW & Wales	CX54EPN	NW & Wales	CX56CEU	NW & Wales
CX07CTE	NW & Wales	CX09BFY	NW & Wales	CX54EPO	NW & Wales	CX56CEV	NW & Wales
CX07CTF	NW & Wales	CX09BFZ	NW & Wales	CX55EAA	Scotland	CX56CEY	NW & Wales
CX07CTK	NW & Wales	CX09BGE	NW & Wales	CX55EAC	Scotland	CX56CFA	NW & Wales
CX07CTO	NW & Wales	CX09BGF	NW & Wales	CX55EAE	Scotland	CX57BZO	NW & Wales
CX07CTU	NW & Wales	CX09BGK	NW & Wales	CX55EAF	NW & Wales	CX57CYO	NW & Wales
CX07CTV	NW & Wales	CX09BGO	NW & Wales	CX55EAG	NW & Wales	CX57CYP	NW & Wales
CX07CTY	NW & Wales	CX09BGU	NW & Wales	CX55EAJ	NW & Wales	CX57CYS	NW & Wales
CX07CTZ	NW & Wales	CX09BGV	NW & Wales	CX55EAK	NW & Wales	CX57CYT	NW & Wales
CX07CUA	NW & Wales	CX09BGW	NW & Wales	CX55EAM	NW & Wales	CX57CYU	NW & Wales
CX07CUC	NW & Wales	CX09BGZ	NW & Wales	CX55EAO	NW & Wales	CX57CYV	NW & Wales
CX07CUG	NW & Wales	CX09BHA	NW & Wales	CX55EAP	NW & Wales	CX57CYW	NW & Wales
CX07CUH	NW & Wales	CX54DKD	NW & Wales	CX55EAW	NW & Wales	CX57CYX	NW & Wales
CX07CUJ	NW & Wales	CX54DKE	NW & Wales	CX55EAY	NW & Wales	CX57CYY	NW & Wales
CX07CUK	NW & Wales	CX54DKF	NW & Wales	CX55EBA	NW & Wales	CX57CZA	NW & Wales
CX07CUU	NW & Wales	CX54DKJ	NW & Wales	CX55EBC	NW & Wales	CX58ETY	NW & Wales
CX07CUV	NW & Wales	CX54DKK	NW & Wales	CX55EBD	NW & Wales	CX58ETZ	NW & Wales
CX07CUW	NW & Wales	CX54DKL	NW & Wales	CX55EBF	NW & Wales	CX58EUA	NW & Wales
CX07CUY	NW & Wales	CX54DKN	NW & Wales	CX55EBG	NW & Wales	CX58EUA	NW & Wales
CX07CVA	NW & Wales	CX54DKO	NW & Wales	CX55EBJ	NW & Wales	CX58EUB	NW & Wales
CX07CVB	NW & Wales	CX54DKU	NW & Wales	CX55FAF	NW & Wales	CX58EUD	NW & Wales
CX08DJJ	NW & Wales	CX54DKV	NW & Wales	CX55FAJ	NW & Wales	CX58EUE	NW & Wales
CX08DJK	NW & Wales	CX54DKY	NW & Wales	CX56CDY	NW & Wales	CX58EUF	NW & Wales
CX08DJO	NW & Wales	CX54DLD	NW & Wales	CX56CDZ	NW & Wales	CX58EUH	NW & Wales
CX08DJU	NW & Wales	CX54DLF	NW & Wales	CX56CEA	NW & Wales	CX58EUJ	NW & Wales
CX09BFM	NW & Wales	CX54DLJ	NW & Wales	CX56CEF	NW & Wales	CX58EUK	NW & Wales
CX09BFN	NW & Wales	CX54DLK	NW & Wales	CX56CEJ	NW & Wales	CX58EUL	NW & Wales
CX09BFO	NW & Wales	CX54EPJ	NW & Wales	CX56CEK	NW & Wales	CX58EUM	NW & Wales
CX09BFP	NW & Wales	CX54EPK	NW & Wales	CX56CEN	NW & Wales	CX58EUN	NW & Wales
CX09BFU	NW & Wales	CX54EPL	NW & Wales	CX56CEO	NW & Wales	CX58EUO	NW & Wales

Arriva Midlands operates the commuter service from Tamworth to Birmingham through Sutton Coldfield. Interurban livery with City Link route lettering is carried by 4201, FJ08LVM, a Volvo B9TL with Wrightbus Eclipse Gemini bodywork, as it sets off on the return journey north. *Mark Bailey*

CX58EUP	NW & Wales	CX58EWG	NW & Wales	CX58EZF	NW & Wales	CX58GBY	NW & Wales
CX58EUR	NW & Wales	CX58EWH	NW & Wales	CX58EZG	NW & Wales	CX58GBZ	NW & Wales
CX58EUT	NW & Wales	CX58EWJ	NW & Wales	CX58EZH	NW & Wales	CX58GCF	NW & Wales
CX58EUU	NW & Wales	CX58EWK	NW & Wales	CX58EZJ	NW & Wales	D170FYM	NW & Wales
CX58EUV	NW & Wales	CX58EWL	NW & Wales	CX58EZK	NW & Wales	D171FYM	NW & Wales
CX58EUW	NW & Wales	CX58EWM	NW & Wales	CX58EZL	NW & Wales	D187FYM	Scotland
CX58EUY	NW & Wales	CX58EWN	NW & Wales	CX58FYU	NW & Wales	D242FYM	NW & Wales
CX58EUZ	NW & Wales	CX58EWO	NW & Wales	CX58FYV	NW & Wales	D553YNO	Original Tour
CX58EVB	NW & Wales	CX58EWP	NW & Wales	CX58FYW	NW & Wales	D675YNO	Original Tour
CX58EVC	NW & Wales	CX58EWR	NW & Wales	CX58FYY	NW & Wales	DD08BCL	Tellings group
CX58EVD	NW & Wales	CX58EWS	NW & Wales	CX58FYZ	NW & Wales	DK55FWY	NW & Wales
CX58EVF	NW & Wales	CX58EWT	NW & Wales	CX58FZM	NW & Wales	DK55FWZ	NW & Wales
CX58EVG	NW & Wales	CX58EWU	NW & Wales	CX58FZN	NW & Wales	DK55FXA	NW & Wales
CX58EVH	NW & Wales	CX58EWV	NW & Wales	CX58FZO	NW & Wales	DK55FXB	NW & Wales
CX58EVJ	NW & Wales	CX58EWW	NW & Wales	CX58FZP	NW & Wales	DK55FXC	NW & Wales
CX58EVK	NW & Wales	CX58EWY	NW & Wales	CX58FZR	NW & Wales	DK55FXD	NW & Wales
CX58EVL	NW & Wales	CX58EWZ	NW & Wales	CX58FZS	NW & Wales	DK55FXE	NW & Wales
CX58EVN	NW & Wales	CX58EXA	NW & Wales	CX58FZT	NW & Wales	DK55FXF	NW & Wales
CX58EVP	NW & Wales	CX58EXB	NW & Wales	CX58FZU	NW & Wales	DK55FXG	NW & Wales
CX58EVR	NW & Wales	CX58EXC	NW & Wales	CX58FZV	NW & Wales	DK55FXH	NW & Wales
CX58EVT	NW & Wales	CX58EXE	NW & Wales	CX58FZW	NW & Wales	DK55FXJ	NW & Wales
CX58EVU	NW & Wales	CX58EXF	NW & Wales	CX58FZY	NW & Wales	DK55FXL	NW & Wales
CX58EVV	NW & Wales	CX58EXG	NW & Wales	CX58FZZ	NW & Wales	DK55FXM	NW & Wales
CX58EVW	NW & Wales	CX58EXH	NW & Wales	CX58GAA	NW & Wales	DK55FXO	NW & Wales
CX58EVY	NW & Wales	CX58EXJ	NW & Wales	CX58GAO	NW & Wales	DK55FXR	NW & Wales
CX58EWA	NW & Wales	CX58EXK	NW & Wales	CX58GAU	NW & Wales	DK55FXS	NW & Wales
CX58EWB	NW & Wales	CX58EXL	NW & Wales	CX58GBE	NW & Wales	DK55FXT	NW & Wales
CX58EWC	NW & Wales	CX58EZA	NW & Wales	CX58GBF	NW & Wales	DK55FXU	NW & Wales
CX58EWD	NW & Wales	CX58EZB	NW & Wales	CX58GBO	NW & Wales	DK55FXV	NW & Wales
CX58EWE	NW & Wales	CX58EZC	NW & Wales	CX58GBU	NW & Wales	DK55FXW	NW & Wales
CX58EWF	NW & Wales	CX58EZE	NW & Wales	CX58GBV	NW & Wales	DK55FXX	NW & Wales

DK55FXY	NW & Wales	FD02UKU	Midlands	FJ08LVO	Midlands	FN52XBG	Midlands
DK55FXZ	NW & Wales	FD52GGO	Midlands	FJ08LVP	Midlands	FY58HYH	Midlands
DK55FYA	NW & Wales	FD52GGP	Midlands	FJ08LVR	Midlands	FY58HYK	Midlands
DK55FYB	NW & Wales	FD52GGU	Midlands	FJ08LVS	Midlands	FY58HYL	Midlands
DK55FYC	NW & Wales	FD52GGV	Midlands	FJ08LVT	Midlands	FY58HYM	Midlands
DK55FYD	NW & Wales	FE51WSU	Midlands	FJ51JYN	Tellings group	FY58HYN	Midlands
DK55FYE	NW & Wales	FE51WSV	Midlands	FJ54OTN	Midlands	FY58HYO	Midlands
DK55FYF	NW & Wales	FE51YWH	Midlands	FJ54OTP	Midlands	FY58HYP	Midlands
DK55FYG	NW & Wales	FE51YWJ	Midlands	FJ54OTR	Midlands	FY58HYR	Midlands
DK55FYH	NW & Wales	FE51YWK	Midlands	FJ54OTT	Midlands	FY58HYS	Midlands
DK55FYJ	NW & Wales	FE51YWL	Midlands	FJ54OTV	Midlands	FY58HYT	Midlands
DK55FYL	NW & Wales	FE51YWM	Midlands	FJ54OTW	Midlands	FY58HYU	Midlands
DK55FYM	NW & Wales	FG56OBF	Midlands	FJ54OTX	Midlands	FY58HYV	Midlands
DK55FYN	NW & Wales	FJ04PFX	Midlands	FJ55BVT	Midlands	FY58HYW	Midlands
DK55FYO	NW & Wales	FJ06ZKK	Tellings group	FJ55BVU	Midlands	FY58HYX	Midlands
DK55FYP	NW & Wales	FJ06ZKL	Tellings group	FJ55BWA	Midlands	G2PGL	Tellings group
DK55FYR	NW & Wales	FJ06ZPV	Midlands	FJ55BWB	Midlands	G35HKY	NW & Wales
DK55FYS	NW & Wales	FJ06ZPW	Midlands	FJ55BWC	Midlands	G37HKY	Tellings group
DK55FYT	NW & Wales	FJ06ZPX	Midlands	FJ55BWD	Midlands	G129YEV	The Shires
DK55FYV	NW & Wales	FJ06ZRL	Midlands	FJ55BWE	Midlands	G130YEV	The Shires
DK55FYW	NW & Wales	FJ06ZRN	Midlands	FJ55BWF	Midlands	G131YWC	The Shires
E52UNE	NW & Wales	FJ06ZRO	Midlands	FJ55BWG	Midlands	G132YWC	The Shires
E224WBG	NW & Wales	FJ06ZRP	Midlands	FJ56KFA	Midlands	G2190TV	Tellings group
E227WBG	NW & Wales	FJ06ZSD	Midlands	FJ56KFC	Midlands	G231VWL	The Shires
E766JAR	Original Tour	FJ06ZSE	Midlands	FJ56KFD	Midlands	G234BRT	Tellings group
E767JAR	Original Tour	FJ06ZSF	Midlands	FJ56KFE	Midlands	G235BRT	Tellings group
E768JAR	Original Tour	FJ06ZSG	Midlands	FJ56KFF	Midlands	G235VWL	The Shires
E769JAR	Original Tour	FJ06ZSK	Midlands	FJ56KFG	Midlands	G254SRG	North East
E770JAR	Original Tour	FJ06ZSL	Midlands	FJ56KFK	Midlands	G257UVK	North East
E771JAR	Original Tour	FJ06ZSN	Midlands	FJ56KFL	Midlands	G258UVK	North East
E772JAR	Original Tour	FJ06ZSO	Midlands	FJ56OBC	Midlands	G283UMJ	The Shires
E773JAR	Original Tour	FJ06ZSP	Midlands	FJ56OBD	Midlands	G286UMJ	The Shires
E774JAR	Original Tour	FJ06ZST	Midlands	FJ56OBE	Midlands	G287UMJ	The Shires
E964JAR	Original Tour	FJ06ZSU	Midlands	FJ56OBG	Midlands	G290UMJ	The Shires
E965JAR	Original Tour	FJ06ZSV	Midlands	FJ56OBH	Midlands	G291UMJ	The Shires
EU05BCL	Tellings group	FJ06ZSW	Midlands	FJ56OBK	Midlands	G292UMJ	The Shires
EU05DVW	Original Tour	FJ06ZSX	Midlands	FJ56OBL	Midlands	G293UMJ	The Shires
EU05DVX	Original Tour	FJ06ZSY	Midlands	FJ56OBM	Midlands	G294UMJ	The Shires
EU08BCL	Tellings group	FJ06ZSZ	Midlands	FJ56OBN	Midlands	G371YUR	Tellings group
F166XCS	Tellings group	FJ06ZTB	Midlands	FJ56OBP	Midlands	G504SFT	Scotland
F246MTW	Tellings group	FJ06ZTC	Midlands	FJ56PCX	Midlands	G613BPH	Tellings group
F572SMG	Scotland	FJ06ZTD	Midlands	FJ56PCY	Midlands	G614BPH	Tellings group
F575SMG	Scotland	FJ06ZTE	Midlands	FJ56PCZ	Midlands	G615BPH	Tellings group
F634LMJ	The Shires	FJ06ZTF	Midlands	FJ56PDK	Midlands	G617BPH	Tellings group
F636LMJ	The Shires	FJ06ZTG	Midlands	FJ56PDO	Midlands	G624BPH	Southern Counties
F637LMJ	The Shires	FJ06ZTH	Midlands	FJ58KXF	Midlands	G629BPH	Southern Counties
F639LMJ	Tellings group	FJ06ZTK	Midlands	FJ58KXG	Midlands	G630BPH	Southern Counties
F640LMJ	Tellings group	FJ06ZTL	Midlands	FJ58KXH	Midlands	G631BPH	Southern Counties
F641LMJ	The Shires	FJ06ZTM	Midlands	FJ58KXK	Midlands	G633BPH	Southern Counties
F643LMJ	The Shires	FJ06ZTN	Midlands	FJ58KXL	Midlands	G634BPH	Southern Counties
F644LMJ	The Shires	FJ06ZTO	Midlands	FJ58KXM	Midlands	G635BPH	Southern Counties
F892BKK	Tellings group	FJ06ZTP	Midlands	FJ58KXN	Midlands	G636BPH	Southern Counties
FD02UKB	Midlands	FJ07TKC	The Shires	FJ58KXO	Midlands	G643BPH	Tellings group
FD02UKC	Midlands	FJ07TKE	The Shires	FJ58KXP	Midlands	G645UPP	The Shires
FD02UKE	Midlands	FJ07TKF	The Shires	FJ58KXR	Midlands	G646UPP	The Shires
FD02UKG	Midlands	FJ08DXG	The Shires	FJ58KXS	Midlands	G647UPP	The Shires
FD02UKJ	Midlands	FJ08DXK	The Shires	FJ58KXT	Midlands	G648UPP	The Shires
FD02UKK	Midlands	FJ08DXL	The Shires	FJ58KXU	Midlands	G649UPP	The Shires
FD02UKL	Midlands	FJ08DXM	The Shires	FJ58KXV	Midlands	G650UPP	The Shires
FD02UKN	Midlands	FJ08DXO	The Shires	FJ58KXW	Midlands	G651UPP	The Shires
FD02UKO	Midlands	FJ08DXP	The Shires	FJ58KXX	Midlands	G652UPP	The Shires
FD02UKP	Midlands	FJ08DXR	The Shires	FJ58KXY	Midlands	G653UPP	The Shires
FD02UKR	Midlands	FJ08LVL	Midlands	FK52MML	Midlands	G654UPP	The Shires
FD02UKS	Midlands	FJ08LVM	Midlands	FL52MML	Midlands	G655UPP	The Shires
FD02UKT	Midlands	FJ08LVN	Midlands	FN04AFJ	Midlands	G656UPP	The Shires

Reg	Operator	Reg	Operator	Reg	Operator	Reg	Operator
G657UPP	The Shires	GN04UDM	Southern Counties	GN06EBB	Southern Counties	GN09AWR	Southern Counties
GB03LLC	Tellings group	GN04UDP	Southern Counties	GN06EBF	Southern Counties	GN09AWU	Southern Counties
GB03TGM	Southern Counties	CN04UDO	Southern Counties	GN06EBG	Southern Counties	GN09AWV	Southern Counties
GB03TGM	Tellings group	GN04UDT	Southern Counties	GN06EBH	Southern Counties	GN09AWW	Southern Counties
GB04BCL	Tellings group	GN04UDU	Southern Counties	GN06EUU	Southern Counties	GN09AWX	Southern Counties
GB04LLC	Tellings group	GN04UDV	Southern Counties	GN06EVG	Southern Counties	GN09AWY	Southern Counties
GB04TGM	Tellings group	GN04UDW	Southern Counties	GN06EVH	Southern Counties	GN09AWZ	Southern Counties
GB05BCL	Tellings group	GN04UDX	Southern Counties	GN06EVJ	Southern Counties	GN09AXA	Southern Counties
GB08BCL	Tellings group	GN04UDY	Southern Counties	GN06EVK	Southern Counties	GN09AXB	Southern Counties
GJ52HDZ	Southern Counties	GN04UDZ	Southern Counties	GN06EVL	Southern Counties	GN09AXC	Southern Counties
GK51SYY	Southern Counties	GN04UEA	Southern Counties	GN06EVM	Southern Counties	GN09AXD	Southern Counties
GK51SYZ	Southern Counties	GN04UEB	Southern Counties	GN06EVP	Southern Counties	GN09AXF	Southern Counties
GK51SZC	Southern Counties	GN04UEC	Southern Counties	GN06EVN	Southern Counties	GN09AXG	Southern Counties
GK51SZD	Southern Counties	GN04UED	Southern Counties	GN06EVT	Southern Counties	GN09AXH	Southern Counties
GK51SZE	Southern Counties	GN04UEE	Southern Counties	GN06EVU	Southern Counties	GN09AXJ	Southern Counties
GK51SZF	Southern Counties	GN04UEG	Southern Counties	GN06EWC	Southern Counties	GN09AXK	Southern Counties
GK51SZG	Southern Counties	GN04UEH	Southern Counties	GN06EWD	Southern Counties	GN09AXM	Southern Counties
GK51SZJ	Southern Counties	GN04UEJ	Southern Counties	GN06EWE	Southern Counties	GN09AXO	Southern Counties
GK51SZL	Southern Counties	GN04UEK	Southern Counties	GN07AVB	Southern Counties	GN54MYO	Southern Counties
GK51SZN	Southern Counties	GN04UEL	Southern Counties	GN07AVC	Southern Counties	GN54MYP	Southern Counties
GK52YUW	Southern Counties	GN04UEM	Southern Counties	GN07AVD	Southern Counties	GN54MYR	Southern Counties
GK52YUX	Southern Counties	GN04UEP	Southern Counties	GN07AVE	Southern Counties	GN54MYT	Southern Counties
GK52YUY	Southern Counties	GN04UER	Southern Counties	GN07AVF	Southern Counties	GN54MYU	Southern Counties
GK52YVB	Southern Counties	GN04UES	Southern Counties	GN07AVG	Southern Counties	GN57BNX	Southern Counties
GK52YVC	Southern Counties	GN04UET	Southern Counties	GN07AVJ	Southern Counties	GN57BNY	Southern Counties
GK52YVD	Southern Counties	GN04UEU	Southern Counties	GN07AVL	Southern Counties	GN57BNZ	Southern Counties
GK52YVE	Southern Counties	GN04UEV	Southern Counties	GN07AVM	Southern Counties	GN57BOF	Southern Counties
GK52YVF	Southern Counties	GN04UEW	Southern Counties	GN07AVN	Southern Counties	GN57BOH	Southern Counties
GK52YVG	Southern Counties	GN04UEX	Southern Counties	GN07AVO	Southern Counties	GN57BOJ	Southern Counties
GK52YVJ	Southern Counties	CN04UEY	Southern Counties	GN07AVP	Southern Counties	GN57BOU	Southern Counties
GK52YVL	Southern Counties	GN04UEZ	Southern Counties	GN07DLE	Southern Counties	GN57BOV	Southern Counties
GK53AOA	Southern Counties	GN04UFA	Southern Counties	GN07DLF	Southern Counties	GN57BPE	Southern Counties
GK53AOB	Southern Counties	GN04UFB	Southern Counties	GN07DLJ	Southern Counties	GN57BPF	Southern Counties
GK53AOC	Southern Counties	GN04UFC	Southern Counties	GN07DLK	Southern Counties	GN57BPK	Southern Counties
GK53AOD	Southern Counties	GN04UFD	Southern Counties	GN07DLO	Southern Counties	GN57BPO	Southern Counties
GK53AOE	Southern Counties	GN04UFE	Southern Counties	GN07DLU	Southern Counties	GN57BPU	Southern Counties
GK53AOF	Southern Counties	GN04UFG	Southern Counties	GN07DLV	Southern Counties	GN57BPV	Southern Counties
GK53AOG	Southern Counties	GN04UFH	Southern Counties	GN07DLX	Southern Counties	GN57BPX	Southern Counties
GK53AOH	Southern Counties	GN04UFJ	Southern Counties	GN07DLY	Southern Counties	GN57BPY	Southern Counties
GK53AOJ	Southern Counties	GN04UFK	Southern Counties	GN07DLZ	Southern Counties	GN58BSO	Southern Counties
GK53AOL	Southern Counties	GN04UFL	Southern Counties	GN07DME	Southern Counties	GN58BSU	Southern Counties
GK53AON	Southern Counties	GN04UFM	Southern Counties	GN07DMF	Southern Counties	GN58BSV	Southern Counties
GK53AOO	Southern Counties	GN04UFP	Southern Counties	GN07DMO	Southern Counties	GN58BSX	Southern Counties
GK53AOP	Southern Counties	GN04UFR	Southern Counties	GN07DMU	Southern Counties	GN58BSY	Southern Counties
GK53AOR	Southern Counties	GN04UFS	Southern Counties	GN07DMV	Southern Counties	GN58BSZ	Southern Counties
GK53AOT	Southern Counties	GN04UFT	Southern Counties	GN08CGO	Southern Counties	GN58BTE	Southern Counties
GK53AOU	Southern Counties	GN04UFU	Southern Counties	GN08CGU	Southern Counties	GN58BTF	Southern Counties
GK53AOV	Southern Counties	GN04UFV	Southern Counties	GN08CGV	Southern Counties	GN58BTO	Southern Counties
GK53AOW	Southern Counties	GN04UFW	Southern Counties	GN08CGX	Southern Counties	GN58BTU	Southern Counties
GK53AOX	Southern Counties	GN04UFX	Southern Counties	GN08CGY	Southern Counties	GN58BTV	Southern Counties
GK53AOY	Southern Counties	GN04UFY	Southern Counties	GN08CGZ	Southern Counties	GN58BTX	Southern Counties
GK53AOZ	Southern Counties	GN04UFZ	Southern Counties	GN09AVV	Southern Counties	GN58BTY	Southern Counties
GKA449L	NW & Wales	GN04UGA	Southern Counties	GN09AVW	Southern Counties	GN58BTZ	Southern Counties
GN04UCW	Southern Counties	GN04UGB	Southern Counties	GN09AVX	Southern Counties	GN58BUA	Southern Counties
GN04UCX	Southern Counties	GN04UGC	Southern Counties	GN09AVY	Southern Counties	GN58BUE	Southern Counties
GN04UCY	Southern Counties	GN04UGD	Southern Counties	GN09AVZ	Southern Counties	GN58BUF	Southern Counties
GN04UCZ	Southern Counties	GN04UGE	Southern Counties	GN09AWA	Southern Counties	GN58BUH	Southern Counties
GN04UDB	Southern Counties	GN04UGF	Southern Counties	GN09AWB	Southern Counties	GN58BUJ	Southern Counties
GN04UDD	Southern Counties	GN04UGG	Southern Counties	GN09AWC	Southern Counties	GN58BUO	Southern Counties
GN04UDE	Southern Counties	GN05ANU	Southern Counties	GN09AWG	Southern Counties	GN58BUP	Southern Counties
GN04UDG	Southern Counties	GN05ANV	Southern Counties	GN09AWH	Southern Counties	GN58BUU	Southern Counties
GN04UDH	Southern Counties	GN05ANX	Southern Counties	GN09AWJ	Southern Counties	GN58BUV	Southern Counties
GN04UDJ	Southern Counties	GN05AOA	Southern Counties	GN09AWM	Southern Counties	GN58LVA	Southern Counties
GN04UDK	Southern Counties	GN05AOB	Southern Counties	GN09AWO	Southern Counties	GN58LVB	Southern Counties
GN04UDL	Southern Counties	GN05AOC	Southern Counties	GN09AWP	Southern Counties	GN59FVB	Southern Counties

GN59FVC	Southern Counties	J315BSH	Original Tour	K506BHN	North East	KE51PUF	Southern Counties
GN59FVD	Southern Counties	J316BSH	Original Tour	K507BHN	North East	KE51PUH	Southern Counties
GN59FVE	Southern Counties	J317BSH	Original Tour	K508BHN	North East	KE51PUJ	Southern Counties
GN59FVF	Southern Counties	J318BSH	Original Tour	K515BHN	North East	KE51PUK	Southern Counties
GN59FVG	Southern Counties	J319BSH	Original Tour	K517BHN	North East	KE51PUO	Southern Counties
GN59FVH	Southern Counties	J320BSH	Original Tour	K538ORH	Scotland	KE51PUU	Southern Counties
GN59FVJ	Southern Counties	J321BSH	Original Tour	K539ORH	Scotland	KE51PUV	Southern Counties
GO02CLA	Tellings group	J322BSH	Original Tour	K540ORH	Scotland	KE51PUY	Southern Counties
GO02STS	Tellings group	J323BSH	Original Tour	K541ORH	Scotland	KE51PVD	Southern Counties
GO03CLA	Tellings group	J324BSH	Original Tour	K542ORH	Midlands	KE51PVF	The Shires
GO08BCL	Tellings group	J325BSH	Original Tour	K542ORH	Midlands	KE51PVZ	The Shires
GO53CLA	Tellings group	J326BSH	Original Tour	K543ORH	Scotland	KE53KBO	The Shires
GO54BCL	Tellings group	J327BSH	Original Tour	K544ORH	Midlands	KE53KBP	The Shires
GO58CHC	Southern Counties	J327VAW	Midlands	K551OHN	Midlands	KE53NEU	The Shires
GO58CHD	Southern Counties	J328BSH	Original Tour	K582MGT	North East	KE53NFA	The Shires
GO58CHF	Southern Counties	J329BSH	Original Tour	K709PCN	North East	KE53NFC	The Shires
GO58CHG	Southern Counties	J330BSH	Original Tour	K717PCN	North East	KE53NFD	The Shires
GO58CHH	Southern Counties	J331BSH	Original Tour	K906SKR	Southern Counties	KE53NFF	The Shires
GS05TGM	Tellings group	J332BSH	Original Tour	K907SKR	Southern Counties	KE53NFG	The Shires
GSU347	Scotland	J334BSH	Original Tour	K908SKR	Southern Counties	KE54HHF	The Shires
GSU348	Tellings group	J335BSH	Original Tour	K909SKR	Southern Counties	KE54LNR	The Shires
GYE456W	NW & Wales	J336BSH	Original Tour	K910SKR	Southern Counties	KE54LPC	The Shires
H74DVM	Midlands	J337BSH	Original Tour	K946SGG	Scotland	KE54LPF	The Shires
H78DVM	NW & Wales	J338BSH	Original Tour	K947OEM	Tellings group	KE54LPJ	The Shires
H79DVM	NW & Wales	J339BSH	Original Tour	K948OEM	Tellings group	KE55CKO	The Shires
H81DVM	Midlands	J340BSH	Original Tour	KC03PGE	The Shires	KE55CKP	The Shires
H85DVM	NW & Wales	J341BSH	Original Tour	KC03PGF	The Shires	KE55CKU	The Shires
H86DVM	NW & Wales	J342BSH	Original Tour	KC06EVN	Tellings group	KE55CTF	The Shires
H87DVM	NW & Wales	J343BSH	Original Tour	KC06EVP	Tellings group	KE55CTK	The Shires
H155PVW	Tellings group	J344BSH	Original Tour	KC51NFO	Southern Counties	KE55CTO	The Shires
H197GRO	The Shires	J345BSH	Original Tour	KC51PUX	Southern Counties	KE55CTU	The Shires
H202GRO	The Shires	J346BSH	Original Tour	KE03OUK	The Shires	KE55CTV	The Shires
H262GEV	Southern Counties	J347BSH	Original Tour	KE03OUL	The Shires	KE55CVA	The Shires
H263GEV	Southern Counties	J348BSH	Original Tour	KE03OUM	The Shires	KE55CVG	The Shires
H264GEV	Southern Counties	J349BSH	Original Tour	KE03OUN	The Shires	KE55CVH	The Shires
H265GEV	Southern Counties	J350BSH	Original Tour	KE03OUP	The Shires	KE55CVJ	The Shires
H278LEF	NW & Wales	J351BSH	Original Tour	KE03OUS	The Shires	KE55CVK	The Shires
H279LEF	NW & Wales	J352BSH	Original Tour	KE03OUU	The Shires	KE55CVL	The Shires
H430XGK	Tellings group	J413NCP	North East	KE03UKK	The Shires	KE55CVM	The Shires
H588DVM	NW & Wales	J433BSH	Original Tour	KE04CZF	The Shires	KE55FBX	The Shires
H683GPF	Tellings group	J465MKL	Yorkshire	KE04CZG	The Shires	KE55FBY	The Shires
H768EKJ	Southern Counties	J468OKP	North East	KE04CZH	The Shires	KE55FDF	The Shires
H769EKJ	Southern Counties	J807KHD	Tellings group	KE04OSU	The Shires	KE55FDG	The Shires
H804RWJ	Tellings group	J926CYL	North East	KE04OSV	The Shires	KE55GVY	The Shires
HDZ2604	The Shires	JJD403D	London	KE04PZF	The Shires	KE55GVZ	The Shires
HDZ2605	The Shires	K1BLU	NW & Wales	KE04PZG	The Shires	KE55GWA	The Shires
HDZ2606	The Shires	K27EWC	NW & Wales	KE05FMM	The Shires	KE55GWC	The Shires
HDZ2607	The Shires	K73SRG	Midlands	KE05FMO	The Shires	KE55GXR	The Shires
HDZ2611	The Shires	K101OHF	Midlands	KE05FMP	The Shires	KE55KPG	The Shires
HIL2148	Scotland	K103OHF	Midlands	KE05FMU	The Shires	KE55KPJ	The Shires
HX04HUH	Tellings group	K105OHF	Midlands	KE05FMV	The Shires	KE55KTC	The Shires
HX04HUK	Tellings group	K107OHF	Midlands	KE05FMX	The Shires	KE55KTD	The Shires
HX51LRK	Tellings group	K108OHF	Midlands	KE05GOH	The Shires	KE55KTJ	The Shires
HX51LRL	Tellings group	K140RYS	North East	KE07EVX	The Shires	KE57EPA	The Shires
HX51LRN	Tellings group	K320CVX	Yorkshire	KE07EVY	The Shires	KE57EPC	The Shires
HX51LRO	Tellings group	K321CVX	The Shires	KE07EWA	The Shires	KJ02JXT	The Shires
HX51LSO	The Shires	K322CVX	The Shires	KE07EWB	The Shires	KL52CWN	The Shires
J6SLT	NW & Wales	K401HWW	Yorkshire	KE07EWC	The Shires	KL52CWO	The Shires
J7SLT	NW & Wales	K402HWW	Yorkshire	KE51PSZ	The Shires	KL52CWP	The Shires
J23GCX	Yorkshire	K403HWW	Yorkshire	KE51PTO	The Shires	KL52CWR	The Shires
J64BJN	The Shires	K410FHJ	The Shires	KE51PTU	The Shires	KL52CWT	The Shires
J201JRP	Tellings group	K415BHN	North East	KE51PTX	The Shires	KL52CWU	The Shires
J220HGY	North East	K422BHN	North East	KE51PTY	Southern Counties	KL52CWV	The Shires
J221HGY	North East	K504BHN	North East	KE51PTZ	Southern Counties	KL52CWW	The Shires
J251KWM	NW & Wales	K505BHN	North East	KE51PUA	Southern Counties	KL52CWZ	The Shires

New to Stephensons of Uttoxeter, L95HRF is now 7458 in the North East fleet, where all of Arriva's Optare Spectra buses have been concentrated. The model is based on the DB250 chassis and was available in both normal and low floor variants. *David Little*

KL52CXA	The Shires	KX09GYC	The Shires	KX54AVE	The Shires	L156UKB	Midlands
KL52CXB	The Shires	KX09GYD	The Shires	KX59ACJ	The Shires	L157YVK	Midlands
KL52CXC	The Shires	KX09GYE	The Shires	KX59ACO	The Shires	L159BFT	Yorkshire
KL52CXD	The Shires	KX09GYF	The Shires	KX59AEE	The Shires	L159GYL	North East
KL52CXE	The Shires	KX09GYG	The Shires	KX59AEF	The Shires	L160GYL	North East
KL52CXF	The Shires	KX09GYH	The Shires	KYV663X	NW & Wales	L161GYL	North East
KL52CXG	The Shires	KX09GYJ	The Shires	KYV689X	NW & Wales	L201TKA	Scotland
KL52CXH	The Shires	KX09GYK	The Shires	L25LSX	Scotland	L201YCU	Southern Counties
KL52CXJ	The Shires	KX09GYN	The Shires	L94HRF	North East	L203YCU	Midlands
KL52CXK	The Shires	KX09GYO	The Shires	L95HRF	North East	L207YCU	Southern Counties
KL52CXM	The Shires	KX09GYP	The Shires	L100SBS	North East	L208YCU	Southern Counties
KL52CXN	The Shires	KX09GYR	The Shires	L102MEH	North East	L210TKA	Scotland
KL52CXO	The Shires	KX09GYS	The Shires	L114YVK	North East	L210YCU	Southern Counties
KL52CXP	The Shires	KX09GYT	The Shires	L115YVK	NW & Wales	L211SBG	NW & Wales
KL52CXR	The Shires	KX09GYU	The Shires	L127YVK	Midlands	L211YCU	Southern Counties
KL52CXS	The Shires	KX09GYV	The Shires	L128YVK	Yorkshire	L212YCU	Southern Counties
KP51SYF	Tellings group	KX09GYW	The Shires	L129YVK	North East	L216TKA	Scotland
KP51UEV	Tellings group	KX09GYY	The Shires	L130YVK	Midlands	L219TKA	North East
KP51UEW	Tellings group	KX09GYZ	The Shires	L131YVK	Midlands	L220TKA	North East
KP51UEX	Tellings group	KX09GZA	The Shires	L135YVK	Midlands	L225TKA	North East
KP51UEY	Tellings group	KX09GZB	The Shires	L136YVK	Yorkshire	L231TKA	Scotland
KP51UEZ	Tellings group	KX09GZC	The Shires	L137YVK	Yorkshire	L232TKA	North East
KS05JJE	The Shires	KX09GZD	The Shires	L140YVK	Yorkshire	L235TKA	Scotland
KU02YUF	Tellings group	KX09GZE	The Shires	L141YVK	North East	L244TKA	North East
KU02YUG	Tellings group	KX09KDJ	The Shires	L142YVK	Midlands	L273FVN	Scotland
KX04HSJ	Tellings group	KX09KDN	The Shires	L143YVK	Midlands	L274FVN	Scotland
KX09GXW	The Shires	KX09KDO	The Shires	L144YVK	Midlands	L300SBS	Midlands
KX09GXY	The Shires	KX09KDU	The Shires	L146YVK	Yorkshire	L301TEM	NW & Wales
KX09GXZ	The Shires	KX09KDV	The Shires	L149BFT	North East	L302TEM	NW & Wales
KX09GYA	The Shires	KX54AVD	The Shires	L152YVK	North East	L303NFA	Midlands
KX09GYB	The Shires			L155YVK	Midlands	L303TEM	NW & Wales

L306HPP	Scotland	LF02PMO	London	LF52UPK	London	LG03MBX	London
L308HPP	Midlands	LF02PMV	London	LF52UPM	London	LG03MBY	London
L311HPP	Midlands	LF02PMX	London	LF52UPN	London	LG03MDE	London
L502TKA	NW & Wales	LF02PMY	London	LF52UPO	London	LG03MDF	London
L505TKA	NW & Wales	LF02PNE	London	LF52UPP	London	LG03MDK	London
L506CPJ	Southern Counties	LF02PNJ	London	LF52UPR	London	LG03MDN	London
L507BNX	Midlands	LF02PNK	London	LF52UPS	London	LG03MDU	London
L507CPJ	Southern Counties	LF02PNL	London	LF52UPT	London	LG03MEV	London
L507TKA	NW & Wales	LF02PNN	London	LF52UPV	London	LG03MFA	London
L508TKA	NW & Wales	LF02PNO	London	LF52UPW	London	LG03MFE	London
L509CPJ	Southern Counties	LF02PNU	London	LF52UPX	London	LG03MFF	London
L510CPJ	Southern Counties	LF02PNV	London	LF52UPZ	London	LG03MFK	London
L513BNX	Midlands	LF02PNX	London	LF52URA	London	LG03MLL	London
L514BNX	Midlands	LF02PNY	London	LF52URB	London	LG03MLN	London
L515BNX	Midlands	LF02POA	London	LF52URC	London	LG03MLV	London
L516BNX	Midlands	LF02POH	London	LF52URD	London	LG03MMU	London
L516CPJ	Midlands	LF02PSO	London	LF52URE	London	LG03MMV	London
L519FHN	North East	LF02PSU	London	LF52URG	London	LG03MMX	London
L521FHN	North East	LF02PSY	London	LF52URH	London	LG03MOA	London
L522BNX	Midlands	LF02PSZ	London	LF52URJ	London	LG03MOF	London
L524FHN	North East	LF02PTO	London	LF52URK	London	LG03MOV	London
L526FHN	North East	LF02PTU	London	LF52URL	London	LG03MPF	London
L527FHN	North East	LF02PTX	London	LF52URM	London	LG03MPU	London
L529FHN	North East	LF02PTY	London	LF52URN	London	LG03MPV	London
L531FHN	North East	LF02PTZ	London	LF52URO	London	LG03MPX	London
L532EHD	North East	LF02PVA	The Shires	LF52URP	London	LG03MPY	London
L533FHN	North East	LF02PVE	London	LF52URR	London	LG03MPZ	London
L536FHN	North East	LF02PVJ	London	LF52URS	London	LG03MRU	London
L537FHN	North East	LF02PVK	London	LF52URT	London	LG03MRV	London
L541FHN	North East	LF02PVL	London	LF52URU	London	LG03MRX	London
L542FHN	North East	LF02PVN	London	LF52URV	London	LG03MRY	London
L544GHN	North East	LF02PVO	London	LF52URW	London	LG03MSU	London
L549GHN	North East	LF52UNV	London	LF52URX	London	LG03MSV	London
L551GHN	North East	LF52UNW	London	LF52URY	London	LG03MSX	London
L588JSG	NW & Wales	LF52UNX	London	LF52URZ	London	LG05BCL	Tellings group
L601EKM	Midlands	LF52UNY	London	LF52USB	London	LG52DAA	London
L605EKM	Southern Counties	LF52UNZ	London	LF52USC	London	LG52DAO	London
L729VNL	North East	LF52UOA	London	LF52USD	London	LG52DAU	London
L735VNL	North East	LF52UOB	London	LF52USE	London	LG52DBO	London
L736VNL	North East	LF52UOC	London	LF52USG	London	LG52DBU	London
L737VNL	North East	LF52UOD	London	LF52USH	London	LG52DBV	London
L738VNL	North East	LF52UOE	London	LF52USJ	London	LG52DBY	London
L739VNL	North East	LF52UOG	London	LF52USL	London	LG52DBZ	London
L740VNL	North East	LF52UOH	London	LF52USM	London	LG52DCE	London
L741VNL	North East	LF52UOJ	London	LF52USN	London	LG52DCF	London
L748VNL	North East	LF52UOK	London	LF52USO	London	LG52DCO	London
L759VNL	North East	LF52UOL	London	LF52USS	London	LG52DCU	London
LB52UYK	Tellings group	LF52UOM	London	LF52UST	London	LG52DCV	London
LF02PKA	London	LF52UON	London	LF52USU	London	LG52DCX	London
LF02PKC	London	LF52UOO	London	LF52USV	London	LG52DCY	London
LF02PKD	London	LF52UOP	London	LF52USW	London	LG52DCZ	London
LF02PKE	London	LF52UOR	London	LF52USX	London	LG52DDA	London
LF02PKJ	London	LF52UOS	London	LF52USY	London	LG52DDE	London
LF02PKO	London	LF52UOT	London	LF52USZ	London	LG52DDF	London
LF02PKU	London	LF52UOU	London	LF52UTA	London	LG52DDJ	London
LF02PKV	London	LF52UOV	London	LF52UTB	London	LG52DDK	London
LF02PKX	London	LF52UOW	London	LF52UTC	London	LG52DDL	London
LF02PKY	London	LF52UOX	London	LF52UTE	London	LJ03MDV	London
LF02PLJ	London	LF52UOY	London	LF52UTG	London	LJ03MDX	London
LF02PLN	London	LF52UPA	London	LF52UTH	London	LJ03MDY	London
LF02PLO	London	LF52UPB	London	LF52UTL	London	LJ03MDZ	London
LF02PLU	London	LF52UPC	London	LF52UTM	London	LJ03MEU	London
LF02PLV	London	LF52UPD	London	LG03MBF	London	LJ03MFN	London
LF02PLX	London	LF52UPG	London	LG03MBU	London	LJ03MFP	London
LF02PLZ	London	LF52UPH	London	LG03MBV	London	LJ03MFU	London

LJ03MFV	London	LJ03MVC	London	LJ04LDC	London	LJ05BJV	London
LJ03MFX	London	LJ03MVD	London	LJ04LDD	London	LJ05BJX	London
LJ03MFY	London	LJ03MVE	London	LJ04LDF	London	LJ05BJY	London
LJ03MFZ	London	LJ03MVF	London	LJ04LDK	London	LJ05BJZ	London
LJ03MGE	London	LJ03MVG	London	LJ04LDL	London	LJ05BKA	London
LJ03MGU	London	LJ03MVT	London	LJ04LDN	London	LJ05BKD	London
LJ03MGV	London	LJ03MVV	London	LJ04LDU	London	LJ05BKF	London
LJ03MGX	London	LJ03MVW	London	LJ04LDX	London	LJ05BKY	London
LJ03MGY	London	LJ03MVX	London	LJ04LDY	London	LJ05BKZ	London
LJ03MGZ	London	LJ03MVY	London	LJ04LDZ	London	LJ05BLF	London
LJ03MHA	London	LJ03MVZ	London	LJ04LEF	London	LJ05BLK	London
LJ03MHE	London	LJ03MWA	London	LJ04LEU	London	LJ05BLN	London
LJ03MHF	London	LJ03MWC	London	LJ04LFB	London	LJ05BLV	London
LJ03MHK	London	LJ03MWD	London	LJ04LFD	London	LJ05BLX	London
LJ03MHL	London	LJ03MWE	London	LJ04LFE	London	LJ05BLY	London
LJ03MHM	London	LJ03MWF	London	LJ04LFF	London	LJ05BMO	London
LJ03MHN	London	LJ03MWG	London	LJ04LFG	London	LJ05BMU	London
LJ03MHU	London	LJ03MWK	London	LJ04LFH	London	LJ05BMV	London
LJ03MHV	London	LJ03MWL	London	LJ04LFK	London	LJ05BMY	London
LJ03MHX	London	LJ03MWN	London	LJ04LFL	London	LJ05BMZ	London
LJ03MHY	London	LJ03MWP	London	LJ04LFM	London	LJ05BNA	London
LJ03MHZ	London	LJ03MWU	London	LJ04LFN	London	LJ05BNB	London
LJ03MJE	London	LJ03MWV	London	LJ04LFP	London	LJ05BND	London
LJ03MJF	London	LJ03MWX	London	LJ04LFR	London	LJ05BNE	London
LJ03MJK	London	LJ03MXH	London	LJ04LFS	London	LJ05BNF	London
LJ03MJU	London	LJ03MXK	London	LJ04LFT	London	LJ05BNK	London
LJ03MJV	London	LJ03MXL	London	LJ04LFU	London	LJ05BNL	London
LJ03MJX	London	LJ03MXM	London	LJ04LFV	London	LJ05GKX	London
LJ03MJY	London	LJ03MXN	London	LJ04LFW	London	LJ05GKY	London
LJ03MKA	London	LJ03MXP	London	LJ04LFX	London	LJ05GKZ	London
LJ03MKC	London	LJ03MXR	London	LJ04LFZ	London	LJ05GLF	London
LJ03MKD	London	LJ03MXS	London	LJ04LGA	London	LJ05GLK	London
LJ03MKE	London	LJ03MXT	London	LJ04LGC	London	LJ05GLV	London
LJ03MKF	London	LJ03MXU	London	LJ04LGD	London	LJ05GLY	London
LJ03MKG	London	LJ03MXV	London	LJ04LGE	London	LJ05GLZ	London
LJ03MKK	London	LJ03MXW	London	LJ04LGF	London	LJ05GME	London
LJ03MKL	London	LJ03MXX	London	LJ04LGG	London	LJ05GMF	London
LJ03MKM	London	LJ03MXY	London	LJ04LGK	London	LJ05GOP	London
LJ03MKN	London	LJ03MXZ	London	LJ04LGL	London	LJ05GOU	London
LJ03MKU	London	LJ03MYA	London	LJ04LGN	London	LJ05GOX	London
LJ03MKV	London	LJ03MYB	London	LJ04LGV	London	LJ05GPF	London
LJ03MKX	London	LJ03MYC	London	LJ04LGW	London	LJ05GPK	London
LJ03MKZ	London	LJ03MYD	London	LJ04LGX	London	LJ05GPO	London
LJ03MLE	London	LJ03MYF	London	LJ04LGY	London	LJ05GPU	London
LJ03MLF	London	LJ03MYG	London	LJ04YWE	London	LJ05GPX	London
LJ03MLK	London	LJ03MYH	London	LJ04YWS	London	LJ05GPY	London
LJ03MLX	London	LJ03MYK	London	LJ04YWT	London	LJ05GPZ	London
LJ03MLY	London	LJ03MYL	London	LJ04YWU	London	LJ05GRF	London
LJ03MLZ	London	LJ03MYM	London	LJ04YWV	London	LJ05GRK	London
LJ03MMA	London	LJ03MYN	London	LJ04YWW	London	LJ05GRU	London
LJ03MME	London	LJ03MYP	London	LJ04YWX	London	LJ05GRX	London
LJ03MMF	London	LJ03MYR	London	LJ04YWY	London	LJ05GRZ	London
LJ03MMK	London	LJ03MYS	London	LJ04YWZ	London	LJ05GSO	London
LJ03MSY	London	LJ03MYT	London	LJ04YXA	London	LJ05GSU	London
LJ03MTE	London	LJ03MYU	London	LJ04YXB	London	LJ07EBO	London
LJ03MTF	London	LJ03MYV	London	LJ05BHL	London	LJ07EBP	London
LJ03MTK	London	LJ03MYX	London	LJ05BHN	London	LJ07EBU	London
LJ03MTU	London	LJ03MYY	London	LJ05BHO	London	LJ07ECF	London
LJ03MTV	London	LJ03MYZ	London	LJ05BHP	London	LJ07ECN	London
LJ03MTY	London	LJ03MZD	London	LJ05BHU	London	LJ07ECT	London
LJ03MTZ	London	LJ03MZE	London	LJ05BHV	London	LJ07ECU	London
LJ03MUA	London	LJ03MZF	London	LJ05BHW	London	LJ07ECW	London
LJ03MUB	London	LJ03MZG	London	LJ05BHX	London	LJ07ECX	London
LJ03MUW	London	LJ03MZL	London	LJ05BHY	London	LJ07ECY	London
LJ03MUY	London	LJ04LDA	London	LJ05BHZ	London	LJ07ECZ	London

LJ07EDC	London	LJ08CWC	London	LJ51DAA	London	LJ51DHZ	London
LJ07EDF	London	LJ08CXR	London	LJ51DAO	London	LJ51DJD	London
LJ07EDK	London	LJ08CXS	London	LJ51DAU	London	LJ51DJE	London
LJ07EDL	London	LJ08CXT	London	LJ51DBO	London	LJ51DJF	London
LJ07EDO	London	LJ08CXU	London	LJ51DBU	London	LJ51DJK	London
LJ07EDP	London	LJ08CXV	London	LJ51DBV	London	LJ51DJO	London
LJ07EDR	London	LJ08CYC	London	LJ51DBX	London	LJ51DJU	London
LJ07EDU	London	LJ08CYE	London	LJ51DBY	London	LJ51DJV	London
LJ07EDV	London	LJ08CYF	London	LJ51DBZ	London	LJ51DJX	London
LJ07EDX	London	LJ08CYG	London	LJ51DCE	London	LJ51DJY	London
LJ07EEA	London	LJ08CYH	London	LJ51DCF	London	LJ51DJZ	London
LJ07EEB	London	LJ08CYK	London	LJ51DCO	London	LJ51DKA	London
LJ07UDD	Original Tour	LJ08CYL	London	LJ51DCU	London	LJ51DKD	London
LJ07XEN	Original Tour	LJ08CYO	London	LJ51DCV	London	LJ51DKE	London
LJ07XEO	Original Tour	LJ08CYP	London	LJ51DCX	London	LJ51DKF	London
LJ07XEP	Original Tour	LJ08CYS	London	LJ51DCY	London	LJ51DKK	London
LJ07XER	Original Tour	LJ09KOE	London	LJ51DCZ	London	LJ51DKL	London
LJ07XES	Original Tour	LJ09KOH	London	LJ51DDA	London	LJ51DKN	London
LJ07XET	Original Tour	LJ09KOU	London	LJ51DDE	London	LJ51DKO	London
LJ07XEU	Original Tour	LJ09KOV	London	LJ51DDF	London	LJ51DKU	London
LJ07XEV	Original Tour	LJ09KOW	London	LJ51DDK	London	LJ51DKV	London
LJ07XEW	Original Tour	LJ09KOX	London	LJ51DDL	London	LJ51DKX	London
LJ08CSO	London	LJ09KPA	London	LJ51DDN	London	LJ51DKY	London
LJ08CSU	London	LJ09KPE	London	LJ51DDO	London	LJ51DLD	London
LJ08CSV	London	LJ09KPF	London	LJ51DDU	London	LJ51DLF	London
LJ08CSX	London	LJ09KPG	London	LJ51DDV	London	LJ51DLK	London
LJ08CSY	London	LJ09KPK	London	LJ51DDX	London	LJ51DLN	London
LJ08CSZ	London	LJ09KPL	London	LJ51DDY	London	LJ51DLU	London
LJ08CTE	London	LJ09KPN	London	LJ51DDZ	London	LJ51DLV	London
LJ08CTF	London	LJ09KPO	London	LJ51DEU	London	LJ51DLX	London
LJ08CTK	London	LJ09KPR	London	LJ51DFA	London	LJ51DLY	London
LJ08CTO	London	LJ09KPT	London	LJ51DFC	London	LJ51DLZ	London
LJ08CTV	London	LJ09KPU	London	LJ51DFD	London	LJ51ORA	London
LJ08CTX	London	LJ09KPV	London	LJ51DFE	London	LJ51ORC	London
LJ08CTY	London	LJ09KPX	London	LJ51DFF	London	LJ51ORF	London
LJ08CTZ	London	LJ09KPY	London	LJ51DFG	London	LJ51ORG	London
LJ08CUA	London	LJ09KPZ	London	LJ51DFK	London	LJ51ORH	London
LJ08CUE	London	LJ09KRD	London	LJ51DFL	London	LJ51ORK	London
LJ08CUG	London	LJ09KRE	London	LJ51DFN	London	LJ51ORL	London
LJ08CUH	London	LJ09KRF	London	LJ51DFO	London	LJ51OSK	London
LJ08CUK	London	LJ09KRG	London	LJ51DFP	London	LJ51OSX	London
LJ08CUO	London	LJ09KRK	London	LJ51DFU	London	LJ51OSY	London
LJ08CUU	London	LJ09KRN	London	LJ51DFX	London	LJ51OSZ	London
LJ08CUV	London	LJ09KRO	London	LJ51DFY	London	LJ53BAA	London
LJ08CUW	London	LJ09KRU	London	LJ51DFZ	London	LJ53BAO	London
LJ08CUY	London	LJ09SSO	London	LJ51DGE	London	LJ53BAU	London
LJ08CVA	London	LJ09SSU	London	LJ51DGF	London	LJ53BAV	London
LJ08CVB	London	LJ09SSV	London	LJ51DGO	London	LJ53BBE	London
LJ08CVC	London	LJ09SSX	London	LJ51DGU	London	LJ53BBF	London
LJ08CVD	London	LJ09SSZ	London	LJ51DGV	London	LJ53BBK	London
LJ08CVF	London	LJ09STX	London	LJ51DGX	London	LJ53BBN	London
LJ08CVG	London	LJ09STZ	London	LJ51DGY	London	LJ53BBO	London
LJ08CVH	London	LJ09SUA	London	LJ51DGZ	London	LJ53BBU	London
LJ08CVK	London	LJ09SUF	London	LJ51DHA	London	LJ53BBV	London
LJ08CVL	London	LJ09SUH	London	LJ51DHC	London	LJ53BBX	London
LJ08CVM	London	LJ09SUO	London	LJ51DHD	London	LJ53BBZ	London
LJ08CVO	London	LJ09SUU	London	LJ51DHF	London	LJ53BCF	London
LJ08CVR	London	LJ09SUV	London	LJ51DHG	London	LJ53BCK	London
LJ08CVS	London	LJ09SUX	London	LJ51DHK	London	LJ53BCO	London
LJ08CVT	London	LJ09SUY	London	LJ51DHL	London	LJ53BCU	London
LJ08CVU	London	LJ09SVA	London	LJ51DHO	London	LJ53BCV	London
LJ08CVV	London	LJ09SVC	London	LJ51DHP	London	LJ53BCX	London
LJ08CVX	London	LJ09SVD	London	LJ51DHV	London	LJ53BCY	London
LJ08CVZ	London	LJ09SVE	London	LJ51DHX	London	LJ53BCZ	London
LJ08CWA	London	LJ09SVF	London	LJ51DHY	London	LJ53BDE	London

LJ53BDF	London	LJ54BAA	London	LJ54LHK	London	LJ57USZ	London
LJ53BDO	London	LJ54BAO	London	LJ54LHL	London	LJ57UTA	London
LJ53BDU	London	LJ54BAU	London	LJ54LHM	London	LJ57UTB	London
LJ53BDV	London	LJ54BAV	London	LJ54LHN	London	LJ57UTC	London
LJ53BDX	London	LJ54BBE	London	LJ54LHO	London	LJ57UTE	London
LJ53BDY	London	LJ54BBF	London	LJ54LHP	London	LJ57UTF	London
LJ53BDZ	London	LJ54BBK	London	LJ54LHR	London	LJ58AUC	London
LJ53BEO	London	LJ54BBN	London	LJ55BPZ	London	LJ58AUE	London
LJ53BEU	London	LJ54BBO	London	LJ55BRV	London	LJ58AUV	London
LJ53BEY	London	LJ54BBU	London	LJ55BRX	London	LJ58AUW	London
LJ53BFA	London	LJ54BBV	London	LJ55BRZ	London	LJ58AUX	London
LJ53BFE	London	LJ54BBX	London	LJ55BSO	London	LJ58AUY	London
LJ53BFF	London	LJ54BBZ	London	LJ55BSU	London	LJ58AVB	London
LJ53BFK	London	LJ54BCE	London	LJ55BSV	London	LJ58AVC	London
LJ53BFL	London	LJ54BCF	London	LJ55BSX	London	LJ58AVD	London
LJ53BFM	London	LJ54BCK	London	LJ55BSY	London	LJ58AVE	London
LJ53BFN	London	LJ54BCO	London	LJ55BSZ	London	LJ58AVG	London
LJ53BFO	London	LJ54BCU	London	LJ55BTE	London	LJ58AVK	London
LJ53BFP	London	LJ54BCV	London	LJ55BTF	London	LJ58AVT	London
LJ53BFU	London	LJ54BCX	London	LJ55BTO	London	LJ58AVU	London
LJ53BFX	London	LJ54BCY	London	LJ55BTU	London	LJ58AVV	London
LJ53BFY	London	LJ54BCZ	London	LJ55BTV	London	LJ58AVX	London
LJ53BGF	London	LJ54BDE	London	LJ55BTX	London	LJ58AVY	London
LJ53BGK	London	LJ54BDF	London	LJ55BTY	London	LJ58AVZ	London
LJ53BGO	London	LJ54BDO	London	LJ55BTZ	London	LJ58AWA	London
LJ53BGU	London	LJ54BDU	London	LJ55BUA	London	LJ58AWC	London
LJ53NFE	London	LJ54BDV	London	LJ55BUE	London	LJ58AWF	London
LJ53NFF	London	LJ54BDX	London	LJ55BUP	London	LJ58AWG	London
LJ53NFG	London	LJ54BDY	London	LJ55BUR	London	LJ59AAE	London
LJ53NFT	London	LJ54BDZ	London	LJ55BUS	London	LJ59AAF	London
LJ53NFU	London	LJ54BEO	London	LJ55BUT	London	LJ59AAK	London
LJ53NFV	London	LJ54BEU	London	LJ55BUU	London	LJ59AAN	London
LJ53NFX	London	LJ54BFA	London	LJ55BUV	London	LJ59AAO	London
LJ53NFY	London	LJ54BFE	London	LJ55BUW	London	LJ59AAU	London
LJ53NFZ	London	LJ54BFF	London	LJ55BUX	London	LJ59AAV	London
LJ53NGE	London	LJ54BFK	London	LJ55BUY	London	LJ59AAX	London
LJ53NGF	London	LJ54BFL	London	LJ55BUZ	London	LJ59AAY	London
LJ53NGG	London	LJ54BFM	London	LJ55BVD	London	LJ59AAZ	London
LJ53NGN	London	LJ54BFN	London	LJ55BVE	London	LJ59ABF	London
LJ53NGU	London	LJ54BFO	London	LJ55BVF	London	LJ59ABK	London
LJ53NGV	London	LJ54BFP	London	LJ55BVG	London	LJ59ABN	London
LJ53NGX	London	LJ54BFV	London	LJ55BVH	London	LJ59ABO	London
LJ53NGY	London	LJ54BFY	London	LJ55BVK	London	LJ59ABU	London
LJ53NGZ	London	LJ54BFZ	London	LJ55BVL	London	LJ59ABV	London
LJ53NHA	London	LJ54BGE	London	LJ55BVM	London	LJ59ABX	London
LJ53NHB	London	LJ54BGF	London	LJ56AOW	London	LJ59ABZ	London
LJ53NHC	London	LJ54BGK	London	LJ56AOX	London	LJ59ACF	London
LJ53NHD	London	LJ54BGO	London	LJ56AOY	London	LJ59ACO	London
LJ53NHE	London	LJ54BJE	London	LJ56APZ	London	LJ59ACU	London
LJ53NHF	London	LJ54BJF	London	LJ56ARF	London	LJ59ACV	London
LJ53NHG	London	LJ54BJK	London	LJ56ARO	London	LJ59ACX	London
LJ53NHH	London	LJ54BJO	London	LJ56ARU	London	LJ59ACY	London
LJ53NHK	London	LJ54BJU	London	LJ56ARX	London	LJ59ACZ	London
LJ53NHL	London	LJ54BKG	London	LJ56ARZ	London	LJ59ADO	London
LJ53NHN	London	LJ54BKK	London	LJ56ASO	London	LJ59ADV	London
LJ53NHO	London	LJ54BKL	London	LJ56ASU	London	LJ59ADX	London
LJ53NHP	London	LJ54BKN	London	LJ56ASV	London	LJ59ADZ	London
LJ53NHT	London	LJ54BKO	London	LJ56ASX	London	LJ59AEA	London
LJ53NHV	London	LJ54BKU	London	LJ57USS	London	LJ59AEA	London
LJ53NHX	London	LJ54BKV	London	LJ57UST	London	LJ59AEB	London
LJ53NHY	London	LJ54BKX	London	LJ57USU	London	LJ59AEC	London
LJ53NHZ	London	LJ54LGV	London	LJ57USV	London	LJ59AED	London
LJ53NJF	London	LJ54LHF	London	LJ57USW	London	LJ59AEE	London
LJ53NJK	London	LJ54LHG	London	LJ57USX	London	LJ59AEF	London
LJ53NJN	London	LJ54LHH	London	LJ57USY	London	LJ59AEG	London

One of the small school buses operating with SAL in Lecco is Iveco 80E15, 3188, AH841CY. Bodywork is by Cacciamali. *Bill Potter*

LJ59AEK	London	LX05KNZ	Original Tour	M157WKA	NW & Wales	M215YKD	NW & Wales
LJ59AEL	London	LX05KOA	Original Tour	M160GRY	Scotland	M216YKD	NW & Wales
LJ59AEM	London	M2SLT	NW & Wales	M160SKR	NW & Wales	M217AKB	NW & Wales
LJ59AEN	London	M20GGY	NW & Wales	M161SKR	NW & Wales	M218AKB	NW & Wales
LJ59AET	London	M20MPS	Midlands	M162SKR	NW & Wales	M218YKC	NW & Wales
LJ59AEU	London	M30GGY	NW & Wales	M163SKR	NW & Wales	M219AKB	NW & Wales
LJ59AEV	London	M65FDS	Scotland	M165GRY	Scotland	M219YKC	NW & Wales
LJ59AEW	London	M67FDS	Scotland	M167WKA	NW & Wales	M220AKB	NW & Wales
LJ59AEX	London	M100PHA	Midlands	M169GRY	Scotland	M221AKB	NW & Wales
LJ59AEY	London	M102RMS	NW & Wales	M170GRY	North East	M223AKB	NW & Wales
LJ59AEZ	London	M103RMS	NW & Wales	M172GRY	North East	M224AKB	NW & Wales
LJ59GTF	London	M104RMS	Scotland	M172YKA	NW & Wales	M225AKB	NW & Wales
LJ59GTU	London	M105RMS	NW & Wales	M176GRY	North East	M226AKB	NW & Wales
LJ59GTZ	London	M106RMS	Scotland	M177GRY	North East	M227AKB	NW & Wales
LJ59GUA	London	M107RMS	Scotland	M178LYP	North East	M228AKB	NW & Wales
LJ59GVC	London	M108RMS	Scotland	M179LYP	North East	M229AKB	NW & Wales
LJ59GVE	London	M109RMS	Scotland	M180LYP	North East	M230AKB	NW & Wales
LJ59GVF	London	M109XKC	NW & Wales	M186YKA	North East	M231AKB	NW & Wales
LJ59GVG	London	M110RMS	Scotland	M186YKA	North East	M232AKB	NW & Wales
LJ59GVK	London	M110XKC	NW & Wales	M189YKA	NW & Wales	M301YBG	NW & Wales
LL04BCL	Tellings group	M112RMS	Scotland	M190YKA	NW & Wales	M302SAJ	North East
LX03HCE	London	M112XKC	NW & Wales	M191YKA	NW & Wales	M302YBG	NW & Wales
LX03HCG	London	M113RMS	Scotland	M192YKA	NW & Wales	M303SAJ	North East
LX03HCL	London	M113XKC	NW & Wales	M195YKA	NW & Wales	M303YBG	NW & Wales
LX03HDE	London	M114RMS	Scotland	M198YKA	NW & Wales	M304SAJ	North East
LX05GDV	Original Tour	M115RMS	Scotland	M200CBB	Southern Counties	M305SAJ	North East
LX05GDY	Original Tour	M117RMS	Scotland	M201YKA	NW & Wales	M370FTY	North East
LX05GDZ	Original Tour	M118RMS	Scotland	M211YKD	North East	M371FTY	North East
LX05GEJ	Original Tour	M119RMS	Scotland	M212YKD	NW & Wales	M372FTY	North East
LX05HRO	Original Tour	M120RMS	Scotland	M213YKD	NW & Wales	M373FTY	North East
LX05HSC	Original Tour	M121RMS	Scotland	M214YKD	NW & Wales	M374FTY	North East

The Italian flag flies by the rail line as 2403, COB53310 pulls into Como town centre on route 50. This is one of three 17.5m BredaMenariniBus articulated buses with Siccar bodywork from the 1980s included with ASF Autolinee. Arriva manages the Como operation and owns approximately half of the share capital through SPT with the remaining portion owned by Omnibus Partecipazioni. *Bill Potter*

M375FTY	North East	M519WHF	NW & Wales	M549WTJ	NW & Wales	M689HPF	North East
M376FTY	North East	M520WHF	NW & Wales	M550WTJ	NW & Wales	M690HPF	North East
M377FTY	North East	M521WHF	NW & Wales	M551WTJ	NW & Wales	M691HPF	North East
M401EFD	Midlands	M523MPF	NW & Wales	M552WTJ	NW & Wales	M692HPF	North East
M402EFD	Midlands	M523WHF	NW & Wales	M553WTJ	NW & Wales	M693HPF	North East
M403EFD	Midlands	M524MPF	NW & Wales	M554WTJ	NW & Wales	M694HPF	Yorkshire
M404EFD	Midlands	M524WHF	NW & Wales	M556WTJ	NW & Wales	M696HPF	Yorkshire
M411UNW	Yorkshire	M525MPM	Southern Counties	M557WTJ	NW & Wales	M697HPF	Yorkshire
M413UNW	Yorkshire	M525WHF	NW & Wales	M558WTJ	NW & Wales	M702HPF	Yorkshire
M415UNW	Yorkshire	M527WHF	NW & Wales	M559WTJ	NW & Wales	M761JPA	NW & Wales
M417UNW	Yorkshire	M528WHF	NW & Wales	M561WTJ	NW & Wales	M762JPA	NW & Wales
M419UNW	Yorkshire	M529WHF	NW & Wales	M562WTJ	NW & Wales	M763JPA	NW & Wales
M420UNW	Yorkshire	M530WHF	NW & Wales	M563WTJ	NW & Wales	M764JPA	Southern Counties
M421UNW	Midlands	M531WHF	NW & Wales	M564YEM	NW & Wales	M770DRG	North East
M422UNW	Midlands	M532WHF	NW & Wales	M565YEM	NW & Wales	M802MOJ	Midlands
M423UNW	Yorkshire	M532WHF	NW & Wales	M566YEM	NW & Wales	M803MOJ	Midlands
M424UNW	Yorkshire	M533WHF	NW & Wales	M567YEM	NW & Wales	M804MOJ	Midlands
M429UNW	Midlands	M534WHF	NW & Wales	M568YEM	NW & Wales	M805MOJ	Midlands
M430UNW	Midlands	M535WHF	NW & Wales	M569YEM	NW & Wales	M812RCP	Yorkshire
M431UNW	Midlands	M536WHF	NW & Wales	M570YEM	NW & Wales	M813RCP	Yorkshire
M501AJC	North East	M537WHF	NW & Wales	M571YEM	NW & Wales	M814RCP	Yorkshire
M502AJC	North East	M538WHF	NW & Wales	M572YEM	NW & Wales	M815RCP	Yorkshire
M503AJC	North East	M540WHF	NW & Wales	M573YEM	NW & Wales	M816RCP	Yorkshire
M504AJC	North East	M541WHF	NW & Wales	M574YEM	NW & Wales	M817RCP	Yorkshire
M514WHF	NW & Wales	M542WHF	NW & Wales	M575YEM	NW & Wales	M818RCP	Yorkshire
M515WHF	NW & Wales	M543WHF	NW & Wales	M617PKP	Southern Counties	M819RCP	Yorkshire
M516WHF	NW & Wales	M544WTJ	NW & Wales	M619PKP	Southern Counties	M831SDA	Scotland
M517KPA	NW & Wales	M545WTJ	NW & Wales	M685HPF	North East	M832SDA	Scotland
M518KPA	NW & Wales	M546WTJ	NW & Wales	M686HPF	North East	M833SDA	Scotland
M518WHF	NW & Wales	M547WTJ	NW & Wales	M687HPF	North East	M847RCP	NW & Wales
M519KPA	NW & Wales	M548WTJ	NW & Wales	M688HPF	North East	M849RCP	NW & Wales

Reg	Region	Reg	Region	Reg	Region	Reg	Region
M911MKM	Southern Counties	MX09LLXR	NW & Wales	MX59FGF	NW & Wales	N131DWM	NW & Wales
M913MKM	Southern Counties	MX09LLXS	NW & Wales	MX59FGG	NW & Wales	N132DWM	NW & Wales
M914MKM	Southern Counties	MX09LXE	NW & Wales	MX59FGJ	NW & Wales	N133DWM	NW & Wales
M915MKM	Southern Counties	MX09LXF	NW & Wales	MX59FHB	NW & Wales	N134DWM	NW & Wales
M916MKM	Southern Counties	MX09LXG	NW & Wales	MX59JJE	NW & Wales	N160VVO	North East
M917MKM	Southern Counties	MX09LXH	NW & Wales	MX59JJF	NW & Wales	N162VVO	North East
M918MKM	Southern Counties	MX09LXJ	NW & Wales	MX59JJK	NW & Wales	N163VVO	North East
M919MKM	Southern Counties	MX09LXT	NW & Wales	MX59JJL	NW & Wales	N164VVO	North East
M920MKM	Southern Counties	MX09LXU	NW & Wales	MX59JJO	NW & Wales	N166PUT	Midlands
M921PKN	NW & Wales	MX09LXV	NW & Wales	MX59JJU	NW & Wales	N168PUT	Midlands
M922PKN	Southern Counties	MX09LXW	NW & Wales	MX59JJV	NW & Wales	N170PUT	Midlands
M923PKN	Southern Counties	MX09LXY	NW & Wales	MX59JJY	NW & Wales	N171PUT	Midlands
M925PKN	Southern Counties	MX09LXZ	NW & Wales	MX59JJZ	NW & Wales	N173PUT	Midlands
M927EYS	NW & Wales	MX09LYA	NW & Wales	MX59JKZ	NW & Wales	N174PUT	Midlands
M928EYS	NW & Wales	MX09LYC	NW & Wales	MX59JZA	NW & Wales	N175PUT	Midlands
M929EYS	NW & Wales	MX09LYD	NW & Wales	MX59JZB	NW & Wales	N177PUT	Midlands
M930EYS	NW & Wales	MX09LYF	NW & Wales	MX59JZC	NW & Wales	N178PUT	Midlands
M931EYS	NW & Wales	MX09LYG	NW & Wales	MX59JZD	NW & Wales	N179PUT	Midlands
M932EYS	NW & Wales	MX09LYH	NW & Wales	MX59JZE	NW & Wales	N1810YH	North East
M933EYS	NW & Wales	MX09LYJ	NW & Wales	MX59JZF	NW & Wales	N1820YH	North East
M934EYS	NW & Wales	MX09OOJ	NW & Wales	MX59JZG	NW & Wales	N1830YH	North East
M936EYS	NW & Wales	MX09OOU	NW & Wales	MX59JZH	NW & Wales	N204LCK	Tellings group
M945LYR	Southern Counties	MX09OOV	NW & Wales	MX59JZJ	NW & Wales	N210TPK	NW & Wales
MF52LYY	NW & Wales	MX09OOW	NW & Wales	MXT179	Original Tour	N211DWM	NW & Wales
MF52LYZ	NW & Wales	MX09OOY	NW & Wales	N4BLU	NW & Wales	N211TPK	NW & Wales
MF52LZA	NW & Wales	MX09OPA	NW & Wales	N32KGS	The Shires	N212TPK	NW & Wales
MF52LZB	NW & Wales	MX09OPB	NW & Wales	N35JPP	The Shires	N213TPK	NW & Wales
MK02BUS	The Shires	MX09OPC	NW & Wales	N36JPP	The Shires	N214TPK	NW & Wales
MK52XNN	NW & Wales	MX09OPD	NW & Wales	N37JPP	The Shires	N215TPK	NW & Wales
MK52XNO	NW & Wales	MX09OPE	NW & Wales	N38JPP	The Shires	N216TPK	NW & Wales
MK52XNP	NW & Wales	MX09OPF	NW & Wales	N39JPP	The Shires	N217TPK	NW & Wales
MK52XNR	NW & Wales	MX09OPG	NW & Wales	N41JPP	The Shires	N218TPK	NW & Wales
MK52XNS	Scotland	MX09OPH	NW & Wales	N42JPP	The Shires	N219TPK	NW & Wales
MM02ZVH	NW & Wales	MX09OPJ	NW & Wales	N43JPP	The Shires	N220TPK	Southern Counties
MM02ZVJ	NW & Wales	MX09OPK	NW & Wales	N45JPP	The Shires	N221TPK	Southern Counties
MM03TGM	Tellings group	MX09OPL	NW & Wales	N46JPP	The Shires	N223TPK	Southern Counties
MV02XYJ	Scotland	MX09OPM	NW & Wales	N51FWU	Yorkshire	N224TPK	Southern Counties
MV02XYK	Scotland	MX09OPN	NW & Wales	N52FWU	Yorkshire	N225TPK	Southern Counties
MX09EKK	NW & Wales	MX09OPO	NW & Wales	N101YVU	NW & Wales	N226TPK	Southern Counties
MX09EKL	NW & Wales	MX54ZVA	Tellings group	N103YVU	NW & Wales	N227TPK	Southern Counties
MX09EKM	NW & Wales	MX59AAE	NW & Wales	N104YVU	NW & Wales	N228TPK	Southern Counties
MX09EKN	NW & Wales	MX59AAF	NW & Wales	N105YVU	NW & Wales	N229TPK	Southern Counties
MX09EKO	NW & Wales	MX59AAJ	NW & Wales	N106DWM	NW & Wales	N230TPK	Southern Counties
MX09EKP	NW & Wales	MX59AAK	NW & Wales	N107DWM	NW & Wales	N231TPK	Southern Counties
MX09EKR	NW & Wales	MX59AAN	NW & Wales	N108DWM	NW & Wales	N233CKA	NW & Wales
MX09EKT	NW & Wales	MX59AAO	NW & Wales	N109DWM	NW & Wales	N233TPK	Southern Counties
MX09EKU	NW & Wales	MX59AAU	NW & Wales	N110DWM	NW & Wales	N234CKA	NW & Wales
MX09EKW	NW & Wales	MX59AAV	NW & Wales	N113DWM	NW & Wales	N234TPK	Southern Counties
MX09EKY	NW & Wales	MX59AAY	NW & Wales	N114DWM	NW & Wales	N235CKA	NW & Wales
MX09JHH	NW & Wales	MX59AAZ	NW & Wales	N115DWM	NW & Wales	N235TPK	Southern Counties
MX09JHK	NW & Wales	MX59ABF	NW & Wales	N116DWM	NW & Wales	N236CKA	NW & Wales
MX09JHL	NW & Wales	MX59ABK	NW & Wales	N117DWM	NW & Wales	N236TPK	Southern Counties
MX09JHO	NW & Wales	MX59ABN	NW & Wales	N118DWM	NW & Wales	N237CKA	NW & Wales
MX09JHU	NW & Wales	MX59FFR	NW & Wales	N119DWM	NW & Wales	N237VPH	Southern Counties
MX09JHV	NW & Wales	MX59FFS	NW & Wales	N120DWM	NW & Wales	N238CKA	NW & Wales
MX09JHY	NW & Wales	MX59FFT	NW & Wales	N121DWM	NW & Wales	N238VPH	Midlands
MX09JHZ	NW & Wales	MX59FFU	NW & Wales	N122DWM	NW & Wales	N239CKA	NW & Wales
MX09JJE	NW & Wales	MX59FFV	NW & Wales	N123DWM	NW & Wales	N239VPH	Southern Counties
MX09JJF	NW & Wales	MX59FFW	NW & Wales	N124DWM	NW & Wales	N240CKA	NW & Wales
MX09JTY	NW & Wales	MX59FFX	NW & Wales	N125DWM	NW & Wales	N240VPH	Midlands
MX09LLXK	NW & Wales	MX59FFZ	NW & Wales	N126DWM	NW & Wales	N241CKA	NW & Wales
MX09LLXL	NW & Wales	MX59FGA	NW & Wales	N127DWM	NW & Wales	N241VPH	Midlands
MX09LLXM	NW & Wales	MX59FGB	NW & Wales	N128DWM	NW & Wales	N242CKA	NW & Wales
MX09LLXN	NW & Wales	MX59FGD	NW & Wales	N129DWM	NW & Wales	N242VPH	Midlands
MX09LLXO	NW & Wales	MX59FGE	NW & Wales	N130DWM	NW & Wales	N243CKA	NW & Wales

Pictured in Southend is Southern Counties' 3252, N252BKK, one of ten Scania L113s with Wright Axcess-ultralow bodywork now running in the town, and the only Scania products with Southern Counties. *Dave Heath*

N243VPH	Midlands	N258CKA	NW & Wales	N288CKB	NW & Wales	N307CLV	NW & Wales
N244CKA	NW & Wales	N259BKK	Southern Counties	N288NCN	North East	N308CLV	NW & Wales
N244VPH	Midlands	N259CKA	NW & Wales	N289CKB	NW & Wales	N322TPK	Southern Counties
N245CKA	NW & Wales	N25FWU	NW & Wales	N289NCN	North East	N357OBC	Midlands
N245VPH	Southern Counties	N260CKA	NW & Wales	N28KGS	The Shires	N358OBC	Midlands
N246CKA	NW & Wales	N261CKA	NW & Wales	N290CKB	NW & Wales	N386OTY	North East
N246VPH	Southern Counties	N262CKA	NW & Wales	N290NCN	North East	N387OTY	North East
N247CKA	NW & Wales	N263CKA	NW & Wales	N291CKB	NW & Wales	N388OTY	North East
N247VPH	Southern Counties	N264CKA	NW & Wales	N292CKB	NW & Wales	N389OTY	North East
N248CKA	NW & Wales	N271CKB	NW & Wales	N293CKB	NW & Wales	N390OTY	North East
N248VPH	Midlands	N272CKB	NW & Wales	N294CKB	NW & Wales	N391OTY	North East
N249CKA	NW & Wales	N273CKB	NW & Wales	N295CKB	NW & Wales	N392OTY	North East
N249VPH	Midlands	N274CKB	NW & Wales	N296CKB	NW & Wales	N393OTY	North East
N24FWU	NW & Wales	N275CKB	NW & Wales	N297CKB	NW & Wales	N429XRC	Midlands
N250BKK	Southern Counties	N276CKB	NW & Wales	N298CKB	NW & Wales	N429XRC	Midlands
N250CKA	NW & Wales	N277CKB	NW & Wales	N299CKB	NW & Wales	N429XRC	Midlands
N251BKK	Southern Counties	N278CKB	NW & Wales	N29KGS	The Shires	N474MUS	Scotland
N251CKA	NW & Wales	N279CKB	NW & Wales	N301AMC	North East	N511XVN	North East
N252BKK	Southern Counties	N281CKB	NW & Wales	N301CKB	NW & Wales	N512XVN	North East
N252CKA	NW & Wales	N281NCN	North East	N301ENX	Midlands	N513XVN	North East
N253BKK	Southern Counties	N282CKB	NW & Wales	N302CKB	NW & Wales	N514XVN	North East
N253CKA	NW & Wales	N282NCN	North East	N302ENX	Midlands	N515XVN	North East
N254BKK	Southern Counties	N283CKB	NW & Wales	N303AMC	North East	N516XVN	North East
N254CKA	NW & Wales	N283NCN	North East	N303CLV	NW & Wales	N517XVN	North East
N255BKK	Southern Counties	N284CKB	NW & Wales	N303ENX	Midlands	N518XVN	North East
N255CKA	NW & Wales	N284NCN	North East	N304CLV	NW & Wales	N519XVN	North East
N256BKK	Southern Counties	N285CKB	NW & Wales	N304ENX	Midlands	N520XVN	North East
N256CKA	NW & Wales	N285NCN	North East	N305AMC	North East	N521MJO	The Shires
N257BKK	Southern Counties	N286CKB	NW & Wales	N305CLV	NW & Wales	N522MJO	The Shires
N257CKA	NW & Wales	N287CKB	NW & Wales	N305ENX	Midlands	N522XVN	North East
N258BKK	Southern Counties	N287NCN	North East	N306CLV	NW & Wales	N523MJO	Southern Counties

Reg	Area	Reg	Area	Reg	Area	Reg	Area
N523MJO	The Shires	N613DWY	NW & Wales	N803BKN	Scotland	NK09BRF	North East
N523XVN	North East	N614CKA	NW & Wales	N804BKN	Scotland	NK09BRV	North East
N524MJO	The Shires	N615CKA	NW & Wales	N806EHA	Midlands	NK09BRX	North East
N524XVN	North East	N616CKA	NW & Wales	N806XHN	North East	NK09BRZ	North East
N525XVN	North East	N617CKA	NW & Wales	N807EHA	Midlands	NK09DFMZ	North East
N527SPA	NW & Wales	N618CKA	NW & Wales	N807XHN	North East	NK09DFNZ	North East
N528SPA	NW & Wales	N619CKA	NW & Wales	N808EHA	Midlands	NK09EJD	North East
N529SPA	NW & Wales	N620CKA	NW & Wales	N808XHN	North East	NK09EJE	North East
N530SPA	NW & Wales	N621CKA	NW & Wales	N809XHN	North East	NK09EJF	North East
N531DWM	NW & Wales	N621KUA	Yorkshire	N810XHN	North East	NK09EJG	North East
N532DWM	NW & Wales	N622CKA	NW & Wales	N908ETM	The Shires	NK09EJJ	North East
N539TPF	Southern Counties	N622KUA	Yorkshire	N912ETM	The Shires	NK09EJL	North East
N540TPF	Tellings group	N623CKA	NW & Wales	N913ETM	The Shires	NK09EJV	North East
N541TPF	Tellings group	N623KUA	Yorkshire	NDZ4521	The Shires	NK09EJX	North East
N543TPK	Southern Counties	N671GUM	NW & Wales	NDZ7918	Southern Counties	NK09EJY	North East
N544TPK	Southern Counties	N672GUM	Scotland	NDZ7918	The Shires	NK09EJZ	North East
N576CKA	NW & Wales	N673GUM	Midlands	NDZ7919	The Shires	NK09EKA	North East
N577CKA	NW & Wales	N674GUM	Midlands	NDZ7926	Southern Counties	NK09EKB	North East
N578CKA	NW & Wales	N675GUM	Scotland	NDZ7933	The Shires	NK09EKC	North East
N579CKA	NW & Wales	N676GUM	NW & Wales	NDZ7935	The Shires	NK09EKD	North East
N580CKA	NW & Wales	N677GUM	Scotland	NK04VMD	Tellings group	NK09FNC	North East
N581CKA	NW & Wales	N678GUM	NW & Wales	NK05GVG	North East	NK09FND	North East
N582CKA	NW & Wales	N679GUM	Midlands	NK05GVX	North East	NK09FNE	North East
N583CKA	NW & Wales	N680GUM	Midlands	NK05GVY	North East	NK09FNF	North East
N584CKA	NW & Wales	N681GUM	Scotland	NK05GWA	North East	NK09FNG	North East
N585CKA	NW & Wales	N682GUM	NW & Wales	NK05GWC	North East	NK09FVR	North East
N586CKA	NW & Wales	N683GUM	Scotland	NK05GWD	North East	NK51ZSR	Tellings group
N587CKA	NW & Wales	N685GUM	Scotland	NK05GWE	North East	NK51ZST	Tellings group
N588CKA	NW & Wales	N686GUM	Scotland	NK05GWF	North East	NK51ZSU	Tellings group
N589CKA	NW & Wales	N687GUM	Scotland	NK05GWG	North East	NK53HHX	North East
N590CKA	NW & Wales	N689GUM	Midlands	NK05GWJ	North East	NK53HHY	North East
N591CKA	NW & Wales	N689GUM	Scotland	NK05GWM	North East	NK53HHZ	North East
N592CKA	NW & Wales	N690GUM	Midlands	NK05GWN	North East	NK53HJA	North East
N593CKA	NW & Wales	N691GUM	Scotland	NK05GWO	North East	NK53VKA	North East
N594CKA	NW & Wales	N697EUR	The Shires	NK05GWU	North East	NK55MYR	North East
N595CKA	NW & Wales	N698EUR	The Shires	NK05GWV	North East	NK55MYS	North East
N596CKA	NW & Wales	N699EUR	The Shires	NK05GWV	North East	NK55MYT	North East
N597CKA	NW & Wales	N701GUM	NW & Wales	NK05GWX	North East	NK56HKV	North East
N598CKA	NW & Wales	N702EUR	The Shires	NK05GWY	North East	NK56HKW	North East
N599CKA	NW & Wales	N703EUR	The Shires	NK05GXA	North East	NK57DXX	North East
N601CKA	NW & Wales	N703GUM	NW & Wales	NK05GXB	North East	NK57DXY	North East
N601DWY	NW & Wales	N704EUR	The Shires	NK05GXC	North East	NK57DXZ	North East
N602DWY	NW & Wales	N704GUM	NW & Wales	NK05GXD	North East	NK57DYA	North East
N603CKA	NW & Wales	N705EUR	The Shires	NK05GXE	North East	NK57GWX	North East
N603DWY	NW & Wales	N705GUM	NW & Wales	NK05GXF	North East	NK57GWY	North East
N604CKA	NW & Wales	N705TPK	NW & Wales	NK05GXG	North East	NK57GWZ	North East
N604DWY	NW & Wales	N706EUR	The Shires	NK05GXH	North East	NK57GXA	North East
N605CKA	NW & Wales	N706GUM	NW & Wales	NK05GXJ	North East	NK57GXB	North East
N605CKA	NW & Wales	N706TPK	NW & Wales	NK05GXL	North East	NK57GXC	North East
N605DWY	NW & Wales	N707GUM	NW & Wales	NK05GXM	North East	NK57GXD	North East
N606CKA	NW & Wales	N707TPK	NW & Wales	NK05GXN	North East	NK57GXE	North East
N606DWY	NW & Wales	N708GUM	Scotland	NK05GXO	North East	NK57GXF	North East
N607CKA	NW & Wales	N708TPK	NW & Wales	NK05GXW	North East	NKJ785	Original Tour
N607DWY	NW & Wales	N709GUM	Scotland	NK07FZC	North East	NL52XZV	Tellings group
N608CKA	NW & Wales	N709TPK	NW & Wales	NK07FZD	North East	NL52XZW	Tellings group
N608DWY	NW & Wales	N710EUR	The Shires	NK07FZE	North East	NL52XZX	Tellings group
N609CKA	NW & Wales	N710GUM	Scotland	NK07FZF	North East	NL52XZY	Tellings group
N609DWY	NW & Wales	N711GUM	Scotland	NK07FZG	North East	NM02DYA	Southern Counties
N610CKA	NW & Wales	N712EUR	The Shires	NK09BPF	North East	P3SLT	NW & Wales
N610DWY	NW & Wales	N712GUM	Scotland	NK09BPO	North East	P10LPG	Yorkshire
N611CKA	NW & Wales	N713EUR	The Shires	NK09BPU	North East	P10TGM	Tellings group
N611DWY	NW & Wales	N713TPK	Southern Counties	NK09BPV	North East	P49MVU	NW & Wales
N612CKA	NW & Wales	N714TPK	Southern Counties	NK09BPX	North East	P52MVU	NW & Wales
N612DWY	NW & Wales	N715TPK	Southern Counties	NK09BPY	North East	P53MVU	NW & Wales
N613CKA	NW & Wales	N716TPK	NW & Wales	NK09BPZ	North East	P56MVU	NW & Wales

P58MVU	NW & Wales	P194LKJ	Southern Counties	P247MKN	Southern Counties	P313FEA	Midlands
P81MOR	Midlands	P194VUA	Yorkshire	P250APM	Southern Counties	P313HEM	NW & Wales
P130RWR	North East	P195LKJ	Southern Counties	P250NBA	NW & Wales	P314FEA	Midlands
P135GND	NW & Wales	P195VUA	Yorkshire	P251APM	Southern Counties	P314HEM	NW & Wales
P136GND	NW & Wales	P196LKJ	Southern Counties	P253APM	Southern Counties	P315FEA	Midlands
P137GND	NW & Wales	P196VUA	Yorkshire	P254APM	Southern Counties	P315HEM	NW & Wales
P138GND	NW & Wales	P197VUA	Southern Counties	P255APM	Southern Counties	P316FEA	Midlands
P139GND	NW & Wales	P197VUA	Yorkshire	P256FPK	The Shires	P316HEM	NW & Wales
P140GND	NW & Wales	P198LKJ	Southern Counties	P257FPK	Southern Counties	P317FEA	Midlands
P170VUA	Yorkshire	P198VUA	Yorkshire	P258FPK	Southern Counties	P317HEM	NW & Wales
P171VUA	Yorkshire	P199LKJ	Southern Counties	P259FPK	Southern Counties	P318FEA	Midlands
P172VUA	Yorkshire	P199VUA	Yorkshire	P259FPK	Southern Counties	P318HEM	NW & Wales
P173VUA	Yorkshire	P201HRY	Midlands	P260NBA	NW & Wales	P319HEM	NW & Wales
P174VUA	Midlands	P201LKJ	Southern Counties	P261FPK	Southern Counties	P319HOJ	Midlands
P175SRO	The Shires	P202HRY	Midlands	P262FPK	Southern Counties	P320HEM	NW & Wales
P175VUA	Yorkshire	P202LKJ	Southern Counties	P263FPK	Southern Counties	P320HOJ	Midlands
P176LKL	Southern Counties	P202RUM	Yorkshire	P264FPK	Southern Counties	P321HOJ	Midlands
P176SRO	The Shires	P203HRY	Midlands	P265FPK	Southern Counties	P322HOJ	Midlands
P176VUA	Yorkshire	P203LKJ	Southern Counties	P266FPK	Southern Counties	P323HOJ	Midlands
P177LKL	Southern Counties	P204HRY	Midlands	P267FPK	Southern Counties	P324HOJ	Midlands
P177SRO	The Shires	P204LKJ	Southern Counties	P270FPK	Southern Counties	P324HVX	Southern Counties
P177VUA	Yorkshire	P205HRY	Midlands	P271FPK	Southern Counties	P325HOJ	Midlands
P178LKL	Southern Counties	P205LKJ	Southern Counties	P271VRG	North East	P326HOJ	Midlands
P178SRO	The Shires	P206HRY	Midlands	P272FPK	Southern Counties	P326HVX	Yorkshire
P179SRO	The Shires	P206LKJ	Southern Counties	P272VRG	North East	P327HOJ	Midlands
P179VUA	Yorkshire	P207LKJ	Southern Counties	P273FPK	Southern Counties	P327HVX	Southern Counties
P180LKL	NW & Wales	P208LKJ	Southern Counties	P274VRG	North East	P328HVX	Southern Counties
P180SRO	The Shires	P209LKJ	Southern Counties	P275FPK	Southern Counties	P329HVX	Southern Counties
P180VUA	Yorkshire	P210LKJ	Scotland	P275VRG	North East	P330HVX	Southern Counties
P181LKL	Southern Counties	P211LKJ	Scotland	P276FPK	Southern Counties	P331HVX	Southern Counties
P181SRO	The Shires	P212LKJ	Scotland	P276VRG	North East	P332HVX	Southern Counties
P181VUA	Yorkshire	P213LKJ	Southern Counties	P277FPK	Southern Counties	P334HVX	The Shires
P182LKL	NW & Wales	P214LKJ	NW & Wales	P277VRG	North East	P410CCU	North East
P182SRO	The Shires	P215LKJ	Southern Counties	P278VRG	North East	P411CCU	North East
P182VUA	Yorkshire	P216LKJ	Southern Counties	P279FPK	Southern Counties	P412CCU	North East
P183LKL	NW & Wales	P218MKL	Southern Counties	P279VRG	North East	P413CCU	North East
P183SRO	The Shires	P219MKL	Southern Counties	P284FPK	Southern Counties	P414CCU	North East
P183VUA	Yorkshire	P220MKL	Southern Counties	P286FPK	Southern Counties	P415CCU	North East
P184LKL	Southern Counties	P221MKL	Southern Counties	P288FPK	Southern Counties	P416CCU	North East
P184SRO	The Shires	P221SGB	Scotland	P289FPK	Southern Counties	P417CCU	North East
P184VUA	Yorkshire	P223MKL	Southern Counties	P290FPK	Southern Counties	P418CCU	North East
P185LKL	Southern Counties	P224MKL	Southern Counties	P291FPK	Southern Counties	P419CCU	North East
P185SRO	The Shires	P225MKL	Southern Counties	P292FPK	Southern Counties	P419HVX	NW & Wales
P185VUA	Yorkshire	P226MKL	Southern Counties	P293FPK	Southern Counties	P41MVU	NW & Wales
P186LKJ	Southern Counties	P227MKL	Southern Counties	P294FPK	Southern Counties	P420CCU	North East
P186SRO	The Shires	P228MKL	Southern Counties	P295FPK	Southern Counties	P420HVX	NW & Wales
P186VUA	Yorkshire	P229MKL	Southern Counties	P296FPK	Southern Counties	P421HVX	Southern Counties
P187LKJ	Southern Counties	P230MKL	Southern Counties	P301HEM	NW & Wales	P422HVX	NW & Wales
P187SRO	The Shires	P231MKL	Southern Counties	P302HEM	NW & Wales	P423HVX	Southern Counties
P187VUA	Yorkshire	P232MKL	Southern Counties	P303HEM	NW & Wales	P425HVX	Southern Counties
P188LKJ	Southern Counties	P233MKN	Southern Counties	P305HEM	NW & Wales	P426HVX	Southern Counties
P188SRO	The Shires	P234MKN	Southern Counties	P306FEA	Midlands	P427HVX	Southern Counties
P188VUA	Yorkshire	P235MKN	Southern Counties	P306HEM	NW & Wales	P428HVX	Southern Counties
P189LKJ	Southern Counties	P236MKN	Southern Counties	P307FEA	Midlands	P429AHR	Tellings group
P189SRO	The Shires	P237MKN	Southern Counties	P307HEM	NW & Wales	P429HVX	Southern Counties
P189VUA	Yorkshire	P238MKN	Southern Counties	P308FEA	Midlands	P42MVU	NW & Wales
P190LKJ	Southern Counties	P239MKN	Southern Counties	P308HEM	NW & Wales	P430HVX	NW & Wales
P190SRO	The Shires	P240MKN	Southern Counties	P309FEA	Midlands	P431HVX	Southern Counties
P190VUA	Yorkshire	P241MKN	Southern Counties	P309HEM	NW & Wales	P43MVU	NW & Wales
P191LKJ	Southern Counties	P242MKN	Southern Counties	P310FEA	Midlands	P456EEF	North East
P191VUA	Yorkshire	P243MKN	Southern Counties	P310HEM	NW & Wales	P458EEF	North East
P192LKJ	Southern Counties	P244MKN	Southern Counties	P311FEA	Midlands	P459EEF	North East
P192VUA	Yorkshire	P244NBA	NW & Wales	P311HEM	NW & Wales	P45MVU	NW & Wales
P193LKJ	Southern Counties	P245MKN	Southern Counties	P312FEA	Midlands	P460EEF	North East
P193VUA	Yorkshire	P246MKN	Southern Counties	P312HEM	NW & Wales	P461EEF	North East

P46MVU	NW & Wales	P823GMS	Scotland	P930MKL	Southern Counties	R91GNW	NW & Wales
P481DPE	Southern Counties	P823RWU	NW & Wales	P931MKL	Southern Counties	R101GNW	London
P514CVO	Midlands	P824GMS	Scotland	P932MKL	Southern Counties	R103GNW	Yorkshire
P525YJO	The Shires	P824RWU	Midlands	P933MKL	Southern Counties	R112GNW	North East
P526YJO	The Shires	P825KES	Scotland	P934MKL	Southern Counties	R113GNW	North East
P527YJO	The Shires	P825RWU	NW & Wales	P935MKL	Southern Counties	R118TKO	Southern Counties
P533MBU	NW & Wales	P826RWU	NW & Wales	P936MKL	Southern Counties	R119TKO	Southern Counties
P534MBU	NW & Wales	P827KES	Scotland	P937MKL	Southern Counties	R120TKO	Southern Counties
P535MBU	NW & Wales	P827RWU	NW & Wales	P938MKL	NW & Wales	R121TKO	Southern Counties
P536MBU	NW & Wales	P828KES	Scotland	P939MKL	NW & Wales	R122TKO	Southern Counties
P537MBU	NW & Wales	P828RWU	NW & Wales	P940MKL	NW & Wales	R129GNW	North East
P538MBU	NW & Wales	P829KES	Scotland	P941MKL	NW & Wales	R130GNW	North East
P539MBU	NW & Wales	P82MOR	Midlands	P942MKL	NW & Wales	R151GNW	NW & Wales
P540MBU	NW & Wales	P830KES	Scotland	P943MKL	NW & Wales	R152GNW	NW & Wales
P541MBU	NW & Wales	P830RWU	NW & Wales	P952RUL	Midlands	R153GNW	NW & Wales
P542MBU	NW & Wales	P831KES	Scotland	P953RUL	NW & Wales	R159UAL	Midlands
P543MBU	NW & Wales	P831RWU	NW & Wales	P954RUL	Midlands	R165GNW	The Shires
P544MBU	NW & Wales	P832KES	Scotland	P955RUL	Midlands	R169GNW	The Shires
P545MBU	NW & Wales	P832RWU	NW & Wales	P956RUL	Midlands	R170GNW	The Shires
P601RGS	The Shires	P833HVX	The Shires	P957RUL	Midlands	R173VBM	The Shires
P607CAY	Midlands	P833KES	Scotland	P958RUL	Midlands	R174VBM	Tellings group
P609CAY	Southern Counties	P833RWU	NW & Wales	P959RUL	NW & Wales	R177TKU	Tellings group
P610CAY	Southern Counties	P834KES	Scotland	P960RUL	NW & Wales	R177VBM	The Shires
P612FHN	North East	P834RWU	NW & Wales	P961RUL	NW & Wales	R179VBM	The Shires
P613CAY	Southern Counties	P835KES	Scotland	P962RUL	Scotland	R180VBM	The Shires
P615PGP	North East	P835RWU	Midlands	P963RUL	Scotland	R191RBM	The Shires
P616PGP	North East	P836KES	Scotland	P964RUL	Scotland	R192RBM	The Shires
P617PGP	North East	P836RWU	Midlands	P965RUL	Scotland	R193RBM	The Shires
P618FHN	North East	P837KES	Scotland	P966RUL	Scotland	R194RBM	The Shires
P61MVU	NW & Wales	P837RWU	Midlands	P967RUL	Scotland	R195RBM	The Shires
P621FHN	North East	P838KES	Scotland	P968RUL	Scotland	R196DNM	The Shires
P627FHN	North East	P838RWU	NW & Wales	PN02HVM	The Shires	R196RBM	The Shires
P631FHN	North East	P839KES	Scotland	PN02HVO	The Shires	R197RBM	The Shires
P633FHN	North East	P839RWU	Midlands	PN02HVP	The Shires	R198RBM	The Shires
P634PGP	North East	P840KES	Scotland	PN02HVR	The Shires	R199RBM	The Shires
P637PGP	North East	P840PWW	Midlands	PN02HVS	The Shires	R201CKO	NW & Wales
P638PGP	North East	P841PWW	Midlands	PN02LZM	Tellings group	R201RBM	The Shires
P6710PP	The Shires	P842PWW	Midlands	PN02LZO	Tellings group	R201VPU	The Shires
P6720PP	The Shires	P843PWW	Midlands	PN02LZP	Tellings group	R202CKO	NW & Wales
P6730PP	The Shires	P844PWW	Midlands	PN02LZR	Tellings group	R202RBM	The Shires
P6740PP	The Shires	P845PWW	Midlands	PN52XBF	Midlands	R203CKO	NW & Wales
P801RWU	Scotland	P846PWW	Midlands	PN52XBH	Midlands	R203RBM	The Shires
P802RWU	Scotland	P847PWW	Midlands	PN52XRJ	Midlands	R204CKO	North East
P803RWU	Scotland	P848PWW	Midlands	PN52XRK	Midlands	R204RBM	The Shires
P804RWU	Scotland	P849PWW	Midlands	PN52XRL	Midlands	R204VPU	The Shires
P805RWU	Scotland	P850PWW	Midlands	PN52XRM	Midlands	R205CKO	North East
P806DBS	Scotland	P851PWW	Midlands	PN52XRO	Midlands	R205RBM	The Shires
P807DBS	Scotland	P852PWW	Midlands	PN52XRP	Midlands	R205VPU	The Shires
P808DBS	Scotland	P853PWW	Midlands	PN52XRR	Midlands	R206CKO	North East
P809DBS	Scotland	P854PWW	Midlands	PN52XRS	Midlands	R206GMJ	The Shires
P810DBS	Scotland	P855PWW	Midlands	PN52XRT	Midlands	R206VPU	The Shires
P811DBS	Scotland	P902DRG	North East	PN52XRU	Midlands	R207CKO	North East
P812DBS	Scotland	P903DRG	North East	PN52XRV	Midlands	R207GMJ	The Shires
P813DBS	Scotland	P904DRG	North East	PN52XRW	Midlands	R207VPU	The Shires
P814DBS	Scotland	P905JNL	North East	PSU969	Midlands	R208CKO	North East
P814VTY	North East	P913PWW	Scotland	PSU988	Midlands	R208GMJ	The Shires
P815DBS	Scotland	P914PWW	Scotland	PSU989	Midlands	R208VPU	The Shires
P816GMS	Scotland	P915PWW	Scotland	R9CLA	Tellings group	R209CKO	North East
P817GMS	Scotland	P916PWW	NW & Wales	R10WAL	Yorkshire	R209GMJ	The Shires
P818GMS	Scotland	P917PWW	NW & Wales	R44BLU	NW & Wales	R209VPU	The Shires
P819GMS	Scotland	P918PWW	NW & Wales	R45VJF	Midlands	R210CKO	Southern Counties
P820GMS	Scotland	P926MKL	Southern Counties	R46VJF	Midlands	R210GMJ	The Shires
P821GMS	Scotland	P927MKL	Southern Counties	R50TGM	Tellings group	R211CKO	Southern Counties
P822GMS	Scotland	P928MKL	Southern Counties	R51XVM	NW & Wales	R211GMJ	The Shires
P822RWU	Scotland	P929MKL	Southern Counties	R69GNW	Yorkshire	R212CKO	Southern Counties

Reg	Area	Reg	Area	Reg	Area	Reg	Area
R212GMJ	The Shires	R315WVR	NW & Wales	R425RPY	North East	R548ABA	NW & Wales
R213CKO	NW & Wales	R317WVR	NW & Wales	R425TJW	Midlands	R549ABA	NW & Wales
R213GMJ	The Shires	R319WVR	NW & Wales	R426COO	North East	R54XVM	NW & Wales
R214GMJ	The Shires	R321WVR	NW & Wales	R426RPY	North East	R550ABA	NW & Wales
R215GMJ	The Shires	R322WVR	NW & Wales	R426TJW	Midlands	R551ABA	NW & Wales
R233AEY	NW & Wales	R324WVR	NW & Wales	R427COO	North East	R552ABA	NW & Wales
R234AEY	NW & Wales	R326WVR	NW & Wales	R427RPY	North East	R553ABA	NW & Wales
R235AEY	NW & Wales	R327WVR	NW & Wales	R427TJW	Midlands	R554ABA	NW & Wales
R236AEY	NW & Wales	R329TJW	Midlands	R428COO	North East	R556ABA	NW & Wales
R237AEY	NW & Wales	R329WVR	NW & Wales	R428TJW	Midlands	R557ABA	NW & Wales
R238AEY	NW & Wales	R330TJW	Midlands	R429COO	North East	R558ABA	NW & Wales
R239AEY	NW & Wales	R000WVR	NW & Wales	R429TJW	Midlands	R559ABA	NW & Wales
R251JNL	North East	R331TJW	Midlands	R430COO	North East	R560ABA	NW & Wales
R255WRJ	NW & Wales	R331WVR	NW & Wales	R430RPY	North East	R561ABA	NW & Wales
R261EKO	Southern Counties	R332TJW	Midlands	R431COO	North East	R561UOT	Midlands
R262EKO	Southern Counties	R332WVR	NW & Wales	R431RPY	North East	R562ABA	NW & Wales
R263EKO	Southern Counties	R334TJW	Midlands	R432RPY	North East	R563ABA	NW & Wales
R264EKO	Southern Counties	R334WVR	NW & Wales	R433RPY	North East	R564ABA	NW & Wales
R265EKO	Southern Counties	R335TJW	Midlands	R434RPY	North East	R565ABA	NW & Wales
R266EKO	Southern Counties	R335WVR	NW & Wales	R435RPY	North East	R566ABA	NW & Wales
R267EKO	Southern Counties	R336TJW	Midlands	R436RPY	North East	R567ABA	NW & Wales
R268EKO	Southern Counties	R336WVR	NW & Wales	R437RPY	North East	R568ABA	NW & Wales
R269EKO	Southern Counties	R337TJW	Midlands	R438RPY	North East	R569ABA	NW & Wales
R270EKO	Southern Counties	R337WVR	NW & Wales	R439RPY	North East	R570ABA	NW & Wales
R271EKO	Southern Counties	R338TJW	Midlands	R440GWY	Yorkshire	R571ABA	NW & Wales
R272EKO	Southern Counties	R339TJW	Midlands	R440RPY	North East	R57XVM	NW & Wales
R28GNW	Yorkshire	R340TJW	Midlands	R441KWT	Yorkshire	R59XVM	NW & Wales
R291KRG	North East	R311TJW	Midlands	R442KWT	Yorkshire	R601MHN	NW & Wales
R292KRG	North East	R342TJW	Midlands	R443KWT	Yorkshire	R602MHN	NW & Wales
R293KRG	North East	R343TJW	Midlands	R445KWT	Yorkshire	R602WMJ	The Shires
R294KRG	North East	R344TJW	Midlands	R446KWT	Yorkshire	R603MHN	NW & Wales
R295KRG	North East	R369TWR	The Shires	R447KWT	Yorkshire	R603WMJ	The Shires
R296CMV	Southern Counties	R370TWR	The Shires	R447SKX	The Shires	R604MHN	NW & Wales
R297CMV	Southern Counties	R371TWR	The Shires	R448KWT	Yorkshire	R604WMJ	The Shires
R298CMV	Southern Counties	R372TWR	The Shires	R448SKX	The Shires	R605WMJ	The Shires
R299CMV	Southern Counties	R381JYS	Scotland	R449KWT	Yorkshire	R606FBU	NW & Wales
R29GNW	Yorkshire	R382JYS	Scotland	R449SKX	The Shires	R606MHN	NW & Wales
R301CMV	Southern Counties	R383JYS	Scotland	R450KWT	Yorkshire	R607MHN	NW & Wales
R301PCW	NW & Wales	R384JYS	Scotland	R450SKX	The Shires	R607WMJ	The Shires
R302CMV	Southern Counties	R385JYS	Scotland	R451KWT	Yorkshire	R608MHN	NW & Wales
R302CVU	NW & Wales	R415TJW	Midlands	R451SKX	The Shires	R608WMJ	The Shires
R303CVU	NW & Wales	R416COO	NW & Wales	R452KWT	Yorkshire	R609MHN	NW & Wales
R304CMV	Southern Counties	R416HVX	The Shires	R452SKX	The Shires	R614MNU	Midlands
R304CVU	NW & Wales	R416TJW	Midlands	R453KWT	Yorkshire	R615MNU	Midlands
R305CMV	Southern Counties	R417COO	NW & Wales	R453SKX	The Shires	R616MNU	Midlands
R305CVU	NW & Wales	R417HVX	The Shires	R454KWT	Yorkshire	R617MNU	Midlands
R307CMV	Southern Counties	R417TJW	Midlands	R455KWT	Yorkshire	R618MNU	Midlands
R308CMV	Southern Counties	R418COO	NW & Wales	R455SKX	The Shires	R619MNU	Midlands
R308CVU	NW & Wales	R418HVX	The Shires	R456KWT	Yorkshire	R620MNU	Midlands
R309CVU	NW & Wales	R418TJW	Midlands	R456SKX	Southern Counties	R621MNU	Midlands
R309WVR	NW & Wales	R419COO	NW & Wales	R457KWT	Yorkshire	R622MNU	Midlands
R310CMV	Southern Counties	R419TJW	Midlands	R458KWT	Yorkshire	R624MNU	Midlands
R310CVU	NW & Wales	R420COO	NW & Wales	R459KWT	Yorkshire	R625MNU	North East
R310NGM	Southern Counties	R420TJW	Midlands	R460KWT	Yorkshire	R636MNU	North East
R310WVR	NW & Wales	R421COO	North East	R461KWT	Yorkshire	R637MNU	Southern Counties
R311CVU	NW & Wales	R421TJW	Midlands	R47XVM	NW & Wales	R638MNU	Midlands
R311NGM	Southern Counties	R422COO	North East	R48XVM	NW & Wales	R639MNU	North East
R311WVR	NW & Wales	R422TJW	Midlands	R503MOT	Midlands	R639MNU	Southern Counties
R312CVU	NW & Wales	R423COO	North East	R504MOT	Midlands	R640MNU	North East
R312NGM	Southern Counties	R423RPY	North East	R505MOT	Midlands	R641MNU	North East
R312WVR	NW & Wales	R423TJW	Midlands	R521UCC	NW & Wales	R685MHN	NW & Wales
R313CVU	NW & Wales	R424COO	North East	R522UCC	NW & Wales	R701KCU	North East
R313NGM	Southern Counties	R424RPY	North East	R524TWR	The Shires	R701MHN	North East
R313WVR	NW & Wales	R424TJW	Midlands	R546ABA	NW & Wales	R702MHN	North East
R314WVR	NW & Wales	R425COO	North East	R547ABA	NW & Wales	R703MHN	Scotland

R705MHN	North East	S157KNK	The Shires	S244JUA	London	S313JUA	London
R710MHN	North East	S158KNK	The Shires	S245JUA	London	S314JUA	London
R711MHN	North East	S159KNK	The Shires	S246JUA	London	S315JUA	The Shires
R716MHN	North East	S160KNK	The Shires	S247JUA	London	S316JUA	London
R723MHN	North East	S161KNK	The Shires	S248JUA	London	S317JUA	The Shires
R758DUB	The Shires	S169JUA	London	S248UVR	NW & Wales	S318JUA	The Shires
R759DUB	The Shires	S170JUA	London	S249JUA	London	S322JUA	London
R792DUB	North East	S171JUA	London	S249UVR	NW & Wales	S341KHN	North East
R792DUB	North East	S172JUA	London	S250JUA	London	S342KHN	North East
R796DUB	North East	S173JUA	London	S250UVR	NW & Wales	S343KHN	North East
R798DUB	North East	S174JUA	London	S251JUA	London	S344KHN	North East
R809WJA	The Shires	S175JUA	London	S251UVR	NW & Wales	S345KHN	North East
R903BKO	Southern Counties	S176JUA	London	S252JUA	London	S345YOG	Midlands
R904BKO	Southern Counties	S177JUA	London	S253JUA	London	S346KHN	North East
R905BKO	Southern Counties	S178JUA	London	S254JUA	London	S346YOG	Midlands
R906BKO	Southern Counties	S179JUA	London	S255JUA	London	S347YOG	Midlands
R907BKO	Southern Counties	S180JUA	London	S256JUA	London	S348KHN	North East
R908BKO	Southern Counties	S181JUA	London	S257JUA	London	S348YOG	Midlands
R910JNL	North East	S182JUA	London	S258JUA	London	S349KHN	North East
R914JNL	North East	S183JUA	London	S259JUA	London	S349YOG	Midlands
R916JNL	North East	S202JUA	London	S260JUA	London	S350KHN	North East
R917JNL	North East	S203JUA	London	S261JUA	London	S350PGA	North East
R918JNL	North East	S204JUA	London	S262JUA	London	S350YOG	Midlands
R919JNL	North East	S205JUA	London	S263JUA	London	S351KHN	North East
R920JNL	North East	S206JUA	London	S264JUA	London	S351PGA	North East
R920RAU	Midlands	S207DTO	Midlands	S265JUA	NW & Wales	S351YOG	Midlands
R922JNL	North East	S207JUA	London	S266JUA	NW & Wales	S352KHN	North East
R923JNL	North East	S208DTO	Midlands	S267JUA	NW & Wales	S352PGA	North East
R929RAU	Midlands	S208JUA	London	S268JUA	NW & Wales	S352YOG	Midlands
R942VPU	Southern Counties	S209JUA	London	S269JUA	NW & Wales	S353KHN	North East
R943VPU	The Shires	S210JUA	London	S270JUA	NW & Wales	S353PGA	North East
R957RCH	Tellings group	S211JUA	London	S271JUA	NW & Wales	S353YOG	Midlands
R958RCH	Tellings group	S212JUA	London	S272JUA	London	S354KHN	North East
R959RCH	Tellings group	S213JUA	London	S273JUA	London	S356KHN	North East
R985FNW	Yorkshire	S214JUA	London	S274JUA	London	S357KHN	North East
R989FNW	Yorkshire	S215JUA	London	S275JUA	London	S358KHN	North East
RDZ1702	NW & Wales	S216JUA	London	S276JUA	London	S401ERP	The Shires
RDZ1705	NW & Wales	S216XPP	The Shires	S277JUA	London	S402ERP	The Shires
RDZ1708	NW & Wales	S217JUA	London	S278JUA	London	S403ERP	The Shires
RDZ1709	NW & Wales	S217XPP	The Shires	S279JUA	London	S404ERP	The Shires
RDZ1710	NW & Wales	S218JUA	London	S280JUA	London	S426MCC	The Shires
RDZ1711	NW & Wales	S219JUA	London	S281JUA	London	S427MCC	The Shires
RDZ1712	NW & Wales	S220JUA	London	S282JUA	London	S428MCC	The Shires
RDZ1713	NW & Wales	S221JUA	London	S283JUA	London	S429MCC	The Shires
RDZ1714	NW & Wales	S223JUA	London	S284JUA	London	S462GUB	Yorkshire
RL51ZKR	Tellings group	S224JUA	London	S285JUA	London	S463GUB	Yorkshire
RL51ZKS	Tellings group	S225JUA	London	S286JUA	London	S464GUB	Yorkshire
RN52EYH	Tellings group	S226JUA	London	S287JUA	London	S465GUB	Yorkshire
RN52EYJ	Tellings group	S227JUA	London	S288JUA	London	S466GUB	Yorkshire
S5CLA	Tellings group	S228JUA	London	S289JUA	London	S467GUB	Yorkshire
S6CLA	Tellings group	S229JUA	London	S290JUA	London	S468GUB	Yorkshire
S10BCL	Tellings group	S230JUA	London	S291JUA	London	S469GUB	Yorkshire
S20BCL	Tellings group	S231JUA	London	S292JUA	London	S470GUB	Yorkshire
S43BLU	NW & Wales	S232JUA	London	S301JUA	London	S471GUB	Yorkshire
S45BLU	NW & Wales	S233JUA	London	S302JUA	London	S472ANW	Yorkshire
S146KNK	The Shires	S234JUA	London	S303JUA	London	S473ANW	Yorkshire
S147KNK	The Shires	S235JUA	London	S304JUA	London	S474ANW	Yorkshire
S148KNK	The Shires	S236JUA	London	S305JUA	London	S475ANW	Yorkshire
S149KNK	The Shires	S237JUA	London	S306JUA	London	S476ANW	Yorkshire
S150KNK	The Shires	S238JUA	London	S307JUA	London	S477ANW	Yorkshire
S151KNK	The Shires	S239JUA	London	S308JUA	London	S478ANW	Yorkshire
S152KNK	The Shires	S240JUA	London	S309JUA	London	S479ANW	Yorkshire
S153KNK	The Shires	S241JUA	London	S310JUA	London	S480ANW	Yorkshire
S154KNK	The Shires	S242JUA	London	S311JUA	London	S481ANW	Yorkshire
S156KNK	The Shires	S243JUA	London	S312JUA	London	S482ANW	Yorkshire

Reg	Area	Reg	Area	Reg	Area	Reg	Area
S483ANW	Yorkshire	S708KFT	North East	SN06BPZ	Southern Counties	T204XBV	Original Tour
S484ANW	Yorkshire	S709KFT	North East	SN06BRF	Southern Counties	T205XBV	Original Tour
S485ANW	Yorkshire	S710KFT	North East	SN53ESG	Midlands	T206XBV	Original Tour
S486ANW	Yorkshire	S711KFT	North East	SN53ESO	Midlands	T207XBV	Original Tour
S487ANW	Yorkshire	S712KRG	North East	SN54GPK	The Shires	T208XBV	Original Tour
S488ANW	Yorkshire	S713KRG	North East	SN54GPO	The Shires	T209XBV	Original Tour
S489ANW	Yorkshire	S714KRG	North East	SN54GPU	The Shires	T209XVO	Midlands
S490ANW	Yorkshire	S715KRG	North East	SN54HWY	Tellings group	T210XBV	Original Tour
S491ANW	Yorkshire	S822MCC	North East	SN54HWZ	Tellings group	T211XBV	Original Tour
S558MCC	NW & Wales	S823MCC	North East	SN54HXA	Tellings group	T212XBV	Original Tour
3559MOO	NW & Waloo	C818RJC	NW & Wales	SN54HXR	Tellings group	T213BBR	Tellings group
S610KHN	NW & Wales	S860OGB	Scotland	SN54HXC	Tellings group	T213XBV	Original Tour
S611KHN	North East	S861OGB	Scotland	SN54HXD	Tellings group	T214XBV	Original Tour
S612KHN	North East	S862OGB	Scotland	SN54HXE	Tellings group	T215XBV	London
S613KHN	North East	S863OGB	Scotland	SN54HXF	Tellings group	T216BBR	Tellings group
S614KHN	NW & Wales	S864OGB	Scotland	SN55HTX	Yorkshire	T216XBV	London
S616KHN	North East	S865OGB	Scotland	SN55HTY	Yorkshire	T217BBR	Tellings group
S617KHN	North East	S866OGB	Scotland	SN55HTZ	Yorkshire	T217XBV	London
S618KHN	North East	S867OGB	Scotland	SN56AXG	The Shires	T218BBR	Tellings group
S619KHN	NW & Wales	S868OGB	Scotland	SN56AXH	The Shires	T218NMJ	Southern Counties
S620KHN	NW & Wales	S869OGB	Scotland	SN58ENX	The Shires	T218XBV	London
S621KHN	North East	S872SNB	NW & Wales	SN58ENY	The Shires	T219BBR	Tellings group
S622KHN	NW & Wales	S873SNB	NW & Wales	SN58EOA	The Shires	T219NMJ	The Shires
S623KHN	NW & Wales	S874SNB	NW & Wales	SN58EOB	The Shires	T219XBV	London
S624KHN	Scotland	S875SNB	NW & Wales	SN58EOC	The Shires	T220XBV	London
S625KHN	North East	S876SNB	NW & Wales	SN58EOD	The Shires	T222MTB	NW & Wales
S626KHN	North East	S877SNR	NW & Wales	SN58EOE	The Shires	T224BBR	Tellings group
S627KHN	NW & Wales	S878SNB	NW & Wales	SN58EOF	The Shires	T225BBR	Tellings group
S628KHN	NW & Wales	S879SNB	NW & Wales	SN58EOG	The Shires	T273JKM	Southern Counties
S629KHN	NW & Wales	S903DUB	The Shires	SN58EOH	The Shires	T274JKM	Southern Counties
S630KHN	NW & Wales	SA52MYT	North East	SN58EOJ	The Shires	T275JKM	Southern Counties
S631KHN	NW & Wales	SCZ9651	Southern Counties	SN58EOK	The Shires	T276JKM	Southern Counties
S632KHN	NW & Wales	SCZ9652	Southern Counties	SN58EOM	The Shires	T277JKM	Southern Counties
S633KHN	NW & Wales	SF04RHA	North East	SN58EOO	The Shires	T278JKM	Southern Counties
S634KHN	NW & Wales	SF09LOD	Scotland	T10BLU	NW & Wales	T279JKM	Southern Counties
S635KHN	North East	SF57NMM	Scotland	T11BLU	NW & Wales	T280JKM	Southern Counties
S636KHN	North East	SF57NMO	Scotland	T42PVM	NW & Wales	T281JKM	Southern Counties
S637KHN	North East	SF57NPK	Scotland	T45KAW	The Shires	T282JKM	Southern Counties
S638KHN	North East	SF57NPL	Scotland	T47WUT	Midlands	T283JKM	Southern Counties
S639KHN	North East	SJ57DDN	Scotland	T48WUT	Midlands	T284JKM	Southern Counties
S640KHN	North East	SJ57DDO	Scotland	T49JJF	Midlands	T285JKM	Southern Counties
S642KHN	Scotland	SJ57DDU	Scotland	T51JJF	Midlands	T286JKM	Southern Counties
S643KHN	Scotland	SJ57DDV	Scotland	T52JJF	Midlands	T287JKM	Southern Counties
S644KJU	Midlands	SJ57DDX	Scotland	T53JJF	Midlands	T288JKM	Southern Counties
S645KJU	Midlands	SJ57DDY	Scotland	T54JJF	Midlands	T289JKM	Southern Counties
S646KJU	Midlands	SJ57DDZ	Scotland	T61JBA	Midlands	T293FGN	London
S647KJU	Midlands	SK52MLE	Midlands	T62JBA	NW & Wales	T294FGN	London
S648KJU	North East	SK52MLF	Midlands	T63JBA	NW & Wales	T295FGN	London
S649KJU	Midlands	SK52MLJ	Midlands	T64JBA	NW & Wales	T296FGN	London
S650KJU	Midlands	SK52MLL	Midlands	T65JBA	NW & Wales	T297FGN	London
S651KJU	Midlands	SK52MLN	Midlands	T74AUA	North East	T298FGN	London
S652KJU	Midlands	SK52MLO	Midlands	T75AUA	North East	T299FGN	London
S653KJU	Midlands	SN03LDV	Midlands	T76AUA	North East	T301FGN	London
S701VKM	Southern Counties	SN03LDX	Midlands	T78AUA	North East	T302FGN	London
S702KFT	North East	SN03LGC	Midlands	T79AUA	North East	T303FGN	London
S702VKM	Southern Counties	SN03LGD	Midlands	T81AUA	North East	T304FGN	London
S703KFT	North East	SN03LGE	Midlands	T82AUA	North East	T305FGN	London
S703VKM	Southern Counties	SN03LGF	Midlands	T83AUA	North East	T306FGN	London
S704KFT	North East	SN06BPE	Southern Counties	T109LKK	Southern Counties	T307FGN	London
S704VKM	Southern Counties	SN06BPF	Southern Counties	T110GGO	London	T308FGN	London
S705KFT	North East	SN06BPK	Southern Counties	T110LKK	Southern Counties	T309FGN	London
S705VKM	Southern Counties	SN06BPU	Southern Counties	T131AUA	Tellings group	T310FGN	London
S706KFT	North East	SN06BPV	Southern Counties	T133AUA	Tellings group	T311FGN	London
S706VKM	Southern Counties	SN06BPX	Southern Counties	T202XBV	Original Tour	T312FGN	London
S707KFT	North East	SN06BPY	Southern Counties	T203XBV	Original Tour	T313FGN	London

Many vehicles in the Milton Keynes element of The Shires retain their yellow and blue livery. Seen in the town in July 2009 is 3003, V393KVY, a rare model with Arriva and found only at Milton Keynes and Colchester.
Dave Heath

T314FGN	London	T424LGP	The Shires	T619PNC	NW & Wales	T827NMJ	The Shires
T314PNB	NW & Wales	T425LGP	The Shires	T620PNC	NW & Wales	T828NMJ	The Shires
T315FGN	London	T490KGB	The Shires	T621PNC	NW & Wales	T829NMJ	The Shires
T315PNB	NW & Wales	T491KGB	The Shires	T622PNC	NW & Wales	T911KKM	Southern Counties
T316FGN	London	T492KGB	The Shires	T623PNC	NW & Wales	T912KKM	Southern Counties
T316PNB	NW & Wales	T494KGB	The Shires	T624EUB	Yorkshire	T913KKM	Southern Counties
T317FGN	London	T495KGB	The Shires	T625EUB	Yorkshire	T914KKM	Southern Counties
T317PNB	NW & Wales	T526AOB	NW & Wales	T626EUB	Yorkshire	T915KKM	Southern Counties
T318FGN	London	T527AOB	NW & Wales	T627EUB	Yorkshire	T916KKM	Southern Counties
T318PNB	NW & Wales	T528AOB	NW & Wales	T628EUB	Yorkshire	T917KKM	NW & Wales
T319FGN	London	T529AOB	NW & Wales	T629EUB	Yorkshire	T918KKM	Southern Counties
T319PNB	NW & Wales	T560JJC	NW & Wales	T630EUB	Yorkshire	T919KKM	Southern Counties
T320FGN	London	T561JJC	NW & Wales	T631EUB	Yorkshire	T920KKM	NW & Wales
T320PNB	NW & Wales	T562JJC	NW & Wales	T632EUB	Yorkshire	T921KKM	Southern Counties
T322FGN	London	T564JJC	NW & Wales	T633EUB	Yorkshire	T922KKM	NW & Wales
T322PNB	NW & Wales	T565JJC	NW & Wales	T634EUB	Yorkshire	TJI1683	Tellings group
T323FGN	London	T566JJC	NW & Wales	T635EUB	Yorkshire	TWY7	Yorkshire
T323PNB	NW & Wales	T567JJC	NW & Wales	T636EUB	Yorkshire	UAR247Y	Original Tour
T324FGN	London	T568JJC	NW & Wales	T637EUB	Yorkshire	UAR250Y	Original Tour
T324PNB	NW & Wales	T569JJC	NW & Wales	T638EUB	Yorkshire	UAR776Y	Original Tour
T325FGN	London	T570JJC	NW & Wales	T639EUB	Yorkshire	UK03LLC	Tellings group
T405ENV	The Shires	T576FFC	Tellings group	T701RCN	North East	UK03TGM	Tellings group
T405SMV	Tellings group	T591CGT	Southern Counties	T702RCN	North East	UK04TGM	Tellings group
T406ENV	The Shires	T592CGT	Southern Counties	T820NMJ	Southern Counties	UK05BCL	Tellings group
T406SMV	Tellings group	T612PNC	NW & Wales	T821NMJ	Southern Counties	UK08BCL	Tellings group
T407ENV	The Shires	T613PNC	NW & Wales	T821PNB	NW & Wales	UOI772	North East
T408ENV	The Shires	T614PNC	NW & Wales	T822NMJ	Southern Counties	V228BLU	NW & Wales
T408LGP	The Shires	T615PNC	NW & Wales	T823NMJ	Southern Counties	V33BLU	NW & Wales
T409ENV	The Shires	T616PNC	NW & Wales	T824NMJ	Southern Counties	V34ENC	NW & Wales
T410ENV	The Shires	T617PNC	NW & Wales	T825NMJ	Southern Counties	V35ENC	NW & Wales
T421GGO	London	T618PNC	NW & Wales	T826NMJ	Southern Counties	V41DJA	NW & Wales

Reg	Region	Reg	Region	Reg	Region	Reg	Region
V82EVU	The Shires	V230KDA	Midlands	V329DGT	London	V435DGT	London
V142EJR	Tellings group	V231HBH	The Shires	V330DGT	London	V501DFT	North East
V201KDA	Midlands	V231KDA	Midlands	V331DGT	London	V502DFT	North East
V201PCX	Yorkshire	V232HBH	The Shires	V332DGT	London	V503DFT	North East
V202KDA	Midlands	V232KDA	Midlands	V334DGT	London	V504DFT	North East
V203KDA	Midlands	V233HBH	The Shires	V335DGT	London	V505DFT	North East
V203PCX	Yorkshire	V233KDA	Midlands	V336DGT	London	V506DFT	North East
V204KDA	Midlands	V234HBH	The Shires	V337DGT	London	V507DFT	North East
V204PCX	Yorkshire	V234KDA	Midlands	V337MBV	Midlands	V508DFT	North East
V205KDA	Midlands	V235HBH	The Shires	V338DGT	London	V509DFT	North East
V205PCX	Yorkshire	V235KDA	Midlands	V338MBV	Midlands	V510DFT	North East
V206DJR	North East	V236HBH	The Shires	V339DGT	London	V511DFT	North East
V206KDA	Midlands	V236KDA	Midlands	V341DGT	London	V512DFT	North East
V206PCX	Yorkshire	V237HBH	The Shires	V342DGT	London	V513DFT	North East
V207DJR	North East	V237KDA	Midlands	V343DGT	London	V514DFT	North East
V207KDA	Midlands	V238HBH	The Shires	V344DGT	London	V515DFT	North East
V207PCX	Yorkshire	V238KDA	Midlands	V345DGT	London	V530GDS	North East
V208DJR	North East	V239HBH	The Shires	V346DGT	London	V531GDS	North East
V208KDA	Midlands	V239KDA	Midlands	V347DGT	London	V532GDS	North East
V208PCX	Yorkshire	V250HBH	The Shires	V348DGT	London	V533GDS	North East
V209DJR	North East	V251HBH	The Shires	V349DGT	London	V534GDS	North East
V209KDA	Midlands	V252HBH	The Shires	V351DGT	London	V535GDS	North East
V209PCX	Yorkshire	V253HBH	The Shires	V352DGT	London	V536GDS	North East
V210DJR	North East	V254HBH	The Shires	V353DGT	London	V553ECC	NW & Wales
V210KDA	Midlands	V255HBH	The Shires	V354DGT	London	V554ECC	NW & Wales
V210PCX	Yorkshire	V256HBH	The Shires	V355DGT	London	V556ECC	NW & Wales
V211DJR	North East	V257HBH	The Shires	V356DGT	London	V557ECC	NW & Wales
V211KDA	Midlands	V258HBH	The Shires	V357DGT	London	V571DJC	NW & Wales
V211PCX	Yorkshire	V259HBH	The Shires	V358DGT	London	V572DJC	NW & Wales
V212DJR	North East	V260HBH	The Shires	V359DGT	London	V573DJC	NW & Wales
V212KDA	Midlands	V261HBH	The Shires	V361DGT	London	V574DJC	NW & Wales
V212PCX	Yorkshire	V262HBH	The Shires	V362DGT	London	V575DJC	NW & Wales
V213DJR	North East	V263HBH	The Shires	V363DGT	London	V576DJC	NW & Wales
V213KDA	Midlands	V264HBH	The Shires	V364DGT	London	V577DJC	NW & Wales
V213PCX	Yorkshire	V265HBH	The Shires	V365DGT	London	V578DJC	NW & Wales
V214DJR	North East	V266HBH	The Shires	V392KVY	The Shires	V579DJC	NW & Wales
V214KDA	Midlands	V267HBH	The Shires	V393KVY	The Shires	V580DJC	NW & Wales
V215KDA	Midlands	V268HBH	The Shires	V404ENC	NW & Wales	V580ECC	NW & Wales
V215PCX	Yorkshire	V270HBH	The Shires	V405ENC	NW & Wales	V581DJC	NW & Wales
V216KDA	Midlands	V271HBH	The Shires	V406ENC	NW & Wales	V582DJC	NW & Wales
V216PCX	Yorkshire	V272HBH	The Shires	V407ENC	NW & Wales	V583DJC	NW & Wales
V217KDA	Midlands	V273HBH	The Shires	V408ENC	NW & Wales	V584DJC	NW & Wales
V217PCX	Yorkshire	V274HBH	The Shires	V409ENC	NW & Wales	V585DJC	NW & Wales
V218KDA	Midlands	V275HBH	The Shires	V410ENC	NW & Wales	V586DJC	NW & Wales
V218PCX	Yorkshire	V276HBH	The Shires	V411ENC	NW & Wales	V587DJC	NW & Wales
V219KDA	Midlands	V280HBH	The Shires	V412ENC	NW & Wales	V588DJC	NW & Wales
V220KDA	Midlands	V281HBH	The Shires	V412UNH	The Shires	V590DJC	NW & Wales
V220PCX	Yorkshire	V282HBH	The Shires	V413ENC	NW & Wales	V591DJC	NW & Wales
V221KDA	Midlands	V283HBH	The Shires	V413UNH	The Shires	V601DBC	Midlands
V221PCX	Yorkshire	V284HBH	The Shires	V414ENC	NW & Wales	V602DBC	Midlands
V223KDA	Midlands	V285HBH	The Shires	V415ENC	NW & Wales	V603DBC	Midlands
V223PCX	Yorkshire	V286HBH	The Shires	V421DGT	The Shires	V604DBC	Midlands
V224KDA	Midlands	V287HBH	The Shires	V422DGT	The Shires	V605DBC	Midlands
V224PCX	Yorkshire	V288HBH	The Shires	V423DGT	Midlands	V606DBC	Midlands
V225KDA	Midlands	V289HBH	The Shires	V424DGT	Midlands	V607DBC	Midlands
V225PCX	Tellings group	V290HBH	The Shires	V425DGT	Midlands	V608DBC	Midlands
V226KDA	Midlands	V291HBH	The Shires	V426DGT	Midlands	V609DBC	Midlands
V226PCX	Yorkshire	V292HBH	The Shires	V427DGT	Midlands	V609LGC	London
V227KDA	Midlands	V293HBH	The Shires	V428DGT	Midlands	V610DBC	Midlands
V227PCX	Yorkshire	V294HBH	The Shires	V429DGT	Midlands	V610LGC	London
V228KDA	Midlands	V311NGD	Scotland	V430DGT	Midlands	V611DBC	Midlands
V228PCX	Yorkshire	V312NGD	Scotland	V431DGT	Midlands	V611LGC	London
V229KDA	Midlands	V313NGD	Scotland	V432DGT	London	V612DBC	Midlands
V229XUB	Yorkshire	V326DGT	London	V433DGT	London	V612DNL	North East
V230HBH	The Shires	V327DGT	London	V434DGT	London	V612LGC	London

Reg	Location	Reg	Location	Reg	Location	Reg	Location
V613LGC	London	V671DVU	NW & Wales	VYJ806	London	W248SNR	Midlands
V614LGC	London	V672DVU	NW & Wales	W3CLA	Tellings group	W249SNR	Midlands
V615LGC	London	V673DVU	NW & Wales	W3CTS	The Shires	W251SNR	Midlands
V616LGC	London	V674DVU	NW & Wales	W12LUE	NW & Wales	W269NFF	NW & Wales
V617LGC	London	V675DVU	NW & Wales	W40BCL	Tellings group	W292PPT	North East
V618LGC	London	V676DVU	NW & Wales	W40TGM	Tellings group	W293PPT	North East
V619LGC	London	V701LWT	London	W50TGM	Tellings group	W294PPT	North East
V620LGC	London	V703DNL	North East	W60TGM	Tellings group	W295PPT	North East
V621LGC	London	V705DNL	North East	W69PRG	North East	W296PPT	North East
V622LGC	London	V706DNL	North East	W72PRG	North East	W297PPT	North East
V623LGC	London	V707DNL	North East	W76PRG	North East	W298PPT	North East
V624DBN	NW & Wales	V708DNL	North East	W78PRG	Scotland	W299PPT	North East
V625DVU	NW & Wales	V709DNL	North East	W79PRG	Scotland	W301PPT	North East
V626DVU	NW & Wales	V710DNL	North East	W80TGM	Tellings group	W302PPT	North East
V627DVU	NW & Wales	V711DNL	North East	W81PRG	North East	W303PPT	North East
V628DVU	NW & Wales	V712DNL	North East	W82PRG	Scotland	W304PPT	North East
V628LGC	London	V713DNL	North East	W83PRG	North East	W307PPT	North East
V629DVU	NW & Wales	V714DNL	North East	W102EWU	Yorkshire	W308PPT	North East
V630DVU	NW & Wales	V715DNL	North East	W103EWU	Yorkshire	W309PPT	North East
V631DVU	NW & Wales	V715LWT	NW & Wales	W104EWU	Yorkshire	W311PPT	North East
V632DVU	NW & Wales	V716DNL	North East	W106EWU	Yorkshire	W312PPT	North East
V633DVU	NW & Wales	V717DNL	North East	W107EWU	Yorkshire	W313PPT	North East
V633LGC	London	V718DNL	North East	W108EWU	Yorkshire	W314PPT	North East
V634DVU	NW & Wales	V719DNL	North East	W109EWU	Yorkshire	W315PPT	North East
V635DVU	NW & Wales	V720DNL	North East	W128XRO	The Shires	W317PPT	North East
V636DVU	NW & Wales	V721DNL	North East	W129XRO	The Shires	W319PPT	North East
V637DVU	NW & Wales	V722DNL	North East	W131XRO	The Shires	W359XKX	The Shires
V638DVU	NW & Wales	V723DNL	North East	W132XRO	The Shires	W361XKX	The Shires
V639DVU	NW & Wales	V724DNL	North East	W134XRO	The Shires	W362XKX	The Shires
V640DVU	NW & Wales	V725DNL	North East	W136VGJ	London	W363XKX	The Shires
V640KVH	Yorkshire	V726DNL	North East	W136XRO	The Shires	W364XKX	The Shires
V640LGC	London	V727DNL	North East	W137VGJ	London	W365XKX	The Shires
V641DVU	NW & Wales	V728DNL	North East	W137XRO	The Shires	W366VGJ	London
V641KVH	Yorkshire	V729DNL	North East	W138VGJ	London	W367VGJ	London
V642DVU	NW & Wales	V730DNL	North East	W138XRO	The Shires	W367XKX	The Shires
V643DVU	NW & Wales	V731DNL	North East	W139XRO	The Shires	W368VGJ	London
V644DVU	NW & Wales	V732DNL	North East	W165HBT	Yorkshire	W368XKX	The Shires
V645DVU	NW & Wales	V733DNL	North East	W166HBT	Yorkshire	W369VGJ	London
V646DVU	NW & Wales	V734DNL	North East	W166PNT	North East	W369XKX	The Shires
V647DVU	NW & Wales	V735DNL	North East	W174CDN	NW & Wales	W371VGJ	London
V648DVU	NW & Wales	V736DNL	North East	W183CDN	Southern Counties	W372VGJ	London
V649DVU	NW & Wales	V737DNL	North East	W191CDN	NW & Wales	W373VGJ	London
V650DVU	NW & Wales	V738DNL	North East	W192CDN	NW & Wales	W374VGJ	London
V650LGC	London	V739DNL	North East	W193CDN	NW & Wales	W376VGJ	London
V651DVU	NW & Wales	V740DNL	North East	W194CDN	NW & Wales	W377VGJ	London
V652DVU	NW & Wales	V741DNL	North East	W198CDN	Southern Counties	W378VGJ	London
V653DVU	NW & Wales	V742DNL	North East	W226SNR	Midlands	W379VGJ	London
V653LWT	North East	V743ECU	North East	W227SNR	Midlands	W381VGJ	London
V654DVU	NW & Wales	V744ECU	North East	W228SNR	Midlands	W382VGJ	London
V655DVU	NW & Wales	V745ECU	North East	W229SNR	Midlands	W383VGJ	London
V656DVU	NW & Wales	V746ECU	North East	W231SNR	Midlands	W384VGJ	London
V657DVU	NW & Wales	V747ECU	North East	W232SNR	Midlands	W385VGJ	London
V658DVU	NW & Wales	V748ECU	North East	W233SNR	Midlands	W386VGJ	London
V659DVU	NW & Wales	V749ECU	North East	W234SNR	Midlands	W387VGJ	London
V660DVU	NW & Wales	V897DNB	The Shires	W235SNR	Midlands	W388VGJ	London
V660LGC	London	VLT5	London	W236SNR	Midlands	W389VGJ	London
V661DVU	NW & Wales	VLT6	London	W237SNR	Midlands	W391VGJ	London
V662DVU	NW & Wales	VLT12	London	W238SNR	Midlands	W392VGJ	London
V663DVU	NW & Wales	VLT27	London	W239SNR	Midlands	W393OUF	Tellings group
V664DVU	NW & Wales	VLT32	London	W241SNR	Midlands	W393VGJ	London
V665DVU	NW & Wales	VLT47	London	W242SNR	Midlands	W394OJC	NW & Wales
V667DVU	NW & Wales	VLT173	London	W243SNR	Midlands	W394VGJ	London
V668DVU	NW & Wales	VLT244	London	W244SNR	Yorkshire	W395RBB	Tellings group
V669DVU	NW & Wales	VLT295	London	W246SNR	Midlands	W395VGJ	London
V670DVU	NW & Wales	VX04JHY	Tellings group	W247SNR	Midlands	W396RBB	Tellings group

Reg	Region	Reg	Region	Reg	Region	Reg	Region
W396VGJ	London	W466XKX	London	W751SBR	North East	X214JOF	NW & Wales
W397RBB	North East	W467XKX	London	W752SBR	North East	X215ANC	NW & Wales
W397VGJ	London	W468XKX	London	W753SBR	North East	X215JOF	NW & Wales
W398RBB	North East	W469XKX	London	W754SBR	North East	X216ANC	NW & Wales
W398VGJ	London	W471XKX	London	W756SBR	North East	X216HCD	Tellings group
W399RBB	North East	W472XKX	London	W757SBR	North East	X216JOF	NW & Wales
W399VGJ	London	W473XKX	London	W758SBR	North East	X217ANC	NW & Wales
W401VGJ	London	W474XKX	London	W759SBR	North East	X217HCD	Tellings group
W402VGJ	London	W475XKX	London	W900BCL	Tellings group	X217JOF	NW & Wales
W403VGJ	London	W476XKX	London	W901UJM	Tellings group	X218ANC	NW & Wales
W101VCJ	London	W477YKX	London	W901LIM	Tellings group	X218HCD	Tellings group
W407VGJ	London	W478XKX	London	W902UJM	Tellings group	X219ANC	NW & Wales
W408VGJ	London	W479XKX	London	W902UJM	Tellings group	X219HCD	Tellings group
W409VGJ	London	W481XKX	London	W903UJM	Tellings group	X221ANC	NW & Wales
W411VGJ	London	W482YGS	The Shires	W904UJM	Tellings group	X221HCD	Tellings group
W412VGJ	London	W483YGS	The Shires	W905UJM	Tellings group	X223ANC	NW & Wales
W413VGJ	London	W484YGS	The Shires	W906UJM	Tellings group	X223HCD	Tellings group
W414VGJ	London	W485YGS	The Shires	W907UJM	Tellings group	X224ANC	NW & Wales
W415KNH	The Shires	W486YGS	The Shires	W986WDS	The Shires	X226ANC	NW & Wales
W416KNH	The Shires	W487YGS	The Shires	WA56ENN	Tellings group	X227ANC	NW & Wales
W421XKX	The Shires	W488YGS	The Shires	WLT348	London	X228ANC	NW & Wales
W422XKX	The Shires	W489YGS	The Shires	WLT372	London	X229ANC	NW & Wales
W423XKX	The Shires	W491YGS	The Shires	WLT385	London	X231ANC	NW & Wales
W424XKX	The Shires	W492YGS	The Shires	WLT554	London	X232ANC	NW & Wales
W425XKX	The Shires	W493YGS	The Shires	WLT664	London	X233ANC	NW & Wales
W426XKX	The Shires	W494YGS	The Shires	WLT676	London	X234ANC	NW & Wales
W427CWX	Tellings group	W495YGS	The Shires	WLT719	London	X235ANC	NW & Wales
W427XKX	The Shires	W496YGS	The Shires	WLT751	London	X236ANC	NW & Wales
W428XKX	The Shires	W497YGS	The Shires	WLT807	London	X237ANC	NW & Wales
W429XKX	The Shires	W498YGS	The Shires	WLT888	London	X238ANC	NW & Wales
W431WGJ	London	W501RBB	North East	WLT892	London	X239ANC	NW & Wales
W431XKX	The Shires	W601YKN	Southern Counties	WLT895	London	X239PGT	London
W432WGJ	London	W602VGJ	London	WLT897	London	X241ANC	NW & Wales
W432XKX	The Shires	W602YKN	Southern Counties	WLT901	London	X241PGT	London
W433WGJ	London	W603VGJ	London	WLT970	London	X242ANC	NW & Wales
W433XKX	The Shires	W603YKN	Southern Counties	WLT997	London	X242PGT	London
W434WGJ	London	W604VGJ	London	WSU476	Scotland	X243HJA	NW & Wales
W434XKX	Southern Counties	W604YKN	Southern Counties	WSV571	Tellings group	X243PGT	London
W435WGJ	London	W605VGJ	London	WSV572	Tellings group	X244HJA	NW & Wales
W435XKX	Southern Counties	W605YKN	Southern Counties	X13LUE	NW & Wales	X244PGT	London
W436WGJ	London	W606VGJ	London	X14LUE	NW & Wales	X246HJA	NW & Wales
W436XKX	Southern Counties	W607VGJ	London	X23BLU	NW & Wales	X246PGT	London
W437WGJ	London	W608VGJ	London	X143WNL	Tellings group	X247HJA	NW & Wales
W437XKX	Southern Counties	W631RNP	North East	X144WNL	Tellings group	X247PGT	London
W438WGJ	London	W651CWX	Yorkshire	X152ENJ	Tellings group	X248HJA	NW & Wales
W438XKX	Southern Counties	W652CWX	Yorkshire	X153ENJ	Tellings group	X248PGT	London
W439XKX	Southern Counties	W653CWX	Yorkshire	X154ENJ	Tellings group	X249HJA	NW & Wales
W441CWX	Tellings group	W654CWX	Yorkshire	X157ENJ	Tellings group	X249PGT	London
W441XKX	Southern Counties	W656CWX	Yorkshire	X158ENJ	Tellings group	X251HJA	NW & Wales
W442XKX	The Shires	W657CWX	Yorkshire	X159ENJ	Tellings group	X252HBC	Midlands
W443XKX	The Shires	W658CWX	Yorkshire	X201ANC	NW & Wales	X252HJA	NW & Wales
W445XKX	The Shires	W659CWX	Yorkshire	X202ANC	NW & Wales	X253HJA	NW & Wales
W446XKX	The Shires	W661CWX	Yorkshire	X203ANC	NW & Wales	X254HJA	NW & Wales
W447XKX	The Shires	W662CWX	Yorkshire	X204ANC	NW & Wales	X256HJA	NW & Wales
W452XKX	The Shires	W663CWX	Yorkshire	X207ANC	NW & Wales	X257HJA	NW & Wales
W453XKX	The Shires	W664CWX	Yorkshire	X208ANC	NW & Wales	X258HJA	NW & Wales
W454XKX	The Shires	W665CWX	Yorkshire	X209ANC	NW & Wales	X259HJA	NW & Wales
W457XKX	The Shires	W667CWX	Yorkshire	X209JOF	NW & Wales	X261OBN	NW & Wales
W458XKX	The Shires	W668CWX	Yorkshire	X211ANC	NW & Wales	X262OBN	NW & Wales
W459XKX	The Shires	W669CWX	Yorkshire	X211JOF	NW & Wales	X263OBN	NW & Wales
W461XKX	London	W671CWX	Yorkshire	X212ANC	NW & Wales	X264OBN	NW & Wales
W462XKX	London	W672CWX	Yorkshire	X212JOF	NW & Wales	X265OBN	NW & Wales
W463XKX	London	W673CWX	Yorkshire	X213ANC	NW & Wales	X266OBN	NW & Wales
W464XKX	London	W674CWX	Yorkshire	X213JOF	NW & Wales	X267OBN	NW & Wales
W465XKX	London	W681DDN	The Shires	X214ANC	NW & Wales	X268OBN	NW & Wales

X269OBN	NW & Wales	X445FGP	London	X685YUG	Yorkshire	Y46TDA	NW & Wales
X2710BN	NW & Wales	X445HJA	NW & Wales	X686YUG	Yorkshire	Y47ABA	NW & Wales
X271RFF	NW & Wales	X446FGP	London	X687YUG	Yorkshire	Y47HBT	The Shires
X2720BN	NW & Wales	X446HJA	NW & Wales	X688YUG	Yorkshire	Y48ABA	NW & Wales
X272RFF	NW & Wales	X447FGP	London	X689YUG	Yorkshire	Y102TGH	London
X273RFF	NW & Wales	X447HJA	NW & Wales	X691YUG	Yorkshire	Y10TGM	Tellings group
X274RFF	NW & Wales	X448FGP	London	X692YUG	Yorkshire	Y184TUK	Midlands
X295MBH	The Shires	X448HJA	NW & Wales	X693YUG	Yorkshire	Y188TDP	Tellings group
X296MBH	The Shires	X449FGP	London	X694YUG	Yorkshire	Y215BGB	North East
X297MBH	The Shires	X449HJA	NW & Wales	X695YUG	Yorkshire	Y241KBU	NW & Wales
X32KON	NW & Wales	X451FGP	London	X696YUG	Yorkshire	Y242KBU	NW & Wales
X351AUX	The Shires	X452FGP	London	X701DBT	NW & Wales	Y243KBU	NW & Wales
X415FGP	London	X453FGP	London	X702DBT	NW & Wales	Y253YBC	Midlands
X416AJA	NW & Wales	X454FGP	London	X703DBT	NW & Wales	Y254YBC	Midlands
X416FGP	London	X457FGP	Southern Counties	X704DBT	NW & Wales	Y256YBC	Midlands
X417AJA	NW & Wales	X458FGP	Southern Counties	X705DBT	NW & Wales	Y257YBC	Midlands
X417BBD	The Shires	X459FGP	Southern Counties	X706DBT	NW & Wales	Y258KNB	North East
X417FGP	London	X471GGO	London	X707DBT	NW & Wales	Y258YBC	Midlands
X418AJA	NW & Wales	X475GGO	London	X708DBT	NW & Wales	Y259KNB	North East
X418BBD	The Shires	X478GGO	London	X709DBT	NW & Wales	Y259YBC	Midlands
X418FGP	London	X481GGO	London	X731DAU	Tellings group	Y261YBC	Midlands
X419AJA	NW & Wales	X485GGO	London	X781NWX	Midlands	Y262YBC	Midlands
X419BBD	The Shires	X501GGO	London	X782NWX	NW & Wales	Y263YBC	Midlands
X419FGP	London	X502GGO	London	X783NWX	Midlands	Y264YBC	Midlands
X421AJA	NW & Wales	X503GGO	London	X801AJA	NW & Wales	Y265YBC	Midlands
X421FGP	London	X504GGO	London	X802AJA	NW & Wales	Y266YBC	Midlands
X422AJA	NW & Wales	X506GGO	London	X803AJA	NW & Wales	Y267YBC	Midlands
X422FGP	London	X507GGO	London	X804AJA	NW & Wales	Y291PDN	Tellings group
X423AJA	NW & Wales	X508GGO	London	X805AJA	NW & Wales	Y291PDN	Tellings group
X423FGP	London	X519GGO	London	X806AJA	NW & Wales	Y291TKJ	Southern Counties
X424AJA	NW & Wales	X521GGO	London	X807AJA	NW & Wales	Y292TKJ	Southern Counties
X424FGP	London	X522GGO	London	X808AJA	NW & Wales	Y293PDN	Tellings group
X425FGP	London	X523GGO	London	X809AJA	NW & Wales	Y293TKJ	Southern Counties
X426AJA	NW & Wales	X524GGO	London	X811AJA	NW & Wales	Y294PDN	North East
X426FGP	London	X526GGO	London	X812AJA	NW & Wales	Y294PDN	Tellings group
X427AJA	NW & Wales	X527GGO	London	X813AJA	NW & Wales	Y294TKJ	Southern Counties
X427FGP	London	X529GGO	London	X814AJA	NW & Wales	Y295PDN	North East
X428FGP	London	X531GGO	London	X815AJA	NW & Wales	Y295PDN	Tellings group
X428HJA	NW & Wales	X532GGO	London	X816AJA	NW & Wales	Y295TKJ	Southern Counties
X429FGP	London	X533GGO	London	X817AJA	NW & Wales	Y296PDN	Tellings group
X429HJA	NW & Wales	X534GGO	London	X818AJA	NW & Wales	Y296TKJ	Southern Counties
X431FGP	London	X536GGO	London	X819AJA	NW & Wales	Y297TKJ	Southern Counties
X431HJA	NW & Wales	X537GGO	London	X821AJA	NW & Wales	Y298TKJ	Southern Counties
X432FGP	London	X538GGO	London	X822AJA	NW & Wales	Y299TKJ	Southern Counties
X432HJA	NW & Wales	X541GGO	London	X956DBT	NW & Wales	Y301TKJ	Southern Counties
X433FGP	London	X546GGO	London	XL04BCL	Tellings group	Y302TKJ	Southern Counties
X433HJA	NW & Wales	X646WTN	North East	Y12CLA	Tellings group	Y303TKJ	Southern Counties
X434FGP	London	X647WTN	North East	Y13CLA	Tellings group	Y30TGM	Tellings group
X434HJA	NW & Wales	X648WTN	North East	Y15TGM	Tellings group	Y346UON	Midlands
X435FGP	London	X649WTN	North East	Y20BLU	NW & Wales	Y347UON	Midlands
X435HJA	NW & Wales	X651WTN	North East	Y20TGM	Tellings group	Y348UON	Midlands
X436FGP	London	X652WTN	North East	Y21BLU	NW & Wales	Y349UON	Midlands
X436HJA	NW & Wales	X653WTN	North East	Y22CJW	NW & Wales	Y351UON	Midlands
X437FGP	London	X654WTN	North East	Y32TDA	NW & Wales	Y352UON	Midlands
X437HJA	NW & Wales	X656WTN	North East	Y36KNB	NW & Wales	Y353UON	Midlands
X438FGP	London	X657WTN	North East	Y36TDA	NW & Wales	Y354UON	Midlands
X438HJA	NW & Wales	X675YUG	Yorkshire	Y37KNB	NW & Wales	Y356UON	Midlands
X439FGP	London	X676YUG	Yorkshire	Y37TDA	NW & Wales	Y357UON	Midlands
X439HJA	NW & Wales	X677YUG	Yorkshire	Y38KNB	NW & Wales	Y358UON	Midlands
X441FGP	London	X678YUG	Yorkshire	Y38TDA	NW & Wales	Y361UON	Midlands
X441HJA	NW & Wales	X679YUG	Yorkshire	Y39TDA	NW & Wales	Y362UON	Midlands
X442FGP	London	X681YUG	Yorkshire	Y42HBT	The Shires	Y363UON	Midlands
X442HJA	NW & Wales	X682YUG	Yorkshire	Y42TDA	NW & Wales	Y364UON	Midlands
X443FGP	London	X683YUG	Yorkshire	Y46ABA	NW & Wales	Y365UON	Midlands
X443HJA	NW & Wales	X684YUG	Yorkshire	Y46HBT	The Shires	Y366UON	Midlands

Y367UON	Midlands	Y504TGJ	Tellings group	Y712KNF	NW & Wales	YG52CFK	Yorkshire
Y451KBU	NW & Wales	Y504UGC	London	Y713KNF	NW & Wales	YG52CFL	Yorkshire
Y451UGC	Southern Counties	Y506TGJ	Tellings group	Y714KNF	NW & Wales	YG52CFM	Yorkshire
Y452KBU	NW & Wales	Y506UGC	London	Y715KNF	NW & Wales	YG52CFN	Yorkshire
Y452UGC	London	Y507TGJ	Tellings group	Y716KNF	NW & Wales	YG52CFO	Yorkshire
Y453KBU	NW & Wales	Y507UGC	London	Y717KNF	NW & Wales	YG52CFP	Yorkshire
Y454KBU	NW & Wales	Y508UGC	London	Y718KNF	NW & Wales	YG52CFU	Yorkshire
Y457KBU	NW & Wales	Y509UGC	London	Y719KNF	NW & Wales	YG52CFV	Yorkshire
Y457KNF	NW & Wales	Y511UGC	London	Y721KNF	NW & Wales	YG52CFX	Yorkshire
Y458KBU	NW & Wales	Y512UGC	London	Y722KNF	NW & Wales	YG52CMU	The Shires
Y458KNF	NW & Wales	Y513UGC	London	Y723KNF	NW & Wales	YJ02FKY	Tellings group
Y459KBU	NW & Wales	Y514UGC	London	Y724KNF	NW & Wales	YJ03PFX	Midlands
Y461KNF	NW & Wales	Y516UGC	London	Y726KNF	NW & Wales	YJ04BKF	Midlands
Y461UGC	Southern Counties	Y517UGC	London	Y727KNF	NW & Wales	YJ04HJC	Yorkshire
Y462KNF	NW & Wales	Y518UGC	London	Y728KNF	NW & Wales	YJ04HJD	Yorkshire
Y462UGC	Southern Counties	Y519UGC	London	Y729KNF	NW & Wales	YJ04HJE	Yorkshire
Y463KNF	NW & Wales	Y521UGC	The Shires	Y733KNF	NW & Wales	YJ04HJF	Yorkshire
Y463UGC	Southern Counties	Y522UGC	London	Y744KNF	NW & Wales	YJ04HJG	Yorkshire
Y464KNF	NW & Wales	Y523UGC	London	Y801DGT	London	YJ05JXU	The Shires
Y464UGC	Southern Counties	Y524UGC	London	Y802DGT	London	YJ05JXV	The Shires
Y465KNF	NW & Wales	Y526UGC	London	Y803DGT	London	YJ05PVT	Midlands
Y465UGC	Southern Counties	Y527UGC	London	Y804DGT	London	YJ06ATK	NW & Wales
Y466KNF	NW & Wales	Y529UGC	London	Y805DGT	London	YJ06FXS	The Shires
Y466UGC	Southern Counties	Y531UGC	The Shires	Y806DGT	London	YJ06FXT	The Shires
Y467KNF	NW & Wales	Y532UGC	London	YD02PXW	Yorkshire	YJ06FXU	The Shires
Y467UGC	Southern Counties	Y533UGC	London	YD02PXX	Yorkshire	YJ06FXV	The Shires
Y468KNF	NW & Wales	Y538VFF	NW & Wales	YD02PXY	Yorkshire	YJ06LDK	The Shires
Y468UGC	Southern Counties	Y539VFF	NW & Wales	YD02PXZ	Yorkshire	YJ06LFE	The Shires
Y469KNF	NW & Wales	Y541UGC	London	YD02PYU	Yorkshire	YJ06LFF	The Shires
Y469UGC	Southern Counties	Y541UJC	NW & Wales	YD02PYV	Yorkshire	YJ06LFG	The Shires
Y471KNF	NW & Wales	Y542UGC	London	YD02PYW	Yorkshire	YJ06LFH	The Shires
Y471UGC	London	Y542UJC	NW & Wales	YD02PYX	Yorkshire	YJ06LFK	The Shires
Y472KNF	NW & Wales	Y543UGC	London	YD02PYY	Yorkshire	YJ06LFL	The Shires
Y472UGC	London	Y543UJC	NW & Wales	YD02PYZ	Yorkshire	YJ06LFZ	Southern Counties
Y473KNF	NW & Wales	Y544UGC	London	YD02RJJ	Yorkshire	YJ06WLX	Yorkshire
Y473UGC	London	Y544UJC	NW & Wales	YD02RJO	Yorkshire	YJ06WLZ	Yorkshire
Y474UGC	London	Y546UGC	London	YD04MFJ	Tellings group	YJ06WMA	Yorkshire
Y475KNF	NW & Wales	Y546UJC	NW & Wales	YD04MFK	Tellings group	YJ06WMC	Yorkshire
Y475UGC	London	Y547UGC	London	YD04MFM	Tellings group	YJ06WMD	Yorkshire
Y476UGC	London	Y547UJC	NW & Wales	YE06HNT	The Shires	YJ06WME	Yorkshire
Y477UGC	London	Y548UGC	London	YE06HNU	The Shires	YJ06WMF	Yorkshire
Y478UGC	London	Y548UJC	NW & Wales	YE06HPA	The Shires	YJ06WMG	Yorkshire
Y479UGC	London	Y549UGC	London	YE06HPC	The Shires	YJ06WMK	Yorkshire
Y481UGC	London	Y549UJC	NW & Wales	YE06HPF	The Shires	YJ06WML	Yorkshire
Y482UGC	London	Y551UJC	NW & Wales	YE06HPJ	The Shires	YJ06WWV	Yorkshire
Y483UGC	London	Y552UJC	NW & Wales	YE06HPK	The Shires	YJ06WWX	Yorkshire
Y484UGC	London	Y581UGC	London	YE06HPL	The Shires	YJ06WWY	Yorkshire
Y485UGC	London	Y685EBR	North East	YE06HPN	The Shires	YJ06WWZ	Yorkshire
Y486UGC	London	Y686EBR	North East	YE06HPO	The Shires	YJ06YRO	NW & Wales
Y487UGC	London	Y687EBR	North East	YE06HPP	The Shires	YJ06YRP	The Shires
Y488UGC	London	Y688EBR	North East	YE06HPU	The Shires	YJ06YRR	The Shires
Y489UGC	London	Y689EBR	North East	YE06HPX	Southern Counties	YJ06YRS	The Shires
Y491UGC	London	Y691EBR	North East	YE06HPY	Southern Counties	YJ06YRT	The Shires
Y492UGC	London	Y692EBR	North East	YE06HPZ	Southern Counties	YJ06YRU	The Shires
Y493UGC	London	Y693EBR	North East	YE06HRA	The Shires	YJ06YRY	NW & Wales
Y494UGC	London	Y694EBR	North East	YE06HRC	The Shires	YJ06YRZ	NW & Wales
Y495UGC	London	Y701XJF	Midlands	YE06HRD	The Shires	YJ07BCZ	The Shires
Y496UGC	London	Y702XJF	Midlands	YE06HRF	The Shires	YJ07BEO	The Shires
Y497UGC	London	Y703XJF	Midlands	YE06HRG	The Shires	YJ07BEU	The Shires
Y498UGC	London	Y704XJF	Midlands	YE06HRJ	The Shires	YJ07JSU	Scotland
Y499UGC	London	Y705XJF	Midlands	YG52CFA	Yorkshire	YJ07JSV	Scotland
Y501TGJ	Tellings group	Y706XJF	Midlands	YG52CFD	Yorkshire	YJ07JSX	Scotland
Y501UGC	London	Y707XJF	Midlands	YG52CFE	Yorkshire	YJ07JSY	Scotland
Y502UGC	London	Y709XJF	Midlands	YG52CFF	Yorkshire	YJ07JSZ	Scotland
Y503UGC	London	Y711KNF	NW & Wales	YG52CFJ	Yorkshire	YJ07JVF	The Shires

YJ07JVU	The Shires	YJ09CUO	Yorkshire	YJ09MLZ	Midlands	YJ57AZP	Midlands
YJ07JVV	The Shires	YJ09CUV	Scotland	YJ09MMA	Midlands	YJ57AZR	Midlands
YJ07JVW	The Shires	YJ09CUW	Scotland	YJ09MME	Midlands	YJ57AZT	Midlands
YJ07JVX	The Shires	YJ09CUX	Scotland	YJ09MMF	Midlands	YJ57AZU	Midlands
YJ07JVY	The Shires	YJ09CUY	Yorkshire	YJ09MMK	Southern Counties	YJ57BEO	Yorkshire
YJ07JVZ	The Shires	YJ09CVA	Yorkshire	YJ09MMO	Southern Counties	YJ57BEU	Yorkshire
YJ07VPW	The Shires	YJ09CVB	Yorkshire	YJ09MMU	Southern Counties	YJ57BKD	Southern Counties
YJ07VPX	The Shires	YJ09CVC	Yorkshire	YJ09OTW	Midlands	YJ57BPZ	Midlands
YJ07VPY	The Shires	YJ09CVD	Yorkshire	YJ09OTY	The Shires	YJ57BRF	Midlands
YJ07VRC	The Shires	YJ09CVE	Yorkshire	YJ09OTZ	The Shires	YJ57BRV	Midlands
YJ07VRD	The Shires	YJ09CVH	Scotland	YJ09OUA	Midlands	YJ57BUA	Midlands
YJ07VRE	The Shires	YJ09CVK	Scotland	YJ09OUB	Midlands	YJ57BUE	Midlands
YJ07VRF	The Shires	YJ09CVL	Scotland	YJ51YWW	Tellings group	YJ57BVB	North East
YJ07VRU	Midlands	YJ09CVM	Scotland	YJ53VFY	Midlands	YJ57BVC	North East
YJ08DVA	Yorkshire	YJ09CVN	Scotland	YJ54CFG	The Shires	YJ57BVD	North East
YJ08DVB	Yorkshire	YJ09CVO	Scotland	YJ54CKE	Midlands	YJ57BVE	North East
YJ08DVC	Yorkshire	YJ09CVR	Scotland	YJ54CKF	Midlands	YJ57BVF	North East
YJ08DVF	Yorkshire	YJ09CXL	The Shires	YJ54CKG	Scotland	YJ57BVG	North East
YJ08DVG	Yorkshire	YJ09CYH	Tellings group	YJ54CKK	Scotland	YJ57BVT	Yorkshire
YJ08DVH	Yorkshire	YJ09EYA	Yorkshire	YJ54CPE	Midlands	YJ57BVU	Yorkshire
YJ08DVK	Yorkshire	YJ09EYB	Yorkshire	YJ54CPF	Midlands	YJ57BVV	Yorkshire
YJ08DVN	Yorkshire	YJ09EYC	Yorkshire	YJ55BKG	NW & Wales	YJ57BVW	Yorkshire
YJ08DVO	Yorkshire	YJ09EYF	Yorkshire	YJ55BKK	NW & Wales	YJ57BVX	Yorkshire
YJ08DVP	Yorkshire	YJ09EYG	Yorkshire	YJ55BKL	NW & Wales	YJ57BVY	Yorkshire
YJ08DVR	Yorkshire	YJ09EYH	Yorkshire	YJ55BKN	NW & Wales	YJ57BVZ	Yorkshire
YJ08DVT	Yorkshire	YJ09EYK	Yorkshire	YJ55BKO	NW & Wales	YJ57BWA	The Shires
YJ08DVU	Yorkshire	YJ09EYL	Yorkshire	YJ55BKU	NW & Wales	YJ57BWB	The Shires
YJ08DZA	Southern Counties	YJ09EYM	Yorkshire	YJ55BKV	NW & Wales	YJ57BWC	The Shires
YJ08DZB	Southern Counties	YJ09EYO	Yorkshire	YJ55KZS	Midlands	YJ57BWD	The Shires
YJ08DZC	Southern Counties	YJ09EYP	Yorkshire	YJ55WOA	The Shires	YJ57BWE	The Shires
YJ08DZD	Southern Counties	YJ09EYR	Yorkshire	YJ55WOB	The Shires	YJ57BWF	The Shires
YJ08DZE	Southern Counties	YJ09EYS	Yorkshire	YJ55WOC	The Shires	YJ57EJD	The Shires
YJ08DZF	Southern Counties	YJ09LBK	Midlands	YJ55WOD	The Shires	YJ57EJE	The Shires
YJ08DZG	Southern Counties	YJ09LBL	Midlands	YJ55WOH	The Shires	YJ57EJF	The Shires
YJ08DZH	Southern Counties	YJ09LBN	Midlands	YJ55WOM	The Shires	YJ57EJG	The Shires
YJ08DZK	Southern Counties	YJ09MJE	Midlands	YJ55WOR	The Shires	YJ57EJK	The Shires
YJ08DZL	Southern Counties	YJ09MJF	Midlands	YJ55WOU	The Shires	YJ57EJL	The Shires
YJ08DZM	Southern Counties	YJ09MJK	Midlands	YJ55WOV	The Shires	YJ57EJN	The Shires
YJ08DZN	Southern Counties	YJ09MJV	Midlands	YJ55WOX	The Shires	YJ57EKA	Midlands
YJ08ECY	Yorkshire	YJ09MJX	Midlands	YJ55WPO	The Shires	YJ57EKB	Midlands
YJ08EEB	Yorkshire	YJ09MJY	Midlands	YJ55WSV	The Shires	YJ57EKC	Midlands
YJ08EEF	Yorkshire	YJ09MKC	Midlands	YJ55WSW	The Shires	YJ57EKD	Midlands
YJ08EEG	Yorkshire	YJ09MKD	Midlands	YJ55WSX	The Shires	YJ57EKE	Midlands
YJ08EEH	Yorkshire	YJ09MKE	Midlands	YJ55WSY	The Shires	YJ57EKF	The Shires
YJ08EEM	Yorkshire	YJ09MKF	Midlands	YJ55WSZ	The Shires	YJ57EKG	The Shires
YJ08EEN	Yorkshire	YJ09MKG	Midlands	YJ55YGV	The Shires	YJ57XWH	The Shires
YJ08EEP	Yorkshire	YJ09MKK	Midlands	YJ55YGW	The Shires	YJ58CAA	North East
YJ08EER	Yorkshire	YJ09MKL	Midlands	YJ56ATK	NW & Wales	YJ58CAE	North East
YJ08EES	Yorkshire	YJ09MKM	Midlands	YJ56ATY	The Shires	YJ58CAO	North East
YJ08EET	Yorkshire	YJ09MKN	Midlands	YJ56ATZ	The Shires	YJ58CAV	North East
YJ08EEU	Yorkshire	YJ09MKO	Midlands	YJ56JYE	Yorkshire	YJ58CAW	North East
YJ08EEV	Yorkshire	YJ09MKP	Midlands	YJ56JYF	Yorkshire	YJ58CAX	North East
YJ08EEW	Yorkshire	YJ09MKU	Midlands	YJ56JYG	Yorkshire	YJ58CBF	North East
YJ08EFT	Tellings group	YJ09MKV	Midlands	YJ56JYH	Yorkshire	YJ58CBO	North East
YJ08XDK	The Shires	YJ09MKX	Midlands	YJ56JYK	Yorkshire	YJ58CBU	North East
YJ09CSU	Yorkshire	YJ09MKZ	Midlands	YJ56JYL	Yorkshire	YJ58CBV	North East
YJ09CTV	Yorkshire	YJ09MLE	Midlands	YJ56JYN	Yorkshire	YJ58CCA	Midlands
YJ09CTX	Yorkshire	YJ09MLF	Midlands	YJ56JYO	Yorkshire	YJ58CCD	Midlands
YJ09CTY	Yorkshire	YJ09MLK	Midlands	YJ56JYP	Yorkshire	YJ58CCE	Midlands
YJ09CTZ	Yorkshire	YJ09MLL	Midlands	YJ57AZD	Midlands	YJ58CCF	Midlands
YJ09CUA	Yorkshire	YJ09MLN	Midlands	YJ57AZF	Midlands	YJ58CCK	Midlands
YJ09CUC	Yorkshire	YJ09MLO	Midlands	YJ57AZG	Midlands	YJ58CCN	Midlands
YJ09CUG	Yorkshire	YJ09MLV	Midlands	YJ57AZL	Midlands	YJ58CCO	Midlands
YJ09CUH	Yorkshire	YJ09MLX	Midlands	YJ57AZN	Midlands	YJ58CCU	Midlands
YJ09CUK	Yorkshire	YJ09MLY	Midlands	YJ57AZO	Midlands	YJ58CDZ	Southern Counties

Arriva Yorkshire received thirteen Optare Tempos of which 1306, YJ09EYK, is one and is seen on the Leeds to Castleford service, for which it carries route lettering. *Andy Jarosz*

YJ58CEA	Southern Counties	YJ58FJY	The Shires	YJ59BUE	Yorkshire	YK07BGF	The Shires
YJ58CEF	Southern Counties	YJ58FJZ	The Shires	YJ59BUH	Yorkshire	YK08ERO	North East
YJ58FFA	North East	YJ58FKA	The Shires	YJ59BUO	Midlands	YK08ERU	North East
YJ58FFB	North East	YJ58PFU	The Shires	YJ59BUP	Midlands	YK08ERV	North East
YJ58FFC	North East	YJ58PFV	The Shires	YJ59BUU	Midlands	YK08ERX	North East
YJ58FFG	North East	YJ58PFX	The Shires	YJ59BUV	Midlands	YK08ERY	North East
YJ58FFV	The Shires	YJ58PFY	The Shires	YJ59BUW	Midlands	YK08ERZ	North East
YJ58FFW	The Shires	YJ58PFZ	The Shires	YJ59BVA	Midlands	YK08ESU	North East
YJ58FHA	Yorkshire	YJ58PHX	Midlands	YJ59BVB	Midlands	YK08ESV	North East
YJ58FHB	Yorkshire	YJ58PKA	The Shires	YJ59BVF	Midlands	YK08ESY	North East
YJ58FHC	Yorkshire	YJ58PKC	The Shires	YJ59BVG	Midlands	YK08ETA	North East
YJ58FHD	Yorkshire	YJ58PKD	The Shires	YJ59BVH	Midlands	YK08ETD	North East
YJ58FHE	Yorkshire	YJ58PKE	The Shires	YJ59BVK	Midlands	YK08ETE	North East
YJ58FHF	Yorkshire	YJ58PKF	The Shires	YJ59BVL	Midlands	YK08ETF	North East
YJ58FIG	Yorkshire	YJ58PKK	The Shires	YJ59BVM	Midlands	YK08ETJ	North East
YJ58FHH	Yorkshire	YJ58PKN	The Shires	YJ59BVN	Midlands	YK08ETL	North East
YJ58FHJ	Yorkshire	YJ58PKO	The Shires	YJ59GJK	North East	YK08ETO	North East
YJ58FHL	Yorkshire	YJ58PKU	The Shires	YJ59GJO	North East	YK08ETR	North East
YJ58FHM	Yorkshire	YJ58PKV	The Shires	YJ59GJV	North East	YK08ETT	North East
YJ58FHN	Yorkshire	YJ58PKX	The Shires	YJ59GJX	North East	YK08ETU	North East
YJ58FHO	Yorkshire	YJ58VCG	The Shires	YJ59GJY	North East	YK08ETX	North East
YJ58FHP	Yorkshire	YJ59BRY	Yorkshire	YJ59GKA	North East	YK08ETY	North East
YJ58FHX	The Shires	YJ59BRZ	Yorkshire	YJ59GKC	North East	YK08XBD	North East
YJ58FHY	The Shires	YJ59BSO	Yorkshire	YJ59GKD	North East	YK08XBE	North East
YJ58FHZ	The Shires	YJ59BTO	Yorkshire	YJ59GKE	North East	YK08XBF	North East
YJ58FJA	The Shires	YJ59BTU	Yorkshire	YJ59GKF	North East	YK08XBG	North East
YJ58FJN	The Shires	YJ59BTV	Yorkshire	YK04KWF	Tellings group	YK08XBH	North East
YJ58FJO	The Shires	YJ59BTX	Yorkshire	YK04KWG	Tellings group	YK08XBK	North East
YJ58FJP	The Shires	YJ59BTY	Yorkshire	YK05CAE	NW & Wales	YK08XBL	North East
YJ58FJU	The Shires	YJ59BTZ	Yorkshire	YK05CBX	NW & Wales		
YJ58FJX	The Shires	YJ59BUA	Yorkshire	YK07BGE	The Shires		

Code	Region	Code	Region	Code	Region	Code	Region
YK08XBM	North East	YN07LHE	Tellings group	YN54ZHM	Tellings group	YT09ZBV	Midlands
YK08XBN	North East	YN08BCL	Tellings group	YN55KZZ	Tellings group	YT09ZBW	Midlands
YK08XBO	North East	YN08HZK	Midlands	YN55PZY	The Shires	YT09ZBX	Midlands
YK08XBP	North East	YN08HZL	Midlands	YN55PZZ	The Shires	YT09ZBY	Midlands
YK08XBR	North East	YN08HZM	Midlands	YN55WSU	Tellings group	YT09ZBZ	Midlands
YK08XBS	North East	YN08HZP	Midlands	YN55WSV	Tellings group	YU04XFB	Tellings group
YK08XBT	North East	YN08HZR	Midlands	YN55WSW	Tellings group	YU04XFC	Tellings group
YK08XBU	North East	YN08HZS	Midlands	YN56NNA	Midlands	YU04XFD	Tellings group
YK08XBV	North East	YN08HZT	Midlands	YN56NNB	Midlands	YX08HWC	Southern Counties
YK08XBW	North East	YN08HZU	Midlands	YN57BKE	Southern Counties	YX08HWD	Southern Counties
YK08XBY	North East	YN08HZV	Midlands	YN57BKF	Southern Counties	YX08HWE	Southern Counties
YK57FHH	The Shires	YN08HZW	Midlands	YN57BKG	Southern Counties	YX08HWF	Southern Counties
YK57FHJ	The Shires	YN08HZX	Midlands	YN58RCF	Midlands	YX08HWG	Southern Counties
YM55RRX	Tellings group	YN08HZY	Midlands	YR58SUH	Scotland	YX08HWH	Southern Counties
YM55RRY	Tellings group	YN08HZZ	Midlands	YR58SUO	Scotland	YX08HWK	Southern Counties
YMB512W	NW & Wales	YN08OBY	Tellings group	YR58SUU	Scotland	YX08HWL	Southern Counties
YN03NCF	The Shires	YN08OBY	Tellings group	YR58SUV	Scotland	YX08HWM	Southern Counties
YN03NEF	The Shires	YN51WGY	Tellings group	YR58SUX	Scotland	YX08HWN	Southern Counties
YN03UXW	North East	YN51WGZ	Tellings group	YR58SUY	Scotland	YX08HWO	Southern Counties
YN04AHA	Midlands	YN53OYW	Tellings group	YR59SRO	Midlands	YX08HWP	The Shires
YN04LXM	The Shires	YN53OYX	Tellings group	YR59SRU	Midlands	YX09EVG	Scotland
YN04XZH	NW & Wales	YN53OYZ	Tellings group	YR59SRV	Midlands	YX09EVH	Scotland
YN05HCG	Midlands	YN53OZH	Tellings group	YR59SRX	Midlands	YX57AOB	Southern Counties
YN05VRT	Tellings group	YN53OZP	Tellings group	YR59SRY	Midlands	YX57AOC	Southern Counties
YN05VRU	Tellings group	YN53OZR	Tellings group	YR59SRZ	Midlands	YX57AOD	Southern Counties
YN06CJU	Tellings group	YN53SVG	The Shires	YR59SSJ	Midlands	YX57CAA	Southern Counties
YN06CJV	Tellings group	YN53VBX	Southern Counties	YR59SSK	Midlands	YX57CAE	The Shires
YN06CJX	Tellings group	YN54AGV	Tellings group	YR59SSO	Midlands	YX57CAO	The Shires
YN06CJY	Tellings group	YN54AJO	Tellings group	YR59SSU	Midlands	YX57CAU	The Shires
YN06CJZ	Tellings group	YN54AKO	Tellings group	YR59SSV	Midlands	YX57CAV	The Shires
YN06JXJ	The Shires	YN54AMO	Tellings group	YR59SSX	Midlands	YX57COE	The Shires
YN06JXK	The Shires	YN54ANU	Tellings group	YR59SSZ	Midlands	YX57COF	The Shires
YN06JXL	The Shires	YN54DDJ	Tellings group	YR59STX	Midlands	YX57COG	The Shires
YN06JXM	The Shires	YN54DDK	Tellings group	YR59STY	Midlands	YX57COR	The Shires
YN06JXO	The Shires	YN54DDL	Tellings group	YS02UBY	The Shires	YX57COS	The Shires
YN06JXP	The Shires	YN54DDO	Tellings group	YT09ZBL	Midlands	YX57COT	The Shires
YN06PFD	Tellings group	YN54OCY	Southern Counties	YT09ZBN	Midlands	YX57CCO	The Shires
YN06PFE	Tellings group	YN54WDE	Southern Counties	YT09ZBO	Midlands	YX57CCU	The Shires
YN06PFF	Tellings group	YN54WWF	Tellings group	YT09ZBP	Midlands	YX57CCV	The Shires
YN06PFG	Tellings group	YN54WWG	Tellings group	YT09ZBR	Midlands	YX57CCY	The Shires
YN06TFZ	Tellings group	YN54ZHK	Tellings group	YT09ZBU	Midlands		

ISBN 9781904875291 © Published by *British Bus Publishing Ltd*, November 2009

British Bus Publishing Ltd, 16 St Margaret's Drive, Telford, TF1 3PH
Telephone: 01952 255669

web; www.britishbuspublishing.co.uk   - e-mail: sales@britishbuspublishing.co.uk